Waste Plastics 2020

Topical Conference at the 2020 AIChE Spring Meeting and 16th Global Congress on Process Safety

Online
17-21 August 2020

ISBN: 978-1-7138-2338-4

Printed from e-media with permission by:

Curran Associates, Inc.
57 Morehouse Lane
Red Hook, NY 12571

Some format issues inherent in the e-media version may also appear in this print version.

Printed with permission by Curran Associates, Inc. (2021)

For permission requests, please contact AIChE
at the address below.

AIChE
120 Wall Street, FL 23
New York, NY 10005-4020

Phone: (800) 242-4363
Fax: (203) 775-5177

www.aiche.org

Additional copies of this publication are available from:

Curran Associates, Inc.
57 Morehouse Lane
Red Hook, NY 12571 USA
Phone: 845-758-0400
Fax: 845-758-2633
Email: curran@proceedings.com
Web: www.proceedings.com

TABLE OF CONTENTS

591161 2020 AIChE Spring Meeting & 16th GCPS AGILE Award Keynote Address - Beyond the Horizon: The Factors and Forces Affecting the Global Petrochemical Industry

Abstract:

Over the past 10 years, LyondellBasell and the global petrochemical industry has undergone a dramatic transformation. In this era of unprecedented disruption, it is likely that the next the decade will present both challenges and opportunities for the petrochemical sector. LyondellBasell CEO Bob Patel will summarize the company's progress to date and provide a forward-looking perspective on key opportunities and risks faced by the industry in 2020 and beyond.

591041 Plastics Circularity Issues and Dow's Response

Abstract:

To fulfill long term growth in demand for plastics while also meeting sustainability expectations, a transformational change is needed. This presentation will review changing expectations within the global hydrocarbon industry and Dow's overall response.

591157 Securing Enterprises via Cybersecurity Understanding, Resiliency, and Education

Abstract:

Securing Enterprises via Cybersecurity Understanding, Resiliency, and Education

Industry 4.0 initiatives are enabled by leveraging the deployment of information technologies into the operational technology environment. While a logical approach in defining the future of smart manufacturing, it remains fret with challenges and risks that must be mitigated assure enterprise resilience to cyber and cyber-physical attacks. As economies recalibrate in the aftermath of the current public health emergency, definitions of what constitutes "critical infrastructure" are broadening while, at the same time, attack surfaces and the threat landscape are expanding. This means companies must place a greater emphasis on taking the steps and making the investments necessary to overcome the challenges and navigate these challenges and risks to maximize their resilience and secure and protect their enterprises. This presentation will explore the link between cybersecurity and process safety in the context of a safety culture. It will also identify cybersecurity challenges in industry 4.0 initiatives, the integration of information technologies (IT) and operational technologies (IT), the delivery of training, and the need for credentialing and ongoing professional development.

AN INTEGRATED PATTERN QUERY METHOD TO ASSIST PLANT MONITORING

Shu Xu, Anthony Amaro Jr., Mark Nixon
Emerson Automation Solutions

Introduction

Live plant behavior is usually recorded and saved as multivariate time series data which can be explored to assist process engineers in process monitoring, abnormality/fault detection and diagnosis. For example, patterns that indicate a trend or a fault might repeat itself over the whole operating period. In addition, faced with sufficient process data, process engineers are challenged to efficiently and effectively establish relationships between specific variables which share patterns with a high degree of correlation. In current work, we demonstrate the adoption of an integrated pattern query framework which combines PCA similarity factor and a fast similarity search algorithm based on dynamic time warping (DTW) to overcome such a challenge. Simulated industrial case studies illustrate the efficacy and efficiency of the proposed framework. Such results demonstrate the successful application of data analytics to cope with real problems in chemical manufacturing within the Digital Transformation era.

The remainder of this paper is organized as follows. Section 2 briefly introduces methodology information regarding PCA similarity factor, dynamic time warping, and an integrated framework. Section 3 provides case study details including the benchmark Tennessee Eastman process simulator and simulation conditions. Section 4 summarizes results obtained in the case study. We conclude the paper and discuss future works in Section 5.

Methods

PCA similarity factor

Principal component analysis (PCA) is a linear dimensional reduction technique that has been widely used in process monitoring applications [1,2]. PCA is intended to optimally capture data variations while minimizing the contaminating effect of random noise. In addition, PCA transforms the original correlated variables $\boldsymbol{x_i}$ into only a few independent principal components which are linear combinations of $\boldsymbol{x_i}$ and represent the directions of major variabilities. The PCA similarity factor metric [3,4] is proposed based on analyzing the structural similarity of such a transformed subspace. For a given two historical data sets, A and B, S_{PCA} can be calculated by:

$$S_{PCA}(A,B) = \frac{1}{k} trace(LM^T ML^T) = \frac{1}{k}\sum_{i=1}^{k}\sum_{j=1}^{k} cos^2\theta_{ij} \qquad (1)$$

where L and M are the matrices that contain the first k principal components of A and B, accounting for most of their variances respectively, and θ_{ij} is the angle between the ith principal component of A and the jth principal component of B. As we can see from Equation (1) the geometric interpretation of S_{PCA} is the sum of the squares of the cosines of the angles between each principal component of L and M.

Pattern-matching based on dynamic time warping

Dynamic time warping (DTW) is an advanced similarity measurement technique originally developed for speech recognition application, and it has been successfully applied to pre-processing chromatography data [5], mass spectroscopy (MS) data, and Nuclear Magnetic Resonance (NMR) spectrum data [6] and Batch Data[10]. In DTW, an optimization problem is formulated to maximize the geometrical similarity between two vectors, and it is commonly solved by dynamic programming. The mathematical framework for warping and alignment in DTW is introduced here. The nomenclature and the terminology follow the convention in Ref. [7].

First, we define reference trajectory R and target trajectory T of length N and M respectively as follows:

$$R := (r_1, r_2, \dots, r_N), T := (t_1, t_2, \dots, t_M) \tag{2}$$

We assume that the two trajectories are sampled at the same sampling interval. To measure the quality of alignment, a local cost function (or local distance measure) c is defined as:

$$c(r_n, t_m) = \|r_n - t_m\| \tag{3}$$

This definition assumes that r_n and t_m is of same dimension (which means same number of variables being measured).

Using the local cost function definition, we can define an alignment cost matrix $C \in \mathbb{R}^{N\times M}$

$$C(n, m) := c(r_n, t_m) \tag{4}$$

The goal of DTW is to find a warping path through the local cost matrix that gives us the minimal cumulative cost between the two trajectories. The warping path P is a series of pair-wise vectors that map sample indices from trajectory T to sample indices on R, and is defined as $P = (p_1, \dots, p_L)$ and $p_l = (n_l, m_l) \in [1{:}N] \times [1{:}M]$. The warping trajectory is constrained by three conditions:

- Boundary condition: $p_1 = (1, 1)$, and $p_L = (N, M)$.
- Monotonicity condition: $n_1 \le n_2 \le \cdots \le n_L$ and $m_1 \le m_2 \le \cdots \le m_L$.
- Step-size condition: $p_{l+1} - p_l \in \{(1,0), (0,1), (1,1)\}$ for $l \in [1{:}L-1]$.

The boundary condition ensures that the first and the last samples map to the beginning and the end of the two trajectories. The monotonicity condition ensures that there will be no backtracking or "folding" of trajectories during warping. The step-size condition ensures that no sample indices will be skipped and limits the amount of warping applied to only three directions $(0, 1)$, $(1, 0)$ and $(1, 1)$.

Given any arbitrary warping path P, the total cost associated with the warping path and the two trajectories in question can be defined:

$$c_p(R, T) := \sum_{l=1}^{L} c(r_{nl}, t_{ml}) \tag{5}$$

With these definitions in place, we can formulate the objective function for DTW warping. The optimal warping path determined through DTW aims to minimize the total cost $c_p(R, T)$.

$$p^*_{DTW} := \arg\min\{c_p(R, T)\} \tag{6}$$

It could be very difficult to find p^* for long trajectories since the search space would be of dimension $\mathbb{R}^{N\times M}$. An efficient procedure to solve this problem in polynomial time $O(M \cdot N)$ based on dynamic programming has been devised.

From the local cost matrix $C(R, T)$, a cumulative cost matrix can be defined as:

$$D(n, m) := DTW(R(1:n), T(1:m)) \quad (7)$$

Where each element in the matrix D represents the cumulative cost accrued arriving at (m, n) from the starting point $(1, 1)$ by taking the optimal path subjected to the three warping constraints.

The elements of the matrix D can be filled out efficiently using the following relationship:

$$D(n, 1) := \sum_{k=1}^{n} c(r_k, t_1), n \in [1:N] \quad (8)$$

$$D(1, m) := \sum_{k=1}^{m} c(r_1, t_k), m \in [1:M] \quad (9)$$

$$D(n, m) := \min\{D(n-1, m-1), D(n-1, m), D(n, m-1)\} + c(r_n, t_m) \quad (10)$$

To perform DTW, the cumulative cost matrix D can first be calculated using the rules shown in Equations (8) ~ (10). It takes $O(M \cdot N)$ iterations to populate the entire cumulative cost matrix. After the D matrix is populated, the optimal warping trajectory p^*_{DTW} can be found by back-stepping from the end (N, M) back to the starting point $(1, 1)$ using the following rule.

$$p_{l-1} := \begin{cases} (1, m-1), & n = 1 \\ (n-1, 1), & m = 1 \\ \arg\min\{D(n-1, m-1), D(n-1, m), D(n, m-1)\}, & otherwise \end{cases} \quad (11)$$

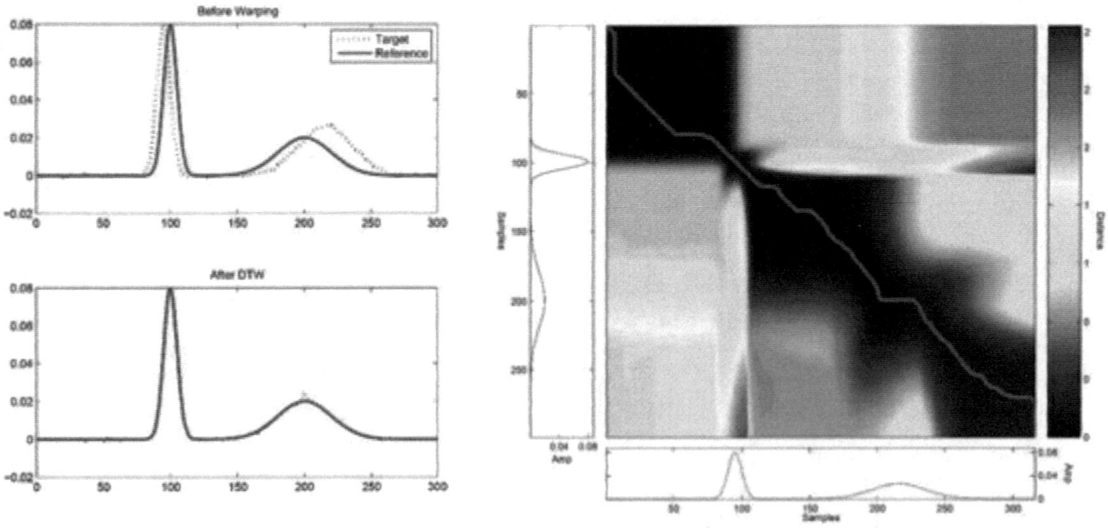

(a) Trajectories from before (top) and after (bottom) warping

(b) The cumulative distance matrix D (heat map) and the optimal path (purple line)

Figure 1 Trajectory plot and diagnostics from before and DTW

Figure 1 shows a result of DTW algorithm on a simple two-trajectory warping problem, where Figure 1(a) shows that the target trajectory (dashed line) has been warped to conform the reference (red line), and Figure 1(b) illustrates the cumulative cost matrix D for this problem. The darker and more blue color indicates areas of lower cumulative cost. The purple line indicates the optimal path that achieves the lowest cumulative cost as defined in Equation (11).

It is interesting to note that the (n,m)th element of the cumulative cost matrix D, $D(n,m)$, can be used as an index to measure the similarity between two univariate signals, and can assist process engineers in pattern matching and identification.

An integrated methodology framework

In this paper, we propose an integrated methodology which combines the PCA similar factor with the DTW-based pattern matching. The flow chart of such a method is illustrated in Figure 2. Given a multivariate data set ***A*** containing a pattern indicating a trend or a fault, we first extract the structure of the data set ***A***, and use it to find periods of data with a high PCA similar factor. Then we identify key contributors to such a trend or a fault via variable contribution plots. Last, we refine our searching based on the DTW-based pattern matching technique, as discussed in the section shown above.

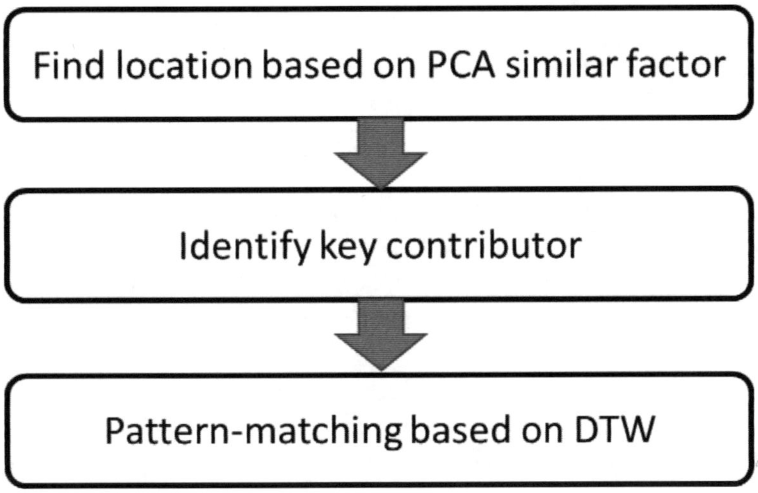

Figure 2 Flow chart of an integrated methodology of pattern query

Case study

The benchmark Tennessee Eastman process simulator

The Tennessee Eastman process model is a realistic simulation program of a chemical plant which is widely accepted as a benchmark for evaluating process control and monitoring methods [1]. Figure 3 shows the flow diagram of the process with five major units, i.e., a reactor, condenser, compressor, separator and stripper. The process has two products from four reactants. Additionally, an inert and a by-product are also present making a total of 8 components denoted as A, B, C, D, E, F, G and H. The process allows for a total 52 measurements out of which 41 are of process variables and 11 are manipulated variables, see Tables 1 and 2:

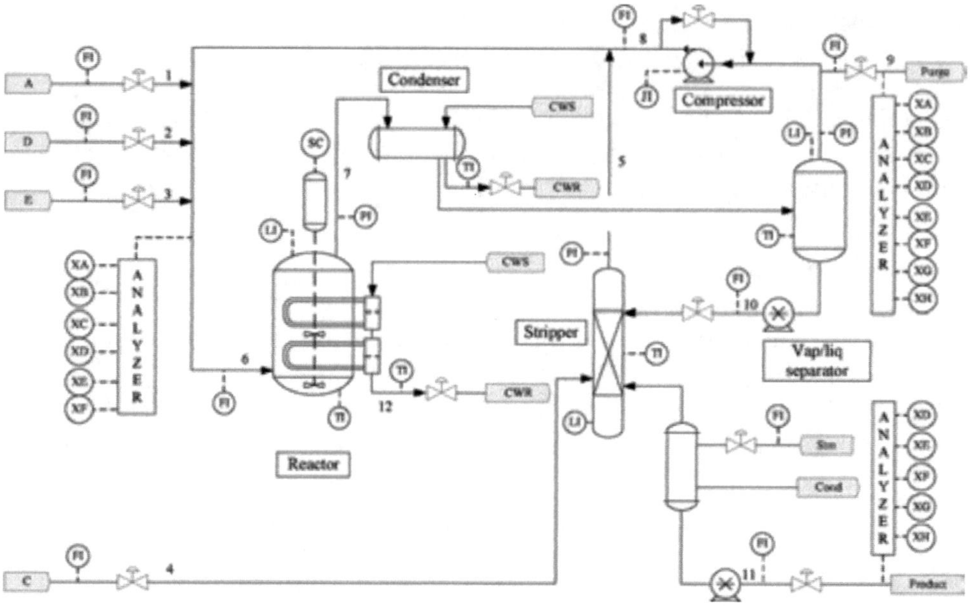

Figure 3 The Tennessee Eastman process

Table 1 Process variables

Block name	Variable name	Number
Input feed	A feed (stream 1)	XMEAS(1)
	D feed (stream 2)	XMEAS(2)
	E feed (stream 3)	XMEAS(3)
	A and C feed	XMEAS(4)
Reactor	Reactor feed rate	XMEAS(6)
	Reactor pressure	XMEAS(7)
	Reactor level	XMEAS(8)
	Reactor temperature	XMEAS(9)
Separator	Separator temperature	XMEAS(11)
	Separator level	XMEAS(12)
	Separator pressure	XMEAS(13)
	Separator underflow	XMEAS(14)
Stripper	Stripper level	XMEAS(15)
	Stripper pressure	XMEAS(16)
	Stripper underflow	XMEAS(17)
	Stripper temperature	XMEAS(18)
	Stripper steam flow	XMEAS(19)
Miscellaneous	Recycle flow	XMEAS(5)
	Purge rate	XMEAS(10)
	Compressor work	XMEAS(20)
	Reactor water temperature	XMEAS(21)
	Separator water temperature	XMEAS(22)

Block name	Variable name	Number
Reactor feed analysis	Component A	XMEAS(23)
	Component B	XMEAS(24)
	Component C	XMEAS(25)
	Component D	XMEAS(26)
	Component E	XMEAS(27)
	Component F	XMEAS(28)
Purge gas analysis	Component A	XMEAS(29)
	Component B	XMEAS(30)
	Component C	XMEAS(31)
	Component D	XMEAS(32)
	Component E	XMEAS(33)
	Component F	XMEAS(34)
	Component G	XMEAS(35)
	Component H	XMEAS(36)
Product analysis	Component D	XMEAS(37)
	Component E	XMEAS(38)
	Component F	XMEAS(39)
	Component G	XMEAS(40)
	Component H	XMEAS(41)

Table 2 Process manipulated variables

Variable name	Number	Base value	Units
D feed flow	XMV(1)	63.053	$kg\,h^{-1}$
E feed flow	XMV(2)	53.980	$kg\,h^{-1}$
A feed flow	XMV(3)	24.644	ks cm h
A and C feed flow	XMV(4)	61.302	ks cm h
Compressor recycle valve	XMV(5)	22.210	%
Purge valve	XMV(6)	40.064	%
Separator pot liquid flow	XMV(7)	38.100	$m^3\,h^{-1}$
Stripper liquid product flow	XMV(8)	46.534	$m^3\,h^{-1}$
Stripper steam valve	XMV(9)	47.446	%
Reactor cooling water flow	XMV(10)	41.106	$m^3\,h^{-1}$
Condenser cooling water flow	XMV(11)	18.114	$m^3\,h^{-1}$
Agitator speed	XMV(12)	50.000	rpm

Downs and Fogel [8] initially defined 20 process faults and an additional valve fault was further introduced in [1], see Table 3.

Table 3 Descriptions of process faults in TE process

Fault number	Process variable	Type
IDV(1)	A/C feed ratio, B composition constant	Step
IDV(2)	B composition, A/C ration constant	Step
IDV(3)	D feed temperature	Step
IDV(4)	Reactor cooling water inlet temperature	Step
IDV(5)	Condenser cooling water inlet temperature	Step
IDV(6)	A feed loss	Step
IDV(7)	C header pressure loss-reduced availability	Step
IDV(8)	A, B, and C feed composition	Random variation
IDV(9)	D feed temperature	Random variation
IDV(10)	C feed temperature	Random variation
IDV(11)	Reactor cooling water inlet temperature	Random variation
IDV(12)	Condenser cooling water inlet temperature	Random variation
IDV(13)	Reaction kinetics	Slow drift
IDV(14)	Reactor cooling water valve	Sticking
IDV(15)	Condenser cooling water valve	Sticking
IDV(16)	Unknown	Unknown
IDV(17)	Unknown	Unknown
IDV(18)	Unknown	Unknown
IDV(19)	Unknown	Unknown
IDV(20)	Unknown	Unknown
IDV(21)	The valve fixed at steady state position	Constant position

Simulation conditions

The data set generated using the TE simulator above have been widely used in process monitoring studies; for example Rieth et al. generated both training and testing data for fault detection evaluation [9]. By considering the time constants of the process in closed loop, the sampling time was selected as 3 minutes. In our study, process data are generated containing 21 randomly generated process faults described in Table 3. A moving window approach is employed to calculate the PCA similarity index and the DTW-based distance, where the window contains a pattern indicating a trend or a fault.

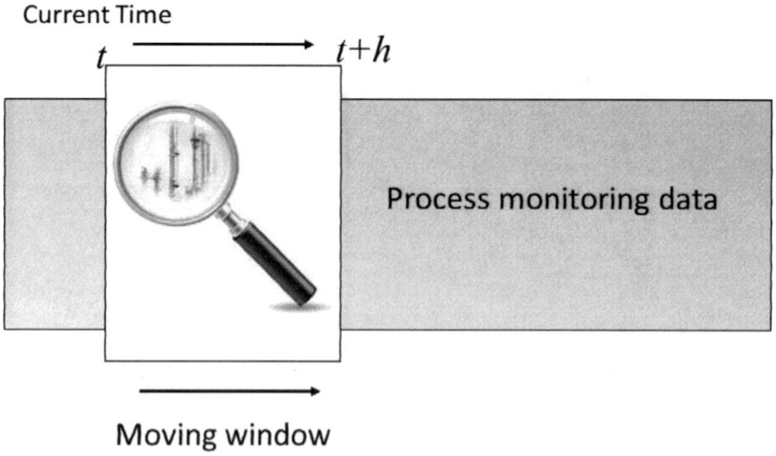

Figure 4 A moving window approach for pattern query

Results and Discussions

Among different types of faults, two of them have been selected for demonstration: a step change (Fault #1) and a sticking valve (Fault #14). The pattern matching results for Faults #1 and #14 are illustrated in Figures 5 and 6. As shown in Figures 5(a) and 6(a), the peaks associated with the locations of Fault #1 and #14 are highlighted in red squares. It is observed that Fault #1 occurs more regularly than Fault #14, and its PCA similarity indices enjoy a smaller standard deviation. After locating where the faults are, we identify the top contributors to both faults through variable contribution plots, and pinpoint their individual patterns using the dynamic time warping (DTW)-based pattern-matching approach, as shown in Figures 5(b) and 6(b). It is observed that the DTW aligns the peaks and valleys of the signal well despite the presence of signal noise. In addition, the Euclidean distance metric used in evaluating the pattern similarity between the two signals obtains a satisfactory result.

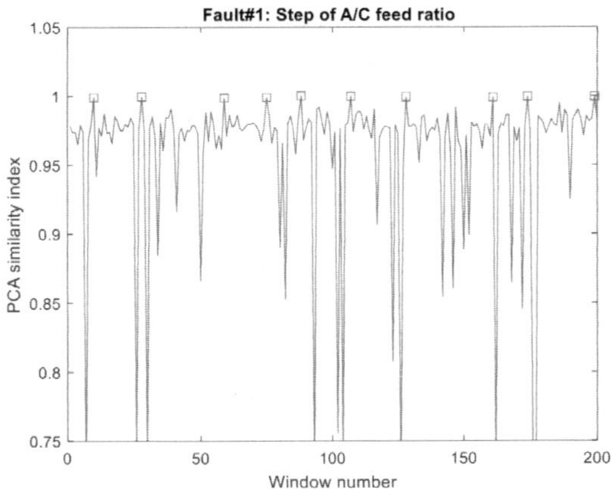

(a) Location found for Fault#1 based on PCA similar index

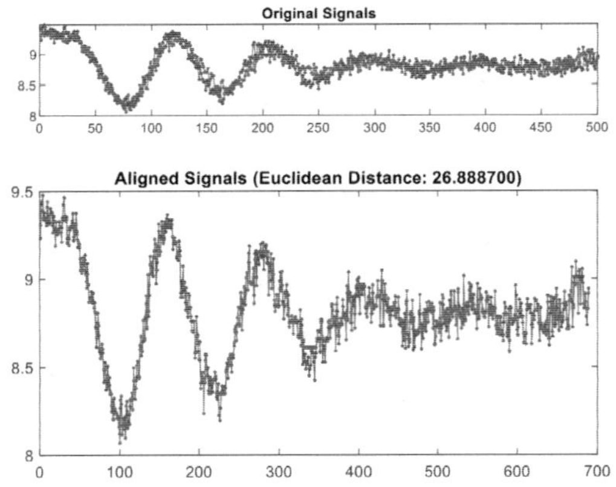

(b) Pattern-matching based on DTW

Figure 5 Pattern query results for Fault #1

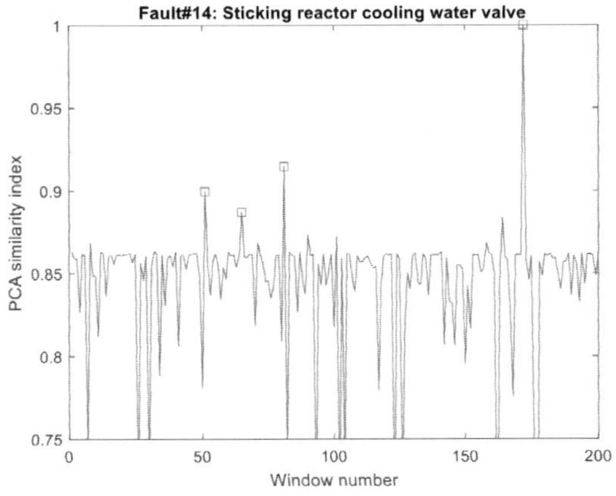

(a) Location found for Fault#14 based on PCA similar index

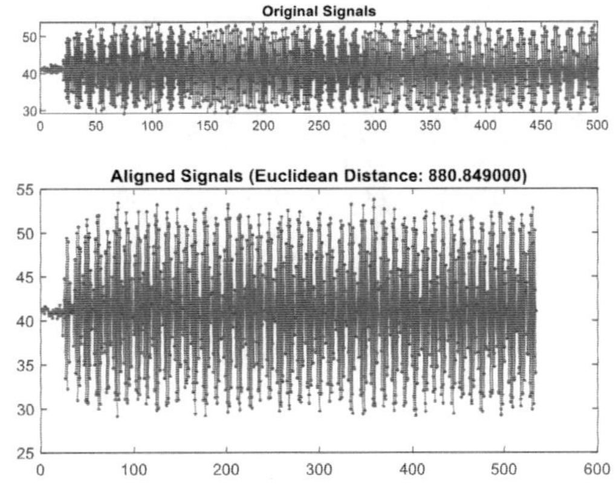

(b) Pattern-matching based on DTW

Figure 6 Pattern query results for Fault #14

Conclusions and Future Work

In the current work, we demonstrate the adoption of an integrated pattern query framework to overcome the challenges faced in process monitoring and fault detection. Such a framework combines PCA similar factor and dynamic time warping (DTW)-based pattern matching approaches, and is tested on data generated from the benchmark Tennessee Eastman process simulator. The method successfully finds periods where process faults occur and pinpoints the measurements of key contributing variables which are similar to the established pattern. Such results illustrate the successful application of data analytics to wrestle with real problems in chemical manufacturing within the Digital Transformation era.

In the future, we will test such a pattern query framework on more industrial data and compare it with other methods.

References

[1] Chiang, L.H., Russell, E.L., and Braatz, R.D. (2000). *Fault Detection and diagnosis in industrial systems*. Springer Science & Business Media.

[2] Kourti, T. and MacGregor, J.F. (1996). "Multivariate SPC Methods for Process and Product Monitoring," *J. Quality Tech.*, 28, 409-428.

[3] Krzanowski, W.J. (1979). Between-Groups Comparison of Principal Components. *J. Amer. Stat. Assoc.*, 74(367), 703-707.

[4] Singhal, A. and Seborg, D.E. (2001) "Matching patterns from historical data using PCA and distance similarity factors," *In Proceedings of the 2001 American Control Conference*. Vol. 2, 1759-1764. IEEE.

[5] Tomasi, G., Van Den Berg, F., Andersson, C.(2004). "Correlation optimized warping and dynamic time warping as preprocessing methods for chromatographic data," *J. Chemom.*, 18, 231-241.

[6] Torgrip,R.J.O, Åberg,M., Karlberg,B., Jacobsson, S.P.(2003). "Peak alignment using reduced set mapping," *J. Chemom.*, 17, 573-582, 2003.

[7] Müller, M. (2007). *Information retrieval for music and motion*.

[8] Downs, J. and Fogel, E.(1993). "A plant-wide industrial process control problem," *Comput. Chem. Eng.* 17, 245-255.

[9] Rieth, C.A.,Amsel, B.D., Tran, R. and Cook, M.B.(2017). "Additional Tennessee Eastman Process Simulation Data for Anomaly Detection Evaluation," *Harvard Dataverse.*

[10] Zhang, Y., Lu, B., and Edgar, T. F. (2013). "Batch Trajectory Synchronization with Robust Derivative Dynamic Time Warping", *ACS Publications*.

589663 Machine Learning Assisted Preventive Maintenance

Abstract:

Machine Learning Assisted Preventive Maintenance

Brian Koren

Graduate Student

Swalm School of Chemical Engineering,

Mississippi State University

bjk187@msstate.edu

Amin Amirlatifi

Assistant professor

Swalm School of Chemical Engineering,

Mississippi State University

amin@che.msstate.edu

ABSTRACT

Majority of manufacturing and chemical plants rely on the knowledge of experienced maintenance staff to plan, schedule, implement, and even predict routine and emergency maintenance needs to avoid unplanned downtime. As many of these knowledgeable staff members begin to reach retirement age, there is a growing demand to document, standardize, and accurately time preventive maintenance, instead of following a predetermined time schedule. This demand is increased by the growing availability of different techniques, such as ultrasonic, vibration and heat analysis, to help predict failures.

The staggering amount of data for predicting maintenance has prompted this research by creating a need for a suite of machine learning methods to analyze and utilize all the different inputs available to predict failures. The present study uses natural language processing (NLP) to read scheduled maintenance plans and analyze maintenance staff notes on how equipment has failed, and what corrective measures were taken afterwards. This information is coupled with time series analysis of data from the time leading up to that failure from the ensemble of sensors in the plant, to define key predictors of the failure using Self Organizing Maps (SOMs) and k-Nearest Neighbors (KNN). A deep neural network (DNN) is trained on the observed trends of key predictors using a combination of Long Short Term Memory (LSTM) and Convolutional Neural Networks (CNN) and system health (healthy/fault prone/failing/failed). The trained system is then used against real-time data from the sensors to predict an upcoming failure, its severity, and prescribe possible remedial actions ahead of failure.

The results from this project showed that many common equipment failures can be defined and predicted by coupling together data from many different inputs surrounding the equipment. The prediction accuracy also improves as more data becomes available to train the model, and when more sensors and inputs are available. The results of this project should encourage industry reliability teams to increase the sensors available for critical equipment and to apply machine learning techniques to handle all the inputs for improved failure prediction.

590246 Data Analytics - a Roadmap to Achieve Operational Excellence

Abstract:

Asset Optimization and Operational performance can be improved significantly by using advanced analytical tools such as Big Data Analytics, IIoT/P, and AI in combination with fundamental models by developing a "**Digital Twin**".

Ingenero uses the "**Digital Twin**" approach to identify operational gaps and effective resolution of issues in timely manner to maximize performance. Ingenero's *Excellence Through Insight* program is successful through combining the right information with in-depth domain expertise and deep data analytics to generate predictive and prescriptive recommendations presented in customized visual displays and dashboards for intelligent decision making.

The success of the process requires the appropriate volume of data. Post collection and mining of data, tool are used for effective reconciliation, validation and consolidation of data to gain insights quickly. This information is utilized as an input for machine learning models and these models convert the raw information into useful statistics/analytics that maximize asset effectiveness.

This paper will discuss Digitalization process utilizing fundamental models combined with data analytic techniques to achieve operational excellence. The appropriate techniques, tools and equipment models that are applied to track the performance and identify the size of deviation and improvement opportunities on real-time basis will be discussed. Additionally the paper will describe how predictive and prescriptive analytics can help determine the impact of deviation, need for corrective action and assist in reaching challenging solutions.

The Industry 4.0 Digitalization initiative is driving industry specific advanced analytics to leverage both historic and real-time data to achieve improvements that are difficult to accomplish using traditional approaches. Substantial economic gains can be achieved when data analytics is coupled with the overall plant operating strategy.

Application of right data analytics process has helped manufacturers to improve their Overall Plant Performance and Capacity by 5 – 15%.

590220 Global Ethylene/Propylene: Refinery Scale Petrochemical Investments and the Impact on Global Markets

Abstract:

Previous announcements have become a reality concerning new refineries built on a chemicals platform with fuel as the byproduct. Long term these type of investments represent a trend that could truly change the market balance. In this presentation we will examine the differing motivations and technologies behind refinery scale olefin investments, how does North America fit into the supply/demand picture, and other long term factors that will play into petrochemical investment decisions.

590229 Disruptive Trends and their Implications

Abstract:

Braskem, known as a leading innovator in petrochemical sustainability, also has a robust transformation agenda preparing for megatrends that could reshape the industry as we know it. In his talk, Kevin will share some real-world examples of exponential change and how disruptive it can be to the most well-established of industries. Exponential change is all around us, but we as humans are often terrible at perceiving and forecasting that change until large disruptions occur. Kevin will talk through four key megatrends impacting the petrochemical industry – Mobility, Energy, Digitization, and Sustainability – and give some examples of how these areas could be disrupted and how Braskem is preparing and embracing these changes.

590251 Panel Discussions - Future of Olefins

Abstract:

Olefins industry has been historically used to cyclic nature of business following the economic cycles. The regional growth has been driven by the regions with demand side or by advantaged feedstock. There are many historical and unique factors at play that will influence the future of olefins industry. In this session, we have invited industry leaders to share their ideas and thoughts about the factors that can significantly impact the industry in mid to long term. Some of the factors that may drive the future include:

- Historical factors like GDP growth, population growth, urbanization trends, quality of life improvements etc.
- Geo-political factors like trade barriers, tensions in Middle East, sanctions etc.
- Environmental pressures like recycling/recovery of plastics, carbon taxation/greenhouse gas emissions, ban of single use plastics etc.
- Changing energy dynamics driven by focus of oil majors away from fuels to petrochemicals
- Large and complex facilities with high capital intensity resulting in challenges to develop, build and operate them safely and achieving desired performance and financial outcomes
- Disruptive technologies like shale developments, transportation systems, sharing platforms, alternate energy sources etc.

589952 Role of Chemicals and Plastics in Quality of Life and Environmental Stewardship: A Look Back over the Past 150 Years

Abstract:

This presentation will look back over the past 150 years of the chemical industry to examine impacts on quality of life, from synthetic nitrogen to address population growth and food supply, to the plastics revolution to address cost reduction and light-weighting of materials to reduce carbon footprints and improve user functionality, and chemicals, catalyst, and materials roles in addressing clean air and water as well as opportunities in future energy. Chemistry and chemicals are key to addressing future energy and enviromental issues.

AIChE Paper Number: 124a

Mixed Plastic Waste To Olefins: From Sorting, Over Impurities To Product Yields And Coke Formation

Marvin Kusenberg
ir.
Ghent University, Ghent, Belgium

Azd Zayoud
Dr. ir.
Ghent University, Ghent, Belgium

Thi Hang Dao
Dr.
Ghent University, Ghent, Belgium

Marko R. Djokic
Dr. ir.
Ghent University, Ghent, Belgium

Nicolas Mys
Dr. ir.
Ghent University, Ghent, Belgium

Uros Kresovic
ir.
Indaver N.V. Belgium, Mechelen, Belgium

Nick Alderweireldt
ir.
Indaver N.V. Belgium, Mechelen, Belgium

Steven De Meester
Prof. Dr. ir.
Ghent University, Ghent, Belgium

Kevin M. Van Geem
Prof. Dr. ir.
Ghent University, Ghent, Belgium

Prepared for presentation at the EPC Technology session: Ethylene Plant Technology Fundamentals, at the 2020 Spring National meeting

Houston, Texas, August 16 – 20, 2020

Mixed Plastic Waste To Olefins: From Sorting, Over Impurities To Product Yields And Coke Formation

Marvin Kusenberg
ir.
Ghent University, Ghent, Belgium

Azd Zayoud
Dr. ir.
Ghent University, Ghent, Belgium

Thi Hang Dao
Dr.
Ghent University, Ghent, Belgium

Marko R. Djokic
Dr. ir.
Ghent University, Ghent, Belgium

Nicolas Mys
Dr. ir.
Ghent University, Ghent, Belgium

Uros Kresovic
ir.
Indaver N.V. Belgium, Mechelen, Belgium

Nick Alderweireldt
ir.
Indaver N.V. Belgium, Mechelen, Belgium

Steven De Meester
Prof. Dr. ir.
Ghent University, Ghent, Belgium

Kevin M. Van Geem
Prof. Dr. ir.
Ghent University, Ghent, Belgium

Abstract:
Plastic waste recycling plays a major role in tackling global challenges such as climate change and an ever growing environmental waste crisis. Mechanical recycling, often claimed to be the most efficient way of utilizing plastic waste, poses limitations for highly mixed plastic waste

streams such as plastic packaging. These obstacles may lead to low quality of recycled products or restrictions for the re-use in, for instance, food packaging. Chemical recycling of plastic waste to base chemicals via pyrolysis and subsequent steam cracking of pyrolysis oils shows great potential to overcome the limitations in present means of plastic waste recycling. In this scenario, the largest concern is the feasibility. Are plastic waste pyrolysis products acceptable steam cracking feedstocks in terms of composition, product yields and radiant coke formation? In this work, steam cracking experiments of pure pyrolysis oil from mixed polyolefins (MPO) as well as of pyrolysis oil from polyethylene (PE)-foils blended with fossil naphtha were performed without prior treatment. Product yields, coke formation and process related issues were benchmarked to fossil naphtha as an industrial feedstock. It was found that plastic waste pyrolysis oils show reasonable product yields in the range of fossil naphtha with roughly twofold coking tendencies. The coke formation could be explained with high olefin contents as well as contaminants in the pyrolysis oils. Possible pre-treatment processes have been suggested to reach feedstock qualities comparable with fossil naphtha. Finally, it has been demonstrated that steam cracking of untreated plastic waste pyrolysis oils shows highly promising results in terms of product yields. It has been concluded that pyrolysis oils of mixed polyolefin waste have the potential to be used as steam cracking feedstock after treatment.

Introduction

During recent years, energy and carbon efficiency became increasingly important in the transition from a linear to a circular economy, which aims to decouple economic growth from the consumption of finite resources. According to the Ellen MacArthur Foundation [1], greenhouse gas emissions could be reduced by 22-44 % in 2050 if the current linear economic path is transitioned to a circular economy.

In this scenario, recycling of plastics plays a key role due to their energy and material intensive production, their abundant use and their long lifetime as a pollutant. Global plastics production exceeded 350 million tons in 2018 and is expected to double within the coming two decades [2]. Roughly one quarter of the produced plastics is used as packaging material, from which only about 5 % is recycled while 95 % is practically single-use material which is either disposed of in landfills or lost to the environment [3]. The enormous volume of plastic waste produced every year creates an increasingly severe waste-handling problem with an estimated 4800-12700 kt of plastics ending up in marine environments [4, 5].

Post-consumer plastic waste is a mixture of several different polymers such as polyethylene (PE), polypropylene (PP), polystyrene (PS), polyvinylchloride (PVC) and polyethylene terephthalate (PET), which further contains a number of additives and auxiliary materials such as stabilizers, flame retardants, plasticizers and colorants. Polyolefins (PE and PP) are the most prominent components found in plastic waste since they are the main ingredients in plastic packaging which represents by far the largest part of the waste streams. [5, 6]

The two main recycling options for plastic waste are mechanical recycling and chemical recycling. Mechanical recycling aims for the conversion of sorted plastic waste streams into products of similar quality and application. For this, single-polymer waste streams are needed which are rarely available due to the highly mixed waste streams, large amounts of additives and the substantial sorting effort required for high purity streams [7, 8]. The large complexity of post-consumer plastic waste makes mechanical recycling of plastic packaging waste challenging due to certain limitations regarding separation and re-use of materials (i.e. food-contact applications or value-added materials) [9]. Additionally, sophisticated multilayer packaging poses considerable challenges to waste sorting facilities. Therefore, mechanically recycled products are often of lower value than the original products. [3]

The highly diverse nature of mixed plastic waste requires a recycling technology which is able to produce high-quality virgin feedstock applicable for any kind of purpose. Therefore, thermochemical decomposition into base chemicals represents a promising recycling technology. Base chemicals such as ethylene and propylene are used to produce new virgin plastics thus closing the material loop [10]. For this purpose, a two-step process can be applied, namely pyrolysis of solid plastic waste and subsequent steam cracking of the pyrolysis oils.

In a pyrolysis reactor, solid plastics are thermally decomposed in an inert atmosphere yielding a distribution of gaseous, liquid and solid products depending on process conditions such as temperature, pressure and residence time [11, 12]. Depending on the polymer, different decomposition mechanisms occur. For instance, polystyrene decomposes in a depolymerization reaction yielding mainly monomers. PE and PP, however, decompose in a random fragmentation reaction yielding a wide range of products. [8, 9, 13] The solid pyrolysis product typically consists of carbon and the inorganic content of the original feedstock [11]. Liquid products of mixed plastic waste pyrolysis contain a complex mixture of hydrocarbons with a wide carbon number distribution. These may be utilized as petrochemical feedstock after upgrading with processes such as hydrotreatment and distillation. Impurities in the pyrolysis oil such as oxygenates, sulfur or nitrogen containing compounds are known to diminish the quality of the liquid product making further treatment necessary in order to obtain petrochemical feedstocks of high quality such as naphtha. [8]

Base chemicals such as ethylene and propylene are produced via steam cracking which is a thermal process utilizing substantial amounts of energy in order to decompose large hydrocarbon molecules into light olefins [6]. Steam cracking of plastic waste pyrolysis oils requires a liquid feedstock that is compatible with industrial steam cracking units in terms of physical properties (e.g. viscosity, boiling point), molecular composition (e.g. olefinic, paraffinic, naphthenic and aromatic contents) and potential contaminant limits (e.g. oxygenates, sulfur, nitrogen, chlorine, metals) [14]. Furthermore, sufficient amounts of high quality feedstock are needed to meet the huge scale of operation of industrial steam crackers. Therefore, it is likely that in an industrial scenario, plastic waste pyrolysis products will be blended with fossil naphtha in order to reach sufficient production scales rather than entirely replacing fossil feedstock [15]. Next to the light olefin yields (i.e. ethylene, propylene, 1,3-butadiene), coke formation is a crucial factor which will determine the economic success of plastic waste pyrolysis oils as steam cracker feedstock, since it is a major factor for the time-on-stream of industrial plants [16].

In this work, waste plastic feedstocks have been pyrolyzed and the liquid products have been steam cracked in a bench-scale steam cracking unit in order to study the feasibility of plastic waste as petrochemical feedstock. As a reference feedstock, fossil naphtha has been steam cracked in a separate experiment to compare product yields as well as coke formation to an industrially accepted feedstock. Plastic pyrolysis oils, as well as steam cracking effluents, have been analyzed using comprehensive two-dimensional gas chromatography coupled to a flame ionization detector (GC × GC-FID). Coke formation was measured using an infrared meter in a controlled decoking procedure after completion of the respective steam cracking experiments.

Materials and methods

Plastic waste samples: Two separate post-consumer plastic waste feedstocks, provided by Indaver N.V. (Belgium), have been used, namely a mixed polyolefin (MPO) sample, mainly consisting of PE and PP and a PE-film sample consisting primarily of PE-foils with a random contamination of other polymers such as PP and PET. Samples provided have been manually sorted, hence no PVC was present. A typical industrial light naphtha was used as a steam cracking reference feedstock.

Analytical setups: Comprehensive two-dimensional gas chromatography coupled to a flame ionization detector (GC × GC-FID) was used to determine the detailed composition of the different feedstocks. The feedstock analyses (off-line), as well as the reactor effluent analyses (on-line), were performed using Thermo Scientific TRACE GC × GC setups (Interscience, Belgium) equipped with a two-stage cryogenic liquid CO_2 modulator [17] and coupled to a flame ionization detector (FID) [18, 19]. As an internal standard, 3-chlorothiophene was used.

Experimental setups and methods: Plastic waste samples have been pyrolyzed in a pyrolysis pilot setup at the Laboratory for Chemical Technology (LCT) at Ghent University, Ghent, Belgium (shown in Figure 1).

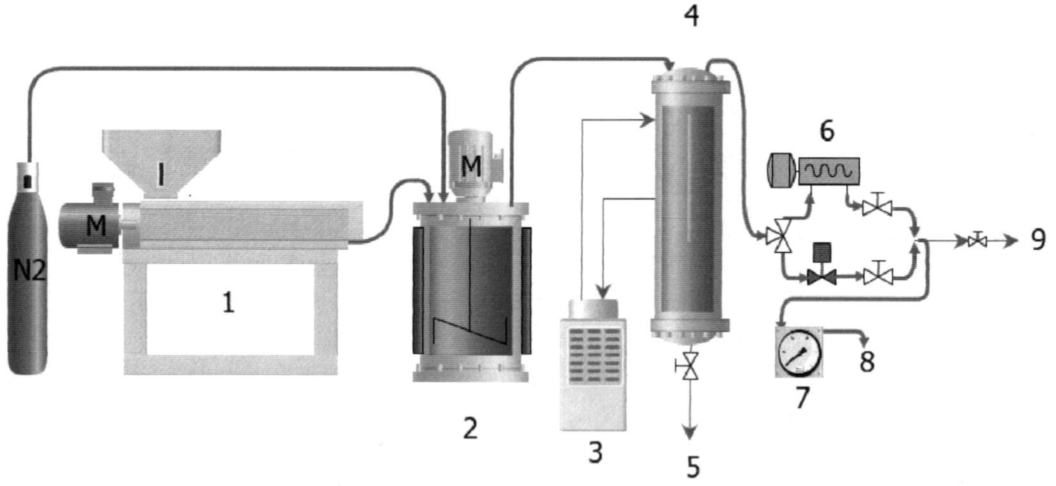

Figure 1: Pyrolysis pilot setup installed at the Laboratory for Chemical Technology (LCT) at Ghent University, Ghent, Belgium: (1) extruder, (2) stirred-tank reactor, (3) cooler, (4) condenser, (5) liquid product sampling, (6) vacuum pump, (7) volumetric gas flow-meter, (8) vent, (9) gaseous product sampling.

The setup consists of an extruder connected to an externally heated stirred-tank reactor vessel. The plastic feedstock is fed in pellet form to the extruder which is used to melt and feed the plastic to the pyrolysis reactor. Plastics are thermally cracked and leave the reactor in vapor form. In the downstream section of the reactor, liquid products are condensed and collected.

The flow of non-condensable gaseous products is measured using a volumetric gas flow-meter. Solid products remain in the reactor and are weighted separately after completion of each experiment in order to close the mass balance. The pyrolysis oils which were steam cracked in this study were produced at 500 °C and 1 bar for PE-film and at 450 °C and 5 bar for MPO. Both experiments were performed with feeding rates of 1 kg/h.

Subsequent steam cracking experiments with pyrolysis oils produced in the plastic waste pyrolysis setup have been carried out at a bench-scale steam cracking setup (BSSC) at the LCT (Ghent University, Belgium). A schematic representation of the bench scale unit is shown in Figure 2.

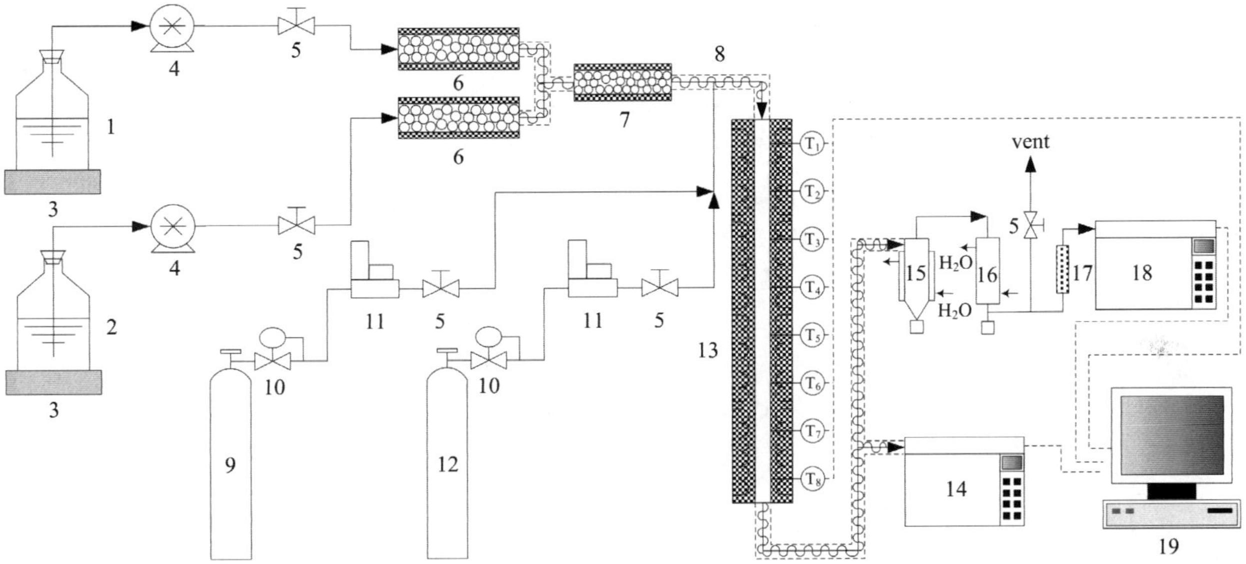

Figure 2: Schematic representation of the bench scale unit: (1) water; (2) hydrocarbon; (3) electronic balance; (4) pump; (5) valve; (6) evaporator; (7) mixer; (8) heater; (9) nitrogen; (10) pressure regulator; (11) mass flow controller; (12) air; (13) reactor; (14) GC × GC; (15,16) condensers; (17) dryrites; (18) RGA; (19) data acquisition. [20]

The setup consists of three sections, namely the feeding section, the furnace/reactor section, as well as the analysis section. Since a detailed description of the setup has already been published elsewhere [20, 21], only a brief description of the setup will be given. The feeding section controls the supply of the feedstock and the water to the reactor coil. The flow of liquids is regulated by Coriolis flow-meter (CORI-FLOW) controlled pumps (Bronkhorst, The Netherlands). Liquid feedstock bottles may be heated using electrical heating mantles in case of more viscous feedstocks to avoid solidification and to guarantee a stable operation of the feeding section. Steam and hydrocarbon feedstock are pre-heated and mixed in electrically heated ovens, filled with quartz beads, enabling an efficient and uniform mixing of feedstock and diluent prior to entering the reactor. The mixture enters the 1.475 m long reactor which is placed vertically in a rectangular electrically heated furnace. There are eight thermocouples positioned along the axial reactor coordinate, measuring the process gas temperature at different positions. The furnace is divided into eight separate sections which can be controlled independently to set a specific temperature profile. The pressure in the reactor is controlled by a back pressure regulator positioned downstream from the outlet of the reactor. Two pressure transducers, placed at the inlet and outlet of the reactor, indicate the coil inlet (CIP) and the coil outlet pressure (COP), respectively. At the reactor outlet, the injection of nitrogen provides an internal standard for analytical measurements and contributes to a certain extent to the quenching.

The reactor effluent is sampled online, i.e. during operation, at high temperature (300 °C). Via a valve-based sampling system and uniformly heated transfer lines, a gaseous sample of the reactor effluent is injected onto the comprehensive two-dimensional gas chromatograph (GC × GC) coupled to a flame ionization detector (FID). The high-temperature 6-port-2-way sampling valve is placed in an oven in which the temperature is kept above the dew point of the effluent sample [18]. Further downstream, the reactor effluent is cooled, water and condensed heavier products (pyrolysis gasoline and pyrolysis fuel oil) are removed, while the remainder of the gaseous effluent stream is sent to the vent. Before reaching the vent, a fraction of the effluent is withdrawn for analysis on the so-called Refinery Gas Analyzer (RGA) using a built-in gas sampling valve system (80 °C).

The reactor is made of Incoloy 800HT (Ni: 30-35, Cr: 19-23, Fe: > 39.5 wt.%). The length of the reactor where the reaction occurs is approximately 1.475 m long, with 0.006 m internal diameter.

The C_{4-} analysis of the quenched effluent gases is performed on the refinery gas analyzer (RGA). Hydrogen, carbon dioxide, carbon monoxide, nitrogen, methane, ethane, ethylene and acetylene are all detected by a thermal conductivity detector (TCD). While the remaining C_1 to C_4 compounds are analyzed by a flame ionization detector (FID) leading to an internal validation. Each detector is mounted on the Interscience Trace GC Ultra. Comprehensive two-dimensional GC × GC-FID is used to detect the C_{5+} compounds. A detailed description of the on-line effluent analysis procedure performed at LCT, Ghent, Belgium has been published elsewhere [18, 22].

Two different coil outlet temperatures (COT) have been tested in the experiments, namely 820 °C and 850 °C. The coil inlet temperature (CIT) was set to 500 °C for all experiments. For both COTs, temperature profiles were set using 8 separate heating zones in the reactor. The coil outlet pressure (COP) was set to 1.7 bar for all experiments using a back pressure regulator downstream of the reactor. A steam dilution of 0.5 was chosen for all experiments. To mimic industrial conditions, 250 ppm of sulfur was continuously added to the steam flow in form of dimethyl disulfide (DMDS), which is known to reduce the CO formation and coking tendency [23]. An overview of the process conditions is given in Table 1.

Table 1: Overview of the performed experiments at the BSSC setup at LCT, Ghent, Belgium.

Temperature profile 1 [°C]	500-550-590-630-670-700-740-780-820
Temperature profile 2 [°C]	500-550-590-630-680-720-770-810-850
Coil outlet pressure [bar]	1.7 bar
Liquid hydrocarbon flow [g/h]	150 g/h
Steam flow [g/h]	75 g/h
Continuous sulfur addition	250 ppm from DMDS

Steam cracking of plastic waste pyrolysis oils was performed without any treatment prior to the experiment. Both pyrolysis oils had a high visible viscosity as well as solid impurities suspended in the liquid. MPO pyrolysis oil, which was in liquid state at ambient conditions, was steam cracked without further mixing with another hydrocarbon. However, due to the viscosity, the feed bottle was heated during operation. PE-film pyrolysis oil had a waxy constitution and was in a solid state at ambient conditions with a melting point around 80 °C. For the steam cracking experiments, it was blended with naphtha for two reasons. First, due to its high melting point it was not practical to feed it pure to the steam cracking unit. Second, plastic waste pyrolysis oils are unlikely to entirely replace fossil feedstock in the future but rather complement it. Therefore, PE-film pyrolysis oil was blended with naphtha in a ratio of 25 wt.% of plastic pyrolysis oil and 75 wt.% of naphtha. The chosen ratio exceeds the expected mixing ratio in industrial steam crackers which is more likely to range around 10 % due to different orders of magnitude in scale between industrial naphtha crackers and plastic waste sorting plants.

Next to steam cracking, controlled decoking experiments were performed after each experiment. For this, the amount of coke which was deposited during each respective steam cracking run was quantified. The quantification was performed using an infrared meter which measures the volumetric concentration of CO and CO_2 in the reactor effluent produced while decoking the reactor coil. For the decoking experiment, air was fed to the reactor which was heated up to 850 °C in all zones. The decoking experiment was finished once CO and CO_2 were no longer detected in the reactor effluent. The coke was quantified using a volumetric drum-type gas meter and referred to the total steam cracking duration per feedstock.

Results and Discussion

Feedstock characterization: Two plastic waste pyrolysis oils (MPO and PE-film) were steam cracked in this study. Prior to the experiments the samples have been analyzed using GC × GC-FID. The molecular composition of both plastic waste pyrolysis oils is shown in Table 2. As a comparison, the molecular composition of fossil naphtha as a benchmark industrial steam cracking feedstock is shown as well. It is to be noted that the composition of plastic waste pyrolysis oils differs significantly depending on the process conditions during pyrolysis as well as on the exact composition of the plastic waste. This has been shown in several studies [9, 11, 13].

Table 2: Molecular composition of the plastic waste pyrolysis oils (MPO, PE-film) and fossil naphtha reference analyzed with GC × GC-FID.

Feedstock	MPO	PE-film	Fossil naphtha
Carbon number range	$C_5 - C_{51}$	$C_5 - C_{47}$	Up to C_9
Paraffins [wt.%]	41	37	79
Olefins [wt.%]	40	55	-
Naphthenes [wt.%]	11	7	17
Aromatics [wt.%]	8	1	4

As an example, Figure 3 provides the GC × GC-FID color plot of MPO pyrolysis oil which underlines the multitude of compounds present. The respective compound families are marked as well.

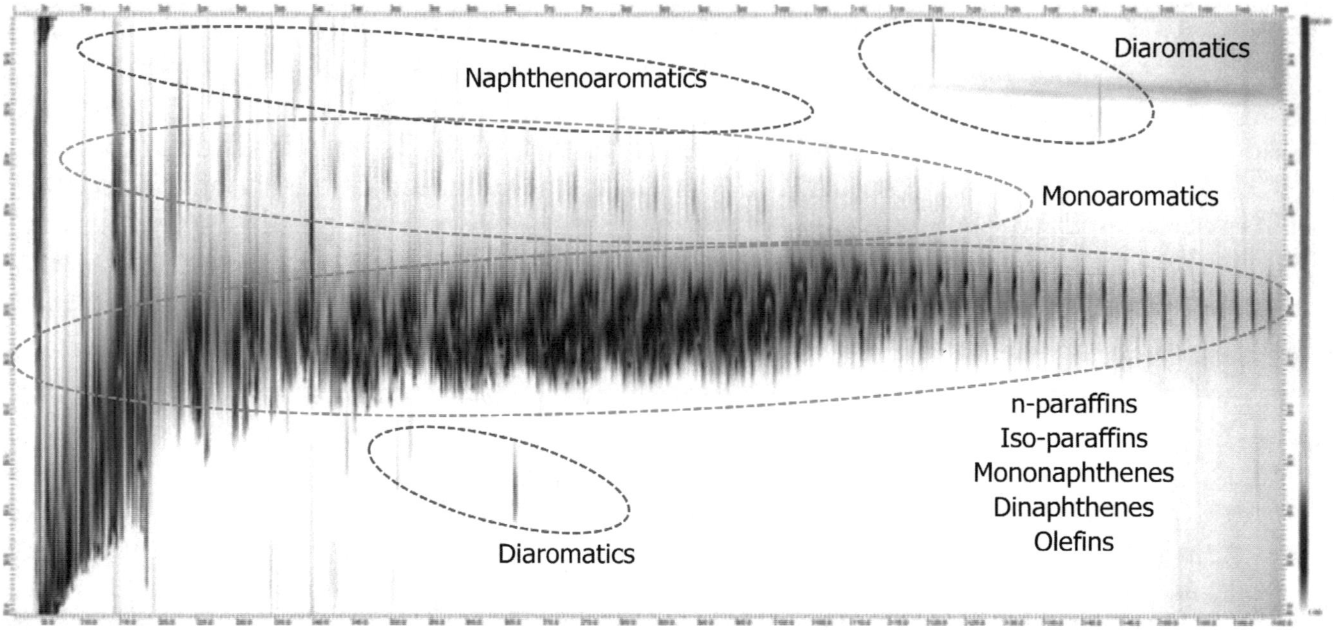

Figure 3: GC × GC-FID color plot of MPO pyrolysis oil. Compound families are marked individually.

It can be seen from the feedstock analysis that the plastic waste pyrolysis oils have a wide carbon range distribution from C_5 up to C_{51} which is in contrast to the carbon range of the light naphtha reference feedstock. This large discrepancy makes the pyrolysis oil problematic for an industrial naphtha cracker without prior fractionation steps. Since the presented bench-scale steam cracking setup used in this study is able to handle feedstocks with boiling points up to 650 °C, no further fractionation steps were performed. Considering the ongoing transition in industry towards heavier petrochemical feedstocks for steam crackers such as crude oil, fractionation into lighter fractions might not be necessary for future use [24, 25].

Regarding the molecular composition, there are large differences between the pyrolysis oils and fossil naphtha especially in the olefin content which would make the pyrolysis oils unattractive for industrial steam crackers due to issues such as high coking and fouling potential as well as the inhibition of paraffin cracking reactions [26, 27]. Processes such as hydrotreatment could be performed to adjust the olefin content as well as to remove nitrogen, sulfur and oxygen-containing compounds which may lead to coke formation, fouling and catalyst poisoning in later upgrading stages [28]. However, the aim of this work is to proof the concept of steam cracking raw plastic pyrolysis oils without further treatment in terms of light olefin yields, coke formation as well as processability in a bench-scale steam cracking unit. It has to be noted that due to the blending of PE-film pyrolysis oil with fossil naphtha, the olefin content is naturally reduced to a far lower content (to ~14 wt.%) by virtue of the absence of olefins in fossil naphtha.

Figure 4: Residue in the feed bottle of PE-film pyrolysis oil.

In the steam cracking experiments observations could be made regarding the processability of the plastic feedstocks. First, due to the very high viscosity of the feedstocks, additional measures were needed such as heating of the feed bottle and tracing of the feeding lines. Second, as already mentioned, both plastic pyrolysis oils contained solid impurities which ultimately caused issues with the feeding pumps since particles in the feedstock eventually precipitated from the liquid (see Figure 4). To overcome this problem, hot filtration for the removal of solids is advisable.

Cracking product yields: Figure 5 shows the product yields of methane and the most important light olefins ethylene, propylene and 1,3-butadiene in wt.% for the two respective COTs 820 °C and 850 °C.

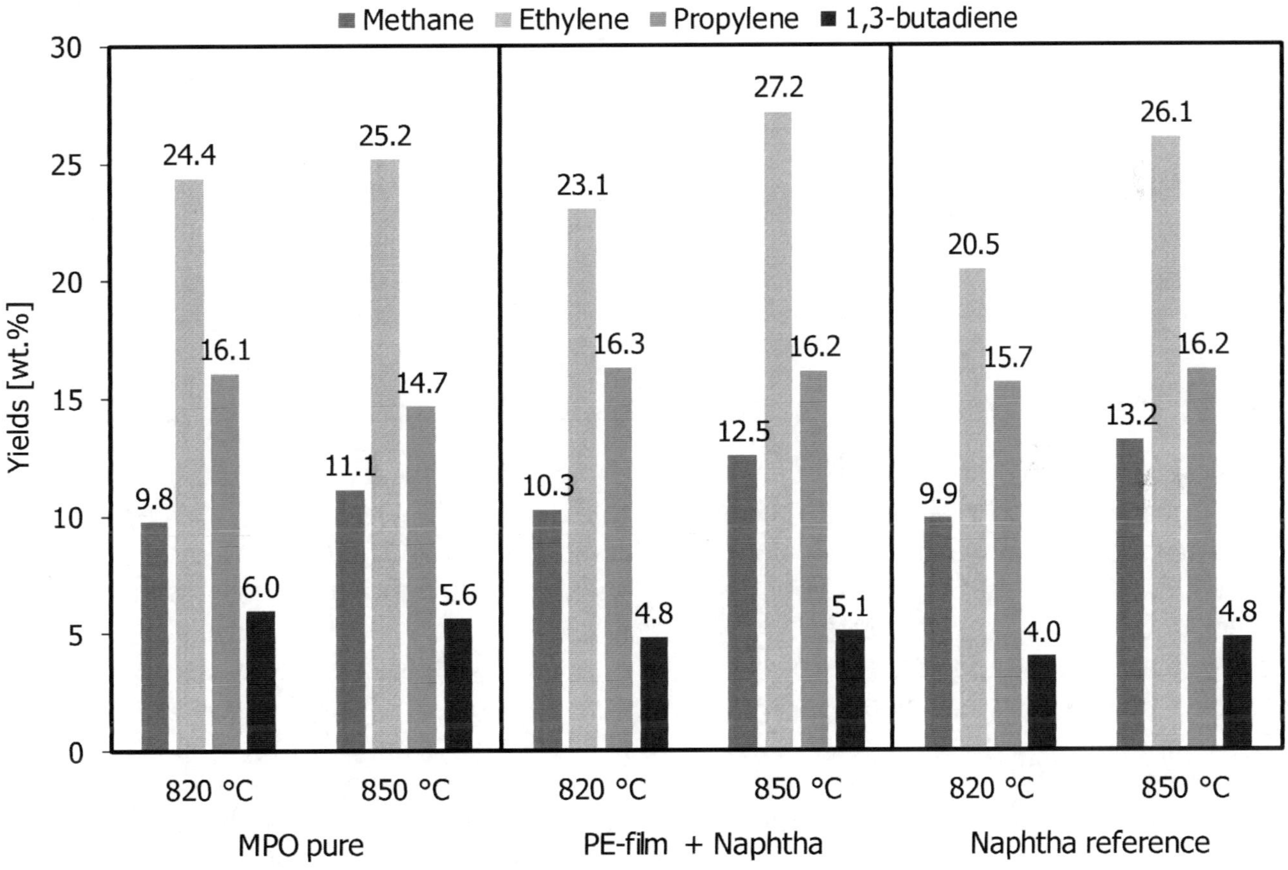

Figure 5: Light olefin yields of steam cracking pure MPO pyrolysis oil, PE-film pyrolysis oil blended with naphtha and naphtha reference in wt.%.

From Figure 5 it can be seen that light olefin yields for all three feedstocks are within the same range with a clearly visible dependency on the temperature. For all three feedstocks, higher cracking severity led to an increase in ethylene yield with a corresponding decrease in propylene yield. This trend is most prominent for fossil naphtha with a sharp increase from 20.5 wt.%

ethylene at a COT of 820 °C to 26.1 wt.% at a COT of 850 °C. The temperature dependency of the yields of pure MPO pyrolysis oil are rather low indicating the influence of the high olefin content of more than 40 wt.% which inhibits paraffin cracking reactions. At a COT of 820 °C, however, the lights' yields of the two pyrolysis oils exceed these of fossil naphtha which indicates that the chosen lower temperature profile is, of course, not optimal for steam cracking of fossil naphtha. Detailed yields of methane, ethylene, propylene and 1,3-butadiene next to the total C_{4-} yields for all experiments are listed in Table 3. The presented values are averaged from at least three stable measurements for each condition, respectively.

Table 3: Result summary of the product yields of the most important light products during steam cracking of the three feedstocks at the chosen process conditions. Numbers given are averaged mass yields of at least three stable measurements for each condition, respectively.

Feedstock	**MPO pure**		**PE-film + Naphtha**		**Naphtha reference**	
COT [°C]	820	850	820	850	820	850
Total C_1-C_4 Species [wt.%]	69.51	68.94	68.02	72.25	68.33	77.45
			Yields [wt.%]			
Methane	9.78	11.14	10.27	12.54	9.89	13.19
Ethylene	24.37	25.17	23.06	27.17	20.48	26.07
Ethane	3.52	3.41	3.17	3.18	2.74	3.05
Propylene	16.10	14.66	16.28	16.15	15.66	16.21
1-butene	2.73	1.76	2.82	2.14	2.82	2.19
Iso-butene	3.30	2.96	3.06	2.75	3.54	3.26
1,3-butadiene	6.00	5.64	4.82	5.09	3.99	4.83

With respect to Table 3, it can be seen that although the plastic waste pyrolysis oils show high ethylene, propylene and 1,3-butadiene yields, the total light olefin yields (C_{4-}) yield lie well below the one of fossil naphtha for a COT of 850 °C. For the lower COT of 820 °C, the values are similar which is confirmed by the fairly low ethylene and propylene yields of fossil naphtha at this cracking severity. The lower total C_1-C_4 species yields of plastic waste pyrolysis oils at 850 °C compared to fossil naphtha can be explained by the respective feedstock molecular composition. Very heavy feedstocks yield more heavy products which are in the range of pyrolysis gasoline (C_5-C_{10}) and pyrolysis fuel oil (C_{10+}) which may induce process fouling. This can be the consequence of high olefin contents which are more prone to secondary reactions yielding heavier products. Additionally, more stable compounds in the feedstock may be unconverted in the cracking experiments [27]. As already mentioned, the olefin content of pure MPO pyrolysis oil is significantly higher than the one of the PE-film – naphtha blend (~40 wt.% vs. ~14 wt.%, respectively) which explains the low total C_{4-} values at a COT of 850 °C compared to the PE-film – naphtha blend and the naphtha reference.

Nevertheless, the presented results underline the feasibility of steam cracking plastic waste pyrolysis oils in terms of product yields. It was shown that steam cracking of untreated plastic waste pyrolysis oils is feasible with light olefin product yields in the range of an industrial naphtha feedstock.

Coking data: A highly important aspect in steam cracking is the formation of coke. Coke depositions on the inner reactor wall act as heat transfer inhibitor increasing the heat demand. Furthermore, the pressure drop along the reactor increases, eventually causing a costly shutdown in order to remove the deposits via a controlled decoking process. [16] In general, coke formation is a direct consequence of feedstock impurities such as heteroatoms but also high olefinic and aromatic contents [26, 29]. Decoking experiments were performed after cracking each respective feedstock according to the described method. Results of the decoking experiments are presented in Figure 6. The total mass of coke was determined using a volumetric drum-type gas meter and put in relation to the total duration of the respective steam cracking runs.

It can be seen from Figure 6 that the plastic waste pyrolysis oils show twofold coke formation compared to fossil naphtha as industrial benchmark. It is further visible that both plastic waste pyrolysis oils show almost equal tendencies (33.4 mg/h for pure MPO pyrolysis oil and 32.8 mg/h for PE-film pyrolysis oil blended with naphtha), which is interesting since pure MPO pyrolysis oil was cracked while PE-film pyrolysis oil was blended with naphtha. This observation implies that the coking behavior of pure PE-film pyrolysis oil would have been substantially higher than the one of MPO pyrolysis oil which indicates that the PE-film pyrolysis oil contains further impurities next to olefins and aromatics which promote coke formation. Considering the mixing ratio of 1:3 of the PE-film + naphtha blend, the theoretical estimated coke formation of pure PE-film pyrolysis oil would be higher than 130 mg/h.

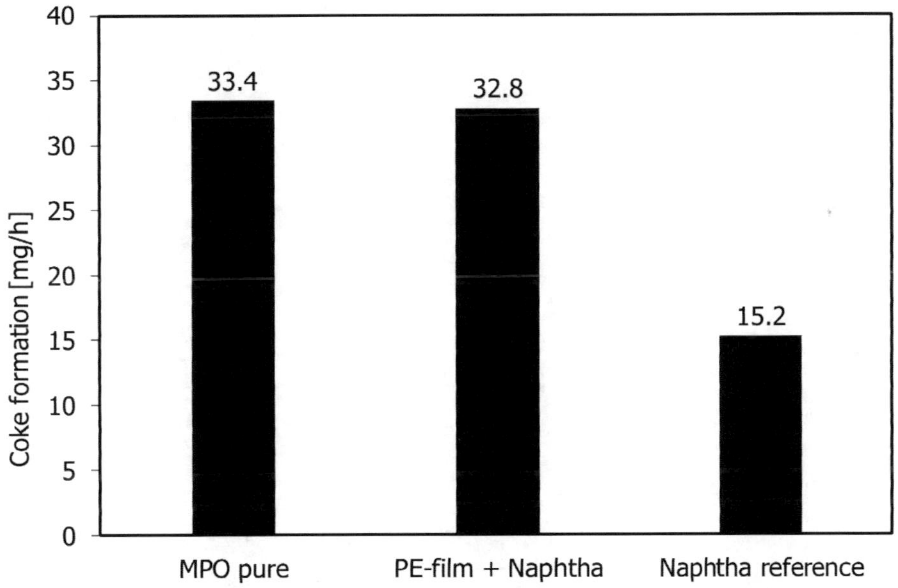

Figure 6: Coke formation in mg/h for MPO pure, PE-film + naphtha as well as fossil naphtha. Data was obtained in a controlled decoking experiment after a total steam cracking duration of at least 4 hours.

Although plastic waste pyrolysis oils show higher coking tendencies than fossil naphtha, the results are still in a reasonable and well explainable range which indicate that steam cracking of these oils is feasible from a coking point of view. The results further suggest that certain treatment steps prior to steam cracking such as hydrotreatment should rather be taken in order to reduce olefin contents, as well as contaminants.

Regarding the PE-film pyrolysis oil – naphtha blend which combined has a reasonably low olefin content of ~14 wt.%, the high coking tendency indicates further impurities which promote coke formation. Further analyses such as elemental analysis (CHNS/O), ICP-AES/OES/MS for metals or ion chromatography for halogens are advised in order to comprehensively understand the composition of post-consumer plastic waste pyrolysis oils. This is especially valid for the plastic foil fractions since these fractions are considered to contain higher amounts of dirt and food residuals as well as metal containing inks [30]. Since contaminants are and will be the determining factor for the industrial use of these products, further research is of paramount importance.

Next to coke formation in the reactor coil, issues such as transfer line exchanger fouling might occur. These issues can be observed by monitoring the pressure difference between reactor inlet and outlet as a function of the reaction time since heavy products condense and deposit in the colder downstream section of the reactor causing a pressure build-up. Fouling issues can furthermore be caused by the presence of sulfur and nitrogen compounds in the feedstock which may react with diolefins in the downstream section. The Δp between the reactor inlet and the reactor outlet it shown in Figure 7 measured during steam cracking pure MPO pyrolysis oil as well as fossil naphtha reference. It can be seen that the pressure difference for fossil naphtha stays constant for a straight cracking duration of 5 hours. MPO, however, shows a steady increase with a sharp rise after 3.5 hours due to a blockage in the downstream section of the reactor. The curve for PE-film pyrolysis oil – naphtha is not shown since the experiment was stopped due to a blockage in the convection section after roughly 4 hours presumably caused by not evaporated feedstock or solid impurities in the liquid feedstock as shown in Figure 4.

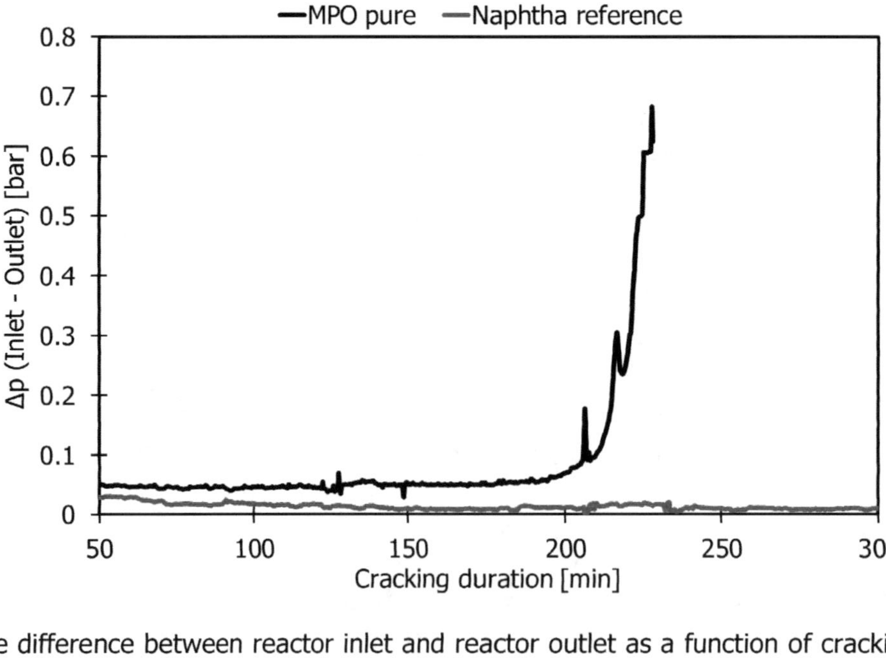

Figure 7: Pressure difference between reactor inlet and reactor outlet as a function of cracking duration for raw MPO pyrolysis oil and fossil naphtha.

The process-related observations suggest that the plastic waste pyrolysis oils require pre-treatment steps such as hot-filtration to remove suspended solids as well as fractionation steps removing heavy residual parts of the oil.

Conclusions and Outlook

Our research clearly shows that steam cracking of post-consumer plastic waste pyrolysis oils is feasible with reasonable product yields and coking tendencies. The pyrolysis oils show potential as steam cracking feedstocks after treatment steps such as hydrotreatment and fractionation via distillation. This has been demonstrated on the basis of steam cracking experiments of raw, untreated plastic pyrolysis oils. Fouling potential leading to blockages in the convection section or the transfer line exchanger has been observed with untreated feedstocks underlining the necessity of pre-treatment steps. These "proof-of-concept" experiments were not intended to ultimately evaluate plastic waste pyrolysis oils as steam cracking feedstocks but rather to demonstrate the feasibility of the streams without prior treatment. The results underline the potential of the feedstocks when pre-treatment steps are applied. Distillation steps for the production of a naphtha range product fraction might reduce the reported fouling and coking issues substantially enabling an even better benchmarking to an industrial naphtha feedstock. In further treatment steps such as hydrotreatment the molecular composition can be tuned to a more naphtha-like composition in respect of olefin and aromatic contents. Additionally, analyses such as elemental analysis (CHNS/O), ICP-AES/OES/MS or ion chromatography will shed light on the concentration of contaminants such as metals and heteroatoms (N, O, S, Cl) which are a further source for coke formation. With these steps further knowledge will be obtained on the influence of the plastic pyrolysis oil composition on steam cracking yields and coke formation as well as process related issues such as transfer line exchanger fouling.

Moreover, thermal pre-treatment of plastic waste could diminish obstacles such as fouling or low quality pyrolysis oils. Post-consumer plastic waste contains a broad range of additives such as stabilizers, colorants and plasticizers as well as chlorinated compounds especially from small amounts of PVC. Due to their high corrosiveness, chlorinated compounds are to be removed as early in the process as possible. By thermally removing volatiles prior to pyrolysis, it is expected to increase the pyrolysis oil quality substantially. Future work, therefore, aims to combine the feeding of the plastic waste to the pyrolysis reactor with a thermal pre-treatment within the extruder by variation of temperature, application of vacuum and screw speed. This way, volatile organic compounds, halogenic contaminants and a range of added compounds may be removed from the plastic melt prior to pyrolysis.

Acknowledgment

The authors acknowledge financial support by the Catalisti-ICON project (HBC.2018.0262) MATTER (Mechanical and Thermochemical Recycling of mixed plastic waste) funded by Flanders Innovation & Entrepreneurship (VLAIO).

References

[1] Ellen MacArthur Foundation, "Completing the picture how the circular economy tackles climate change," 2019, vol. V.3.

[2] Plastics Europe - Association of Plastics Manufacturers, "Plastics - The Facts 2019," 2019.

[3] Ellen MacArthur Foundation, "The New Plastics Economy: Rethinking the future of plastics," 2016,

[4] Jambeck, J. R., Geyer, R., Wilcox, C., Siegler, T. R., Perryman, M., Andrady, A., Narayan, R., and Law, K. L., "Plastic waste inputs from land into the ocean," *Science,* vol. 347, no. 6223, pp. 768-771, 2015, doi: 10.1126/science.1260352.

[5] Geyer, R., Jambeck, J. R., and Law, K. L., "Production, use, and fate of all plastics ever made," *Science Advances,* vol. 3, no. 7, p. e1700782, 2017, doi: 10.1126/sciadv.1700782.

[6] Thunman, H., Berdugo Vilches, T., Seemann, M., Maric, J., Vela, I. C., Pissot, S., and Nguyen, H. N. T., "Circular use of plastics-transformation of existing petrochemical clusters into thermochemical recycling plants with 100% plastics recovery," *Sustainable Materials and Technologies,* vol. 22, p. e00124, 2019, doi: 10.1016/j.susmat.2019.e00124.

[7] Brems, A., Baeyens, J., and Dewil, R., "Recycling and recovery of post-consumer plastic solid waste in a European context," *J Thermal Science,* vol. 16, no. 3, 2012, doi: 10.2298/TSCI120111121B.

[8] Al-Salem, S. M., Lettieri, P., and Baeyens, J., "The valorization of plastic solid waste (PSW) by primary to quaternary routes: From re-use to energy and chemicals," *Progress in Energy and Combustion Science,* vol. 36, no. 1, pp. 103-129, 2010, doi: 10.1016/j.pecs.2009.09.001.

[9] Ragaert, K., Delva, L., and Van Geem, K., "Mechanical and chemical recycling of solid plastic waste," *Waste Management,* vol. 69, pp. 24-58, 2017, doi: 10.1016/j.wasman.2017.07.044.

[10] Garforth, A. A., Ali, S., Hernández-Martínez, J., and Akah, A., "Feedstock recycling of polymer wastes," *Current Opinion in Solid State and Materials Science,* vol. 8, no. 6, pp. 419-425, 2004, doi: 10.1016/j.cossms.2005.04.003.

[11] Buekens, A. G., "Introduction to Feedstock Recycling of Plastics," in *Feedstock Recycling and Pyrolysis of Waste Plastics,* J. Scheirs and W. Kaminsky, Eds.: John Wiley & Sons, Ltd, 2006, pp. 1-41.

[12] Lopez, G., Artetxe, M., Amutio, M., Bilbao, J., and Olazar, M., "Thermochemical routes for the valorization of waste polyolefinic plastics to produce fuels and chemicals. A review," *Renewable and Sustainable Energy Reviews,* vol. 73, pp. 346-368, 2017, doi: 10.1016/j.rser.2017.01.142.

[13] Angyal, A., Miskolczi, N., and Bartha, L., "Petrochemical feedstock by thermal cracking of plastic waste," *Journal of Analytical and Applied Pyrolysis,* vol. 79, no. 1, pp. 409-414, 2007, doi: 10.1016/j.jaap.2006.12.031.

[14] Sundaram, K. M. and Stancato, B., "How much is too much? Feed contaminants and their consequences," presented at the AIChE 2018 Spring National Meeting, Orlando, Florida, 22-26 April 2018.

[15] Amghizar, I., Vandewalle, L. A., Van Geem, K. M., and Marin, G. B., "New Trends in Olefin Production," *Engineering,* vol. 3, no. 2, pp. 171-178, 2017, doi: 10.1016/J.ENG.2017.02.006.

[16] Patil, M., Djokic, M. R., Marin, G. B., Reyniers, M.-F., and Van Geem, K. M., "Effect of Contaminants on Coke Formation during Steam Cracking of Propane," in *2019 Spring Meeting and 15th Global Congress on Process Safety*, 2019: AIChE.

[17] Adahchour, M., Beens, J., and Brinkman, U. A. T., "Recent developments in the application of comprehensive two-dimensional gas chromatography," *Journal of Chromatography A,* vol. 1186, no. 1, pp. 67-108, 2008, doi: 10.1016/j.chroma.2008.01.002.

[18] Van Geem, K. M., Pyl, S. P., Reyniers, M.-F., Vercammen, J., Beens, J., and Marin, G. B., "On-line analysis of complex hydrocarbon mixtures using comprehensive two-dimensional gas chromatography," *Journal of Chromatography A,* vol. 1217, no. 43, pp. 6623-6633, 2010, doi: 10.1016/j.chroma.2010.04.006.

[19] Pyl, S. P., Schietekat, C. M., Van Geem, K. M., Reyniers, M.-F., Vercammen, J., Beens, J., and Marin, G. B., "Rapeseed oil methyl ester pyrolysis: On-line product analysis using comprehensive two-dimensional gas chromatography," *Journal of Chromatography A,* vol. 1218, no. 21, pp. 3217-3223, 2011, doi: 10.1016/j.chroma.2010.12.109.

[20] De Bruycker, R., Anthonykutty, J. M., Linnekoski, J., Harlin, A., Lehtonen, J., Van Geem, K. M., Räsänen, J., and Marin, G. B., "Assessing the Potential of Crude Tall Oil for the Production of Green-Base Chemicals: An Experimental and Kinetic Modeling Study," *Industrial & Engineering Chemistry Research,* vol. 53, no. 48, pp. 18430-18442, 2014, doi: 10.1021/ie503505f.

[21] Harper, M. R., Van Geem, K. M., Pyl, S. P., Marin, G. B., and Green, W. H., "Comprehensive reaction mechanism for n-butanol pyrolysis and combustion," *Combustion and Flame,* vol. 158, no. 1, pp. 16-41, 2011, doi: 10.1016/j.combustflame.2010.06.002.

[22] Pyl, S. P., Schietekat, C. M., Reyniers, M.-F., Abhari, R., Marin, G. B., and Van Geem, K. M., "Biomass to olefins: Cracking of renewable naphtha," *Chemical Engineering Journal,* vol. 176-177, pp. 178-187, 2011, doi: 10.1016/j.cej.2011.04.062.

[23] Dhuyvetter, I., Reyniers, M.-F., Froment, G. F., Marin, G. B., and Viennet, D., "The Influence of Dimethyl Disulfide on Naphtha Steam Cracking," *Industrial & Engineering Chemistry Research,* vol. 40, no. 20, pp. 4353-4362, 2001, doi: 10.1021/ie001131b.

[24] Corma, A., Sauvanaud, L., Mathieu, Y., Al-Bogami, S., Bourane, A., and Al-Ghrami, M., "Direct crude oil cracking for producing chemicals: Thermal cracking modeling," *Fuel,* vol. 211, pp. 726-736, 2018, doi: 10.1016/j.fuel.2017.09.099.

[25] IHS Markit. (2019, 19/02/2020). *Crude Oil-to-Chemicals (COTC).* Available: https://cdn.ihs.com/www/pdf/0519/COTCBrochurePEP.pdf

[26] Kopinke, F. D., Zimmermann, G., Reyniers, G. C., and Froment, G. F., "Relative rates of coke formation from hydrocarbons in steam cracking of naphtha. 2. Paraffins, naphthenes, mono-, di-, and cycloolefins, and acetylenes," *Industrial & Engineering Chemistry Research,* vol. 32, no. 1, pp. 56-61, 1993, doi: 10.1021/ie00013a009.

[27] Albright, L. F., Crynes, B. L., and Corcoran, W. H., *Pyrolysis: Theory and Industrial Practice.* New York: Academic Press Inc, 1983.

[28] Toraman, H. E., Dijkmans, T., Djokic, M. R., Van Geem, K. M., and Marin, G. B., "Detailed compositional characterization of plastic waste pyrolysis oil by comprehensive two-dimensional gas-chromatography coupled to multiple detectors," *Journal of Chromatography A,* vol. 1359, pp. 237-246, 2014, doi: 10.1016/j.chroma.2014.07.017.

[29] Kopinke, F. D., Zimmermann, G., Reyniers, G. C., and Froment, G. F., "Relative rates of coke formation from hydrocarbons in steam cracking of naphtha. 3. Aromatic

hydrocarbons," *Industrial & Engineering Chemistry Research,* vol. 32, no. 11, pp. 2620-2625, 1993, doi: 10.1021/ie00023a027.

[30] Brouwer, M., Picuno, C., Thoden van Velzen, E. U., Kuchta, K., De Meester, S., and Ragaert, K., "The impact of collection portfolio expansion on key performance indicators of the Dutch recycling system for Post-Consumer Plastic Packaging Waste, a comparison between 2014 and 2017," *Waste Management,* vol. 100, pp. 112-121, 2019, doi: 10.1016/j.wasman.2019.09.012.

AIChE Paper Number: 124b

Oxidative Dehydrogenation of Ethane

*Gunther Kracker**
Linde Engineering

Prepared for Presentation at the 2020 AIChE Spring National Meeting
EPC Technology Session: Ethylene Plant Technology Fundamentals

Houston, TX, March 30 – April 1, 2020

**Presenter*
Co Authors: Rainer Kemper, Desislava Tota, Mathieu Zellhuber, Martin Schubert

Oxidative Dehydrogenation of Ethane

Gunther Kracker
Linde Engineering

Abstract:

Currently steam cracking represents the state-of-the-art technology for olefin production, especially for ethylene and propylene. These base petrochemicals are used within a wide range of downstream processes, and as such continuing growth is expected. The steam cracking process is a mature and well-established technology.

The steam cracking feedstock shift to lighter hydrocarbons, primarily ethane, in the last decade has prompted research and development of alternative on-purpose production route to ethylene. In contrast to propylene (e.g. PDH technology), no relevant on-purpose production route has been developed to date for ethylene at commercial scale.

As a promising on-purpose approach the Oxidative Dehydrogenation of Ethane (ODH-E) as a new process technology is introduced in this paper. Linde, partnering with Clariant for the catalyst supply, developed an ODH-E process demonstrating a high selectivity to valuable products with increased profitability compared to steam cracker technology and with a significant reduction potential of CO2 footprint. This paper discusses the development of this promising process technology, its current status, and relevant performance information.

1 Introduction

The majority share of the global olefin production is still today provided by steam cracking which is a mature and well-established technology, but which shows limited development potential.

Now the oxidative dehydrogenation of ethane (ODH-E) offers an interesting alternative to the conventional steam cracking. By catalytic selective conversion of ethane and oxygen, ethylene and acetic acid are formed as commercial high value products under mild conditions at low pressure and temperatures of max. 400°C.

Principle of the ODH-E Process (main reactions to ethylene and acetic acid):

$C_2H_6 + 0.5\ O_2 \rightarrow C_2H_4 + H_2O \qquad \Delta H_R^0 = -105\ kJ/mol$

$C_2H_6 + 1.5\ O_2 \rightarrow CH_3COOH + H_2O \qquad \Delta H_R^0 = -642\ kJ/mol$

Significant steps have been achieved towards the first-of-its-kind commercial realization of this innovative technology. Presently, no comparable commercial process is available for licensing. The competitive advantage compared to steam cracking mainly comes from the reaction section, which can be operated continuously without decoking or regeneration due to the advantage of oxidative process.

The paper will discuss the ongoing development of this promising ODH-E process technology, its current status, and relevant performance information.

2 Technology Development – Review and Status

Linde´s ODH-E technology is being developed in close cooperation with partner Clariant as catalyst supplier. After successful laboratory and pilot plant tests, a Linde demonstration plant was built up and intensively operated under commercial conditions. A schematic overview of the demo plant arrangement is given in Figure 1. The reactor in the demo plant has been realized as a molten salt cooled, single-tube, fixed-bed reactor with commercial size tube geometry (w.r.t. length and inner diameter).

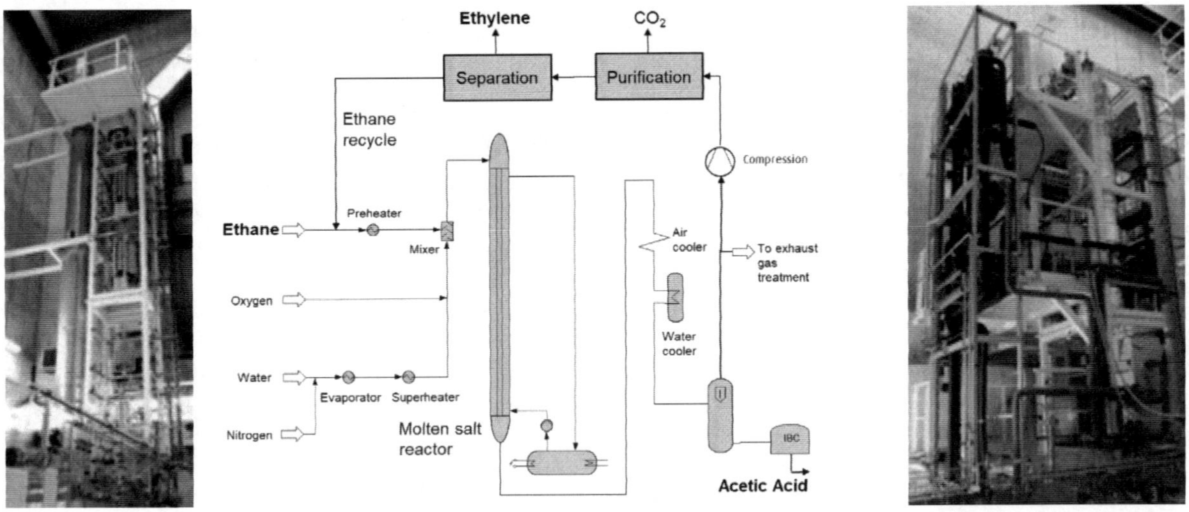

Figure 1: Linde ODH-E Demo Plant

The main focus of the first demonstration phase was on the reactor. Tests have been conducted with various ethane/oxygen/steam feed composition under variation of reaction conditions. The results confirmed laboratory findings regarding high selectivity to valuable products ethylene and acetic acid, with a wide and adjustable range of acetic acid/ethylene production ratios (see Figure 2).

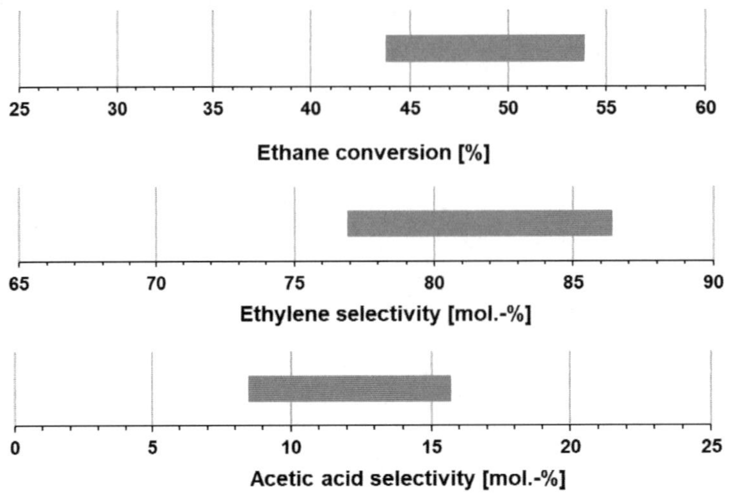

Figure 2: Demo Plant Performance Figures - Ethane Conversion and Product Selectivity

A significant contribution to the demonstration of the new technology has been achieved by operating the demo plant with recycled unconverted ethane. Extending the demo plant with the recycle, the performance data have been validated in a closed-loop plant setup including gas compression, CO2 removal, drying, C2-splitting and closed ethane recycle.

Considering the measured ethane conversion and the product selectivity, overall process performance was confirmed at commercial conditions. At a target ethane conversion at 50% and above, a total selectivity to commercial products (sum of selectivity to ethylene and acetic acid) above 95% was achieved. By choosing different process parameter sets (temperature, pressure, ethane/steam ratio), the desired product ratio of ethylene to acetic acid can be adjusted in a wide range, which is a very attractive feature for potential applications of the ODH-E technology.

As an example shown in Figure 3, a stable plant operation over 1000 hrs accumulated time was realized under commercial conditions. During the long operation campaign with and without closed loop operation, a nearly constant catalyst performance has been observed for all key reaction parameters.

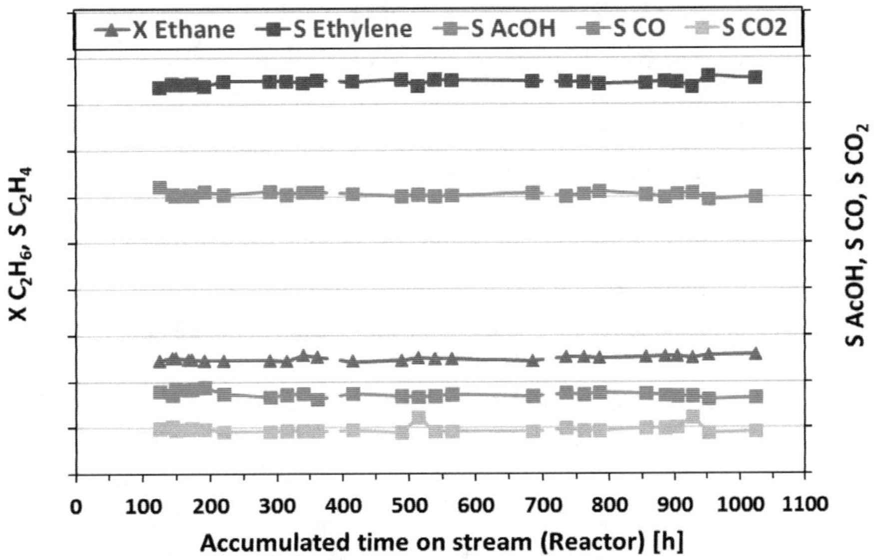

Figure 3: Stable plant operation over 1000 hours (accumulated time)

3 Commercial Technology Layout

The ODH-E technology can be applied as stand-alone facility and takes advantage of integration with a steam cracker. The stand-alone plant concept combines the new developed catalytic process with common well-known process units for purification and gas separation. The process design including a process gas dilution with steam and a cryogenic gas separation train was confirmed to be the most advantageous concept during a systematic evaluation study based on process simulation data. A conceptual design was executed for this preferred process option, resulting in a co-production of ethylene and acetic acid as value products.

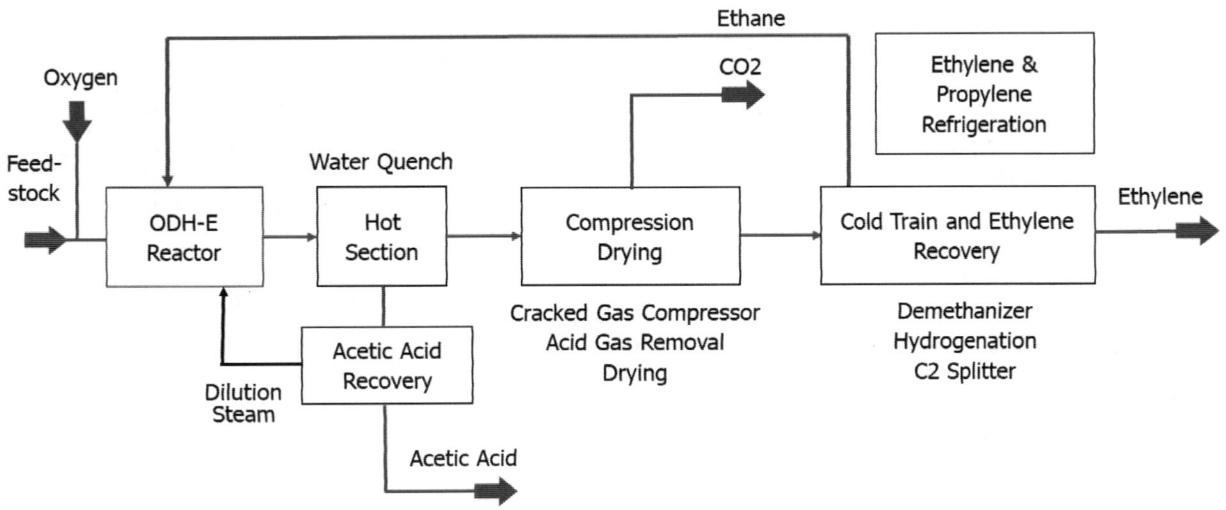

Figure 4: Process Flow Scheme of the Linde ODH-E Technology

Figure 4 provides a process flow overview for a stand-alone ODH-E plant. The fresh feedstock and the recycle streams are converted in the reaction section after mixing with dilution steam and oxygen. The process can take commercially available ethane qualities. An air separation unit upstream of the reactor is foreseen for providing the purified oxygen feed. By catalytic selective conversion of ethane and oxygen under mild conditions, ethylene and acetic acid are formed as main products. Due to parallel and series reactions, the reactor effluent also contains small amounts of by-products CO and CO2, which are removed from the process gas. The reactor effluent is cooled and fed to the water quench column to remove process water and acetic acid. The condensate is routed to a state-of-the-art acetic acid extraction unit to produce high purity acetic acid and recover water for dilution steam.

After separation of water and acetic acid, the process gas leaves the top section of the quench tower to be compressed in a multistage compressor. The compressed gas is then routed for CO2 removal in a purification system comprising an amine wash unit. Here the CO2 product is recovered with high purity for further utilization or storage. Downstream of the amine wash unit, remaining CO2 traces are removed in the caustic wash system.

After CO2 removal the process gas is dried to adsorb water traces before gas processing in the cold separation section. The dried process gas is fed to the cold separation train. After cooling and partial condensation of the C2 products in the low temperature section, the process gas and the condensate stream rich in C2 components are routed to the demethanizer column to separate C2 product from CO and any residual traces of Oxygen. The bottom product is processed in the C2 splitter to recover the ethylene product and recycle the unconverted ethane back to the ODH-E reactor section.

The ODH-E technology can also be integrated with a steam cracker, as shown in Figure 5, which gives an additional flexibility for the application of the ODH-E process. Various combinations, also with existing facilities are possible.

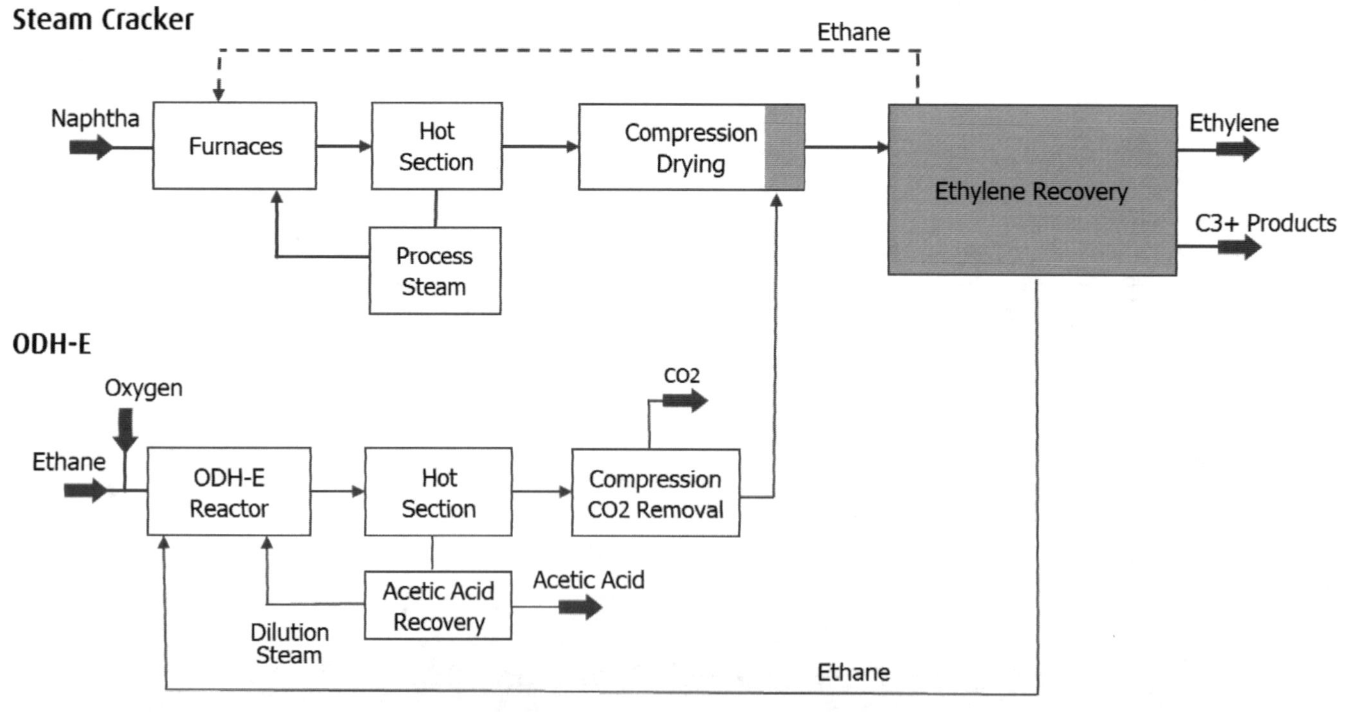

Figure 5: Linde ODH-E Technology Integration with a Steam Cracker

Add-on capacity of ODH-E reactor in existing steam cracker facilities is considered as an attractive business case which can provide additional ethylene production with low investment costs and potential for reducing the CO2 footprint of the combined facility.

4 Reactor Concept of the ODH-E Process

Using the performance data successfully demonstrated by the Linde demo plant and by detailed reactor simulations, a reactor design has been performed for a commercial ODH-E plant with a capacity of 250 kta ethylene.

The new exothermic ODH-E process utilizes an industrially known and well-proven multi-tubular fixed-bed reactor design with molten salt cooling. Figure 6 shows a process design of salt-cooled, fixed-bed reactor. Examples for other industrial applications of the ODH-E reactor system are the synthesis of acrolein/acrylic acid, phthalic and maleic anhydride.

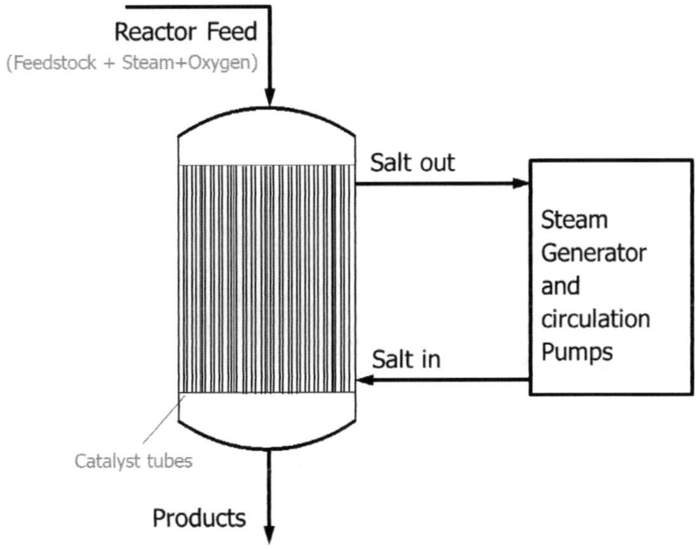

Figure 6: Multitube fixed bed reactor with liquid cooling

The reactor contains a tube bundles filled with the catalyst particles. Within the surrounding reactor jacket flows molten salt as heat transfer medium. The liquid heat carrier is circulated by pumps around the tube bundle to remove the reaction heat of the exothermic oxidative dehydrogenation reaction and to ensure near isothermal conditions. Due to the effective heat transfer a favorable temperature profile in the reactor can be maintained. The salt circulation system includes pumps and a steam generator to use the reaction heat for HP Steam production used for the compressor turbines. The reactor concept ensures highest process safety due to the direct cooling of the tubes.

5 Process Performance Figures

The techno-economic optimization of the commercial scale plant was performed based on rigorous process simulations. Applying the performance figures for a full scale commercial plant layout, an overall material balance and utility consumption have been calculated for a typical stand-alone ODH-E plant with a capacity of 250 kta ethylene.

For the calculated case, the ODH-E plant delivers approximately 0.36 tons of acetic acid per ton of ethylene product, requiring merely 15% higher ethane consumption compared to a steam cracker with the same ethylene capacity. Considering the carbon efficiency[1], the ODH-E process offers a 5% advantage compared to the steam cracking at the same capacity of 250 kta ethylene. This principal advantage of the exothermic ODH-E technology compared to the endothermic steam cracker technology is mostly due to the strongly reduced carbon losses in firing exhaust streams.

Further, the ODH-E technology includes already the carbon capture process step for delivering the CO2 generated in the ODH-E reactor as a by-product in the piping to battery limits for potential utilization in CO2 conversion technologies.

Table 1: Specific CO2 Emissions of the ODH-E Plant (ton CO2 per ton produced ethylene and acetic acid)

	ODH-E Plant including ASU (compressor with E-motor and green power) or ODH-E Plant with import of oxygen		ODH-E Plant including ASU (compressor with E-motor and conventional generated power)		Steam Cracker
	CO2 as carbon captured product	CO2 as emission	CO2 as carbon captured product	CO2 as emission	
Specific CO2 emissions, %	37	46	68	77	100

For the sustainability analysis of the new technology, characteristic values for the greenhouse impact have been calculated to compare the ODH-E process with the steam cracker process. In terms of CO2 emission, even in case of applying conventionally generated electrical power, the ODH-E process offers a considerable improvement compared to the steam cracker processes. For an ODH-E plant including the oxygen production in an Air Separation Unit (ASU) within the battery limits of the ODH-E unit, the specific CO2 emission is reduced by approx. 32% compared to conventional steam cracker in case the captured CO2 is utilized in a downstream application. If no CO2 processing is feasible and CO2 has to be vented, still a 23% CO2 reduction compared to the steam cracker can be claimed.

A further large improvement potential appears considering ODH-E plant including an ASU and utilizing electrical power originating from CO2 neutral or renewable sources. A significant beneficial aspect of the ODH-E technology is a process integration using "green" electricity combined with a "storage" of electrical energy in the form of liquid oxygen generated in the

[1] carbon from feed and fuel gas to carbon from commercial products

ASU. Such integration together with a downstream processing of captured CO2 reduces the specific CO2 emissions by 63% as indicated in Table 1 compared to conventional production with steam crackers.

The application of such a concept would enable an industrial base-chemical production adjusting its energy demand to the availability of renewable electricity, resulting in minimized CO2 footprint. Unlike Power-to-Gas, the objective is not to convert electrical energy to chemical energy for pure energy storage, but to establish more flexible steady production processes and to enable to balance fluctuations in the electrical grid without additional process steps.

6 Benchmarking with Steam Cracking

For the economic analysis, Linde ODH-E technology was compared to state-of-the-art ethane steam cracking technologies. As a reference case, a plant capacity of 250 kta ethylene production was considered. The ODH-E setup considered maximum ethylene production, with corresponding low acetic acid co-production.

A Total Installed Cost (TIC) estimation has been performed for an ethane steam cracker and an ODH-E plant resulting in a verifiable advantage for the ODH-E plant. Feedstock and product prices for the region USA Gulf Coast served as basis for the economic evaluation. Using the Internal Rate of Return (IRR) as indicator for the economic performance, the calculation shows that the ODH-E process profitability increases by 25% compared to the steam cracker. Based on detail analysis, the Linde ODH-E Plant provides higher profitability compared with the steam cracker at same ethylene capacity (at 250 kta ethylene) and is a competitive solution particularly for ethylene and acetic acid demand, allowing for cost-effective and self-sufficient production.

7 Markets

Linde's ODH-E process is hence an on-purpose ethylene technology, which besides ethylene also co-produces a certain amount of acetic acid, with a product split depending on the process conditions. Within the operating window the ODH-E process offers the advantage of a high overall selectivity to valuable products (ethylene + acetic acid selectivity > 95%).

Ethylene consumers, as well as consumers of both, ethylene and acetic acid are the potential target markets. Analysts [1] forecast an increase in the global acetic acid market value to around US$ 16.4 billion by 2026. The market value is estimated to grow at CAGR above 6.2% over the forecast time frame.

As depicted in Figure 7, the acetic acid is mainly converted in synthesis processes to vinyl acetate monomer (VAM), purified terephthalic acid (PTA), acetic anhydride and ethyl acetate.

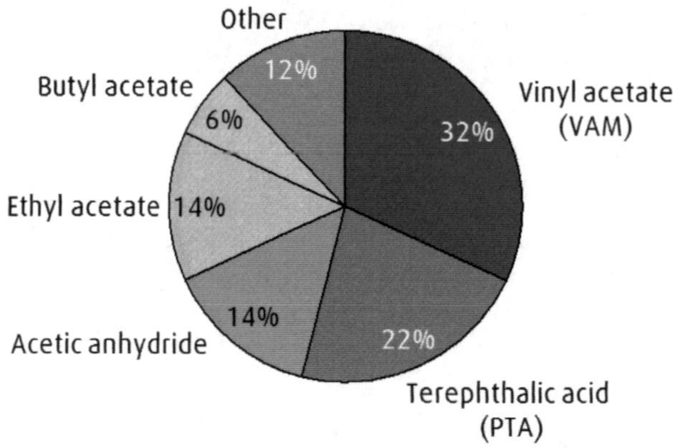

Figure 7: Acetic acid world consumption by end use [2]

VAM and ethyl acetate are products that consume also ethylene feedstock in their manufacturing process. These two products already account for about 46% of the world demand of acetic acid. Moreover, PTA is a product mainly used for the manufacture of polyethylene terephthalate (PET), which consumes ethylene in its value chain to polyethylene terephthalate (PET). The world consumption of acetic acid for PTA accounts for approx. 22%. These three products (VAM, ethyl acetate and PTA), which feature the consumption of ethylene and acetic acid in their production or value chain, account for approx. 60% of the world demand of acetic acid. Therefore, most of the acetic acid demand is linked to products consuming also ethylene. In Table 2 are shown some examples of the required ethylene to acetic acid ratio for these applications.

Table 2: Examples for acetic acid and ethylene consumption in downstream applications

	Acetic acid consumption (t/t product)	**Ratio acetic acid/ ethylene (t/t)**
Vinyl Acetate Monomer (VAM)	≈ 0.7	≈ 2
Terephthalic Acid (PTA)	≈ 0.05	≈ 0.17[1)]
Ethyl acetate	≈ 0.7	≈ 2
Ethyl Vinyl Acetate (EVA)	≈ 0.2	≈ 0.22

1) For PET production via PTA

On these grounds, the co-production of acetic acid with ethylene allows for a new potential promising market for applications simultaneously demanding ethylene and acetic acid.

8 Summary

A new ODH-E technology has been developed by Linde as an alternative to the conventional ethylene production by steam cracking. The performance of the new process has already been successfully demonstrated by Linde demo plant tests, confirming stable reaction conditions and very high selectivity to the valuable products ethylene and acetic acid, furthermore with an easily adjustable product ratio.

Conceptual design of a commercial ODH-E plant and an economic analysis have been performed for comparison with the conventional steam cracking technology. The Linde ODH-E plant provides higher profitability than steam cracker at same ethylene capacity (at 250 kta ethylene) and is a competitive solution for simultaneous ethylene and acetic acid demand, allowing for cost effective and self-sufficient production.

In terms of CO2 emission and greenhouse impact, the ODH-E process offers a strong improvement potential compared to the conventional steam cracker processes, as it delivers CO2 in the process gas carbon captured and it can directly and significantly benefit from the application of green electricity based on renewables. Therefore, the new ODH-E technology represents a significant step forward on the way to economically and environmentally sustainable solutions for base chemicals production.

References

1. Acumen Research and Consulting (2019), "Acetic Acid Market – Global Industry Size, Share, Trends and Forecast 2019-2026"

2. Chemical Economics Handbook (2013), IHS Chemical

Ab initio studies of the mechanism and kinetics of polymerization fouling in ethylene plants

Ying Yu, Dow, Freeport, TX
Mark Davis, Dow, Freeport, TX
Jorge H. Pazmino, Dow, Freeport, TX

Abstract

Polymerization fouling of butadiene and/or styrene can happen under favorable conditions in downstream separation sections of ethylene plants and is a major cause of shutdowns and safety issues. The general move of USGC crackers to lighter feedstocks and the corresponding longer residences times of many reactive chemicals in the plant can lead to higher fouling potential in different areas of the plant. In this work, the mechanism and kinetics of initiation and propagation reactions of the butadiene-styrene system were investigated using density functional theory (DFT) methods. The overwhelmingly dominant initiation mechanism under typical plant conditions was found to be peroxide decomposition over thermal initiation at ppm levels of peroxide impurities. The similarity between isoprene and butadiene in fouling chemistry was unveiled, such as monomer reactivity ratios in the copolymerization with styrene. The results suggest that isoprene is a good analogue of butadiene in lab studies for copolymerization fouling with the advantages of reducing the risk of chemical exposure and fire/explosion. Furthermore, DFT predictions of the instantaneous relative monomer consumption rates of isoprene-styrene system were validated by experimental results at different initial monomer compositions. It is predicted that in a butadiene-styrene system, butadiene will be consumed faster than styrene even at 70% of styrene in monomer composition, and the overall propagation rate constant will increase with higher styrene fractions. We recommend that anti-fouling treatments should focus on eliminating initiators such as peroxides, and aggressively tailoring additive programs based on radical-forming monomers such as styrene. Overall, this study built a basis for integrating computational study with experimental efforts in understanding the fundamentals of polymerization fouling, which could eventually help provide more effective anti-fouling strategies in the production of light hydrocarbons.

1. Introduction

Fouling is a major cause for shutdowns and safety issues in steam crackers and accounts for substantial economic losses. Polymerization fouling is considered to be the main mechanism of foulant formation in downstream separation sections, where favorable conditions for polymerization of butadiene and/or styrene are present. The shift to share gas has escalated the value of cracking lighter for USGC ethylene plants, leading to higher concentrations and/or longer residences times of many reactive chemicals in different areas of the plant that higher fouling potential. Currently, effective online monitoring or predicting tools for polymerization fouling are not yet available. Analytical studies of fouling face great challenges due to the high complexity of foulant composition. Consequently, rational design of anti-fouling treatments is limited by the lack of fundamental understanding of mechanism and kinetics. Given the variety of plant operating conditions, it is critical to obtain fundamental knowledge of polymerization fouling that could serve as a basis to assess and improve anti-fouling treatments and ultimately increase asset utilization. While experimental studies of the polymerization of hazardous monomers (e.g. butadiene) are presented with strict safety constraints, computational studies on such systems offer a safer and more efficient alternative to provide comprehensive mechanistic insights that help guide ongoing and future experiments, as well as the rational design of anti-fouling strategies.

Polymerization fouling falls in the category of free radical polymerization in which polymers form by the successive addition of free radical building blocks. This process is complex and typically consists of initiation, propagation, chain transfer, and termination reactions. Among them, initiation and propagation reactions are key in this study to both identify initiation mechanisms and elucidate co-propagation kinetics under typical operating conditions. With this information, optimum run conditions can be developed and effective inhibition strategies can be identified to reduce process fouling.

Initiation reactions are responsible for generating free radicals that kick off other reactions. In this study, two initiation mechanisms were proposed for the butadiene-styrene system under typical plant conditions with a temperature from 85 °C to 120 °C: thermal initiation and peroxide decomposition. Thermal initiation mechanism of styrene has been studied both experimentally and computationally, as summarized in Figure 1.[1-11] Two mechanisms were previously proposed for the formation of a key Diels-Alder dimer (**A_2H**) intermediate: either a concerted mechanism[2-3] through transition state **$TS_{2\text{-}3}$**, or a stepwise mechanism via a diradical intermediate (**DR_2**).[4] However, there is no consensus yet as to the actual **A_2H** formation mechanism. Therefore, both mechanisms were considered in this study. Molecule-assisted homolysis[10, 12] between **A_2H** and another styrene molecule results in monoradical species, **$A_2\bullet$** and **$HM_2\bullet$**, which initiate polymerization. Two main side reactions include: a **DR_2** ring-closure to form 1,2-diphenylcyclobutane (**DCB**); and an **A_2H** reaction with styrene (**M_2**) to generate an inactive trimer, **A_2M_2**. The thermal initiation mechanism of butadiene has not been reported in the literature. However, dimerization mechanisms of butadiene through Diels-Alder reactions have been widely studied,[13-19] with a simplified dimerization mechanism shown inside the box in Figure 2. The Diels-Alder dimer (**VCH**) is the main product, while a side product, 1,2-divinylcyclobutane (**DVCB**), can be formed by ring closure of the diradical intermediate (**DR_{1tt}**) and may undergo further rearrangement (not shown in Figure 2 as it is beyond the scope of this work).[14, 17-18, 20] In this study, the complete thermal initiation mechanism of butadiene beyond dimerization reactions was mapped as shown in Figure 2 by adapting the mechanism of styrene. **VCH** can react with a butadiene monomer and generate either two monoradicals (**$A_1\bullet$** and **$HM_1\bullet$**) that initiate polymerization, or an inactive trimer (**A_1M_1**). Considering that the conversion from *trans*-butadiene to *cis*-butadiene requires an enthalpy change of 3.14 kcal/mol and a barrier height of 7.43 kcal/mol,[21] we used the energy of *trans*-butadiene as the reference point.

The second proposed initiation mechanism is peroxide decomposition. As possible impurities in the cracked gas, organic peroxides can readily undergo decomposition reactions and form radicals to initiate polymerization. Cumene hydroperoxide (**CHP**) was chosen as a representative peroxide in this study. The mechanism and kinetics of **CHP** decomposition to generate two monoradicals has been reported, with a 0.5 reaction order on **CHP** concentration and rate coefficients listed in Table 1.[22] To determine the dominant initiation mechanisms under typical plant conditions, we reported, for the first time, the predicted thermal initiation energetics of styrene and butadiene systems from the same computational framework, and the comparison of initiation rates between thermal initiation and **CHP** decomposition. Plant conditions were approximated using the compositions of liquid samples from separation units, as listed in Table 2. These results led to quantitatively understanding initiation kinetics and to recommendations for rational design of anti-fouling treatments by inhibiting radical formation.

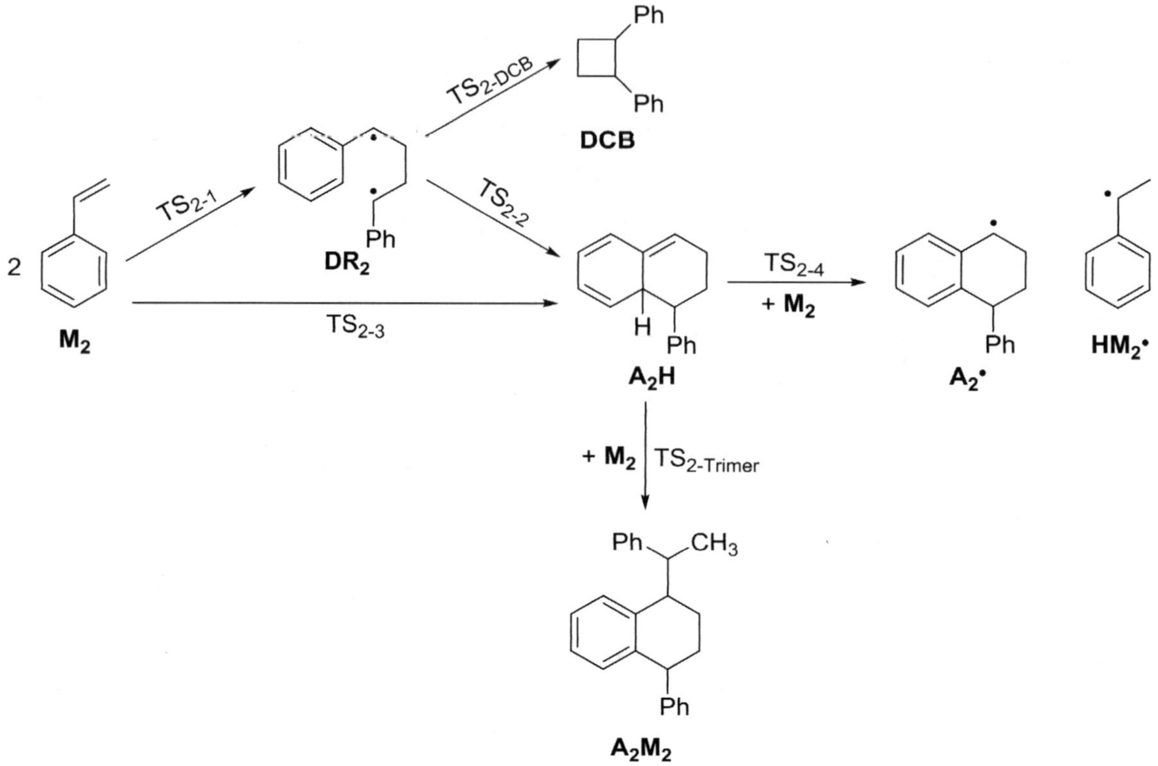

Figure 1. Thermal initiation mechanism of styrene (M_2)

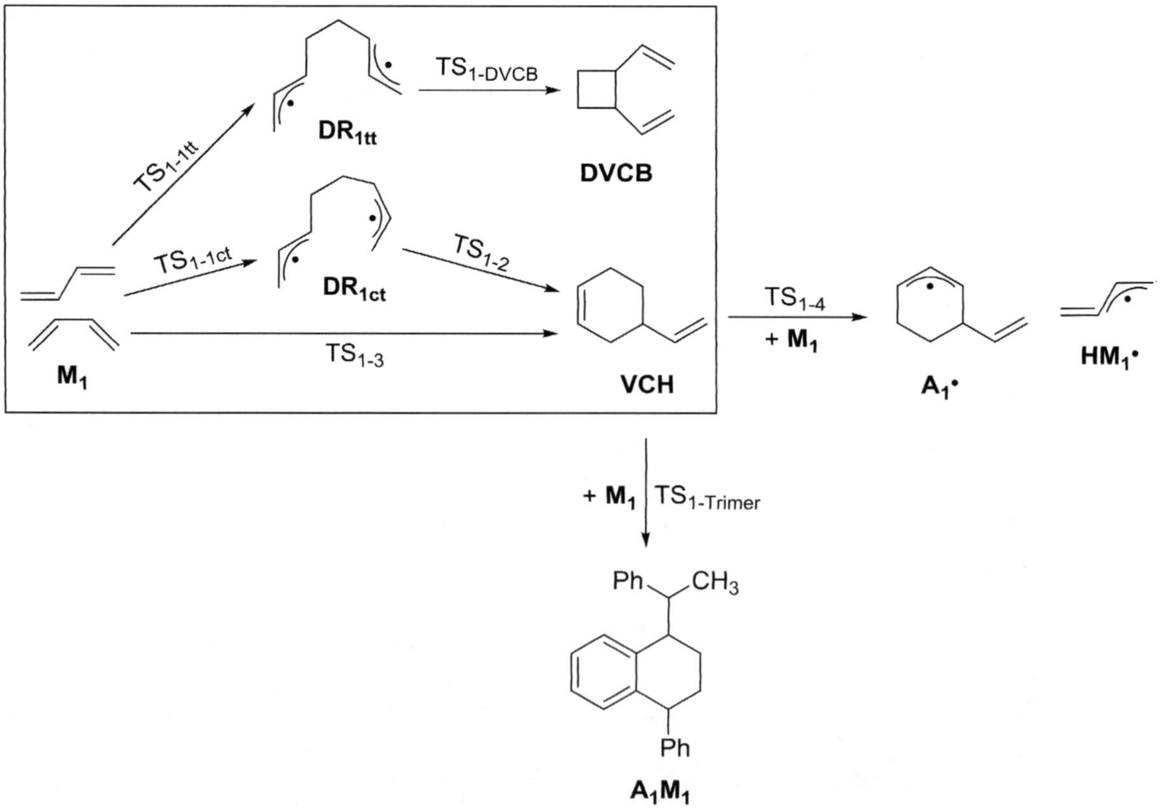

Figure 2. Proposed thermal initiation mechanism of butadiene (M_1)

Table 1. Rate coefficients of CHP decomposition

Overall reaction	**E$_a$** (kcal*mol^{-1})	**A** (mol$^{0.5}$*L$^{-0.5}$*s^{-1})
CHP → 2R•	29.16	1.78E+11

Table 2. Typical composition of process liquid.

Species	**Weight Percentage (%)**
Butadiene	1-10
Styrene	1-20

Propagation is the main reaction to increase chain length, in which long chain radicals add to unsaturated bonds of the monomer. The terminal model is the most commonly used and efficient model in the study of propagation kinetics in copolymerization. It assumes that chemical reactivity of the growing chain only depends on the unit at the chain end (the ultimate unit). The terminal model has been found to predict reactivity ratios, copolymer composition and sequence distribution accurately.[23] In this study, we adopted this model to describe co-propagation and listed the four propagation reactions of butadiene (**M$_1$**) and styrene (**M$_2$**) system in Scheme 1 with rate coefficients k_{ij}, where i and j denote the indexes of the ultimate unit and the monomer, respectively. Extensive experimental studies have been conducted on homo-propagation kinetics of styrene and butadiene, with the measured kinetic parameters listed in Table 3.[24-28] However, considerable variances exist in the reported values, possibly due to differences in reaction conditions. When cross-propagation is considered, monomer reactivity ratios quantify the relative rates of a radical chain reacting with its own type of monomer than with the other monomer, as defined in equations 1 and 2. Table 4 summarizes reported reactivity ratios of butadiene-styrene system from literature. Experimental results cover wide ranges: 1.40-2.20 for r_1, and 0.38-0.7 for r_2, presumably due to the difference in experimental methods and temperature range. Considering the large variances in the origins and values of reported kinetic parameters, a major goal of this study is to investigate co-propagation kinetics of butadiene-styrene system within the same computational framework to ensure consistency in results and enable rational comparison of propagation rates and reactivity ratios. In addition, when the same computational method is used for both initiation and propagation reactions, it will provide a basis for better understanding of the overall co-polymerization kinetics.

$$M_1 \bullet + M_1 \xrightarrow{k_{11}} M_1 \bullet$$

$$M_1 \bullet + M_2 \xrightarrow{k_{12}} M_2 \bullet$$

$$M_2 \bullet + M_1 \xrightarrow{k_{21}} M_1 \bullet$$

$$M_2 \bullet + M_2 \xrightarrow{k_{22}} M_2 \bullet$$

Scheme 1. Terminal model of free radical co-propagation of butadiene (M_1) and styrene (M_2). M_1• and M_2• represent radical chains with the ultimate unit of M_1 or M_2, respectively.

$$r_1 = \frac{k_{11}}{k_{12}} \qquad (1)$$

$$r_2 = \frac{k_{22}}{k_{21}} \qquad (2)$$

Table 3. Reported experimental kinetic parameters of homo-propagation of butadiene (M_1) and styrene (M_2) systems.

	E$_a$ (kcal*mol^{-1})	**A** (L*mol^{-1}s^{-1})
	6.5±1	1.02E+06
M_2• + M_2	5.75	9.72E+06
	7.77	4.27E+07
M_1• + M_1	8.53	8.05E+07
	9.30	1.20E+08

Table 4. Reported monomer reactivity ratios of butadiene-styrene system

r_1	**r_2**
1.83[29]	0.65[29]
1.4-2.2[30]	0.5-0.7[30]
1.40[31-32]	0.38[31-32]

To further simplify the mechanistic model for butadiene-styrene co-polymerization, the following assumptions were applied.

1) Propagation rates are independent of radical chain length;

2) Steady state approximation is applied to radical concentration;
3) Chain transfer reactions are not included.

This co-polymerization kinetic model enables us to calculate key kinetic properties in addition to monomer reactivity ratios. Instantaneous relative monomer consumption rate according to Mayo-Lewis equation[33], $\frac{d[M_1]}{d[M_2]}$, and average co-propagation rate coefficient, $<k_p>$, can both be determined as in equations 3 and 4,

$$\frac{d[M_1]}{d[M_2]} = \frac{[M_1](r_1[M_1]+[M_2])}{[M_2]([M_1]+r_2[M_2])} = \frac{1+r_1\frac{f_1}{f_2}}{1+r_2\frac{f_2}{f_1}} \tag{3}$$

$$< k_p > = \frac{r_1 f_1^2+2f_1 f_2+r_2 f_2^2}{\frac{r_1 f_1}{k_{11}}+\frac{r_2 f_2}{k_{22}}} \tag{4}$$

where f_1 and f_2 are the molar fractions of monomer $\mathbf{M_1}$ and $\mathbf{M_2}$, respectively. Predictions of these properties under diverse monomer fractions lead to suggestions of anti-fouling treatments. The fact that these properties can be monitored experimentally provides handles to valid computational predictions and can facilitate integrated computational and experimental studies in the future.

Another goal of this *ab initio* study is to evaluate the similarity of isoprene to butadiene in fouling chemistry for the design of safer lab experiments on polymerization fouling. While butadiene is highly volatile and hazardous, isoprene poses less danger to lab settings and can serve as a good analogue of butadiene. We used computational methods to quantify the similarity between butadiene and isoprene chemistry by comparing their initiation mechanisms and rationalizing their co-propagation kinetics with styrene. Furthermore, DFT predictions of co-polymerization kinetics of isoprene-styrene system was validated by in-house experiments. Ultimately, this study provided a basis for the integration of computational and experimental studies to better understand polymerization fouling and develop anti-fouling strategies.

2. Computational Methods

All DFT calculations were performed in the Gaussian 09 software package.[34] Geometries of all species were optimized in the gas phase using B3LYP[35-36] functional, and 6-31+G(d,p) basis set. This functional was chosen based on its successful application in predicting thermal initiation and propagation energetics of styrene free radical polymerization,[10, 37-39] and the dimerization energetics of butadiene.[16, 40] All TSs were verified by the presence of a single imaginary frequency and the intrinsic reaction coordinate (IRC) was followed in both directions.[41] The lowest free energy ground states and transition states (TS) were found through a combination of conventional geometry optimization and one-dimensional potential energy dihedral scans to improve accuracy. In the case of TS, the bond length corresponding to the vibration of imaginary frequency was kept frozen during the dihedral scans. Diradical species and transition states that lead to diradicals were treated with unrestricted B3LYP, which involves HOMO-LUMO mixing in the initial guess leading to unrestricted wave functions that had higher stability than the corresponding restricted wave functions. (Note: HOMO is short for highest occupied molecular orbital, LUMO is short for lowest unoccupied molecular orbital.) The temperature correction to the free energy was calculated using the frequencies from full optimization at B3LYP/6-31+G(d,p) level of theory with harmonic oscillator (HO) approximation based on standard formulae from statistical thermodynamics.[42] Free energy calculations also included the zero-point energy (ZPE) correction. To improve the accuracy of calculated electronic energy, single-point (SP) energy calculations were performed at the M06-2X/6-311+G(3df,2p)//B3LYP/6-31+G(d,p) level of theory, where M06-2X[43] is the hybrid metageneralized gradient approximation (hybrid-metaGGA) within DFT. Note that the same level of theory was used throughout this study to investigate thermal initiation and propagation reactions of styrene, butadiene and isoprene, which guaranteed the consistency in all DFT calculation results for further comparison and analysis.

Rate coefficients are calculated by combining electronic structure calculations with transition state theory (TST). Individual rate coefficients are calculated using transition state theory, via the expression shown below:

$$k(T) = \kappa(T)\sigma \frac{k_B T}{h} c^{(1-n)} e^{\frac{-\Delta G^{\neq}}{RT}} \quad (5)$$

where k_B is the Boltzmann's constant, T is the temperature at which the polymer synthesis occurs, h is Planck's constant, c is the standard state concentration, n is the molecularity of the reaction, R is the universal gas constant, and $\Delta G^{\neq}$ is the Gibbs free energy barrier of the reaction. The parameter σ represents the reaction path degeneracy, and $\kappa(T)$ is the tunneling correction factor which accounts for quantum mechanical tunneling. One approach to quantify this factor is given by the second order Wigner correction to the reaction rate,[44] in which the correction factor depends on the magnitude, $|v|$, of the imaginary frequency along the reaction coordinate:

$$\kappa(T) = 1 + \frac{1}{24}\left(\frac{h|v|}{k_B T}\right)^2 \quad (6)$$

Rate constants, k(T) were calculated under different temperatures from 25°C to 150°C, covering a wide range of possible operation conditions as well as experimental conditions. Linear regression was done to obtain Arrhenius rate parameters, activation energy E_a, and the pre-exponential factor, A.

3. Results and Discussion

The organization of this section is as follows. First, the thermal initiation mechanisms and energetics are described for styrene, butadiene and isoprene systems, respectively. The thermal initiation rate of butadiene-styrene system under plant conditions is estimated and compared with **CHP** decomposition rate to determine the dominant initiation mechanism. The similarity between isoprene and butadiene in thermal initiation chemistry is then concluded. Next, the co-propagation kinetics of the butadiene-styrene system is presented. Propagation rate parameters and monomer reactivity ratios are reported and validated against experimental results. Similarly, the isoprene-styrene system is examined to reveal its resemblance to the butadiene-styrene system. Finally, two key properties, instantaneous relative monomer consumption rate, $\frac{d[M_i]}{d[M_j]}$, and average propagation rate coefficient, $\langle k_p \rangle$, are predicted for both the butadiene-styrene and isoprene-styrene systems. Experimental validation of the instantaneous relative monomer consumption rate of the isoprene-styrene system is shown to prove the accuracy of DFT predictions. Recommendations for experimental design and anti-fouling treatments are proposed accordingly.

3.1. Initiation Mechanism and Kinetics

3.1.1. Styrene Thermal Initiation

DFT calculations were conducted for each elementary step with a comprehensive conformational search based on the reported mechanism of styrene thermal initiation. Free energy profiles are shown in Figure 3. The main path to produce monoradicals (in black line) and major side reactions (in colored lines) were captured.

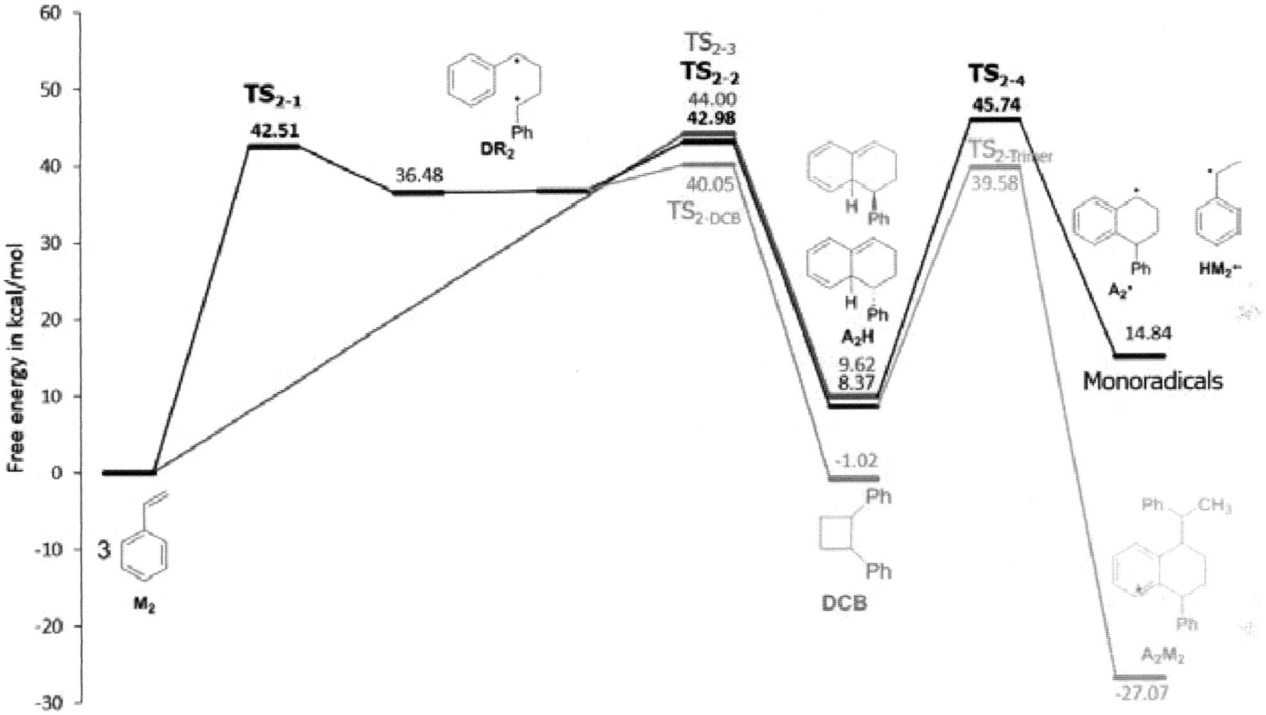

Figure 3. Free energy profiles of styrene thermal initiation at 100°C. Energies are calculated at M06-2X/6-311+G(3df,2p)//B3LYP/6-31+G(d,p) level of theory and are reported in kcal/mol.

Two competing reaction pathways for the formation of the Diels-Alder dimer, **A_2H**, show similar energetics in term of TS heights. The stepwise pathway (black line) is slightly more favorable than the concerted pathway (red line), which agrees with previous literature results.[10] The main side product, **DCB,** competes with **A_2H** in the later stage of stepwise pathway with a TS energy approximately 3 kcal/mol lower. This is in line with experimental evidence of **DCB** formation and the relative TS energies in literature, although a larger difference in TS energies (5.7 kcal/mol) was reported.[10] **A_2H** intermediates are unstable and can react with another styrene monomer to form two monoradicals (**A_2•** and **HM_2•**) or an inactive trimer (**A_2M_2**). The formation of **A_2M_2** is more favorable than monoradicals by a 6 kcal/mol difference in barrier heights. The rate-determining barrier in thermal initiation was found to be the barrier from **A_2H** to monoradicals (**TS_{2-4}**), which is approximately 3 kcal/mol higher than the barrier of **A_2H** formation (**TS_{2-2}**). Given the complexity of the mechanism with side reactions, it is not feasible to obtain a simple analytical rate expression for the rate of monoradical formation, i.e., the thermal initiation rate of styrene. On the other hand, the rate of **A_2H** formation reaction can be easily determined and can serve as an upper bound of thermal initiation rate. Considering the small difference in the free energies between **$TS_{2\text{-}4}$** and **$TS_{2\text{-}2}$**, an approximation was made that the rate of thermal initiation can be represented by the rate of dimerization reaction as an upper limit. The approximated Arrhenius parameters of styrene thermal initiation were obtained: activation energy E_a=29.84 kcal/mol; and pre-exponential factor A=9.94×10^5 L*$mol^{-1}s^{-1}$.

3.1.2. Butadiene Thermal Initiation

DFT calculations were conducted for each elementary step with a comprehensive conformational search based on the proposed mechanism of butadiene thermal initiation, and the free energy profiles are shown in Figure 4. The main path to produce monoradicals (in black line) and major side reactions (in colored lines) were captured. Free energies of all species are relative to that of *trans*-butadiene monomer reactants.

Unlike the thermal initiation mechanism of styrene, the formation of the Diels-Alder dimer (**VCH**) from two butadiene molecules is found to be dominated by the concerted reaction pathway via **$TS_{1\text{-}3}$**. The stepwise pathway has a rate-determining TS (**$TS_{1\text{-}1ct}$**) which is more than 4 kcal/mol higher in free energy than **$TS_{1\text{-}3}$**. The formation of **DVCB** faces a higher barrier than **VCH**, making **VCH** the dominant dimerization product. In contrast to the low stability of **A_2H** intermediates in styrene thermal initiation, **VCH** dimers are highly stable with the free energies more than 20 kcal/mol lower compared with starting monomers. Although **VCH** can react with another butadiene monomer to form two monoradicals (**A_1•** and **HM_1•**) via a transition state (**$TS_{1\text{-}4}$**) lower in free energy than **$TS_{1\text{-}3}$**, the extraordinary stability of VCH results in an actual free energy barrier of this elementary step as high as 52.34 kcal/mol which effectively shuts down the reaction. The formation of the inactive trimer, **A_1M_1,** is also inhibited by high energy barriers, although less than the formation of monoradicals. Indeed, **VCH** serves as a "sink species" that traps most of the butadiene dimers in a deep well from further reactions. Considering the low temperature range in this study, it can be concluded that butadiene thermal initiation reactions make negligible contribution to radical formation compared with styrene thermal initiation. However, dimerization of butadiene proceeds much faster than that of styrene due to a free energy barrier approximately 7 kcal/mol lower.

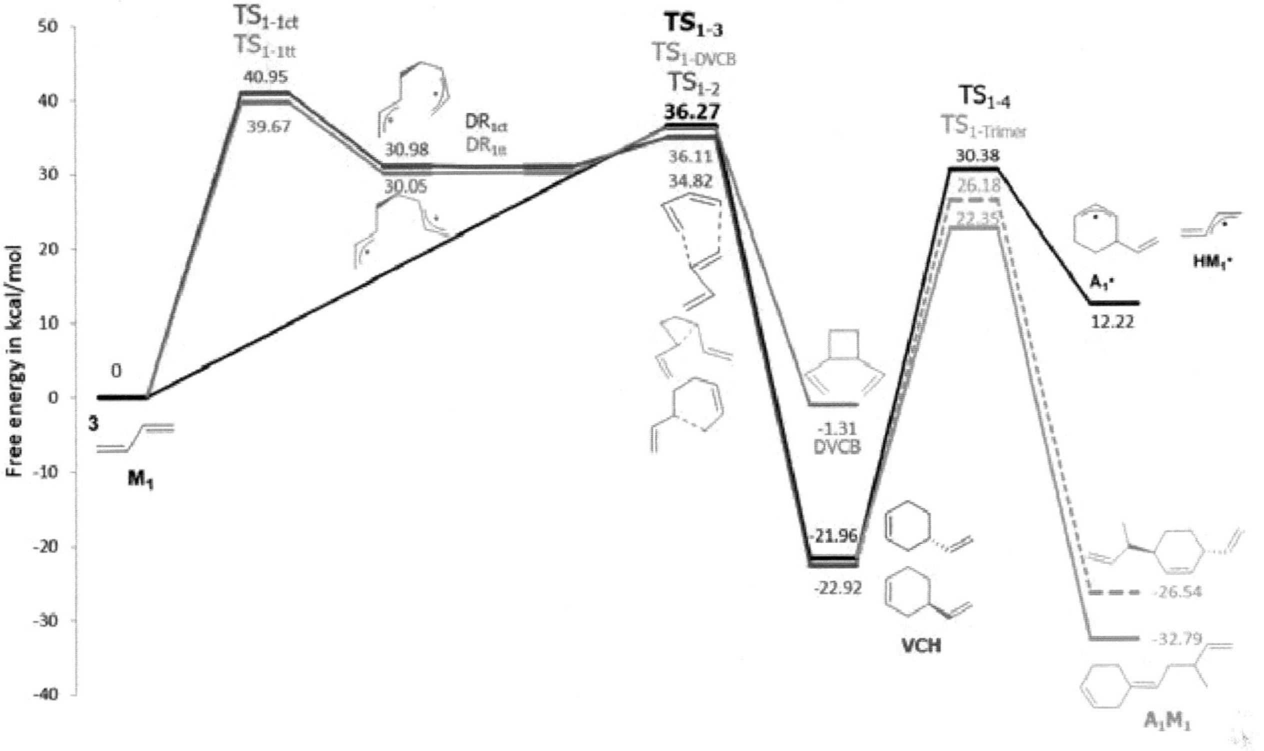

Figure 4. Free energy profiles of butadiene thermal initiation at 100°C. Energies are calculated at M06-2X/6-311+G(3df,2p)//B3LYP/6-31+G(d,p) level of theory and are reported in kcal/mol.

3.1.3. Isoprene Thermal Initiation

DFT calculations were conducted for isoprene thermal initiation following similar mechanism as in the case of butadiene. Free energy profiles are shown in Figure 5, including the main path towards monoradicals (in black line) and major side reactions (in colored lines). The free energies for all species are relative to that of *trans*-isoprene monomer reactants.

Thermal initiation energetics of isoprene show high similarity to those of butadiene. Concerted reaction pathway wins over stepwise reaction pathway towards the formation of Diels-Alder dimers from two isoprene molecules. Formation of Diels-Alder dimers is more favorable than that of the side product, 1,2-divinyl-1,2-dimethylcyclobutane (**DVMCB**). Like in the case of butadiene, Diels-Alder dimers of isoprene are highly stable with the free energies more than 20 kcal/mol lower compared with starting monomers, leading to a free energy barrier as high as 54.27 kcal/mol to form monoradicals. Consequently, reactions that generate monoradicals from Diels-Alder dimers are significantly inhibited, leading to negligible yield to radicals compared with styrene system. It can be concluded that isoprene resembled butadiene in terms of reaction mechanism and energetics of thermal initiation, and is a good analogue of butadiene in the regime of thermal initiation.

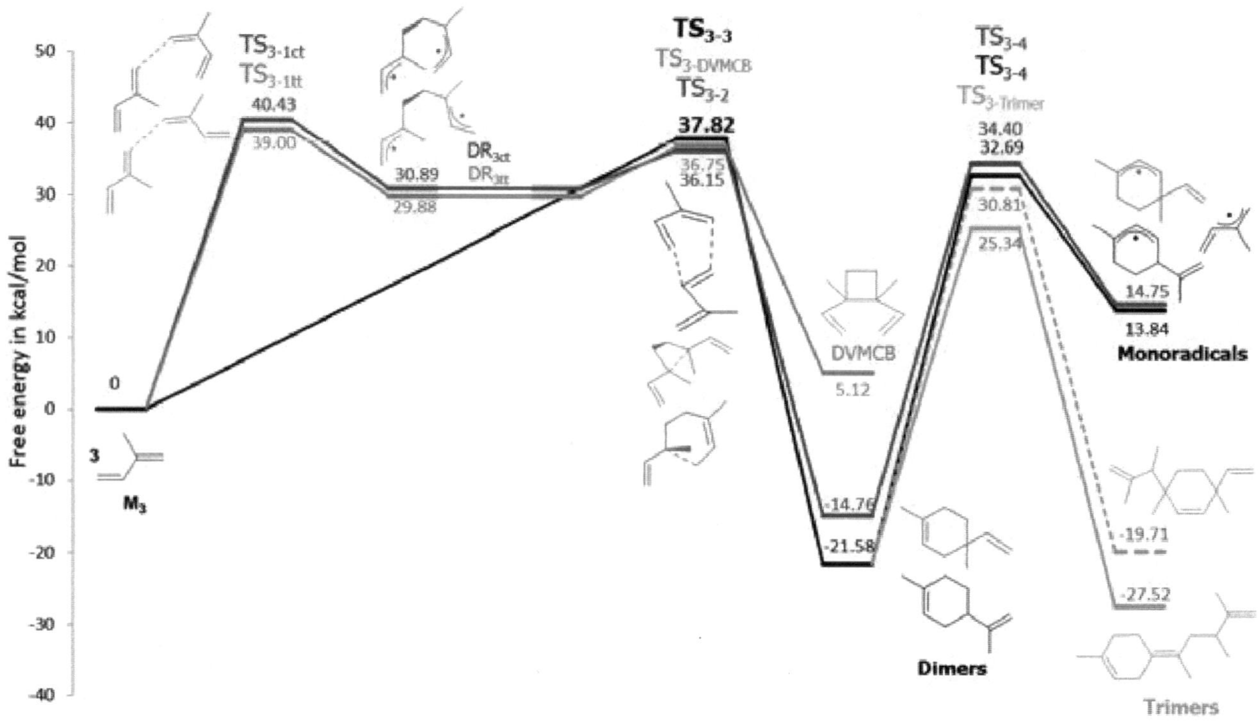

Figure 5. Free energy profiles of isoprene thermal initiation at 100°C. Energies are calculated at M06-2X/6-311+G(3df,2p)//B3LYP/6-31+G(d,p) level of theory and are reported in kcal/mol.

3.1.4. Determine the Dominant Initiation Mechanism

With the energetics of thermal initiation mapped and plant conditions determined (Table 2), a comparison was then made between thermal initiation rate and **CHP** decomposition rate at a representative temperature, 100 °C. The concentration of **CHP** is set to 1 ppm to represent low levels of peroxide within the detection limit. Based on previous results, the major pathway to generate radicals is styrene thermal initiation for both butadiene-styrene and isoprene-styrene systems. Therefore, the overall thermal initiation rate can be represented by the thermal initiation rate of styrene, which is approximated by the rate of stepwise dimerization reaction of styrene. Table 5 summarizes key kinetic parameters of the two initiation mechanisms and the corresponding initiation rates. At such a low **CHP** level, the initiation rate of **CHP** decomposition is three orders of magnitude faster than the thermal initiation rate. In fact, for **CHP** decomposition rate to match thermal initiation rate, only 1.92 ppt **CHP** is needed, which is beyond the limitations of current detecting methods. The stark contrast between initiation rates of the two mechanisms clearly points to peroxide decomposition as the dominant initiation mechanism. To inhibit radical formation from peroxide decomposition, two treatment methods can be proposed. First is to carefully dose proper free radical scavenger into the separation units; second is to lower the impurity level of peroxide or more generally, the level of oxygenates in the cracked gas to slow down the formation of free radicals.

Table 5. Comparison of thermal initiation kinetics between CHP decomposition (1 ppm) and styrene thermal initiation (10 wt% styrene in toluene) at 100 °C

Reaction	**E_a (kcal*mol^{-1})**	**A (mol$^{0.5}$*L$^{-0.5}$*s^{-1})**	**k (100°C) (mol$^{0.5}$*L$^{-0.5}$*s^{-1})**	**r (100°C) (mol*L^{-1}s^{-1})**
CHP Decomposition	29.16	1.78×10^{11}	1.48×10^{-6}	3.48×10^{-9}
	E_a (kcal*mol^{-1})	**A (L*mol^{-1}s^{-1})**	**k (100°C) (L*mol^{-1}s^{-1})**	**r (100°C) (mol*L^{-1}s^{-1})**
Styrene Thermal Initiation	29.84	9.94×10^{5}	3.30×10^{-12}	2.24×10^{-12}

3.2. Propagation Mechanism and Kinetics

3.2.1. Butadiene-styrene system

Upon the generation of radical species, propagation reactions start and monomers are being consumed at fast rates. Propagation kinetics of butadiene-styrene system (Scheme 1) were first investigated using the same level of theory as in the study of initiation kinetics.

DFT predicted kinetic parameters and rate constants are listed in Table 6. Styrene homo-propagation reaction ($M_2\bullet + M_2$) is almost ten times faster than butadiene homo-propagation reaction ($M_1\bullet + M_1$). However, in cross-propagation reactions ($M_2\bullet + M_1$ and $M_1\bullet + M_2$), butadiene monomer reacts more readily than styrene monomer. The monomer reactivity ratios are 2.04 for butadiene, and 0.41 for styrene at 100°C, which fall in the range of experimental results from literature as listed in Table 4, indicating good agreement with experimental measurements.

Table 6. Propagation kinetics of butadiene-styrene copolymerization

Reaction	**E_a** (kcal*mol^{-1})	**A** (L*mol^{-1}s^{-1})	**k_p** at **100**°C (L*mol^{-1}s^{-1})
$M_2\bullet + M_2$	6.63	1.69×10^{6}	2.21×10^{2}
$M_1\bullet + M_1$	10.24	2.43×10^{7}	2.44×10^{1}
$M_2\bullet + M_1$	6.86	5.57×10^{6}	5.34×10^{2}
$M_1\bullet + M_2$	9.08	2.49×10^{6}	1.20×10^{1}

3.2.2. Isoprene-styrene system

Propagation kinetics of isoprene-styrene system were examined to evaluate how well isoprene resembles butadiene in polymerization chemistry. DFT-predicted kinetic parameters and rate constants are summarized in Table 7. Similar to the butadiene-styrene system, the styrene homo-propagation reaction ($M_2\bullet + M_2$) is more than one order of magnitude faster than isoprene homo-propagation ($M_3\bullet + M_3$). In cross-propagation reactions ($M_2\bullet + M_3$ and $M_3\bullet + M_2$), isoprene monomers react more readily than styrene monomers. The monomer reactivity ratios of 2.27 for isoprene and 0.48 for styrene at 100°C reveal the similarity of isoprene to butadiene in co-polymerization with styrene. Because isoprene and butadiene are similar in thermal initiation chemistry and in reactivity ratio, it can be concluded that isoprene is a good analogue of butadiene for the study of polymerization fouling.

Table 7. Propagation kinetics of isoprene-styrene copolymerization

Reaction	E_a (kcal*mol^{-1})	A (L*mol^{-1}s^{-1})	k_p at 100^{0}C (L*mol^{-1}s^{-1})
$M_2 \bullet + M_2$	6.63	1.69×10^6	2.21×10^2
$M_3 \bullet + M_3$	10.17	7.97×10^6	8.80×10^0
$M_2 \bullet + M_3$	6.59	3.33×10^6	4.60×10^2
$M_3 \bullet + M_2$	10.07	3.06×10^6	3.87×10^0

3.3. Key Kinetic Properties of Co-polymerization Reactions

Based on the results of propagation kinetics, two additional key kinetic properties, instantaneous relative monomer consumption rate, $\frac{d[M_i]}{d[M_j]}$, and co-propagation rate coefficient, $\langle k_p \rangle$, can be determined for variant monomer fractions (equations 3 and 4).

Predictions of the instantaneous relative monomer consumption rate as a function of styrene molar fraction for butadiene-styrene and isoprene-styrene systems are shown in Figures 6 and 7, respectively. In both cases, the monomer consumption rate of butadiene or isoprene is higher than that of styrene up to approximately 70% styrene in monomer composition. For butadiene-styrene system, the ratio of monomer consumption rates, $\frac{d[M_1]}{d[M_2]}$, is approximately 4 at typical plant conditions (Table 2) where styrene fraction is 57%. For isoprene-styrene system, the ratio of monomer consumption rates, $\frac{d[M_3]}{d[M_2]}$, is 4.3 when the weight fraction is 5% for isoprene and 10% for styrene. Note that the instantaneous relative monomer consumption rate is an experimentally measurable property which can be used to validate computational predictions. Experiments were conducted for isoprene-styrene system at four different styrene fractions and were plotted in Figure 7 in comparison to DFT predictions. Experimental details are recorded in the supporting information. Across the styrene fraction range from 0.22 to 0.95, both DFT predictions and experimental measurements show decreasing trends of $\frac{d[M_3]}{d[M_2]}$ as a function of styrene fraction. In addition, the absolute values of $\frac{d[M_3]}{d[M_2]}$ from DFT predictions are in close proximity of experimental measurements, which validated the strong predictive power of DFT calculations on isoprene-styrene system to facilitate integrated computational and experimental studies in the future. Considering the similarity between isoprene and butadiene in polymerization fouling, computational results of butadiene-styrene system could be used with more confidence under different plant conditions.

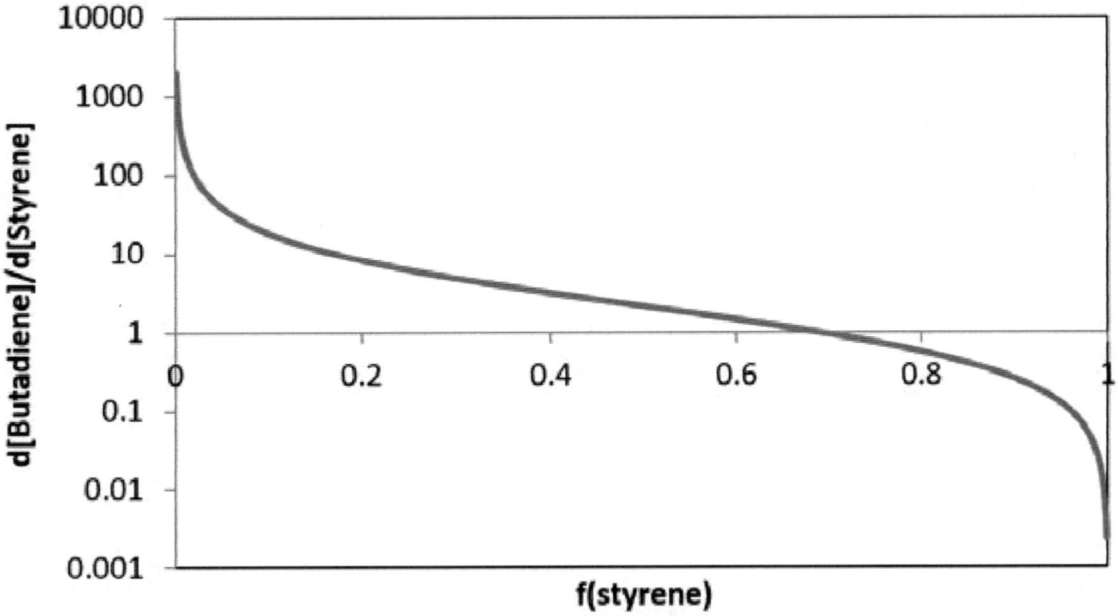

Figure 6. Relationship between styrene (M_2) fraction and the relative monomer consumption rate of butadiene (M_1)/styrene (M_2) system at 100°C.

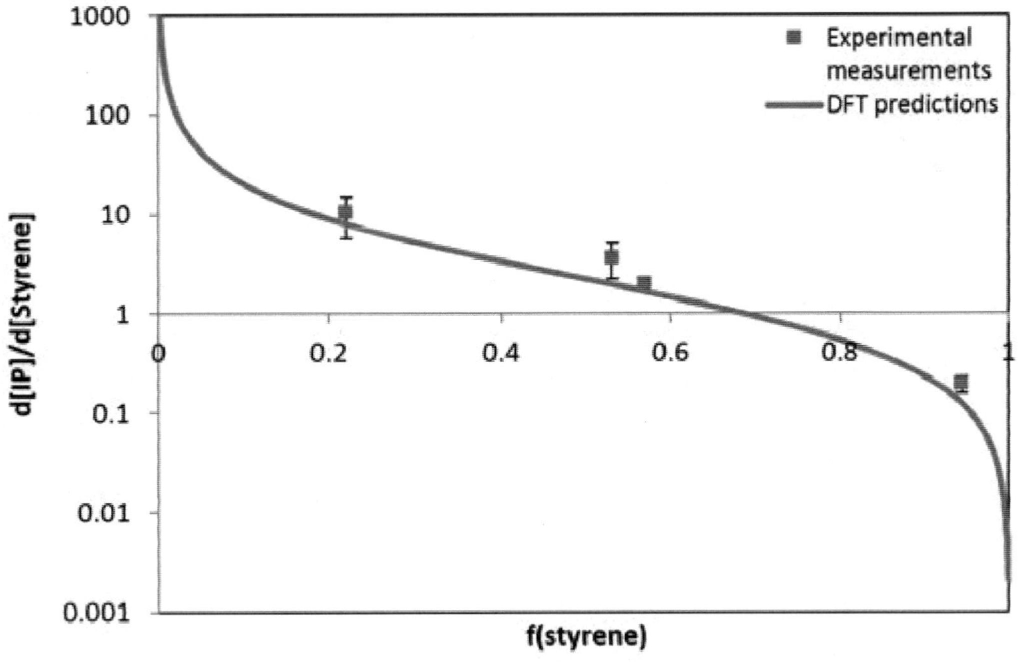

Figure 7. Relationship between styrene (M_2) fraction in monomer feed and the relative monomer consumption rate of isoprene (M_3)/styrene (M_2) system at 100°C. Experimental error bars show standard deviation.

The average co-propagation rate coefficient, <k_p>, as a function of styrene fraction were calculated for both butadiene-styrene and isoprene-styrene systems, as shown in Figure 8 and 9, respectively. In both cases, the co-propagation rate coefficient increases with styrene fraction, leading to faster formation of polymer foulant. The average co-propagation rate coefficient hits the highest value when the monomer composition is pure styrene, which is in line with the fact that styrene homo-propagation has the largest rate constants as in Tables 6 and 7. Correspondingly, a treatment method based on this relationship is to lower the fraction of styrene in the cracked gas. It may be achieved by lowering the temperature of the gas mixture and separating styrene at earlier stages.

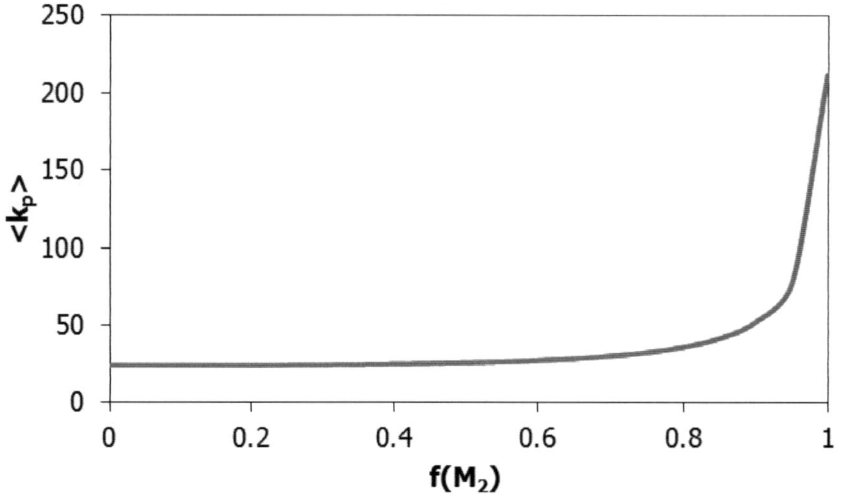

Figure 8. Relationship between styrene fraction and the overall propagation rate of butadiene-styrene system at 100 °C.

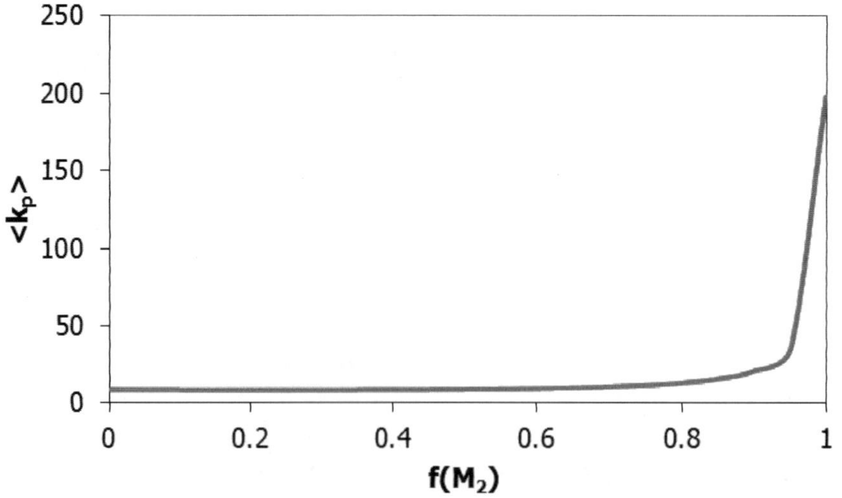

Figure 9. Relationship between styrene fraction and the overall propagation rate of isoprene-styrene system at 100 °C.

4. Conclusions

This study presents a broad yet in-depth investigation of the fundamentals of polymerization fouling using computational methods. Mechanism and kinetics of initiation and propagation reactions of the butadiene-styrene system were rationalized within the same computational framework. The dominant initiation mechanism at typical plant conditions was identified as peroxide decomposition over thermal initiation at ppm level of **CHP** impurity. Although both being extremely slow processes, thermal initiation of styrene is much more favorable than that of butadiene. This is due to the unusually high stability of butadiene Diels-Alder dimers that prevent them from further reactions. Results of co-propagation kinetics of butadiene-styrene system achieved good agreement with literature reports. The predicted reactivity ratios are 2.04 for butadiene and 0.41 for styrene, indicating a higher reactivity of butadiene monomers in propagation reactions. In addition, isoprene was validated as a good analogue to butadiene in polymerization fouling for lab experimental studies given safety considerations. Further analysis provided two more key kinetic properties that can be readily measured. One is the instantaneous relative monomer consumption rate, and the DFT predictions of this property were validated by experimental measurements for isoprene-styrene system; the other is the average co-propagation rate coefficient, which leads to a prediction that the overall polymerization would go faster with higher styrene fraction. Based on these results, we suggest that anti-fouling treatment should focus on lowering oxygenate levels, inhibiting free radical activity, and controlling styrene content in the cracked gas. This work also suggests future directions in elucidating the origins of trace peroxide species to better monitor and control polymerization fouling. Overall, this study provided a basis for integrating computational study with experimental efforts in understanding the fundamentals of polymerization fouling, which could eventually help provide more effective anti-fouling strategies for light hydrocarbon crackers.

References

1. Russell, K. E.; Tobolsky, A. V., Thermal Initiation of Styrene Polymerization. *J Am Chem Soc* **1953,** *75* (20), 5052-5054.
2. Mayo, F. R., Dimerization of Styrene. *J Am Chem Soc* **1968,** *90* (5), 1289-&.
3. Mayo, F. R., Chain Transfer in the Polymerization of Styrene .8. Chain Transfer with Bromobenzene and Mechanism of Thermal Initiation. *J Am Chem Soc* **1953,** *75* (24), 6133-6141.
4. Flory, P. J., The mechanism of vinyl polymerizations. *J Am Chem Soc* **1937,** *59*, 241-253.
5. Bengough, W. I.; Park, G. B., Thermal Polymerization of Styrene in Diethyladipate at Temperatures up to 160degrees. *Eur Polym J* **1976,** *12* (7), 431-433.
6. Barr, N. J.; Bengough, W. I.; Beveridge, G.; Park, G. B., Measurement of Rates of Initiation in Thermal Polymerization of Styrene. *Eur Polym J* **1978,** *14* (4), 245-250.
7. Chong, Y. K.; Rizzardo, E.; Solomon, D. H., Confirmation of the Mayo Mechanism for the Initiation of the Thermal Polymerization of Styrene. *J Am Chem Soc* **1983,** *105* (26), 7761-7762.
8. Buzanowski, W. C.; Graham, J. D.; Priddy, D. B.; Shero, E., Spontaneous Polymerization of Styrene in the Presence of Acid - Further Confirmation of the Mayo Mechanism. *Polymer* **1992,** *33* (14), 3055-3059.
9. Devonport, W.; Michalak, L.; Malmstrom, E.; Mate, M.; Kurdi, B.; Hawker, C. J.; Barclay, G. G.; Sinta, R., "Living" free-radical polymerizations in the absence of initiators: Controlled autopolymerization. *Macromolecules* **1997,** *30* (7), 1929-1934.
10. Khuong, K. S.; Jones, W. H.; Pryor, W. A.; Houk, K. N., The mechanism of the self-initiated thermal polymerization of styrene. Theoretical solution of a classic problem. *J Am Chem Soc* **2005,** *127* (4), 1265-1277.
11. Alsharaeh, E. H.; El-Shall, M. S., Ion mobility study of the mechanism of the gas phase thermal polymerization of styrene and the structures of the early oligomers. *Polymer* **2011,** *52* (24), 5551-5559.
12. Pryor, W. A.; Coco, J. H.; Daly, W. H.; Houk, K. N., Radical Generation from Polymolecular Reactions of Closed Shell Molecules - Molecule-Assisted Homolysis (Mah) - Hydrogen-Atom Transfer from a Diels-Alder Adduct to an Alkene. *J Am Chem Soc* **1974,** *96* (17), 5591-5593.
13. Caramella, P.; Quadrelli, P.; Toma, L.; Romano, S.; Khuong, K. S.; Northrop, B.; Houk, K. N., The three corrugated surfaces of 1,4-divinyltetramethylene diradical intermediates and their connections to 1,2-divinylcyclobutane, 4-vinylcyclohexene, 1,5-cyclooctadiene, and two butadienes. *J Org Chem* **2005,** *70* (8), 2994-3008.
14. Vaughan, W. E., The homogeneous thermal polymerization of 1,3-butadiene. *J Am Chem Soc* **1932,** *54*, 3863-3876.
15. Scarborough, D. L. A.; Kobayashi, R.; Thompson, C. D.; Izgorodina, E. I., Active space and basis set effects in CASPT2 models of the 1,3-butadiene-ethene cycloaddition and the 1,3-butadiene dimerization. *Int J Quantum Chem* **2015,** *115* (15), 989-1001.
16. Quadrelli, P.; Romano, S.; Toma, L.; Caramella, P., Merging and bifurcation of 4+2 and 2+4 cycloaddition modes in the archetypal dimerization of butadiene. A case of competing bispericyclic, pericyclic and diradical paths. *Tetrahedron Lett* **2002,** *43* (48), 8785-8789.
17. Aldeeb, A. A.; Rogers, W. J.; Mannan, M. S., Evaluation of 1,3-butadiene dimerization and secondary reactions in the presence and absence of oxygen. *J Hazard Mater* **2004,** *115* (1-3), 51-56.
18. Duncan, N. E.; Janz, G. J., The Thermal Dimerization of Butadiene, and the Equilibrium between Butadiene and Vinylcyclohexene. *J Chem Phys* **1952,** *20* (10), 1644-1645.
19. Stephenson, L. M.; Gemmer, R. V.; Current, S., Dimerization of Butadiene. *J Am Chem Soc* **1975,** *97* (20), 5909-5910.

20. Reed, H. W. B., The Formation of Trans-1-2-Divinylcyclobutane from Butadiene. *J Chem Soc* **1951,** (Mar), 685-687.
21. Durig, J. R.; Bucy, W. E.; Cole, A. R. H., Torsional Potential Function of Glyoxal. *Can J Phys* **1975,** *53* (19), 1832-1837.
22. Duh, Y. S.; Kao, C. S.; Hwang, H. H.; Lee, W. W. L., Thermal decomposition kinetics of cumene hydroperoxide. *Process Saf Environ* **1998,** *76* (B4), 271-276.
23. Odian, G. G., *Principles of polymerization*. 4th ed.; Wiley: Hoboken, N.J., 2004; p xxiv, 812 p.
24. Bamford, C. H.; Dewar, M. J. S., Studies in Polymerization .1. A Method for Determining the Velocity Constants Polymerization Reactions and Its Application to Styrene. *Proc R Soc Lon Ser-A* **1948,** *192* (1030), 309-328.
25. Deady, M.; Mau, A. W. H.; Moad, G.; Spurling, T. H., Evaluation of the Kinetic-Parameters for Styrene Polymerization and Their Chain-Length Dependence by Kinetic Simulation and Pulsed-Laser Photolysis. *Makromol Chem* **1993,** *194* (6), 1691-1705.
26. Buback, M.; Gilbert, R. G.; Hutchinson, R. A.; Klumperman, B.; Kuchta, F. D.; Manders, B. G.; Odriscoll, K. F.; Russell, G. T.; Schweer, J., Critically Evaluated Rate Coefficients for Free-Radical Polymerization .1. Propagation Rate Coefficient for Styrene. *Macromol Chem Physic* **1995,** *196* (10), 3267-3280.
27. Deibert, S.; Bandermann, F.; Schweer, J.; Sarnecki, J., Propagation Rate Coefficient of Free-Radical Polymerization of 1,3-Butadiene. *Makromol Chem-Rapid* **1992,** *13* (7), 351-355.
28. Morton, M.; Salatiello, P. P.; Landfield, H., Absolute Propagation Rates in Emulsion Polymerization .2. Butadiene in Hydroperoxide-Polyamine Systems. *J Polym Sci* **1952,** *8* (2), 215-224.
29. Gilbert, R. D.; Williams, H. L., Reactivity Ratios of Conjugated Dienes Copolymerized in Emulsion at 5-Degrees. *J Am Chem Soc* **1952,** *74* (16), 4114-4118.
30. Mitchell, J.; Williams, H. L., LABORATORY STUDY OF LOW RATIO BUTADIENE–STYRENE COPOLYMERS. *Canadian Journal of Research* **1949,** *27* (2), 35-46.
31. Orr, R. J., Thermochemical Aspects of Butadiene Styrene Copolymerization. *Polymer* **1961,** *2* (1), 74-82.
32. Kalfus, M.; Grzywa, E., Modeling and Simulation of Chemical Homogeneity of Butadiene-Styrene Copolymer in High-Conversion Emulsion Co-Polymerization. *Polym J* **1978,** *10* (6), 583-590.
33. Mayo, F. R.; Lewis, F. M., Copolymerization I A basis for comparing the behavior of monomers in copolymerization, the copolymerization of styrene and methyl methacrylate. *J Am Chem Soc* **1944,** *66*, 1594-1601.
34. Frisch, M. J.; Trucks, G. W.; Schlegel, H. B.; Scuseria, G. E.; Robb, M. A.; Cheeseman, J. R.; Scalmani, G.; Barone, V.; Mennucci, B.; Petersson, G. A.; Nakatsuji, H.; M., C.; Li, X.; Hratchian, H. P.; Izmaylov, A. F.; Bloino, J.; Zheng, G.; Sonnenberg, J. L.; Hada, M.; Ehara, M.; Toyota, K.; Fukuda, R.; Hasegawa, J.; Ishida, M.; Nakajima, T.; Honda, Y.; Kitao, O.; Nakai, H.; Vreven, T.; Montgomery, J., J. A.; E., P. J.; Ogliaro, F.; Bearpar, M.; Heyd, J. J.; Brother, E.; Kudin, K. N.; Staroverov, V. N.; Keith, T.; Kobayashi, R.; Normand, J.; Raghavachari, K.; Rendell, A.; Burant, J. C.; Iyengar, S. S.; Tomasi, J.; Cossi, M.; Rega, N.; Millam, J. M.; Klene, M.; Knox, J. E.; Cross, J. B.; Bakken, K.; Adamo, C.; Jaramillo, J.; Gomperts, R.; Stratmann, R. E.; Yazyev, O.; Austin, A. J.; Cammi, R.; Pomelli, C.; Ochterski, J. W.; Martin, R. L.; Morokuma, K.; Zakrzewski, V. G.; Voth, G. A.; Salvador, P.; Dannenberg, J. J.; Dapprich, S.; Daniels, A. D.; Farkas, O.; Foresman, J. B.; Ortiz, J. V.; Cioslowski, J.; Fox, D. J. *Gaussian 09*, Revision D.01; Gaussian, Inc.: 2013.
35. Becke, A. D., Density-Functional Thermochemistry .3. The Role of Exact Exchange. *J Chem Phys* **1993,** *98* (7), 5648-5652.
36. Lee, C. T.; Yang, W. T.; Parr, R. G., Development of the Colle-Salvetti Correlation-Energy Formula into a Functional of the Electron-Density. *Phys Rev B* **1988,** *37* (2), 785-789.

37. Coote, M. L., Quantum-Chemical Modeling of Free-Radical Polymerization. *Macromol Theor Simul* **2009,** *18* (7-8), 388-400.
38. Mavroudakis, E.; Liang, K.; Moscatelli, D.; Hutchinson, R. A., A Combined Computational and Experimental Study on the Free-Radical Copolymerization of Styrene and Hydroxyethyl Acrylate. *Macromol Chem Physic* **2012,** *213* (16), 1706-1716.
39. Dossi, M.; Storti, G.; Moscatelli, D., A Quantum Chemistry Study of the Free-Radical Copolymerization Propagation Kinetics of Styrene and 2-Hydroxyethyl Acrylate. *Polym Eng Sci* **2011,** *51* (10), 2109-2114.
40. Quadrelli, P.; Romano, S.; Toma, L.; Caramella, P., A bispericyclic transition structure allows for efficient relief of antiaromaticity enhancing reactivity and endo stereoselectivity in the dimerization of the fleeting cyclopentadienone. *J Org Chem* **2003,** *68* (15), 6035-6038.
41. Martinez, E. P.; Gonzalez, C. S.; Reyes, A. R., Variability and Increase in the Reaction-Time in the Early Diagnosis of Subclinical Hepatic-Encephalopathy. *Rev Esp Enferm Dig* **1990,** *77* (1), 29-32.
42. McQuarrie, D. A., *Statistical Mechanics*. University Science Books: 2000.
43. Zhao, Y.; Truhlar, D. G., The M06 suite of density functionals for main group thermochemistry, thermochemical kinetics, noncovalent interactions, excited states, and transition elements: two new functionals and systematic testing of four M06-class functionals and 12 other functionals. *Theor Chem Acc* **2008,** *120* (1-3), 215-241.
44. Wigner, E., The transition state method. *T Faraday Soc* **1938,** *34* (1), 0029-0040.

591159 When Disasters Happen: A Collaborative Approach Towards the Community's Recovery Process

Abstract:

The 2020 Virtual AIChE Spring Meeting & 16th GCPS Wednesday keynote "When Disasters Happen: A Collaborative Approach Towards the Community's Recovery Process" will be presented by Thomas Munoz, Emergency Manager and Homeland Security Director with Texas City. As the Emergency Manager and Homeland Security Director with Texas City, the home of 19 chemical plants, he is responsible for the safety of all its citizens along with managing the crisis communication plan with local industry and major stakeholders.

Safety and disaster planning is vital to the recovery process after a major incident. As key critical infrastructure industries, issues may arise that have not been planned for in advance, such as the recovery from an incident. Developing and maintaining partnerships with local partners in your community will support addressing unforeseen challenges. Recovery efforts include planning with key partners and understanding the phases of recovery, which also consider the mental health and well of everyone involved in the response, as well as the community impacted. In this keynote address a discussion of pre-incident planning considerations and strategies as well as lessons learned/after action improvements post-recovery will be discussed. Additionally, collaboration considerations regarding local resources to promote recovery in the event of a workplace incident will be discussed.

C0-presenter Amanda Groller, MS.Ed., LPC will discuss the mental health impacts of a disaster.

589942 Successes and Failures of Pilot Plants for Technology Commercialization in Chemicals and Fuels: A Look Back over 3 Decades

Abstract:

Pilot plants for commercial deployments in Shell over the past three decades have included fully integrated units, and virtual integration using modeling to integrate piloting of select unit operations, as well as data from bench- scale units and cold-flow experiments. The successes and issues for these pilot plants and demonstration units will be reviewed, including challenges in scale-down and solids handling which impact trade-offs between risk-reduction accuracy, timing, and development cost. Adaptations of existing pilots to new or modified projects and use of pilot plants and bench scale units to address discrepancies or problems learned after commercial startup will also be presented for assessment of best practices and learnings for the use of pilot plants in process development and technology commercialization.

589015 Lessons Learned in Use of Small-Scale Distillation Columns

Abstract:

Success in piloting new process developments involves integrating process models, technology fundamentals, and defining critical operating parameters for the experimental plan. While it is important that test plans are developed to address key parameters, understanding the limitations of the pilot unit helps to identify how the obtained data can be applied for final industrial scale design. The results of the studies presented here provided insight into using small pilot distillation columns to validate process models and provided important design considerations for commercial scale operation. Key learnings in operating small scale pilot distillation columns and applying data for design considerations is the focus of this presentation using a propylene oxide purification method as an example.

OPTIMAL DESIGN OF A LIQUID-LIQUID EXTRACTION SYSTEM

Authors:

Donald J. Glatz
Brendan Cross

Koch Modular Process Systems, LLC

Prepared for Presentation at

2020 AIChE Spring Meeting
August 2020

INTRODUCTION

In most separation applications involving liquid-liquid extraction (LLE), the process will require one or more distillation columns downstream of the extraction column. Thus, optimization of the process requires a focus on the overall integration of extraction and distillation columns in order to provide the most economical solution. Typically, the solvent used for the extraction step must be distilled from the extracted solute and then recycled back to the extraction column. Also, if the solvent has some solubility in the feed, as is often the case for systems that include an aqueous feed and an organic solvent, then a distillation column will be required to recover and recycle the solvent from the aqueous phase (raffinate). As discussed in previous papers [1] [2], design of the extraction column almost always requires pilot plant testing in order to provide data for accurate scale-up and design of the commercial column. Often the distillation columns can be designed by first simulating the process and then designing the distillation column internals using known design criteria, hydraulic rating software and process engineering experience. However, in some instances, such as foaming, and unknown interactions between various components in the system, pilot plant testing of downstream distillation steps may also be required in order to provide a firm basis for the commercial process design. This paper will present three case studies involving the design of complete LLE systems with downstream distillation. In each case, the extraction column was sized via pilot plant testing. In two of the applications, the distillation columns were designed via simulations only, and in the other application a distillation column was designed via pilot plant testing plus simulation. All three system have been successfully operating in the commercial installations.

Case Study 1 – Extraction of Acetic Acid from Fermentation Broth

This application involved the extraction of acetic acid from an aqueous fermentation broth generated in a cellulosic ethanol process. Ethyl acetate (EA) was the selected solvent. The overall process for this application was expected to be as shown in Figure 1. The first column is an extraction column where EA is used to counter currently contact the broth and extract the acetic acid. The extract phase (EA) leaving the top of the column is directed to a distillation column (the solvent recovery column) to regenerate EA for recycle back to the extraction column, and to remove and recover the acetic acid that was extracted from the EA. Aqueous raffinate leaving the bottom of the extraction column will contain the soluble amount of EA and thus must be sent to a stripping column to recover the EA and also to recycle this solvent back to the extraction column. As shown in Figure 1, the overheads from each distillation column enter a common decanter which produces an organic, EA rich phase and a heavier aqueous rich phase. A portion of the EA rich phase is recycled back to the extraction column, with the balance going back to the solvent recovery column as reflux. The aqueous rich phase is sent back to the stripping column as reflux. As in most 3-column extraction process such as this, the majority of the operating cost for the system is in the extract distillation column. Thus, operating cost is minimized by optimizing the process such that the amount of EA used in the extraction step is minimized and the distillation column is designed to operate in a manner that minimizes the utilities required. Using less EA in the extraction step also minimizes the capitol cost of the solvent recovery column.

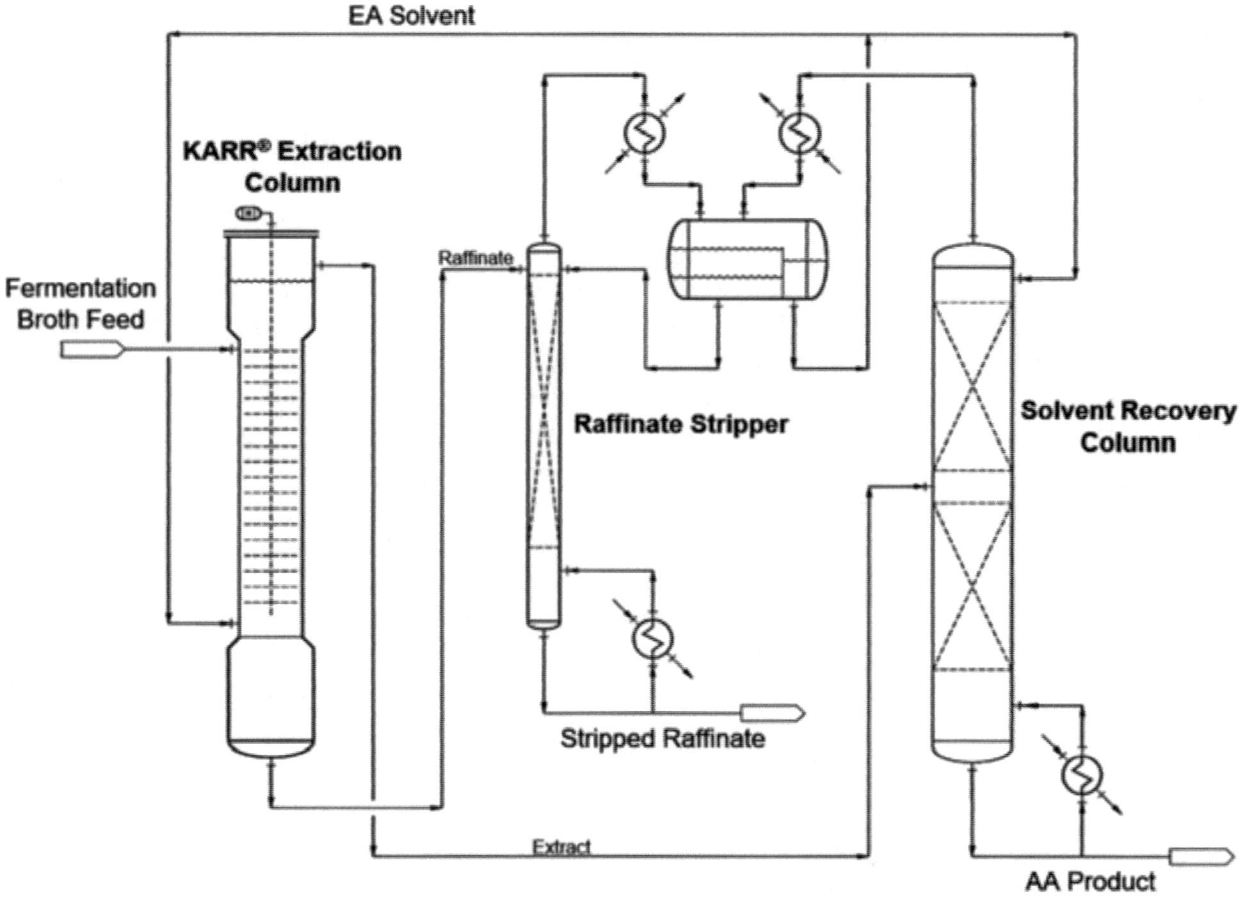

Figure 1 Extraction of Acetic Acid from Fermentation Broth Flow Diagram

Initial investigation into the parameters for this process indicated that sufficient vapor-liquid equilibrium (VLE) data was available for the components in the system. Also, the process engineers involved had the design experience and the appropriate modeling software in order to effectively simulate and design the distillation and stripping columns. The key design information required was the material balance around the extraction column. Selection of the appropriate extraction column technology and pilot plant testing was required in order to optimize the performance of the column and develop the corresponding material balance.

Optimization of the extraction process has been reported in a previous paper [3]. Due to a strong tendency toward emulsification, the KARR® Column was selected for this process [4]. Results from pilot plant testing generated the following design criteria:

- Scale-up was based upon a 1" diameter x 12' plate stack height, pilot KARR® Column with metal perforated plates. The column was tested with the solvent phase dispered.
- The solvent to feed ratio was 1.5 on an overall, weight basis. I.e. the flow rate of EA was 1.5 times the flow rate of the fermentation broth in order to achieve 98-99% acetic acid extraction.
- The design temperature was 40-45 °C.

- The design capacity was 650 GPH/ft^2 (combined flow of feed + solvent divided by the cross-sectional area of the column).

The material balance developed around the KARR® Column, as a result of the pilot test, for the commercial scale system is shown in Table 1. Based upon this material balance, simulations were performed and optimized for the process design of the distillation and stripping columns and then the mechanical designs for these columns were performed. Thus, the process flow diagram shown in Figure 1 is essentially the same as the process flow diagram that was created for the industrial scale acetic acid extraction process. A complete modular system [5] containing all of the extraction and distillation columns, tanks, heat exchangers, pumps, piping, instrumentation and controls was designed and fabricated off site and delivered to the end user.

	Feed	**Solvent**	**Raffinate**	**Extract**
$\dot{m}$ (g/min)	90	135	87	134
x_{AA}	0.052	0.0010	0.0029	0.035
x_{EA}	0	1.0	0.081	0.93
x_{H2O}	0.95	0	0.92	0.040

Table 1 Extraction of Acetic Acid from Fermentation Broth – Pilot Test Material Balance

Case Study 2 – Extraction of Acids from Wastewater

The application involved the extraction of a mixture of short chain organic acids from a wastewater stream using an alkane solvent which has essentially no solubility in the water, and thus a raffinate stripping column would not be required. The process is shown schematically below in Figure 2. The first column is an extraction column where the alkane is used to counter currently contact the aqueous feed containing approximately 4% acids and extract these into the solvent phase. The desired raffinate acid concentration was < 800 ppm, however from the standpoint of a process performance guarantee, <1000 ppm would be acceptable. The extract phase (solvent) leaving the top of the column is directed to a distillation column to recover the solvent overhead for recycle back to the extraction column. Aqueous raffinate leaving the bottom of the extraction column will have a sufficiently low level of solvent that it can be discharged as treated wastewater from the system.

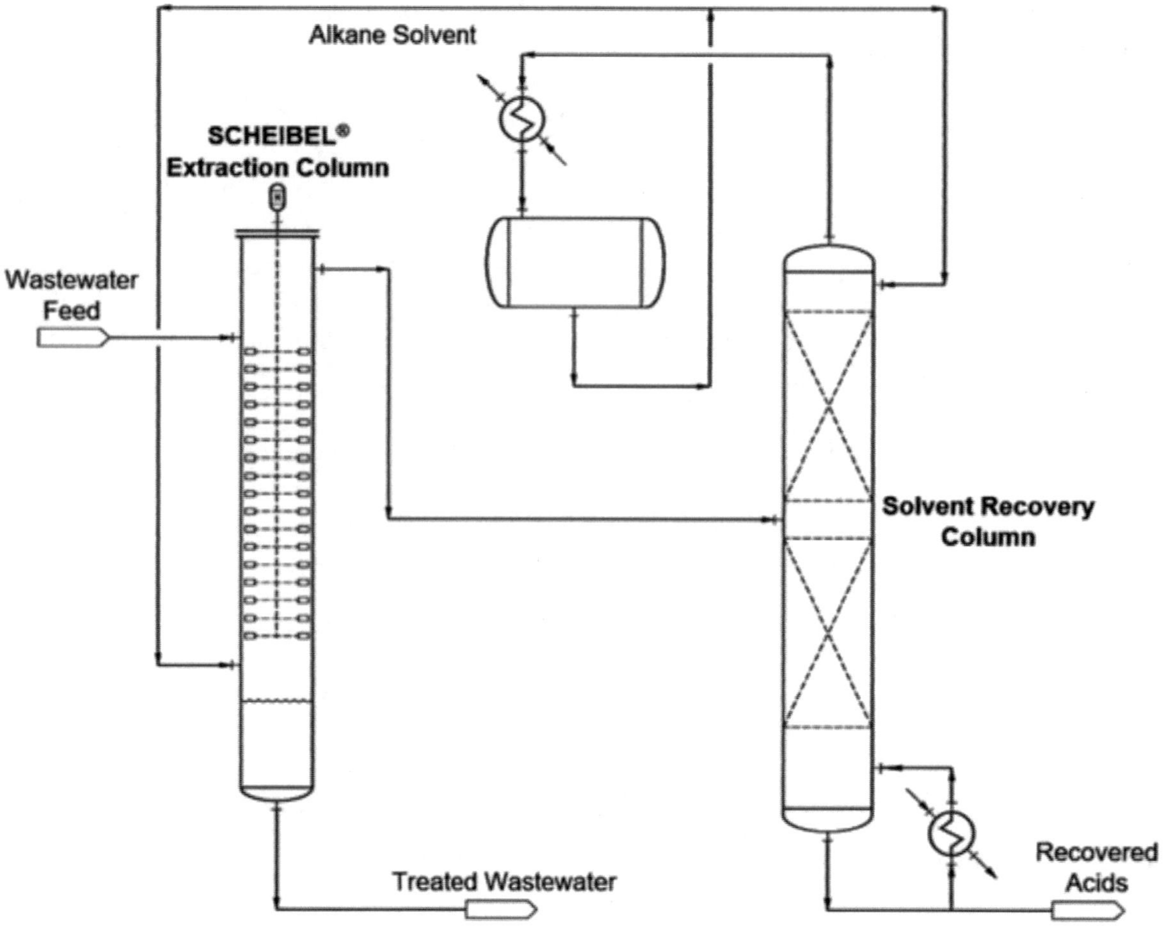

Figure 2 Extraction of Acids from Wastewater Flow Diagram

Both liquid-liquid equilibrium data (for water – organic acids – alkane) and VLE data for the organic acids/alkane separation were obtained from the end user based upon testing in their specified laboratory. The liquid-liquid equilibrium data was confirmed via additional laboratory tests using the actual feed to be fairly accurate. Using the VLE data provided, simulations demonstrated that the overhead solvent could be distilled to achieve very low ppm levels of acid in the recycled solvent. As in Case Study 1, minimizing the amount of alkane solvent used in the extraction column would result in the most economical operating cost for the process in addition to reducing the capitol cost for the downstream solvent recovery column.

Due to a high difference in density between the two phases (approximately 0.3 difference in the specific gravities) and the expectation that the interfacial tension would also be high, a 3" diameter x 60-stage SCHEIBEL® Column [4] was selected to pilot test the extraction step. Initial testing using fresh solvent (no acid) demonstrated successful performance of the column with scale-up [6] capacities up to 600 GPH/ft^2. The minimum solvent to feed ratio that could achieve a raffinate acid concentration below 1000 ppm was 3.5 on an overall, weight basis.

In order to confirm the VLE data received for extract phase separation, pilot plant distillation testing was conducted. The distillation was performed in a 4" diameter x 30' glass column. The column contained 21 feet of 316SS Koch Glitsch BX wire gauze structured packing. There were numerous potential locations for the feed point such that the heights of both the stripping section and the rectification section could be varied. The column was traced with electric heat tape to overcome any heat losses, and then insulated.

Pilot plant testing of the solvent recovery column was unable to achieve the results expected via simulation. It was apparent that the vapor-liquid equilibrium data provided did not reflect the true nature of the separation at the purity required. The test program demonstrated that the distillation column could produce a recovered solvent distillate at several reflux ratios with an acid content between 50-100 ppm acid in the alkane distillate, a level that was significantly higher than simulations predicted.

Based upon these results, additional extraction tests were performed in the pilot SCHEIBEL® Column using solvent spiked with 100 ppm acid. These tests demonstrated that the desired raffinate concentration of < 1000 ppm acid could not be achieved. The concentration of acid leveled out at just above 1000 ppm in the raffinate for testing with S/F ratios between 3.5 and 5.0. Thus, there is a pinch point at this concentration of acid in the solvent. As such a lower level of acid in the recycled solvent is required for this process to achieve < 1000 ppm in the raffinate. Additional tests were then performed using 50 ppm acid in the solvent. The performance guarantee level of acid in the raffinate was achieved at S/F = 1.5, but a ratio of 3.5 was required to achieved less than the desired concentration of 800 ppm.

Like Case Study 1, the process was modeled using a simulation software based upon the pilot test results, and the distillation column was designed using distillation internals hydraulic rating software. A complete modular system was designed and fabricated. The full system was eventually installed and started up with excellent results. A modular system including the SCHEIBEL® Column for extraction and a distillation column with structured packing was fabricated. Since a recycle solvent with 50 ppm acid could not be achieved in the test distillates, and upon agreement with the end user, the design was based upon a raffinate guarantee of <1100 ppm acid and a S/F of 3.5. This system has successfully operated for a number of years. Future plans were to further optimize the process in order to reduce the solvent to feed ratio and/or to achieve 800 ppm or lower levels of acid in the aqueous effluent.

Case Study 3 – Phenol Extraction System

Prior to working on this project, the engineers involved had previous experience with the design, supply and operation of phenol extraction systems. These systems involve the typical, 3-column design as follows:

- An extraction column using a low boiling solvent to extract phenol
- A distillation column for recovery of the phenol (bottoms stream) and recovery of the low boiling overhead solvent.

- A stripping column to strip soluble solvent out of the aqueous raffinate steam, recover this solvent and recycle it to the extraction column, generating wastewater with <1 ppm solvent.

Therefore, the flow diagram for this system as shown in Figure 3 below is analogous to Figure 1 for Case Study 1.

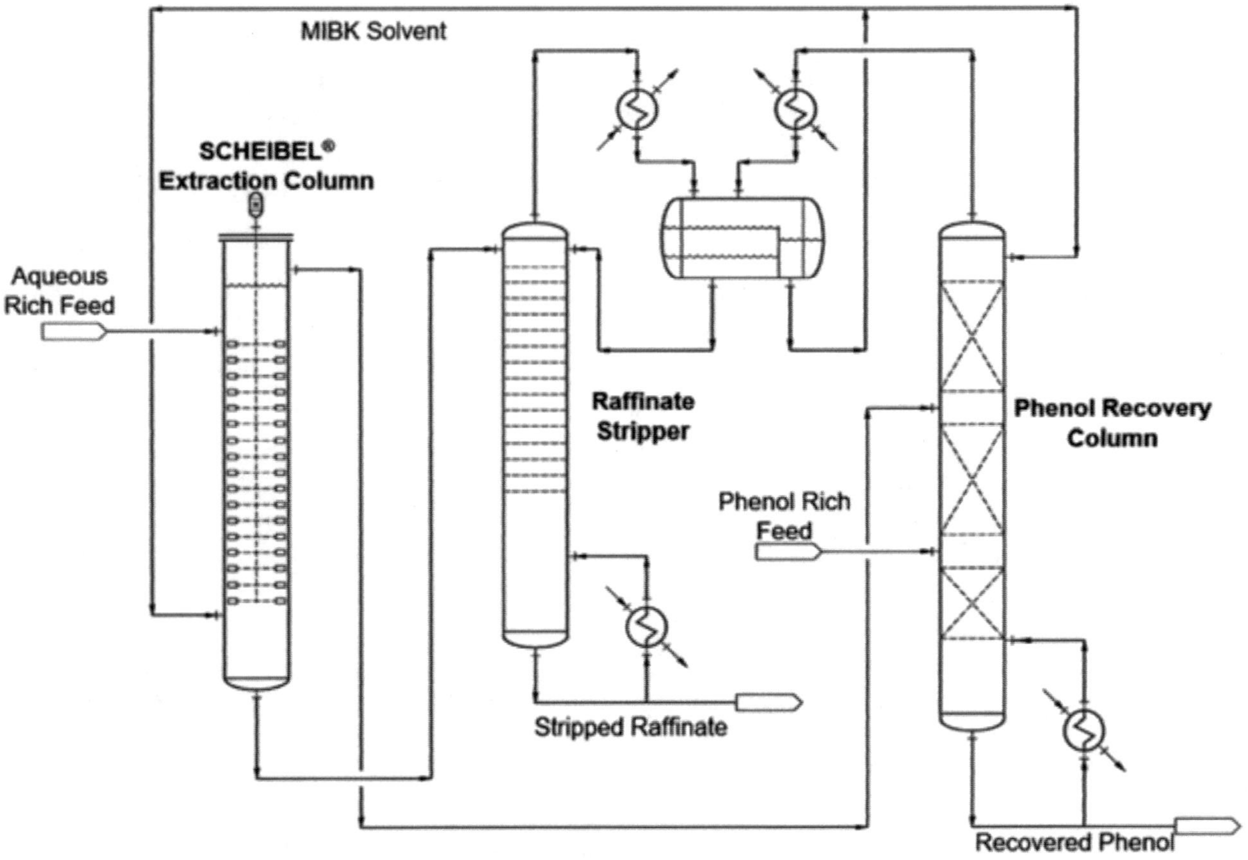

Figure 3 Phenol Extraction System Flow Diagramp

Most phenol extraction systems utilize methyl isobutyl ketone (MIBK) as the solvent due to a very high distribution coefficient for aqueous steams with pH below 7. The design of the distillation and stripping columns are relatively straight forwards via simulations and scale up. However, the design of extraction column can be significantly affected by trace components that are in the feed. As such, in order to accurately design the extraction column and develop a good material balance around this column, pilot plant testing using actual plant feed material is required.

The aqueous feed for this application was unique in that it was a distillate stream from a resin manufacturing process. The distillate stream was actually two phases; a phenol rich phase and an aqueous rich phase that contained the soluble amount of phenol which was analyzed at room temperature to be 7.4%. The phenolic phase by-passed the extraction column and was directed to the distillation column for phenol recovery in the bottom of the column. The aqueous phase was directed to the extraction column for counter current contact with MIBK. The target for the raffinate concentration was < 50 ppm phenol.

Prior to pilot plant testing the extraction process, LLE tests [7] with the actual feed were performed to generate LLE data and finalize the extractor configuration for pilot plant set-up. These tests demonstrated that 6 theoretical stages were required to achieve the <50 ppm phenol in the raffinate at S/F = 0.2 “as is” basis.

Pilot plant testing was performed utilizing a 3” diameter x 60-stage SCHEIBEL® Column as shown in Figure 4 below. Prior simulation for both this feed and previous phenol extraction systems, plus performance in actual commercial systems demonstrated the < 5ppm phenol could easily be achieved in the MIBK distillate for recycle to the extraction column. Thus, no phenol spiking of the MIBK solvent for the extraction tests was required. During testing, the design capacity for scale up was determined to be 600 GPH/ft^2. Based upon raffinate analyses, < 50 ppm raffinate phenol concentration could easily be achieved using 42 - 60 agitated stages and S/F = 0.2. The results indicated that lower S/F with 60 stages could also be used. In the end, the SCHEIBEL® extraction column was sized for 60-stages and S/F = 0.2 in order to be conservative. Like the other case studies discussed, process simulation software and distillation internals hydraulic rating software were used to design the phenol recovery and wastewater stripping columns. A complete modular system including all of the unit operations and other equipment, piping, instrumentation and controls was designed and fabricated designed and fabricated off site and delivered to the end user.

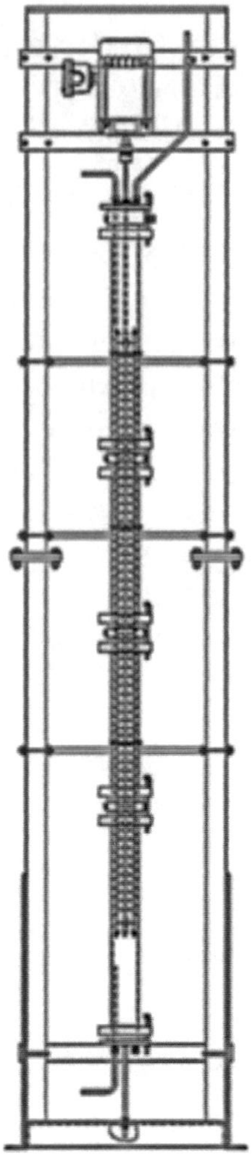

Figure 4 3" Diameter / 60 Stage Scheibel Column Used for Phenol Extraction Pilot Test

CONCLUSION

Designing a liquid-liquid extraction process with a solvent that boils lower than water is used to extract, and possibly recover, higher boiling compounds from aqueous streams requires optimized design of not only the extraction column, but also the interaction between the extraction column and the downstream distillation columns required for the entire system. The three case studies presented in the paper show that first it is important to select the correct extraction column technology and then to have some understanding as to the amount of solute that will be present in the recycled solvent. Pilot plant testing the extraction column is almost always required in order to accurately design the column itself and to determine the corresponding material balance, which in turn allows for downstream distillation design. When sufficient VLE data is available for the known components in the system and the design engineers have experience with process simulation and distillation internals design, then these downstream columns can be designed without the need for pilot plant testing. However, if VLE data is not known or if there are concerns with multicomponent interactions that may not be accurately simulated, or if there are other factors that inhibit confident distillation design, then pilot plant testing the downstream distillation columns can be very useful for their design and de-risking the entire process.

One note of caution, pilot plant testing in the manner discussed in the paper, often cannot account for possible build-up of low boiling contaminants in the solvent, especially when these components are in very low concentration in the aqueous feed. The only way to really determine this would be to set up a full pilot plant system with all the required columns operating in recycle mode and to operate this system for a sufficiently long period of time. Unfortunately, this is not really practical for most applications.

REFERENCES

1. Cusack, R., Karr, A., "A Fresh Look at Liquid-Liquid Extraction, ", Chemical Engineering, April 1991
2. Cusack, R., "Pilot Plants Confirm Process Validity", Chemical Engineering, June 1998
3. Glatz, D., Parker, W., "Recovery Of Valuable Products From Biomass Using A KARR® Column", International Solvent Extraction Conference (ISEC 2008), September 2014
4. Cusack, R., Fremeaux, P., A Fresh Look at Liquid-Liquid Extraction, The Range of Available Extractors is Wide", Chemical Engineering, March 1991
5. Glatz, D., Cross, B., Lightfoot, T., "Pilot Plant Testing of Liquid-Liquid Extraction", CEP Magazine, February 2018
6. Glatz, D., Parker, W., "Successful Design of Agitated Extraction Liquid-Liquid Extraction Columns Via Pilot Plant Testing", International Solvent Extraction Conference (ISEC 2014), September 2014
7. Glatz, D., Cross, B., Lightfoot, T., "Liquid-Liquid Extraction: Generating Equilibrium Data", Chemical Engineering, October 2018

AIChE Paper Number 124d

Reducing greenhouse gas emissions of steam cracking furnaces by oxy-fuel combustion, renewable fuels and high emissivity coatings

Stijn Vangaever
ir.
Laboratory for Chemical Technology, Ghent University, Ghent, Belgium

Yu Zhang
Dr. ir.
Laboratory for Chemical Technology, Ghent University, Ghent, Belgium

Marko R. Djokic
Dr. ir.
Laboratory for Chemical Technology, Ghent University, Ghent, Belgium

Joost Van Thielen
Mr.
CRESS B.V., Breskens, The Netherlands

John Olver
Dr.
Emisshield Inc., Blacksburg, Virginia, USA

Gilles Theis
Mr.
John Zink Hamworthy Combustion, Dudelange, Luxemburg

Kevin M. Van Geem
Prof. Dr. ir.
Laboratory for Chemical Technology, Ghent University, Ghent, Belgium

Prepared for Presentation at the 2020 Spring National Meeting
Houston, Texas (TX), March 29 – April 2, 2020

Reducing greenhouse gas emissions of steam cracking furnaces by oxy-fuel combustion, renewable fuels and high emissivity coatings

Stijn Vangaever
_ir._
Laboratory for Chemical Technology, Ghent University, Ghent, Belgium

Yu Zhang
_Dr. ir._
Laboratory for Chemical Technology, Ghent University, Ghent, Belgium

Marko R. Djokic
_Dr. ir._
Laboratory for Chemical Technology, Ghent University, Ghent, Belgium

Joost Van Thielen
_Mr._
CRESS B.V., Breskens, The Netherlands

John Olver
_Dr._
Emisshield Inc., Blacksburg, Virginia, USA

Gilles Theis
_Mr._
John Zink International Luxembourg, Dudelange, Luxemburg

Kevin M. Van Geem
Prof. Dr. ir.
Laboratory for Chemical Technology, Ghent University, Ghent, Belgium

Abstract: Steam cracking furnaces are responsible for the majority of the emissions of a steam cracking plant. Therefore the energy efficiency and the furnace technology will need to change drastically, preferably in a cost effective way, bearing in mind that more strict regulations will be imposed over time. Several novel technologies were tested at various technology readiness levels before implementing the most promising technologies at a commercial scale in the context of the H2020 IMPROOF project. John Zink Hamworthy Combustion (JZHC) is in the possession of a test furnace where several burners can be tested simultaneously to full scale to simulate part of an ethylene cracker furnace. Two JZHC COOLStar™ burners were tested to see the effect of biogas and oxyfuel combustion on the heat flux profile. The experimental results were validated with Reynolds Averaged Navier-Stokes Computational Fluid Dynamics simulations in ANSYS Fluent 19 using a 25 species mechanism for combustion in CO_2-rich environments. The results show that existing furnaces can easily be retrofitted to run on biogas or to implement oxy-fuel combustion. Steam cracking pilot plant experiments performed at the Laboratory for Chemical Technology show that applying an Emisshield high emissivity coating on the furnace refractory wall can lead to energy savings of approximately 5 %. These energy savings are validated using radiative heat transfer simulations in a simplified geometry. A parametric study is performed to check the effect of boundary wall emissivity on the temperatures inside a so-called tube-in-box enclosure.

Introduction: Novel technologies

Renewable fuels & oxy-fuel combustion

Biogas is the most straightforward bioenergy alternative in terms of production process and purification involved[1]. Several commercial scale biogas plants are already running successfully in different parts of the world. In these plants, biogas is produced through anaerobic fermentation of biomass, and it is mainly constituted of CH_4 (55 – 75 %) and CO_2 (25 – 45 %), with traces of other components such as H_2 (0 – 3 %), N_2 (1 – 5 %), CO (0 - 0.3 %) and H_2S (0.1 - 0.5 %)[2]. Due to the presence of carbon dioxide the caloric value of biogas is significantly lower compared to natural gas. This makes the removal or transformation of carbon dioxide, preferably combined with the purification from other contaminants, essential for increasing the calorific value of biogas in order to act as a one-to-one replacement for natural gas[3]. However, biogas upgrading and cleaning makes the process economically less feasible and puts an additional stress on the environment due to the emissions during processing. For these reasons there has been an increased interest in the direct combustion of unpretreated biogas. One of the essential questions in this work is whether the use of renewable biogas can actually be considered in steam cracking furnaces or by extension to be considered in gas-fired furnaces in general.

Oxy-fuel combustion is a promising technology where fuel is burned with pure oxygen rather than air as oxidant. Due to the absence of the nitrogen component in air, higher flame temperatures can be obtained with quasi non-existing NO_x emissions. A portion of the flue gas is often recycled to temper the flame temperature. The main advantage of the oxy-fuel combustion process is that the flue gas stream is nearly entirely composed of carbon dioxide and water instead of the large amount of nitrogen diluent present in conventional air-combustion scenarios. This means that water vapor can easily be separated from the exhaust stream via low-temperature dehydration, resulting in an near pure carbon dioxide product stream. This carbon dioxide stream can be used for carbon capture, utilization and storage. The main drawback of the oxy-fuel combustion technology is the high energy requirement related to producing high purity oxygen. For this reason, the broader perspective offered by life cycle assessment is often called upon to validate the environmental impact of oxy-fuel combustion combined with carbon capture, utilization and storage.

High emissivity coatings

Interest in high emissivity coatings originated in the aerospace industry. The association with spacecraft and NASA surrounded the research area in an aura of 'high tech' and 'mystery'[4]. Nonetheless, the underlying physical principle is straightforward. During the atmospheric re-entry of space vehicles, friction between the space vehicle and the atmospheric gases drastically heats up the surface of the spacecraft. Since the emissive properties of the radiating surface determine the amount of heat that can be dissipated back to the surroundings, increasing the emissive properties reduces the surface temperature[5]. Emisshield licensed the high emissivity coating technology from NASA in 2001 to include all applications except space and space vehicle applications. For the last decades, high emissivity coatings started to make their way into the chemical industry. High emissivity refractory and metal coatings promise to reduce downtime, minimize coking, increase productivity and to reduce the plant's overall carbon footprint by providing a uniform temperature distribution and an enhanced radiative heat transfer[6].

Methodology and procedures

Renewable fuels & oxy-fuel combustion

Test furnace

The John Zink Hamworthy Combustion test furnace, representing a section of a cracking furnace, is used to validate the feasibility of biogas and oxy-fuel combustion. The test furnace allows for full-scale tests of two floor burners simultaneously at their rated capacity with furnace temperatures up to 1623 K. The rectangular firebox has internal dimensions representative of an industrial furnace with vertical tubes acting as a heat sink. Side by side, two COOLStar™ burners are installed in the test furnace. The main benefit of a full-scale test is that emissions and flame stability with high inert content, shape and height can be studied in a detailed manner. Several openings along the furnace height allow for heat flux measurements.

Both biogas and oxy-fuel firing conditions are tested in this furnace. In both cases the fuel and oxidizer are artificially created by mixing pure CO_2, H_2O and O_2 combined with natural gas. This implies that for the biofuel tests, natural gas was gradually diluted with CO_2 to mimic biogas. For the oxy-fuel tests, the oxidant was composed of O_2 combined with CO_2 and H_2O. The ratio of CO_2 to H_2O was in correspondence with the stoichiometric combustion ratio of natural gas in air similar to operation with recycled flue gas. The O_2 content in the oxidant was initially kept at 21 vol% which is comparable to air. The main question is whether existing industrial furnaces can easily be revamped or retrofitted to run on biogas and or oxy-firing conditions.

Computational Fluid Dynamics model set-up

Three-dimensional steady-state global mass, momentum, energy and species conservation equations are solved for a compressible, reacting fluid using ANSYS Fluent 19. The RNG-based $k - \varepsilon$ two-equation model is used to model turbulence. Initially, the Westbrook and Dryer two-step combustion mechanism was used to obtain the combustion reaction rates[7]. In a later stage, the POLIMI mechanism is reduced to a 25 species mechanism with 154 reactions[8]. The abundance of CO_2 in the mixture has chemical effects resulting in a reduced mechanism that is larger than a conventional methane-in-air combustion mechanism. The Eddy Dissipation Concept is used to model the turbulence-chemistry interaction. The discrete ordinates radiation model is selected to solve the radiative transfer equation with an angular discretization of 2 by 2. A weighted-sum-of-gray gasses model accounts for the gas phase radiative properties.

High emissivity coatings

Steam cracking pilot plant

To experimentally assess the effects of applying a high emissivity coating on the energy efficiency of a gas-fired furnace, the steam cracking pilot plant at the Laboratory for Chemical Technology of Ghent University is used. A schematic representation of the steam cracking pilot plant is shown in Figure 1. For a detailed description of the unit, including all the analytical equipment, the reader is referred to previous work[9, 10]. The pilot plant is divided in different cells that can be fired independently. Twelve wall burners in each cell allow the plant operator to impose any required temperature or heat flux profile over the reactor tube length. In the first two cells 10 kg/h of propane mixed with 4 kg/h of dilution steam is preheated to the desired coil inlet temperature of 917 K. A SCOPE reactor coil, made of Centralloy® HT-E with nine fins and an internal diameter ranging from 35 to 39.1 mm, is installed in the two cracking cells[11]. At the outlet of the pilot plant a coil outlet temperature of 1111 K is maintained to guarantee the same cracking severity for each experiment, corresponding to 90 % propane conversion. At the start of every experiment, the reactor is heated with steam until the desired reactor coil temperature profile is reached. Next, the flow rate of propane and steam is set to the desired operating conditions. Cracking starts and is maintained for 6 h. Finally, the experiment is halted and the reactor is decoked. The experimental procedure, as described, is repeated two times to establish a base case with uncoated furnace walls. Next, the two cracking cells are coated with a high emissivity coating and the experiment is repeated twice. Differential pressure transmitters are installed to measure the premixed burner gas flow rate to the two cracking cells. Additionally, the total natural gas flow rate to the entire furnace is monitored. Several thermocouples are installed to measure the process gas temperature, flue gas temperature, tube metal wall temperature and refractory wall temperature, the latter both inside and outside the furnace.

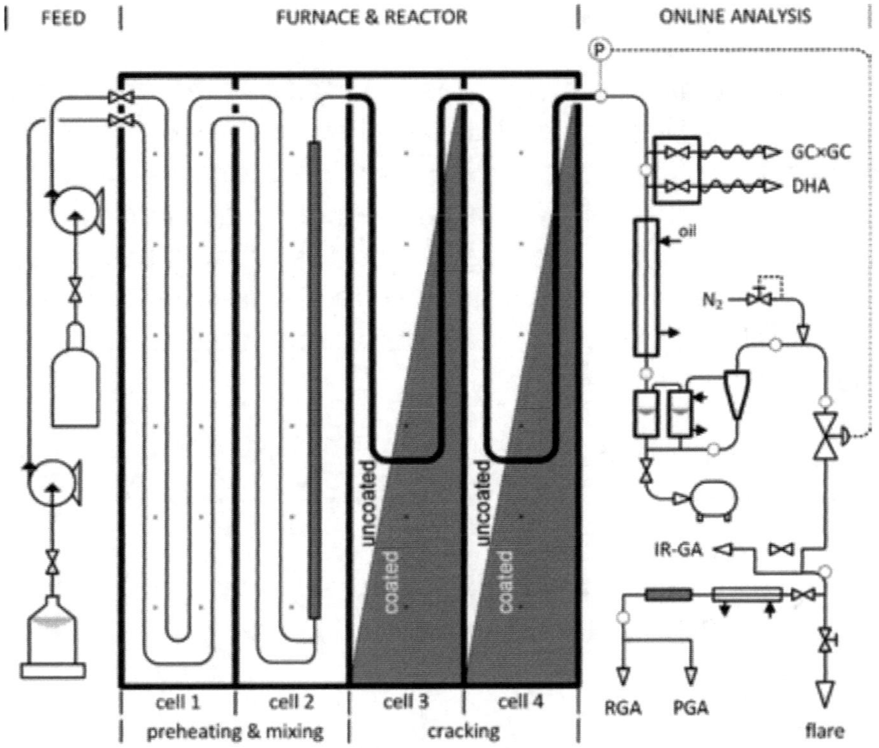

Figure 1: Schematic overview of the pilot plant setup. The pilot plant is divided into a feed section, a furnace-reactor section and an online analysis section. Cracking reactions mainly occur in cells 3 and 4. These cells are coated with a high emissivity coating. (Adapted from Van Geem et al.[10])

Radiative heat transfer tube-in-box model

In this work, a simplified two-dimensional geometry based on a horizontal cross-section of one of the pilot plant cracking cells, is studied . This so-called tube-in-box model, shown in Figure 2, consists of a central cylindrical surface, acting as a heat sink, and an enclosing adiabatic refractory surface. The enclosing adiabatic box has a length and height of 0.380 m. A circle with 0.103 m diameter, equivalent to the pilot plant heat sink area, is positioned in the center. A perfectly stirred flue gas with a uniform temperature, T_G, is located between the adiabatic enclosure and the central heat sink. The flue gas is composed of CO_2, H_2O and N_2 in the stoichiometric combustion ratio of methane in air.

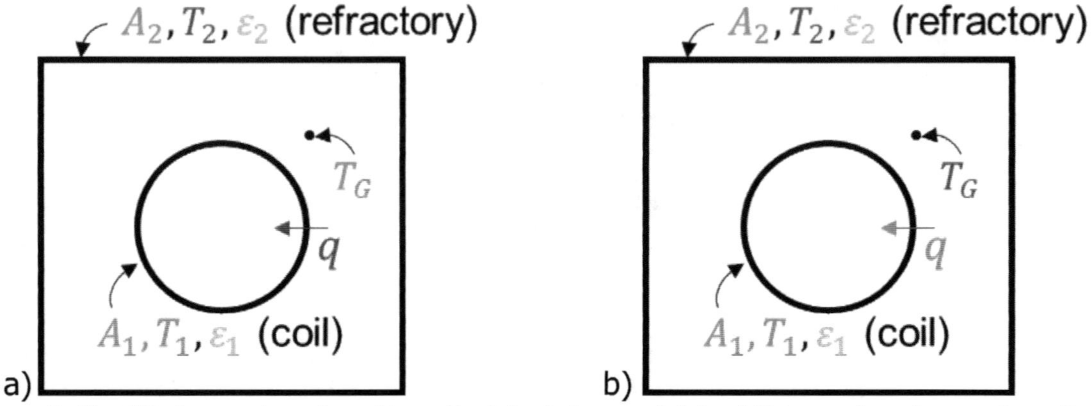

Figure 2: Simplified geometry referred to as the tube-in-box model. In a) the gas phase temperature, T_G, is assumed to be constant for different tube, ε_1, and adiabatic surface, ε_2, emissive properties. In b) the gas phase temperature, T_G, is adjusted to get the same heat flux, q.

Radiative heat transfer from the hot flue gas to the heat sink will be modeled using Hottel's zoning method[12] with varying emissive properties for both the central heat sink, ε_1, and the furnace refractory wall, ε_2.The geometry is determined by the surface areas, A_1 and A_2. The central heat sink temperature, T_1, is kept constant based on the averaged experimentally observed tube metal temperature. Initially, the gas phase temperature , T_G, is kept constant, resulting in an unknown adiabatic wall temperature, T_2, and an unknown heat flux, q, to the central heat sink, visualized in Figure 2 a). In a second simulation, shown in Figure 2 b), the gas phase temperature, T_G, is adjusted to obtain a constant heat flux, q, for varying wall emissive properties. The radiative heat transfer problem will initially be solved with a gray gas assumption. Later, this solution will be compared to a non-gray gas approach.

Results & Discussion

Renewable fuels & oxy-fuel combustion

Mechanism reduction

Conventional global combustion mechanisms such as the two-step Westbrook and Dryer mechanism[7] or the four-step Jones and Lindstedt mechanism[13] are not valid in CO_2-rich conditions[14]. The present CO_2 will compete with O_2 for atomic hydrogen leading to the formation of CO. Similarly, CO_2 might also react with the hydrocarbon radicals contributing to the formation of CO[15]. So instead of refining global combustion mechanisms to be used under oxy-fuel conditions[16], the POLIMI natural gas combustion mechanism developed by the CRECK modelling group is reduced to include the appropriate radical species[8]. The resulting mechanism consists of 25 species with 154 reactions. The mechanism has a maximum error on ignition delay time lower than 20 % in the tested range (equivalence ratio of 0.5 to 1.7 with up to 40 vol% CO_2 for biogas and under oxy-fuel conditions). Figure 3 shows a maximum error of 7 % on the laminar flame speed.

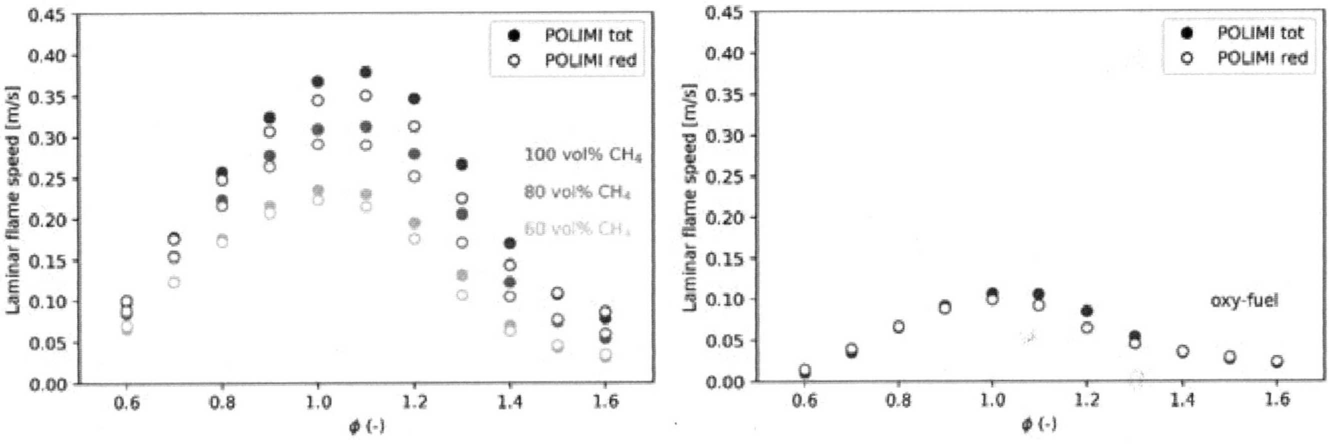

Figure 3: Laminar flame speed for the complete POLIMI mechanism versus the reduced POLIMI mechanism for biogas with varying CO_2 additions (left) or under oxy-firing conditions (right).

Flame envelop & heat flux profile

Figure 4 gives a visualization of the temperature and vertical velocity of a cross-section of the test furnace for different CO_2 dilutions. The peak flame temperature occurs at an elevation of approximately 4.5 m and increases with the addition of CO_2.

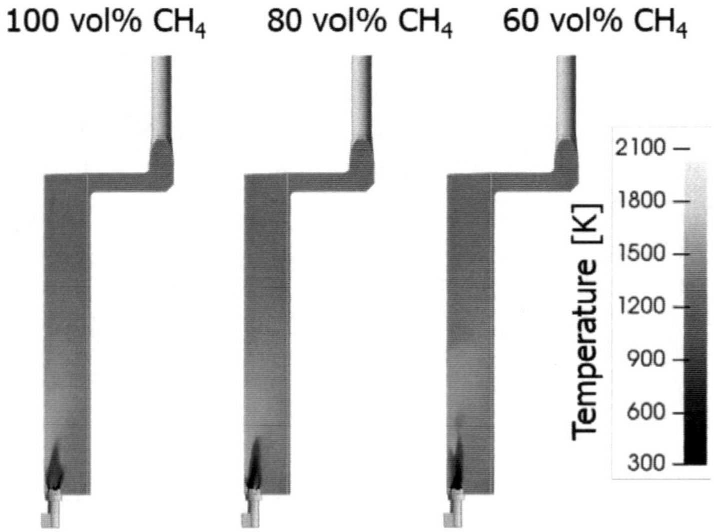

Figure 4: Temperature distribution for conventional combustion (left) and for biogas combustion with 20 vol% CO_2 (middle) and 40 vol% CO_2 (right).

The heat flux profile shows where most heat is transferred to a process coil in function of the furnace height. This visualization is important for licensors of steam cracking furnaces. In order to retrofit an existing furnace with biogas or oxy-fuel combustion, the heat flux profile should not deviate too much from the predefined shape. Experimentally, this is the case, as shown in Figure 5. The initial two-step Westbrook and Dryer combustion mechanism gives a heat flux profile that closely matches the experimental results. In a later stage, the reduced POLIMI mechanism was implemented resulting in an upward shift of the maximum heat flux. This increased flame height could eventually result in flame lift-off when the CO_2 concentration is further increased. However, Large Eddy Simulations are typically required to correctly model the flame detachment. The oxy-fuel simulations have not yet converged.

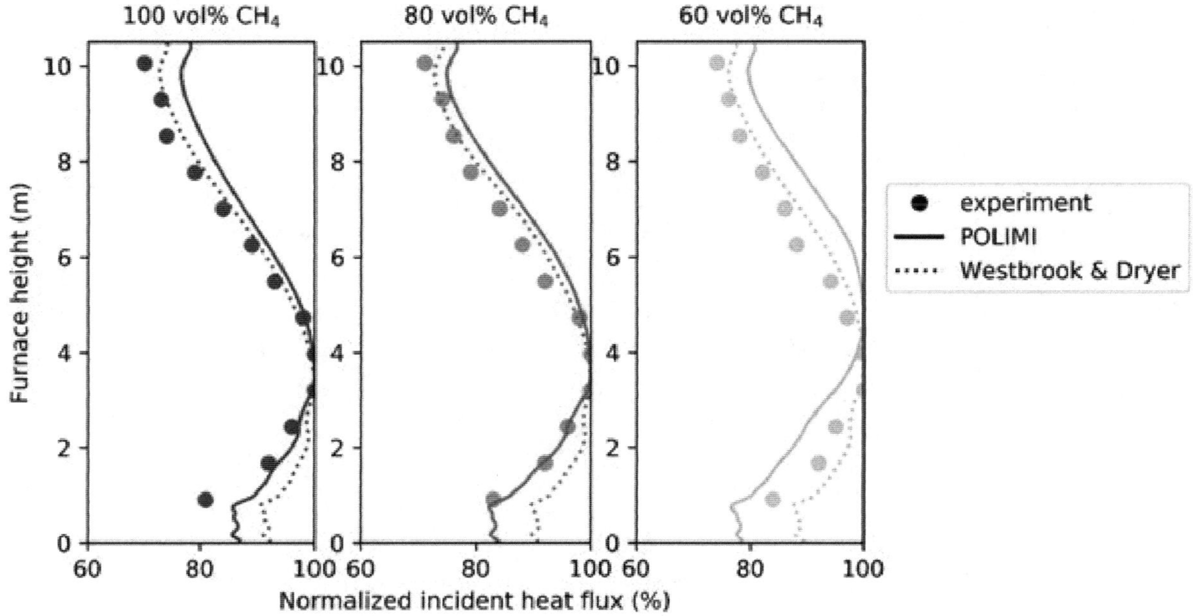

Figure 5: Normalized incident radiative heat flux to the process coils for varying CO_2 additions. The experimental measurements are compared to the numerical simulations using the simplified POLIMI mechanism and the two-step Westbrook & Dryer mechanism.

High emissivity coatings

Steam cracking pilot plant experiments

The fuel consumption of the steam cracking pilot plant before and after the high emissivity coating was applied is shown in Figure 6. For all experiments the natural gas consumption initially decreases to reach a steady-state after almost 6 h of cracking. The decrease can be attributed to the fact that the furnace needs to heat-up and needs to find an equilibrium with the surrounding environment. The fuel consumption at steady-state is gives an energy efficiency increase of 4.77 % using averaged values for the uncoated and coated runs.

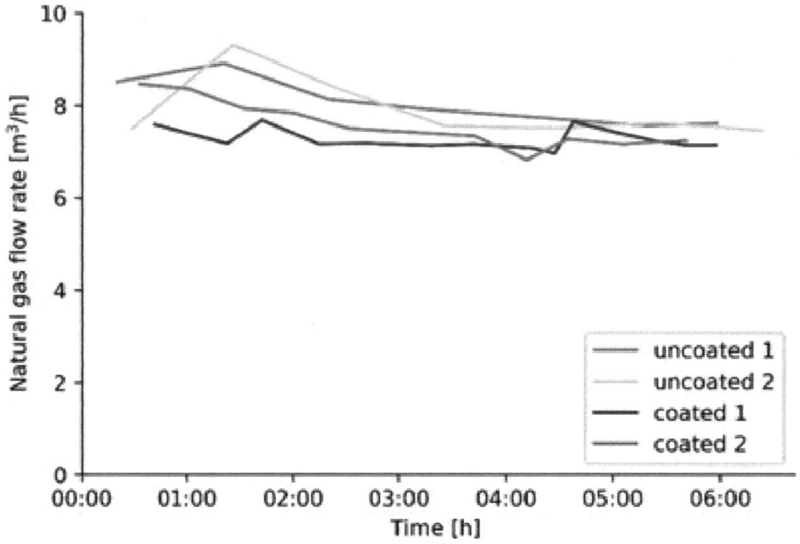

Figure 6: Natural gas flow rate to the entire pilot plant over time for the different experiments before and after applying an Emisshield high emissivity coating.

Radiative heat transfer tube-in-box model

Initially, the total surface-to-surface exchange areas with a participating gray gas and the total gas-to-surface exchange areas are derived for the tube-in-box model. The "exchange areas" approach was introduced by Hottel and Sarofim in their zonal method for radiative heat transfer[12]. The total exchange areas are functions of the geometry and radiative properties of the entire enclosure[17]. The resulting equations for a gray gas are independent of the adiabatic wall emissivity. With the gray gas approximation, all radiation coming from the gas will either be reflected or absorbed and re-emitted by the enclosing surface. Since there is no clear distinction between the 'two types of radiation', the enclosing surface can be regarded as a blackbody independent of its emissivity; redirecting all radiation towards the heat sink.

A non-gray gas model needs to be introduced to make the distinction between reflected and re-emitted radiation. An uncoated furnace, having a high probability of reflecting the incoming radiation, tends to reflect the radiation in an absorption band where it is re-absorbed by the gas. A higher emissivity will increase the probability that the incoming energy will be re-emitted in transparent window where it is more likely to heat the load.

Additionally, radiative transfer characteristics of an opaque wall can often be described with good accuracy by gray emission, absorption and reflection. The radiative properties of a molecular gas vary strongly and rapidly across the wavelength spectrum which makes that the assumption of a gray gas is never a good one[18, 19]. Hence, a non-gray gas modelling technique is certainly required to accurately capture the radiative heat transfer characteristics. In the gray-plus-clear gas model the general weighted sum of gray gasses model is reduced to the combination of one lumped absorption band where the total exchange areas derived for a gray gas are still applicable and one clear band where the gas absorbs no radiation.

Based on the pilot plant experiments, a reactor tube wall temperature of 1173 K will be used as heat sink boundary condition. For the first simulation a constant gas phase temperature of 1373 K was assumed, shown in Figure 2 a). In the next step the gas phase temperature is changed to have the same net heat flux to the inner tube, shown in Figure 2 b). The resulting net radiative heat flux for a constant gas phase temperature is shown in Figure 7 a). The gas phase temperature required to get the same heat flux is shown in Figure 7 b).The refractory emissivity was varied from 0.4 to 1.0 (y-axis) and the reactor tube emissivity was varied from 0.7 to 1.0 (x-axis) to give a complete overview of the effect of the emissive properties of the boundary surfaces on the temperature inside the geometry.

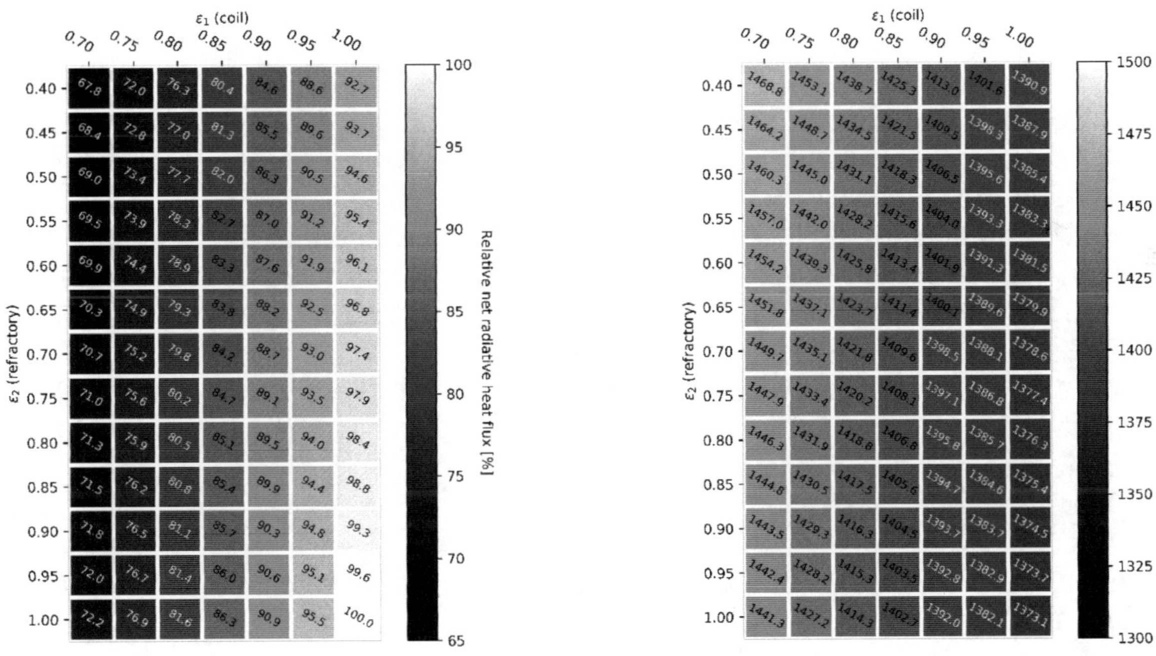

a) b)

Figure 7: Radiative heat flux towards the coil for a constant gas phase temperature a) and gas phase temperature required to get a constant net radiative heat flux b).

Conclusions

In the IMPROOF project, the effect of various novel technologies including biogas, oxy-fuel combustion and high emissivity coatings on the greenhouse gas emissions of steam cracking furnaces has been studied. Experiments performed at the John Zink test furnace shows that existing furnaces can be retrofitted to run on biogas or on oxy-fuel combustion. Numerical simulations support this claim, validating the experimentally obtained heat flux profiles. Steam cracking pilot plant experiments performed at the Laboratory for Chemical Technology show energy savings up to 5 % after an Emisshield high emissivity coating was applied. Radiative heat transfer simulations show the necessity of a non-gray gas model to accurately describe the effect of high emissivity coating in a tube-in-box enclosure. In the following step of the IMPROOF project, 3D reactor technologies and high emissivity coatings will be implemented on an industrial scale. Extensive measurements combined with a thorough life cycle analysis should be able to accurately describe the effect of different technologies on the greenhouse gas emissions of steam cracking furnaces.

Acknowledgment

This project has received funding from the European Union's Horizon 2020 research and innovation programme (H2020-SPIRE-04-2016) under grant agreement No 723706.

References

1. Gnansounou, E.; Raman, J. K., Chapter 7 - Wide Scope Environmental Assessment of Biofuels. In *Biofuels: Alternative Feedstocks and Conversion Processes for the Production of Liquid and Gaseous Biofuels (Second Edition)*, Pandey, A.; Larroche, C.; Dussap, C.-G.; Gnansounou, E.; Khanal, S. K.; Ricke, S., Eds. Academic Press: 2019; pp 163-196.
2. Karellas, S.; Boukis, I.; Kontopoulos, G., Development of an investment decision tool for biogas production from agricultural waste. *Renewable and Sustainable Energy Reviews* **2010,** 14, (4), 1273-1282.
3. Angelidaki, I.; Xie, L.; Luo, G.; Zhang, Y.; Oechsner, H.; Lemmer, A.; Munoz, R.; Kougias, P. G., Chapter 33 - Biogas Upgrading: Current and Emerging Technologies. In *Biofuels: Alternative Feedstocks and Conversion Processes for the Production of Liquid and Gaseous Biofuels (Second Edition)*, Pandey, A.; Larroche, C.; Dussap, C.-G.; Gnansounou, E.; Khanal, S. K.; Ricke, S., Eds. Academic Press: 2019; pp 817-843.
4. Stendera, J.; Bonsall, S., High Emissivity Coatings - Do They Really Work? *Refractories Worldforum* **2012,** 4, (1), 67-72.
5. He, X.; Li, Y.; Wang, L.; Sun, Y.; Zhang, S., High emissivity coatings for high temperature application: progress and prospect. *Thin Solid Films* **2009,** 517, (17), 5120-5129.
6. Emisshield website - Hydrocarbon & Chemical refractory and metal coatings. http://www.emisshield.com/industrial-applications/hydrocarbon-chemical/ (14/01/2020),
7. Westbrook, C. K.; Dryer, F. L., Simplified reaction mechanisms for the oxidation of hydrocarbon fuels in flames. *Combustion science and technology* **1981,** 27, (1-2), 31-43.

8. Stagni, A.; Frassoldati, A.; Cuoci, A.; Faravelli, T.; Ranzi, E., Skeletal mechanism reduction through species-targeted sensitivity analysis. *Combustion and Flame* **2016,** 163, 382-393.
9. Dierickx, J. L.; Plehiers, P. M.; Froment, G. F., On-line gas chromatographic analysis of hydrocarbon effluents: Calibration factors and their correlation. *Journal of Chromatography A* **1986,** 362, 155-174.
10. Van Geem, K. M.; Pyl, S. P.; Reyniers, M.-F.; Vercammen, J.; Beens, J.; Marin, G. B., On-line analysis of complex hydrocarbon mixtures using comprehensive two-dimensional gas chromatography. *Journal of Chromatography A* **2010,** 1217, (43), 6623-6633.
11. Symoens, S.; Dokic, M.; Zhang, J.; Bellos, G.; Jakobi, D.; Weigandt, J.; Klein, S.; Battin-Leclerc, F.; Heynderickx, G.; Van Thielen, J. In *IMPROOF: Integrated model guided process optimization of steam cracking furnaces*, 30th Ethylene producers' conference (30th EPC), 2018; 2018.
12. Hottel, H. C.; Sarofim, A. F., *Radiative Transfer*. McGraw-Hill: 1967.
13. Jones, W.; Lindstedt, R., Global reaction schemes for hydrocarbon combustion. *Combustion and flame* **1988,** 73, (3), 233-249.
14. Frassoldati, A.; Cuoci, A.; Faravelli, T.; Ranzi, E.; Candusso, C.; Tolazzi, D. In *Simplified kinetic schemes for oxy-fuel combustion*, 1st International conference on sustainable fossil fuels for future energy, 2009; 2009; pp 6-10.
15. Glarborg, P.; Bentzen, L. L. B., Chemical Effects of a High CO2 Concentration in Oxy-Fuel Combustion of Methane. *Energy & Fuels* **2008,** 22, (1), 291-296.
16. Andersen, J.; Rasmussen, C. L.; Giselsson, T.; Glarborg, P., Global combustion mechanisms for use in CFD modeling under oxy-fuel conditions. *Energy & Fuels* **2009,** 23, (3), 1379-1389.
17. Green, D. W.; Perry, R. H., *Perry's Chemical Engineers' Handbook, Eighth Edition*. McGraw-Hill Education: 2007.
18. Edwards, D., Molecular gas band radiation. In *Advances in heat transfer*, Elsevier: 1976; Vol. 12, pp 115-193.
19. Modest, M. F., *Radiative heat transfer*. Academic press: 2013.

Session 124: Ethylene Plant Technology Fundamentals
Paper 124f.

Separation of Olefin-Paraffin Mixtures with Optiperm™ Customized Amorphous Fluoropolymer Membranes

Dr. Hannah Murnen

Chief Technology Officer

Compact Membrane Systems

Dr. Sudip Majumdar

Membrane Expert

Compact Membrane Systems

William Charlton

VP Operations

Compact Membrane Systems

Prepared for Presentation at the 2020 Spring National Meeting

Houston, TX, April 1, 2020

Separation of Olefin-Paraffin Mixtures with Optiperm™ Customized Amorphous Fluoropolymer Membranes

Hannah Murnen

Chief Technology Officer

Compact Membrane Systems

Sudip Majumdar

Membrane Expert

Compact Membrane Systems

William Charlton

VP Separations

Compact Membrane Systems

Abstract: The separation of light olefins from paraffins is one of the most energy intensive in the world. The state of the art separation method is distillation, a technology that has remained unchanged over decades. Membranes hold the promise of a more modular, energy efficient and capital efficient method for separation. However, they have historically been hampered by low performance and stability.

Compact Membrane Systems has developed a novel membrane material using a silver embedded fluoropolymer material to transport olefins across the membrane. This material has demonstrated over 2 years of longevity in laboratory operation and has shown high selectivity and permeance under realistic operating conditions. Additionally Optiperm™ membranes were trialed at the Delaware City Refining Company in over 300 days of field operations validating the robustness and capability of the technology in an operating plant environment.

During the past year, CMS has developed and installed a laboratory recycle test systems to allow for the continuous testing of spiral wound modules with areas of ~20 square feet. The recycle test unit enables CMS to test larger scale membrane modules under field operating conditions using actual process streams from end users. Furthermore the recycle rig design incorporates our innovative design for hydrating the process feed within the membrane housing, significantly reducing the complexity of the membrane systems operations.

We will share the latest learnings from our recycle rig testing as well as provide updates on the scale up of our flat sheet spiral wound membrane cartridges. We will also discuss the current state of our hollow fiber membrane configuration development and the progress on our joint project with Dow Chemical through RAPID, a manufacturing institute funded by the Department of Energy. The goal of the work with Dow is to produce a chemical grade propylene from a process feed stream with a propylene concentration of greater than 60%.

CMS will also discuss results on laboratory testing of C_4 streams. This includes mixed gas separation of n-butane and 1-butene with permeances between 200-500 GPU and selectivity greater than 20. We will share results for a number of operating conditions and will highlight several exciting C_4 separation applications including those with isomer streams. In addition, longevity results will be shared demonstrating the stability of the membrane in C_4 separations.

Background

Ethylene and propylene are major chemical industry raw materials and consume a great deal of energy in their production. The separations of these materials are some of the costliest, most energy intensive, and most technically challenging separations in the industry due to the very similar size and nature of the molecules being separated. The technology currently employed for the separation of ethylene and propylene from paraffins is distillation and is estimated to consume 250 trillion BTU/year of energy.

In addition, distillation columns to separate these molecules are very capital-intensive. A membrane based olefin-paraffin separation technology has the potential to provide substantial economic benefit to the petrochemical industry and to drastically reduce the energy required. In addition, membrane systems are scalable, allowing their implementation for smaller, stranded streams where the investment in distillation columns have not previously been installed due to limited economic value. Membrane processes utilizing facilitated transport membranes have been extensively studied for separating olefin/paraffin mixtures. While separations have been demonstrated in the laboratory, problems with membrane stability have prevented development of commercial systems.

Membrane Technology

Compact Membrane Systems (CMS) has developed Optiperm™ membrane, a customized amorphous fluoropolymer (CAF) containing silver ions that selectively transport olefin molecules from a mixture of olefin and paraffin. CMS Optiperm™ membranes are based on a silver-containing fluoropolymer material. The technology exploits a mass-transfer mechanism known as facilitated transport. Silver incorporated into the membrane polymer backbone as a binding site for the double bond in the olefins transporting them across the membrane as depicted in the cartoon below. Saturated paraffins do not have the same interaction. This difference in affinity allows for olefins to pass through the membrane at a significantly higher rate than do paraffins.

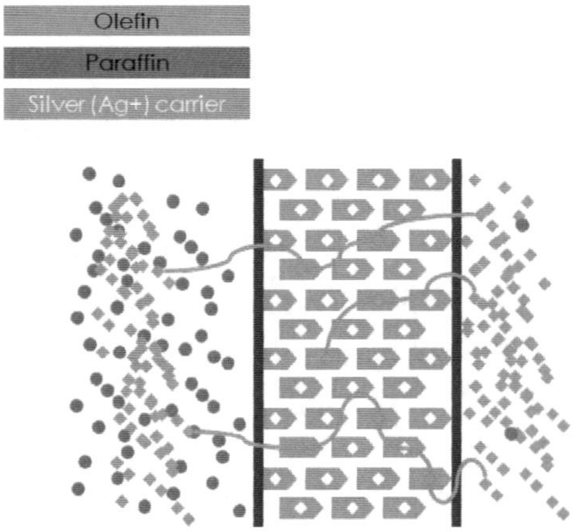

Figure 1. *Illustration of Facilitated Transport*

The two key membrane properties that characterize the effects of facilitated transport are **permeance** and **selectivity**.

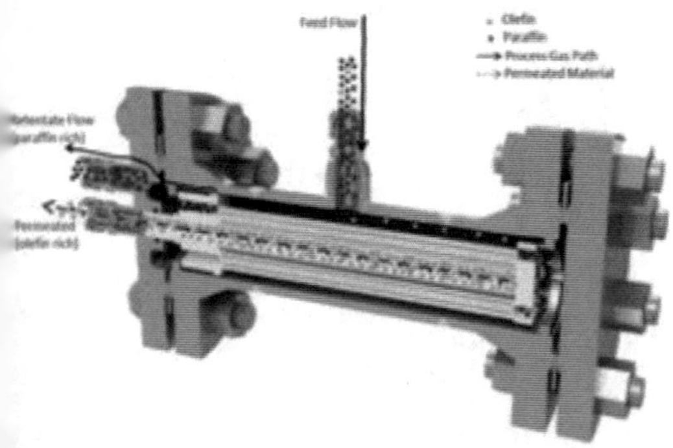

- **Permeance** is defined as the rate at which one component travels through a given surface area of membrane for a given driving force. This performance metric is typically expressed in GPU (gas permeation unit).
- The **selectivity** of one component to another is simply the ratio of the permeance of the two components. This tells you how much olefin is passing through the membrane as compared to paraffin.

Figure 2. *Illustration of Optiperm™ membrane module*

The Optiperm™ membrane has both high permeance and high selectivity, meaning high product purities can be reached with a reasonable membrane area. These characteristics yield very attractive process economics and the membrane can perform separations of olefins from paraffins over a wide range of operating temperatures, pressures, and compositions. CMS has demonstrated olefin removal at temperatures up to 70°C (158°F) and pressures up to 250 psig in the lab. For an efficient separation to occur, the olefin-paraffin mixture must be in the gaseous state. And optimal separation performance requires a highly humidified feed stream as water vapor improves the mobility of the olefin through the membrane. The membrane can remove olefin in both concentrated and dilute streams for compositions ranging 5-95% by volume of olefin. The membrane can process mixtures of C_2, C_3, and C_4 olefin and paraffin mixture.

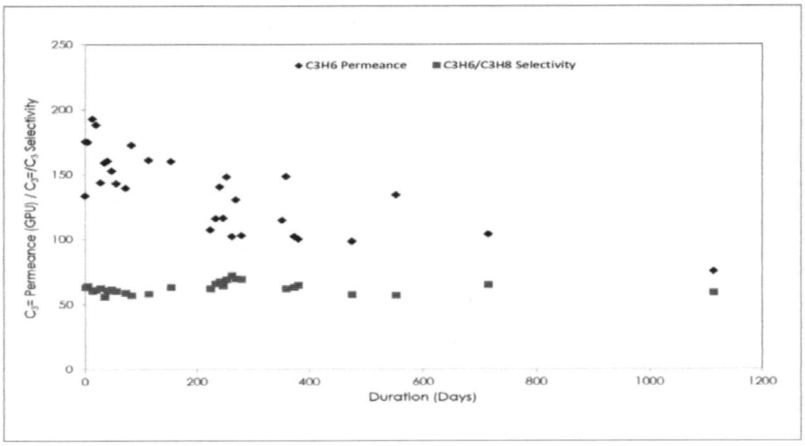

The chart displayed to the left shows a performance profile for a long term lab ageing study of an Optiperm™ membrane separating a 20% propylene /80% propane mixture at 60 psig and room temperature. As you can see, the membrane has shown stable performance over >1100 days, a vast improvement over current state of the art materials.

Figure 3. *Optiperm™ membrane long term aging study*

Pilot Units

The purpose of the CMS Pilot Units is to provide a platform for field demonstration of Compact Membrane Systems' (CMS) new Optiperm™ membrane technology for separation of olefins from paraffins in the gas state. The membranes utilize silver as a facilitating agent to improve the selective transport of olefins through the membrane relative to paraffins. These membranes provide a high olefin transport rate, which reduces the required membrane surface area, and high olefin/paraffin selectivity, which improves product split. Although various membranes for olefin/paraffin separation have been evaluated for many years, the CMS membranes are different in that they exhibit stable performance over a much longer time period than was ever achieved with earlier membranes. The improvement in membrane stability is made possible by incorporating the silver ion into backbone of the fluoropolymer which makes up the bulk of the membrane structure. As with most fluoropolymers, the fluoropolymer used in the CMS olefin/paraffin separation membrane is very resistant to chemical attack. It is this property that has been successfully utilized to create a membrane that provides stable separation performance in olefin/paraffin separations.

Prior to the refinery field trials all of the development tests done with CMS Optiperm™ membranes have been carried out in the laboratory at small scales. The DCRC Pilot Test Unit enabled testing of membranes that are an order of magnitude larger in scale with actual refinery process streams. Moreover, membrane performance could be tested at operating conditions that are difficult to achieve in the laboratory. In particular, membrane performance in the separation of propylene/propane mixtures at feed pressures above 90 psig can be readily tested.

Building upon the learnings of or membranes performance in the lab environment we completed over 300 days of operations that spanned a period of approximately 15 calendar months at the Delaware City Refining Company facility with our Optiperm™ membrane technology utilizing a paraffin-rich mixed C_3 feed stream to gauge the technical performance and economic viability of a full scale system. Prior to our field trials all of our membrane ageing and performance testing have been done in our lab facilities under controlled conditions with bottled gases and one of the things we did not know when we began the pilot trials was weather our membranes would be stable in a real process stream.

DCRC Pilot Test Findings:

DCRC Pilot tests key learnings:

- **Field membrane performance is stable**. Membrane pre-qualification and post-field performance tests confirm steady membrane performance parameters with permeance and selectivity matching the pre field testing values.
- **Membrane performance matches laboratory data** at similar conditions.
- We were able to achieve our primary goal of **producing an HD5 quality propane product**. A commercial scale system design will incorporate all of the design and control elements to continually produce on spec products including HD5 propane and propylene-rich streams.
- **Our spiral module membrane design and materials of construction performed as designed** with no mechanical failures noted in 300 days of field testing using multiple spiral wound membrane modules.

To give a little bit more detail around the first key learning, presented below in Figure 3, is field data from a membrane test conducted over a 50 day period. In this test the membrane was operating at a low stage cut with a 9% propylene feed to produce a permeate stream with a propylene concentration of 25%. Although there is some scatter in the data, the permeate stream composition is consistent over the 50 days of operation.

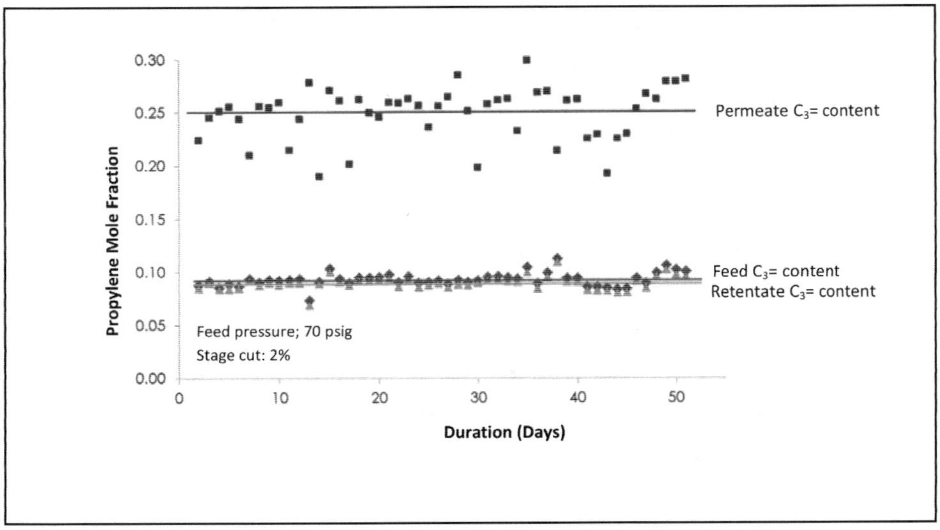

Figure 4. The results of one of our long term aging studies conducted over a 50 day period

Building upon the success of the DCRC Pilot Unit, CMS designed and built a new testing facility at our Newport, Delaware site to enable testing of larger scale spiral wound membranes at field conditions. To give a sense of scale, the testing at DCRC was done on a ~2ft^2 membrane cartridge. This test unit will allow testing of >20ft^2 cartridges. The test unit was designed as a closed loop system which allows for continual circulation, separation and recovery of a mixed olefin paraffin streams at plant operating conditions. The system will run with 24/7 autonomous operation to allow for the long term performance testing of membrane cartridges. The test system provides the research team with the capability to simultaneously test 2 large scale modules at different process conditions and has proven to be irreplaceable tool for developing and testing new membrane designs and operating concepts. Finally, the system also incorporates one our latest innovations, a novel method of humidification that ensures uniform humidification throughout the entire module which we envision to be incorporated into all pilot and commercial scale designs.

The CMS design and engineering team is currently developing three pilot scale units to be deployed in 2020 at customer sites to further enhance our understanding of the Optiperm™ membranes performance in process environments.

2020 Pilot 1 - For the first project, CMS has partnered with Dow through the Department of Energy's RAPID institute to scale-up and test large scale spiral wound membrane modules at a Dow process facility. The CMS and Dow project team has designed and engineered a single stage membrane system capable of testing membrane cartridge with up to 20 ft^2 of active area. The Dow system design will also include CMS' new cutting edge hydration system that was conceptualized during the DCRC pilot trials and proven with our newly installed test systems. The pilot is scheduled to be installed in the 3rd Quarter 2020 and will be deployed in a mixed C_3 feed stream at the Dow Freeport Texas site to separate and upgrade the olefin component of the feed stream to a chemical grade propylene product.

2020 Pilot 2 - To date all of the development tests done with CMS Optiperm™ membranes have been carried out at smaller scales with membranes having less than 5 ft^2 of active surface area. These earlier separation tests were also limited to single stage membrane processes. For the second pilot trial underway in 2020, CMS is partnering with an industry major to develop a Pilot Test Unit utilizing a two stage separation process with membranes having up to 100 ft^2 of active area. Moreover, membrane performance will be tested at actual field operating conditions that are difficult to fully duplicate in the laboratory at this scale. Finally, and probably most importantly, through demonstration of the membrane technology to produce relatively high purity product streams in the field, the Pilot Test Unit will generate increased interest in the technology and promote its acceptance in streams where high purity is required. In its initial deployment, the system will be tested in the recovery of 90+% propylene and 95+% propane streams from polypropylene reactor process effluent stream containing 40% to 70% propylene. The Pilot Test Unit has been designed to enable demonstration of acceptable product split across a wide range of operating conditions (pressures, temperatures and feed compositions). The testing with the two stage pilot unit will commence in the 4th quarter of 2020.

2020 Pilot 3 - The separation and recovery of C_4 olefins is one of the most challenging separations in the petrochemicals industries as they require several integrated stepwise process to isolate the C4 isomers. These separation processes often include extractive distillation, fraction and absorption to effectively separate and concentrate the valuable components. With increasing customer inquiries around Butane separations, CMS has accelerated the study of Optiperm™ effectiveness at separating C_4 olefins from paraffins and C_4 isomers. To further investigate these separations via membrane, CMS is refurbishing and reconfiguring the pilot unit previously used at the DCRC refinery to complete field testing at a process site in Europe to demonstrate Optiperm™ capability in separating C_4 olefin paraffin streams. This will be our first test using a spiral wound module configuration outside of our lab facilities to separate these valuable olefin components of the mixed C_4 streams. Our goal is to commence field testing with C_4s early in the 3rd quarter of 2020.

Modelling of spiral wound elements

With the CMS operations team now fully focused on commercialization of our Optiperm™ membrane technology an enhanced modeling tool with predictive properties was needed to help with the design and optimization of the membrane modules in a variety of operating conditions. To accomplish this goal, CMS partnered with the University Of Minnesota Department Of Chemical Engineering[1] through the RAPID program to develop a fundamental model of spiral wound elements in olefin/paraffin applications. This work developed a mathematical model for the separation of gases with spiral-wound membrane modules. The model accounts for variation of feed and permeate-side variables in two

---

[1] A Mathematical Model for Gas Separations with Spiral-Wound Membranes: Validation and Design in the Presence of Uncertainty

dimensions, and is implemented with open-source software packages. The uncertainties in membrane parameters are propagated into model calculation outputs by Monte Carlo simulation. The results from the model are used to understand the transport phenomena occurring in both dimensions, and the importance of the pressure drop along the permeate channel is realized.

The charts below shows a comparison of the models predicted performance versus the actual values recorded on the CMS recycle test unit for a mixed C_3 stream with an approximate 65% propylene concentration operating at 160 psig, 40 deg F over a 6 day period. As can be seen in the graph the results achieved in this test (solid lines), are very close to that predicted by the model (dashed lines) giving good confidence that the model will enable us to predict performance at a range of operating conditions. Please note that the decrease in permeance for the final day in this test is related to a increase in feed olefin composition which results in a lower GPU due to the facilitated transport nature of this membrane.

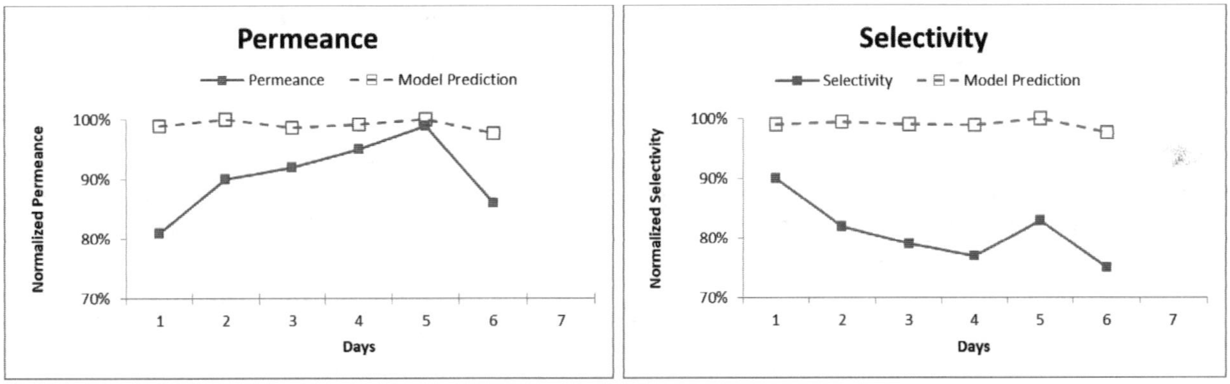

Figure 5. Comparison of the models predicted performance versus the actual values

Additional areas of development

The research team is also investigating other interesting applications of our Optiperm™ membranes both to improve process efficiencies and enable the capture and conversion of low value streams to desirable products. One example is research in the area of Enhanced Oxidative Dehydrogenation of Ethane (ODH) where the Optiperm™ membrane is partnered with an ODH reactor to produce a low cost ethylene product directly from ethane. Commercial applications of the technology may include localized small scale manufacturing of plastic components, and the direct upgrading of ethane streams to ethylene at NGL plants to name a few.

590472 Development of Novel Alloys for Use in Ethylene Cracking Furnaces

Abstract:

Cast chromia-forming austenitic stainless steels such as HP-type (25Cr-35Ni) and 34Cr-45Ni alloys are used in a wide range of industrial applications such as high temperature furnace components, radiant burner tubes, and ethylene cracking furnaces that demand high temperature microstructural stability, corrosion resistance, and creep strength. In particular, alloys used in ethylene cracking furnaces should possess good high temperature mechanical properties, coking resistance, must be fabricable using centrifugal/static casting, and must be weldable. Alloys such as cast HP-type alloys and 35Cr-45Ni achieve their environmental protection through the formation of a chromium-rich oxide scale. Recent work has shown that alloys that form a thin alumina-scale show better coking resistance in ethane steam cracking environments [1]. Although alloys that form alumina scales offer better corrosion protection and coking resistance at relevant temperatures, developing cast austenitic alloys that form a stable alumina scale and achieve creep strength comparable to existing cast chromia-forming alloys is challenging.

Cast alumina-forming austenitic stainless steels have recently been developed at ORNL for use from about 750°C to 1150°C [2]. In a recent study Muralidharan et al discussed the effect of composition on the alumina-forming capability and creep properties of cast alloys of composition Fe- 25 wt.% Ni-14 wt.% Cr-3.5 wt.% Al and with varying levels of Nb, Si, and C [3,4]. It was observed that alloys in this composition range containing about 25 wt. % Ni may be limited to operation at temperatures of about 800°C - 850°C depending on the water vapor levels in the environment due to inadequate oxidation resistance at higher temperatures. In a more recent study, Brady et al. [5] studied the ability of Fe–(25–45)Ni– (10–25)Cr–(4–5)Al–1Si–0.15Hf–0.07Y–0.01B wt% base alloys with and without Nb, Ti, and C additions to form a protective alumina scale at 1100 °C in air with 10% water vapor. It was found that Fe–35Ni–25Cr–4Al with additions of Nb, C showed promise as an alloy base for future development to achieve a cast alumina-forming austenitic alloy for use at temperatures exceeding 1100°C.

Oak Ridge National Laboratory has been working with industrial partners Duraloy Technologies, and MetalTek International and University of Wisconsin, Milwaukee to develop cast austenitic stainless steel compositions for use in ethylene cracking furnace applications. This presentation will outline the methodologies used in the development of the cast Fe-Ni-Cr-Al austenitic stainless steels, their microstructure, and highlight their high temperature oxidation resistance. In addition, the ability of these alloys to resist coking in an ethane steam cracking environment has been evaluated. This talk will outline the experimental setup, testing methodologies, and results from cyclic coking-decoking tests conducted on the developmental alloys. These results were compared to results from baseline 35Cr-45Ni alloys.

Research sponsored by ARPA-E, US Department of Energy, under Contract DE-AC05-00OR22725 with UT-Battelle, LLC.

1. E. Munoz Gandarillas, K. M. Van Geem, Marie-Francoise Reyniers, Guy B. Marin, "Influence of the Reactor Material Composition on Coke Formation during Ethane Steam Cracking," Ind. Eng. Chem. Res. 2014, 53, 6358-6371.
2. G. Muralidharan, Y. Yamamoto, and M. P. Brady, "Cast Alumina Forming Austenitic Stainless

Steels," U.S. Patent 8, 431, 072 B2, April 30, 2013.
3. G. Muralidharan, Y. Yamamoto, M. P. Brady, D. Leonard, B. A. Pint, D. Voke and R. Pankiw, (2015) "Development of Cast Alumina-forming Austenitic Stainless Steel Alloys for use in High Temperature Process Environments," NACE Paper C2015-6114, Houston, TX, presented at NACE Corrosion 2015, Dallas, TX, March 2015.
4. G. Muralidharan, Y. Yamamoto, M. P. Brady, L. R. Walker, H. M. Meyer II, and D. N. Leonard, "Development of Cast Alumina-Forming Austenitic Stainless Steels" JOM, First Online: September 6, 2016, DOI: 10.1007/s11837-016-2094-8.
5. M.P. Brady, G. Muralidharan, Y. Yamamoto, and B. Pint, "Development of 1100°C Capable Alumina-Forming Austenitic Alloys," Oxid Met, DOI 10.1007/s11085-016-9667-3.

590872 Horsehead Corporation Zinc Explosion

Abstract:

A fatal explosion occurred on July 22nd, 2010 at the Horsehead Corporation Zinc Refinery in Monaca, Pennsylvania. The accident resulted in the deaths of two workers and serious injury to a third. The United States Chemical Safety Board conducted a thorough investigation to determine the cause of the disaster and provide recommendations on how to prevent future accidents. From the investigation, it was determined that a cadmium distillation column, a process unit used to separate cadmium and other relatively low boiling impurities, exploded. It was determined that the sump, the structure through which the bottom product drained, became clogged; this resulted in liquid zinc flooding the distillation unit. The large accumulation of liquid led to the development of pressures and temperatures above the unit's design specifications and ultimately the failure of the system. Though this was the physical initiating event of the actual accident process, many other factors played a role in causing the accident.

The design itself used for the distillation column, specifically the sump, had specifications that were determined to be hazardous in prior investigations across the globe. The company failed to address issues with the design even though multiple nonfatal explosions had occurred at the same facility in the years prior to the disaster. The facility considered such accidents as normal and never addressed the problem. If proper safety protocol had been followed and these near misses investigated, the 2010 explosion may have never occurred. The Horsehead facility also did not follow proper management of change (MOC) protocol in the construction of the cadmium distillation column, and it did not educate employees on how to identify and evaluate hazards while operating the process. All these factors led to the eventual explosion on July 22nd, 2010.

585879 Development of a Comprehensive Procedure Writers' Guide Framework

Abstract:

Procedures are considered to be an integral part of safe operations within the process industry, as they serve as safeguards to inform workers of proper operating methods and prevent workers from engaging in unsafe and/or undesired actions. A number of industrial incidents stemming from poorly written procedures have highlighted pitfalls within one of the most critical elements of procedure development, the writers' guides. Previous studies have shown that there are many inconsistencies between writers' guides of different companies, with only a few elements, such as general procedure goals and proper writing styles, appearing in a majority of them. Many other critical elements, particularly those related to human factors and safety/environmental responsibilities, appear in few writers' guides or are poorly addressed. This work seeks to provide a comprehensive framework for the development of writers' guides across process industries, so as to have consistent content within procedures that addresses the issues that can lead to unsafe actions and incidents. As part of this framework, a procedure life-cycle has been developed and proposed based on a literature review of system development life-cycles from a number of industries, including software development and architecture. This life-cycle illustrates the stages of the procedure development process that are influenced by the writer's guide as well as the elements that are used to develop the writer's guide. Using the procedure development life-cycle as a guide, this framework has been developed based on regulatory elements and industry best practices as well as human factors guidelines.

586526 Justifying Improved Management of Loading Hoses

Abstract:

Justifying Improved Management of Loading Hoses

Abstract

There are several sources of loading hose failure rate data. However in general these data sources provide failure rates without regard to the situation in which the hose is used. One of the most vexing aspects of this subject is that there are several variables associated with loading hose specification, usage, maintenance, replacement and more that are known or suspected to have a dramatic effect on the failure rate, but there is no method for quantifying these variables. For this reason it would be useful to develop failure rates that are situation-specific, which in turn plants could use to justify improvements in their hose management policies. In the absence of actual data, this paper intends to solicit expert opinion from industry prior to or during the conference, and publish the results of the survey and suggested failure rate modifiers at the 2020 conference or in an issue of Process Safety Progress.

The conference presentation will present the current published data, as well as the proposed modification factors. To the extent that it is possible to do so without prejudicing 'real time' responses to the survey, the presentation will discuss general features of any survey results received in advance of the conference.

NOTE: This is being submitted for poster session only. BakerRisk has already submitted its allotted number of abstracts for the LPS presentation slots, and this abstract is not intended to be taken in preference to either of those two slots. In its current form, it is probably more suited as a poster.

587021 Machine Learning Based Quantitative Structure-Property Relationship Prediction of Lower Flammability Limit

Abstract:

This study used lower flammability limit (LFL) data and ten calculated molecular descriptors data of 78 pure chemical compounds to construct Quantitative structure-property relationship (QSPR) models. Four machine learning methods, k-nearest neighbors (*k*-NN), support vector machine (SVM), random forest (RF) and boosting, are applied to QSPR models to improve prediction accuracy. Prediction errors and accuracy are compared with traditional multiple linear regression (MLR) models. A novel cross validation method, 10-fold cross validation method, is also used to increase the data usage and prediction reliability. Result shows that models generated by machine learning methods have a significantly lower root mean square error (RMSE) than traditional MLR method in the test dataset. Machine learning based models can be used as substitution methods to improve UFLs predictability of chemical compounds

Using API RP571 in PHAs

Michael H. Oliver
EN Engineering
8308 Lake Bonita Road, Catlettsburg, KY 41129
moliver@enengineering.com

Prepared for Presentation at
American Institute of Chemical Engineers
2020 Spring Meeting and 16th Global Congress on Process Safety
Houston, TX
March 30 – April 1, 2020

Using API RP571 in PHAs

Michael H. Oliver
EN Engineering
8308 Lake Bonita Road, Catlettsburg, KY 41129
moliver@enengineering.com

Keywords: PHA, API RP571, Event-based failure

Abstract

API-RP571, Degradation Mechanisms Affecting Fixed Equipment in the Refining Industry, addresses approximately 60 degradation mechanisms that can affect equipment in petroleum refineries. As the title suggests, the mechanisms are well known in refineries and several of them can also occur in other industries. The overwhelming majority of these mechanisms are "time-based" but there are at least 3 "event-based" mechanisms that should be taken into consideration when conducting a PHA. Those three are low temperature brittle fracture, chloride stress corrosion cracking (Cl-SCC) and polythionic acid stress corrosion (PASCC).

Since all three of these mechanisms are likely to occur during startup, shutdown, upset conditions or maintenance activities they need to be taken into consideration when conducting a PHA with special attention to the actual materials of construction.

With the enormous increase in hydroprocessing capacity in the refining industry in recent decades, there has been a corresponding increase in the quantity of piping and equipment fabricated from austenitic stainless steels which are susceptible to both Cl-SCC and PASCC.

1 Introduction

API Recommended Practice 571 addresses Damage Mechanisms Affecting Fixed Equipment in the Refining Industry and covers 60+ damage mechanisms, the clear majority of which are time based but there are a handful of event-based mechanisms and as such may arise during a PHA depending on the nature of the process being reviewed RP571 is a very well organized document. Damage mechanisms common to a variety of industries are addressed in section 4.0 while those specific to refining and petrochemical industries are in section 5.0. Additionally, section 5.2 includes PFDs showing where the various damage mechanisms may be found in the process.

It is not the intention of this paper to delve into the technical details of the three event-based mechanism but rather the need for heightened awareness of these relatively common failures so they can be addressed during a PHA.

2 Low Temperature Brittle Fracture

A brittle fracture is the fracture of a metallic object without prior plastic deformation.

Brittle fracture typically occurs in carbon steels (CS) below the ductile-brittle transition temperature. Probably the most common CS used in refinery and petrochemical plant construction is A106 for piping which along with the flange equivalent (A105) and the plate equivalent for vessels (A285) have virtually the same composition.

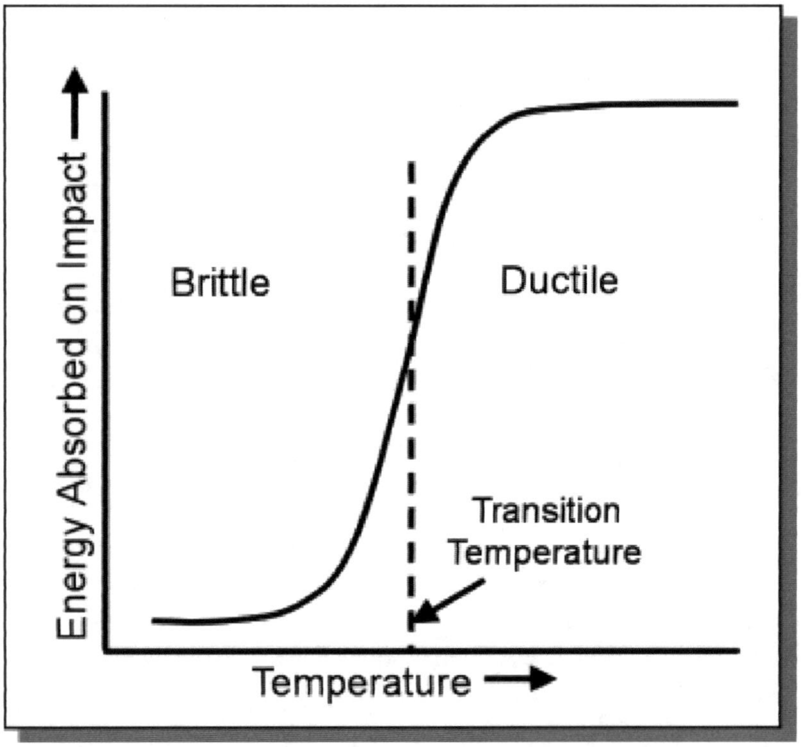

Figure 1. Temperature vs. Energy Absorbed on Impact

The actual ductile to brittle transition temperature for A106 starts at about 50 deg F (10 deg C). Compare that to the following atmospheric boiling points for some light hydrocarbons:

Table 1. Atmospheric Boiling Point of Some Light Hydrocarbons

Hydrocarbon	BP deg F	BP deg C
Propane	-44	-42.1
n-Butane	31	-0.5
i-Butane	11	-11.7

It is not unreasonable to conclude that in abnormal operating conditions, especially in the case of auto-refrigeration (when light hydrocarbons might boil off at near atmospheric pressure) that piping components constructed of A106 material will easily reach temperatures below ductile-brittle transition thus being susceptible to brittle fracture.

A notable low temperature brittle fracture incident in recent history is the Esso Longford (Australia) gas plant incident in September, 1998 which resulted in 2 fatalities plus 8 injured. In that incident, a heat exchanger was cooled to sub transition temperatures and then quickly exposed to high temperature resulting in a major loss of containment.

A review of the piping specification for 3 major oil companies reveals the following:

- Company A allows A106 and A105 flanges down to -20 deg F (-29 deg C) which is consistent with ASME non-shock ratings
- Company B allows A106/A105 only down to 32 deg F (0 deg C), more conservative than ASME
- Company C allows A106/A105 only down to 32 deg F (0deg C), also more conservative than ASME
- Companies B and C also require additional impact test specs for components with wall thickness greater than 0.5 inch

Companies B and C have factored their experience into the ASME piping code and their internal piping spec reflects that experience with more conservative low temperature constraints on A106/A105 components. Those 2 companies (B and C) specify "fine grained steels" for operation below 32 deg F (0 deg C).

What are "fine grained steels"? Those are steels with a finer grain structure and are much less susceptible to low temperature brittle fracture down to -58 deg F (-50 deg C). Typically for piping that would be A333, A350 for flanges and A516 plate for pressure vessels.

Learning point: if during a PHA, it is determined that equipment components could be cooled below the ductile-brittle transition temperature due to startup, shutdown or upset conditions, a review should be conducted of materials of construction to determine if the materials are susceptible to low temperature brittle fracture and if so, recommendations should be made to avoid failure such as heat up before pressure up or material replacement. Age of the equipment may also be a factor. Equipment manufactured to ASME B&PV Code, Section VIII, Div 1 prior to December, 1987 may be more susceptible.

Historical note: during World War 2, the US mass produced 2700 "Liberty Ships" to transport goods to the UK and Europe. The construction time for these ships was typically 42 days. About a third suffered (sometimes catastrophic) hull fractures and more than a few actually cracked in half. This was because the steel used had a relatively high ductile to brittle transition temperature (worse than A106) and in the cold water of the North Atlantic, the hulls became brittle and cracked usually starting at a stress point like a port hole.
The Titanic suffered a similar fate.

3 Chloride Stress Corrosion Cracking

Chloride stress corrosion cracking (Cl-SCC) like Polythionic Acid Stress Corrosion Cracking (PASCC) are included in the broader category of Environmental Cracking. Both these degradation mechanisms are event-based and usually result in sudden failure often with catastrophic results.

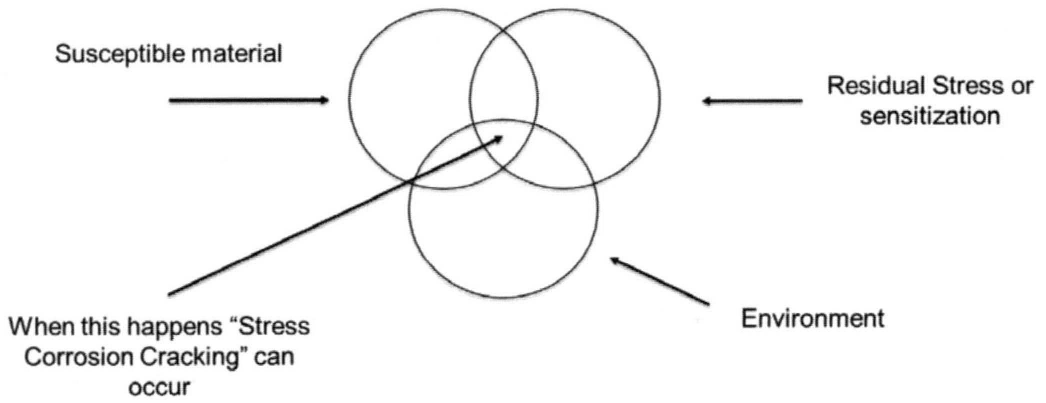

Figure 2. Stress Corrosion Cracking Model

In the case of Chloride Stress Corrosion Cracking and Polythionic Acid Stress Corrosion Cracking, the susceptible materials are the same: austenitic stainless steels including the common 304 family (304, 321 and 347) and 316.

Critical factors for Cl-SCC are chloride content, pH, temperature, stress, O2 content and alloy composition. Drains are particularly susceptible during start up and shutdown if not purged.

The "stress or sensitization" component of the mechanism can be induced stress for example in piping around supports or in heat exchangers where tubing passes through baffles or tube sheets or more commonly, it is sensitization of the metal due to exposure to high temperatures especially in the heat affected zone (HAZ) of welds.

Technically, sensitization is due to chromium carbide precipitates which form resulting in depletion of chromium around grain boundaries. In conditions which may allow formation of water in the presence of chlorides, cracking tends to occur along the grain boundaries and it can happen rapidly.

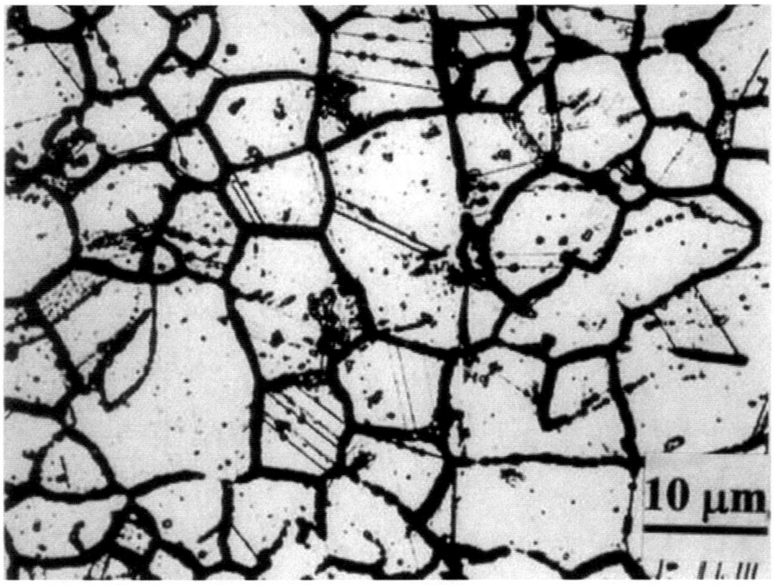

Figure 3. Chromium Carbide Precipitates Around Grain Boundaries

Referring to the SCC model in Figure 2, two of the three components, susceptible material and stress/sensitization are almost certainly not variables. That leaves the "environment" component of the mechanism. That's where particular attention must be devoted during the PHA. Recommendations during the PHA must exclude the possibility of that particular environment from forming internally and externally, especially during startup, shutdown and upsets.

4 Polythionic Acid Stress Corrosion Cracking

There are similarities between Cl-SCC and PASCC in that the susceptible material and stress/sensitization are common to the two, but the environment is very different. In the presence of sulfur containing compounds and/or H2S, a sulfide film forms inside piping and equipment. These are conditions typically found in virtually all hydroprocessing units that remove sulfur. If that film is exposed to air, the combination of O2 and moisture react with that sulfide film to form polythionic acid which can cause stress corrosion cracking in austenitic stainless steels. As with Cl-SCC, these failures can occur hours after exposure to an ambient air environment. This environment is unlikely to occur during any phase of operation, but it is certainly a concern when maintenance activities are underway that may expose sensitized piping and equipment to ambient air. Fortunately, the solution to this problem is potentially simpler than for the other two damage mechanisms discussed in this paper. NACE has produced a recommended practice RP0170 that describes a soda ash wash procedure that should be used before exposing austenitic SS piping and equipment to air during maintenance activities. This procedure has been successfully applied for many years and should be a recommendation from any PHA on a hydroprocess unit containing susceptible materials.

5 Summary and Conclusion

API RP571 is a valuable reference when conducting PHAs especially when concerned about event-based failures. Three common event-based failures include low temperature brittle fracture, chloride stress corrosion cracking and polythionic acid stress corrosion cracking. Recent experiences with EN Engineering clients highlight the diversity regarding company piping specifications and while the 3 examples cited may not be representative of the industry they highlight the need to always consider materials of construction when reviewing abnormal operations such as startup, shutdown and upsets.

Process Safety Capability Building as a Strategic Priority within Organizations

Kristle M. Thompson, PMP, Trinity Consultants, Inc., Houston, TX, USA

Abstract

Capability building provides organizations with the opportunity to develop skill-building approaches, that, if executed strategically, provide a powerful framework for organizational sustainability. This paper is designed for individuals and organizations interested in learning how to build and maintain expertise in process safety. Trinity Consultants will share best practices on how Process Safety Management (PSM) and Risk Management Program (RMP) capability is achieved via targeted recruitment, intensive onboarding and training, well-defined career development tracks, program element standardization, industry best practices, knowledge management, and employee recognition.

Introduction

Nearly 30 years ago the Occupational Safety and Health Administration (OSHA) published the Process Safety Management (PSM) Standard (1910.119). The goal of the PSM standard is to provide guidance on the management of hazards associated with processes using highly hazardous chemicals that have the potential to cause catastrophic incidents in the workplace and surrounding environment.

As we fast-forward to the present, you will find that many organizations still struggle with effectively identifying, evaluating, and controlling process hazards and risks to ensure safe and reliable operations. To address these process hazards, organizations need to develop the necessary expertise, judgement, and initiative to properly implement and maintain an effective PSM program.[1] One of the primary contributing factors that relates to this industry challenge is the level at which process safety is viewed as a strategic priority within organizations by senior leadership. Ensuring senior leadership actively supports process safety through its investment strategy is imperative to the success of a PSM/RMP program. [2]

Trinity's leadership actively supports PSM within the organization and has gained, through direct facility experience nationwide, the depth and breadth of PSM/RMP knowledge. Trinity will share best practices acquired as a result of working with client PSM/RMP programs that are staffed by process safety teams. The teams vary in size, technical expertise, and organizational structures across a broad cross-section of industries.

Specifically, Trinity will discuss how to develop a solid process safety team through:

- Targeted recruitment of process safety professionals;
- Intensive onboarding and training programs;
- Well-defined tailored career development tracks;
- Development of internal standards;

- Development and use of adopted templates from Recognized and Generally Accepted Good Engineering Practices (RAGAGEP);
- Extensive coaching through high-frequency project peer reviews; and
- Employee recognition programs.

1 Targeted Recruitment

Targeted recruitment of safety professionals with global and domestic industry-specific PSM expertise is the first step to building PSM/RMP capability in an organization. Over the past five years, there has been a shift in how organizations prioritize PSM and recruit for PSM candidates, particularly with catastrophic incidents capturing more attention of the public and media outlets. Within organizational structures, what may have once been lumped in with the occupational safety group is now distinguished as its own standalone PSM group within the Environmental, Health, and Safety (EHS) organization. Additionally, more stringent PSM competency requirements have been established by organizations, given the weight of responsibility in a PSM role now requiring science or engineering degrees, and some level of experience in process operations, process hazard analysis (PHA), and PSM/RMP program implementation. Lastly, you will find that there is a competitive market in Process Safety, where salaries are typically higher for Process Safety professionals in comparison to general EHS professionals.

Now that we understand how the landscape of PSM recruiting has shifted in organizations, let's focus on some of the burning questions around targeted recruitment.

How does my organization's HR department target solid process safety candidates for new-hire consideration?

Trinity's reputation for excellence has been shaped by their consultants' intimate knowledge and expertise of industry and EHS regulatory requirements [3], which makes targeted recruitment critical for its organization. Candidates are typically found through the LinkedIn recruiter job boards. Additionally, referrals play an important role in how the organization secures top talent, as referrals tend to be more reliable and improve the quality of candidates. Regardless of how the candidates are initially targeted, each must move through a rigorous recruitment process to include a series of probing technical phone interviews, face-to-face screenings with senior leadership, and full day on-site interviews with technical PSM/RMP staff and leadership.

The ideal candidate must be able to demonstrate competence, preferably across diverse industries domestically and globally, as indicated in the PSM Hiring Competency Checklist below (Figure 1). The candidate must also convey an energy and drive to succeed in the role; the ability to meticulously execute PSM/RMP projects; and have the capability to quickly learn and master any identified PSM/RMP gaps that may have been noted during the recruitment process.

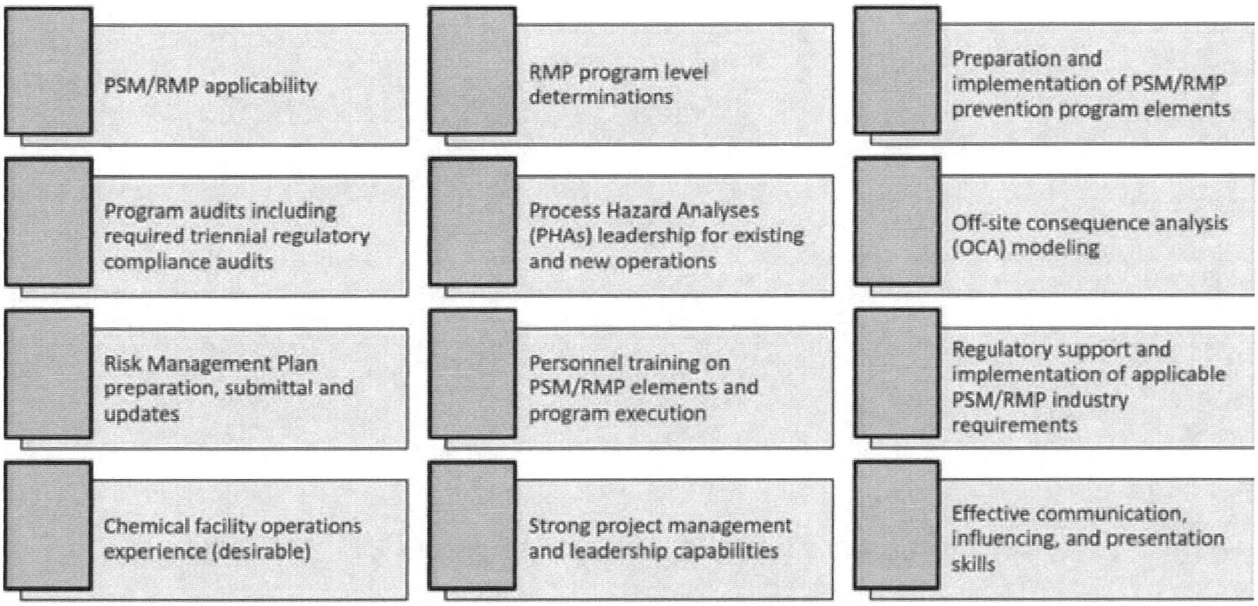

Figure 1. PSM Hiring Competency Checklist

One of the recruiting challenges for this role is identifying candidates who not only know how to lead and delegate work, but also know how to technically execute PSM/RMP work. An effective way of evaluating a candidate's depth and breadth of technical and leadership capabilities, in addition to the conventional interview questioning model, is through demonstration. This can be in the form of a case study or career presentation.

Trinity is an organization of Doers, Leaders, and Sellers. Each consultant regardless of seniority must demonstrate flexibility and be able to perform technical work, effectively lead projects, and create business development opportunities. The world of consulting is fast-paced and often requires varied skills and deliverables.

The concept of Doers, Leaders, and Sellers in the consulting industry is analogous to PSM/RMP-related industry roles as described in Figure 2 below.

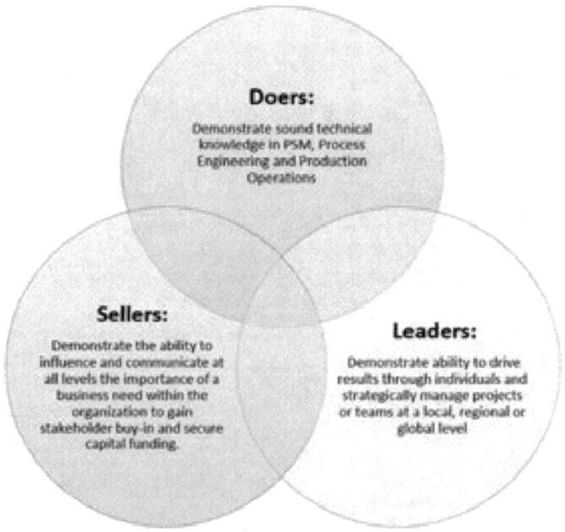

Figure 2. Description of Industry Doers, Leaders, and Sellers

The consulting recruitment model may not directly align with your organization's recruitment strategy but serves as a good starting point to landing solid candidates. A key step for internal or external recruitment is to identify which of the above roles is **required** for the current position, and which skills can be **developed** for career growth from that role. This ensures both a good initial fit for the company and ongoing career satisfaction for the candidate.

What if my organization relies on external EHS/Engineering placement firms for recruitment of PSM Candidates?

Trispyrian is a technical search firm that focuses on placement of engineering, operations, and EHS professionals, and has developed the technical knowledge and network to efficiently find the right candidates for their client organization's need. [4]

One obstacle that often appears in the PSM/RMP recruitment process is the recruiter's (internal or external) failure to understand the differences between roles supporting *Process Safety* versus *Occupational Safety*. Technical search firms ensure that they include the following aspects in their recruiting process.

- Technical screening – The unique ability to screen a candidate's technical knowledge to narrow your interview group to a handful of highly qualified professionals that meet the exact criteria of the role. Ensure the recruiters have been trained to understand the requirements associated with OSHA PSM and EPA RMP regulations, their project work types, technical requirements, and your internal standards to increase candidate fit.

- Simplified hiring process – Understanding facility operations, both simple and complex, provides insight into the skills and qualities desired by hiring managers, thereby simplifying the hiring process.

- Candidate network – Obtains a broad and deep network of potential candidates with proven skill sets, and ongoing partnerships with industry clients for more than a decade.

- Search option flexibility – Offers multiple options to meet your organization's specific staffing needs, ranging from direct hire to long-term contract assignments. For direct hire placements, the placement firm screens and presents a list of the most qualified candidates, and coordinates phone and face-to-face interviews to explore the fit between the candidate and the position. When direct hire is not feasible, the placement firm associates can be placed in long-term contract assignments, and at any point, should an organization decide to extend an offer for permanent employment, the placement firm will facilitate the hiring process.[4]

Selecting an experienced PSM/RMP savvy search firm that values client relationships, experienced in applicable regulations, and understands your operations will allow your organization to find and recruit top, qualified candidates.

What if my organization is small and doesn't have the PSM resources or finds it difficult to execute PSM?

Implementation of the PSM/RMP elements can be quite complex and resource-intensive, particularly for smaller organizations. By evaluating the critical technical skills (similar process experience, PSM/RMP program knowledge) and roles (Doer, Seller, Leader) based on your specific organization and its applicable processes, you can determine the technical areas that may benefit from third-party expertise.

Let's use the Process Hazard Analysis (PHA) element as an example. OSHA states that "an effective and comprehensive process hazard analysis shall be performed by a team with expertise in engineering and process operations, and the team shall include at least one employee who has experience and knowledge specific to the process being evaluated. Also, one member of the team must be knowledgeable in the specific process hazard analysis methodology being used". Given this requirement, the agency acknowledges that some organizations, particularly smaller ones, may not have the staff expertise to perform a PHA, and therefore, the organization may need to hire an engineering or consulting company to perform the analysis. [5]

If your organization does not have the staff to address all PSM/RMP compliance areas, establish a good working relationship with a consulting firm who specializes in PSM/RMP compliance, remains abreast of key regulatory developments, and is highly skilled in PSM/RMP program implementation for companies varying in size, skillset, and organizational structure. Site resources can focus on maintaining compliance with the PSM/RMP program once established for ongoing site changes, and can utilize third-party resources for larger task such as PHA review, compliance audits, and program updates based on regulatory changes.

2 Intensive Onboarding and Training Programs

An intensive onboarding and training program inclusive of on-the-job exposure, as well as internal and external competency training requirements is the second step in developing PSM/RMP capability. Once the qualified candidate has been selected and hired via the targeted recruitment process, it is imperative that the employee is immersed in a well-structured and rigorous onboarding and training program, and that training continues to assist the employee in growing through the life of their career. A key in determining your critical training needs is to identify the tasks that internal staff will perform and start there with your training efforts. Tasks that will be performed by external resources may only require a high level of knowledge by internal staff to review completed work products.

Employee Orientation

All new hires, not just the PSM/RMP team, should participate in an initial orientation. The orientation process should include a general overview of the facility, review PSM job-specific roles and responsibilities, and provide an overview of the organization's mission and safety culture. For example, employees may be informed of emergency exits, muster points, how to access process safety information and PSM-related procedures, and the proper safety-reporting protocols, such as reporting an injury or unsafe condition. [6]

On-the-Job PSM Competency Training

Competency training will be used to address gaps between critical skills, the candidate's experience, and the organization's specific procedures. The process by which internal and external training is provided for a PSM employee may vary by organization, but let's take the training of a PSM Engineer as an example.

Critical on-the-job training for a PSM Engineer would involve both skill-based training and awareness-based training. Skill-based training demonstrates the actual hands-on procedures necessary to perform a specific task [6], such as conducting an incident investigation or facilitating a PHA utilizing an industry recognized methodology. A great way for PSM Engineers to get up to speed with their role is to shadow senior staff. For example, a new Process Safety Engineer may support a lead auditor in the documentation review for the facility compliance audit, or she/he may take advantage of any PHA scribing opportunities to gain competency exposure to the PHA process. Facilitating a Management of Change (MOC) team will familiarize staff with how the PSM program elements are interrelated, and how small, simple process changes can impact many program elements.

Awareness-based training includes general policies, hazard recognition and expectation for maintaining a safe work environment [6]. For example, a Process Safety Engineer should have a high-level familiarity of the intent of the OSHA PSM and EPA RMP regulations, particularly at the start of a role.

The ongoing combination of skill and awareness-based training will ultimately build a PSM Engineer who is well-versed in PSM/RMP regulations, and capable of delivering a solid PSM/RMP program.

Internal PSM Competency Training

Some organizations may have specific internal competencies that must be mastered before carrying out certain PSM element tasks. In keeping with the same PSM Engineer example used above, a PSM Engineer may have to complete a rigorous PHA certification process to facilitate a PHA for their assigned process unit/operation. A PHA certification process may include a rigorous multi-day internal

PHA course and exam covering company-specific hazards, hazard methodologies, risk matrices, and risk acceptance criteria. Additionally, the PSM Engineer may be required to participate in at least two (2) PHAs as a scribe/observer and/or have shared facilitation responsibility with guidance from a more experienced PHA facilitator for a set number of complex operations before securing the PHA certification. As a precursor to or in lieu of an internal PHA course, some facilities utilize a formal multi-day PHA course from a reputable industry trade association, which leads into the next topic of External PSM Competency Training.

External PSM Competency Training

The Center for Chemical Process Safety's (CCPS) HAZOP Studies and Other PHA Techniques for Process Safety and Risk Management course at the American Institute of Chemical Engineers (AICHE) Academy is an industry-recognized course for PHA facilitators. After this rigorous three-day course, PSM engineers and technical professionals will: understand the obligations under the OSHA PSM and EPA RMP regulations; examine various hazard evaluation methods, including HAZOPs, how to select the hazard methodology appropriate for your facility; and lastly, how to determine the adequacy of safeguards and report findings and recommendations. [7]

Another great external course for PSM Engineers and technical professionals is CCPS' Process Safety Boot Camp. This is AICHE Academy's most popular course, and was developed for organizations who require quick ramp-up of staff knowledge in the fundamental concepts of Process Safety. The four-day course is intensive and covers key process areas, including risk-based process safety, process hazard analysis, regulatory compliance and process safety in plant and design operations. [7]

If there are state-specific PSM/RMP regulations for your locations, such as the New Jersey Toxic Catastrophe Prevention Act or the California Accidental Release Prevention and Cal/OSHA PSM programs, invest in external courses based on those regulations to ensure all federal and state requirements are also met.

3 Well-defined Tailored Career Development Tracks

The third step to building PSM/RMP capability is establishing well-defined, tailored career development tracks that reinforce subject matter expertise and support advancement within the organization. To this point, we have discussed what it takes to recruit, onboard, and train employees to be successful in their roles. Eventually, after achieving and maintaining success in their role, these employees will want to understand where the next opportunity lies within the organization, particularly those who are ambitious and expect to see career development opportunities to remain engaged and motivated in their role. It is imperative for organizations to play their part in supporting employees' career path by ensuring transparency and access to resources necessary for career development and progression, specifically:

- Job descriptions
- Well-defined competencies required for advancement
- Clear internal job application process

- On-the-job development opportunities
- Job shadowing and training
- Mentoring and supervisor coaching
- Transfer or lateral moves
- Formal succession planning [8]

A solid career program model should provide a roadmap for employees to strategically ascend in their career and capabilities. The program should promote growth vertically and laterally and ensures transparency in clearly defined career channels, grade levels, and job positions with associated responsibilities and competency requirements. To go a step further, organizations should have a promotion review board process dedicated to the advancement of its employees. Each employee should be provided with a promotion metrics summary matrix tailored to their selected career channel of interest, and this document should serve as a go-to reference for PSM professionals to identify and target any outstanding competency gaps, and potentially enable more expeditious advancement within the organization.

Career development program models and tracks will vary by organization based on company type and organizational structure. However, when employees have access to career development support resources, organizations will reap the benefits of employee engagement, success in their role and retention. Please reference Figure 3 below for a high-level example of two possible PSM career development tracks with potential cross-over, and cross training opportunities if PSM interests shift over the course of their career.

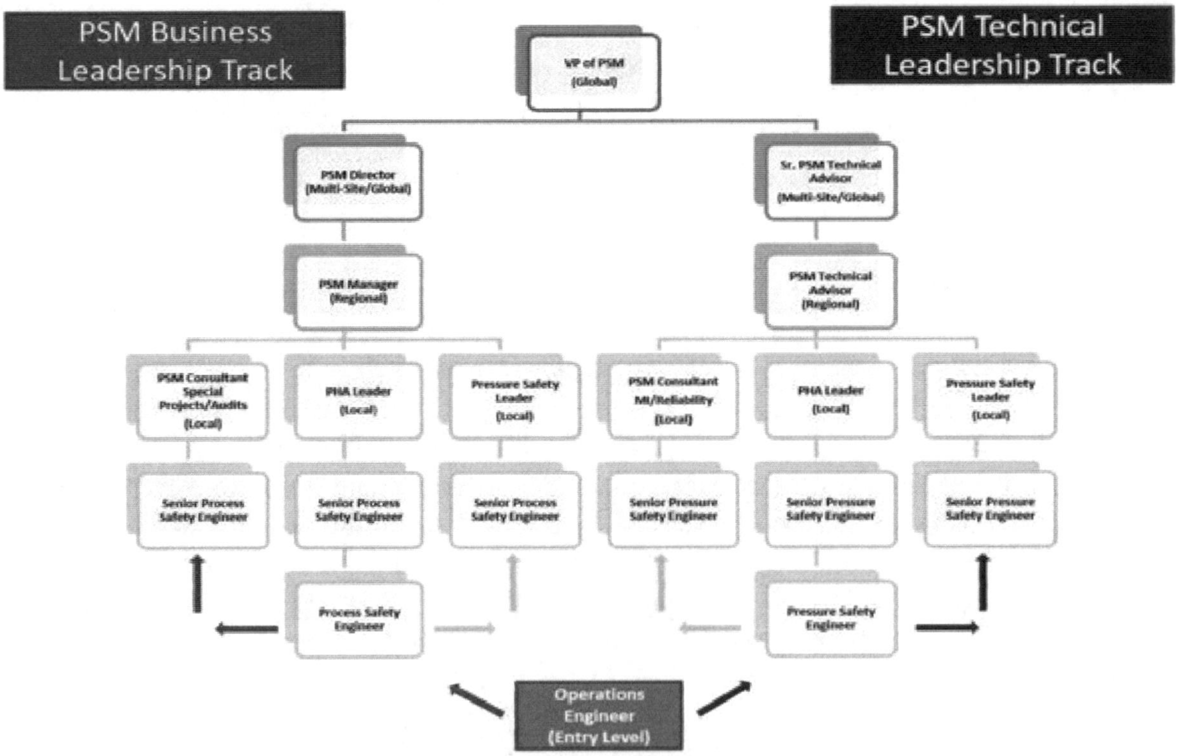

Figure 3. PSM Business Leadership Track vs. PSM Technical Leadership Track

4 Development of Internal Standards

Development of internal standards to produce a consistent quality work product across the PSM team is the fourth step in building capability. PSM/RMP personnel will find great value in the development of internal standardization within their team, primarily because it will ensure delivery of a quality work product, minimize project ambiguity and promote productivity within the organization. PSM elements recommended for internal standardization include the PHA, MOC, Pre-Startup Safety Review (PSSR), and Compliance Audit processes. Organizations that invest time to develop an internal standard on these processes will reap many benefits to include internal consistency in processes and risk evaluation.

For example, a robust PHA Standard Operating Procedure (SOP) should define requirements for:

- Personnel that facilitate, document, and approve PHAs
- Methodologies used to systematically identify, evaluate, and control hazards for covered processes subject to OSHA's PSM standard (codified in 29 CFR 1910.119) and/or EPA RMP standard (40 CFR Part 68)
- Methodologies for PHAs conducted under other scopes
- Quality Assurance / Quality Control (QA/QC) procedures for PHA reports

The SOP should also include a PHA Workflow to align with the steps of the PHA procedure as a best practice, a company-specific risk matrix, and risk acceptance criteria. Please reference Figure 4 for an example of a PHA Workflow.

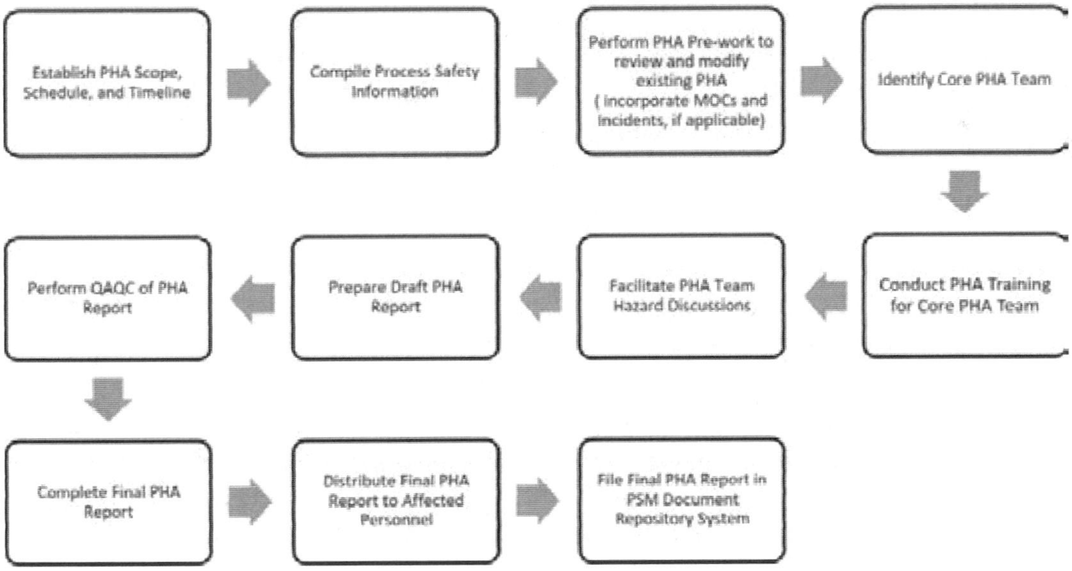

Figure 4. PHA Workflow

PSM Organizations that invest in internal standardization, where applicable, will find that it makes PSM information more accessible to employees, promotes transparency, and creates more confidence within the team [9] that a quality PSM/RMP work product will be achieved.

5 Development and use of Adopted Templates from Recognized and Generally Accepted Good Engineering Practices (RAGAGEP)

The fifth step to develop Process Safety capability is the use of adopted templates from Recognized and Generally Accepted Good Engineering Practices (RAGAGEP) to ensure all PSM projects are executed based on sound, safe, and reliable industry practices. According to OSHA, "the PSM standard allows employers to select the RAGAGEP that apply in their covered processes. Examples of RAGAGEP include widely adopted codes, consensus documents, non-consensus documents, and internal standards."[10]

Let's take an Ammonia Refrigeration facility for the first RAGAGEP example, where industry standard practices and templates are readily available for your organization's use. The International Institute for Ammonia Refrigeration (IIAR), the world's leading advocate for ammonia refrigerants, provides the most up-to date technical industry guidance to the ammonia refrigeration community [11], and serves as an acceptable source of RAGAGEP for non-refrigeration facilities that store or use anhydrous ammonia for process operations. IIAR provides technical expertise and comprehensive guidance in the area of PHAs. As a member of the IIAR community, your organization would have full access to the IIAR What-If/Checklist for Ammonia Refrigeration operations. This PHA template can be customized to accurately capture the site-specific hazards, and will serve to be an extremely valuable compliance tool as your organization implements OSHA's PSM PHA requirements.

For a second RAGAGEP example, look at CCPS's Guidelines for Auditing Process Safety Management Systems as a key industry guidance document to deliver a consistent, comprehensive, and quality PSM/RMP compliance audit for organizations that range in size, industry-type, or chemical process. Whether your organization performs internal audits or utilizes an external consultant to assess PSM/RMP compliance at the site, it is imperative that the compliance audit addresses the facility's level of compliance with the regulatory requirements of 29 CFR 1910.119 (PSM) and 40 CFR 68 (RMP), including evaluating compliance with incorporated standards such as 1910.252(a), 1910.38, and 1910.120 relating to fire prevention and emergency planning.

Additionally, the audit should include clear, unambiguous findings of compliance or non-compliance based on audited evidence, and document references to applicable regulations, RAGAGEP, or best practices for each finding and recommendation. For instance, if an audit was performed on the Mechanical Integrity (MI) program element, the auditor may want to confirm if the facility's pressure vessels are included in the MI program as cited in PSM 1910.119(j)(1)(i) and RMP 68.73(a)(1) regulations. Through discussion and evidence review, the auditor observes that the facility has not carried out pressure vessel inspections and is unaware of the appropriate RAGAGEP to reference for pressure vessel inspections. An appropriate recommendation based on this finding may be for the facility to develop an inspection schedule for all pressure vessels based on consensus RAGAGEP, or manufacturer recommendations, and ensure the pressure vessel inspection is completed by a certified inspector whose qualifications are in accordance with API-510 standard or another regulatory/consensus standard (i.e. certified National Board pressure vessel inspectors). Figure 5 below highlights the key components of a quality compliance audit.

Figure 5. Key Compliance Audit Components

A final consideration for delivering quality audits is to ensure and/or confirm, if externally performed, that the audit checklist template has been reviewed and updated on a defined frequency to ensure the checklist content remains in alignment with the ever-changing world of PSM/RMP compliance. A relevant example would be remaining informed of the many iterations of the EPA's RMP Rule and ensuring that the regulatory language and audit questions in an RMP Prevention Program Level 3 audit checklist template have been updated to reflect the requirements of the recently published December 2019 Final RMP Reconsideration Rule. Oversight of ongoing rule changes should be used to ensure your organizations PSM/RMP program procedures are current and require the facility to incorporate any new or updated regulations, such as the recent RMP emergency response coordination, notification, and exercise requirements.

Relying on RAGAGEP tools for your organization's covered processes is vital to the success of an PSM/RMP Program. Use of recognized industry practices provides a level of certainty in quality and processes and ensures that organizations are less susceptible to missteps [9] in implementing their PSM/RMP program.

6 Extensive Coaching via High-Frequency Peer Reviews

Extensive coaching through high-frequency peer reviews enabling effective transfer of knowledge and continuous improvement within the PSM team is step six. A peer review is a quality control measure, and is defined as a process in which professionals review each other's work to confirm quality, accuracy, and relevance.[12] All PSM organizations should encourage routine peer reviews of work products per internal Quality Management Systems (QMS) to promote continuous improvement. Peer feedback can also be focused on "how" tasks are performed to encourage staff growth in PHA, MOC and audit facilitation roles.

For example, Trinity has a robust QMS and its mission is "to develop and implement intelligent solutions and to provide high quality consulting services that efficiently achieve clients' business objectives. All employees are involved in our efforts to continuously improve our ability to meet and exceed customer expectations."[13]

The organization ensures high quality deliverables and guarantees continuous improvement in its staff's learning and development through peer reviews. Each project is subject to a detailed task review (Tier 1) at the peer level, and a high-level review (Tier 2) at the senior leadership level. The two tiers of quality checks, in the context of deliverables and services, are described below, and can be applied to your teams.

Tier 1: How "correct" is it?

- Are technical calculations correct?
- Are regulatory determinations correct?
- Have the correct procedures been applied or implemented?
- Is the work product free of spelling, grammar, and formatting errors?

Tier 2: How "good" is it?

- Does the work product present insightful and innovative solutions?
- Does the work product provide added value to the client, helping them meet business objectives?
- Is the work product well-written?
- Is the style, organization, and aesthetic presentation of the work product appropriate?
- What distinguishes this work product from competitors?
- Does the work product make a strong impression?[13]

Trinity's Knowledge Management System (KMS) is another critical piece that ties into ongoing staff coaching and peer reviews. The KMS provides consultants with the ability to collaborate, leverage expertise, and effectively capture, store, and access knowledge content on a software platform to enable more prompt, insightful solutions to clients.[14] This tool is similar to LinkedIn, Facebook and Twitter in that each consultant maintains their own online profile highlighting education, experience and specialized skills. Additionally, areas of expertise can be identified via hashtags, which are organized within a structured taxonomy. Other key benefits of the KMS system are listed below:

- Business process, technical, and regulatory questions can be directed to relevant peers through use of hash tags and/or targeting to specific communities
- Discussion threads occur in an open collaborative space, then are stored, and searchable
- Subject matter experts, who are validated and weighted by the software per their KMS participation, can be identified by hashtag-based searches
- Each consultant controls the content flowing into their newsfeed based on their connections to colleagues, membership in communities, or subscription to hashtags
- Consultants can also receive instant email notifications and/or daily digest emails of their KMS newsfeed content
- Consultants can earn recognition badges based on levels of KMS participation and contributions
- Community sites covering the organization's overhead departments, businesses, locations, and important subject-matter topics.

- Keyword based search capabilities to identify documents, people, and relevant past conversation threads within the KMS [14]

Implementation of a Quality Management System and Knowledge Management System are effective avenues for providing ongoing coaching and support. Without these systems, you run the risk of remaining stagnant in employee competency growth and miss out on continuous improvement opportunities to promote greater performance within the organization.

7 Employee Recognition Programs

The final step to developing PSM/RMP capability in your organization is the fostering of employee recognition programs that provide timely recognition through personal acknowledgement of positive behaviors. Employee recognition must be an inherent part of an organization's culture. Otherwise, you risk losing your best and most skilled assets, the employees, and this is critical when you think about the specialized skill-set of PSM professionals. Losing a top performer can create a major competency gap within in your organization and potentially impact its ability to effectively execute PSM/RMP work.

Employee acknowledgement serves as a key motivator for staff, which can lead to improved productivity and cultivate a culture of self-improvement. Employee recognition promotes employee retention, as staff may feel a greater sense of loyalty to an organization that consistently values employees. Lastly, employee recognition is a great morale booster, fostering a happier and more collaborative team environment. [15]

One of the most effective ways to recognize employees is by implementing a formal employee recognition program. An employee recognition program should provide both manager and employee initiated recognition. The program, as an example, can be a point-based recognition system, where each employee is allocated a set amount of points annually to distribute to peers. Employee-received points can be assigned a monetary value and can be redeemed in the calendar year or rolled over to the next calendar year. The recognition can be presented in the form of a personalized recognition card, highlighting the employee's stellar performance. As an extra incentive, your organization could allow top-point earning employees to enter into a drawing to win an additional bonus. [16]

While structured employee recognition programs are a great way to demonstrate appreciation, employee recognition doesn't always have to be as sophisticated or even in the form of a monetary reward. It can be as simple as a fun day away from the office, or a thank you email acknowledging an employee's outstanding work effort. The level of energy that your organization's leaders invest in employee recognition will make all the difference in the success and sustainability of your PSM/RMP team and program.

Conclusion

Capability building should be one of the highest strategic priorities within an organization, primarily because it is a process that is not always straight-forward or easy to tackle. This is especially evident for the PSM/RMP regulated industry, where poor judgement and lack of sound technical experience and leadership could be catastrophic not only for the business, but also for the employees and surrounding community. OSHA and EPA developed their respective PSM/RMP standards as guiding lights to aid organizations in the proper implementation and maintenance of an effective process safety program, but these standards alone are not enough. Senior leadership must demonstrate an unwavering commitment to building process safety capability and keep it at the forefront of their organization's strategic agenda to ensure PSM/RMP program success.

The following steps pave the path for capability building to successfully propel your PSM/RMP organization forward:

- Targeted recruitment of safety professional with global and domestic industry-specific PSM/RMP expertise
- Intensive onboarding and training programs inclusive of on-the-job exposure, internal and external competency training requirements
- Well-defined, tailored career development tracks designed to reinforce subject matter expertise and support advancement within the organization
- Development of internal standards to produce a consistent quality work product across the PSM/RMP team
- Development and use of adopted templates from RAGAGEP to ensure all work is executed based on sound, safe, and reliable industry practices
- Extensive coaching through high-frequency peer reviews allowing for effective transfer of knowledge and continuous improvement within the PSM team
- Employee Recognition programs that provide PSM teams with timely recognition through personal acknowledgement of positive behaviors

While your PSM/RMP team may vary in size, technical expertise, or organizational structures, these best practices will provide a well-documented roadmap and a high level of assurance that PSM/RMP capability is building within your organization.

References

1. Occupational Safety and Health Administration, Process Safety Management Guidelines for Compliance, Available at https://www.osha.gov/Publications/osha3133.html, Accessed on January 9, 2020.
2. Chemical Processing, Process Safety Begins in the Board Room, Available at https://www.chemicalprocessing.com/articles/2013/process-safety-begins-in-the-board-room/, Accessed on January 9, 2020.
3. Trinity Consultants, Statement of Qualifications for Process Safety Management and Risk Management Program, Available via Trinity Consultants Intranet System, Accessed on January 9, 2020.
4. Trinity Consultants, Trispyrian Services Sheet, Available via Trinity Consultants Intranet System, Accessed on January 9, 2020.
5. Occupational Safety and Health Administration, Process Safety Management for Small Business, Available at https://www.osha.gov/Publications/OSHA3908.pdf, Accessed on January 9, 2020.

6. Travelers, How to Onboard and Train Employees into a Safety Culture, Available at https://www.travelers.com/resources/workplace-safety/onboarding-and-training-employees-into-a-safety-culture, Accessed on January 9, 2020.
7. American Institute of Chemical Engineers, 2019 AICHE Academy Course Catalog, Available at www.aiche.org/academy, Accessed on January 9, 2020.
8. The Balance Careers, How to Make a Career Path Plan, Available at https://www.thebalancecareers.com/career-pathing-1918080, Accessed on January 10, 2020.
9. Planon Software, Five Reasons why Organizations can no longer avoid Standardization, Available at https://planonsoftware.com/us/whats-new/blog/five-reasons-why-organizations-can-no-longer-avoid-standardization/ , Accessed on January 10, 2020.
10. Occupational Safety and Health Administration, Process Safety Management for Petroleum Refineries: Lessons Learned from the Petroleum Refinery Process Safety Management National Emphasis Program, Available at https://www.osha.gov/Publications/OSHA3918.pdf , Accessed on January 10, 2020.
11. International Institute of Ammonia Refrigeration, Technology and Standards, Available at https://www.iiar.org/iiar/technology_and_standards/technology_and_standards.aspx, Accessed on January 10, 2020.
12. Medical News Today, What to know about Peer Review, Available at https://www.medicalnewstoday.com/articles/281528.php , Accessed on January 11, 2020.
13. Trinity Consultants, Quality Management System, Available via Trinity Consultants Intranet System, Accessed on January 11, 2020.
14. Trinity Consultants, Knowledge Management System, Available via Trinity Consultants Intranet System, Accessed on January 12, 2020.
15. Efront Learning, Employee Recognition in the Workplace: The Why and How, Available at https://www.efrontlearning.com/blog/2018/04/employee-recognition-workplace-benefits-ways.html, Accessed on January 12, 2020.
16. Trinity Consultants, Employee Recognition Program, Available via Trinity Consultants Intranet System, Accessed on January 12, 2020.

A human's perspective on human factors in bowtie modeling

Two visions

Both the bowtie method as well as human factors have become increasingly popular topics in risk management. The challenge remains, however, to discover the best way to include human factors in the bowtie method. The study of human factors occupies itself with researching the interaction between the user and his/her working environment. In this white paper two visions will be examined on how and why human factors can be included in the bowtie method. The two visions draw from Hollnagel's System-1 and System-2 thinking. Vision 1 (which is more closely related to Safety-I thinking) is in favor of the idea that human factors can be included as a threat as well as an escalation factor. Vision 2 (which is more closely related to Safety- II thinking) is in favor of the idea that human factors can only be included as an escalation factor. This article concerns itself first and foremost with qualitative bowties and not quantitative bowties, because bowtie thinking is, at its core, scenario-driven. This paper is intended for people who use bowtie modeling in their jobs, but who do not have any background in cognitive psychology. By the end of this white paper, we hope you have gained a better understanding of human factors and how to include these in the bowtie method.

The challenge remains

The challenge of including human factors in bowties can best be illustrated by means of an incident. Imagine the following scenario, a recurring operation of many industries is that their employees must travel on the road for work. In this outlined situation the driver and his passengers are driving from site to another location to drop off the work car after a long shift. The driver is tired and speeding to get home as he needed to catch his plane to go on vacation. However, he is driving on gravel roads that have gotten wet from rain. At one point, the driver is distracted and does not notice a large pool of water on the road. The right front wheel hits the pool of water causing the steering wheel to sharply pull to the right. The driver overcorrects, which results in the car going sideways. Control is lost, and the car impacts some bushes. No one is hurt, the car is damaged in the process, but can still be driven back to the drop off location.

How would the risks in this incident fit within the bowtie method? Which human factors are present in this incident setting and where do they fit in the bowtie model? How should these risks and human factors be described? What happens in case human factors are 'wrongly' placed in the bowtie model? To answer these questions and fully understand their answers, both the foundation of the method as well as the building blocks of the bowtie must be examined.

'A' bowtie method?

Barrier management is being used in multiple industries (McLeod et al., 2016), with the bowtie methodology on the rise in popularity. Bowtie methodology is applied mostly in high-hazard industries (nuclear, oil and gas, aviation etc.) but is relevant for nearly any industry that relies on barriers to prevent losses (healthcare, banking, public services, public utilities etc.). Differences in bowtie practices, however, persist and yet there is no agreed upon standard on "the" bowtie method. De Ruijter and Guldenmund (2016) suggest there are many different bowtie methods, and we should perhaps speak of "a" bowtie method. The method takes its name from the shape of the diagram that you create, which looks like a men's bowtie.

One of these methods is the Shell-bowtie risk evaluation method, which is the bowtie method that will be used in this paper. The Shell-bowtie is used to analyze and demonstrate multi-causal relationships in high risk scenarios. The main goal of using bowties is to communicate as simply and visually as possible where an organization's risks lay. A bowtie diagram does two things. First, a bowtie provides a visual summary of all plausible accident scenarios that could exist around a certain hazard. The knot in the middle is called a top event, which is the first 'loss of control moment' of the hazard.

Secondly, and most important, by identifying control measures or barriers, the bowtie displays what a company does to control those scenarios. The idea of the barriers is based on James Reason's Swiss Cheese model. Reason (1997) hypothesized that hazards are prevented from leading to losses by a series of barriers. According to him, these barriers are never 100% effective. Each barrier has unintended intermittent weaknesses and, when they line up, a hazard can lead to losses. Herein lies the reason for why multiple barriers are necessary.

Figure 1: The Swiss Cheese Model

A metaphor commonly used to explain the barrier concept is to compare the barriers to slices of Swiss cheese with holes. The holes in the barriers are dynamic and continuously change in size and location. When holes in all the slices line up, the hazard (the arrow) can pass through these deficient barriers, leading to an accident (losses).

Bowtie elements

To answer the question of where human factors fit in the bowtie method, the bowtie terminology must be defined, and the bowtie elements must be examined more closely.

Hazard: a hazard is the process that an organization needs to do to run a business. If hazards are under control, they will not cause harm, but they introduce the potential for damage/injury.

Top Event: when control over a hazard is lost, it is often possible to identify the moment when a normal situation changes to an abnormal situation. The top event is not a catastrophe yet, but the company is now exposed to the potential harm of the hazard. It is still possible for the organization to control the situation.

It is important to realize that the top event is not an absolute event. Events are usually part of a longer chain of events: drinking and driving can lead to a loss of control over a vehicle, which can lead to crashing into open water, which can lead to occupants becoming entrapped in an enclosed space, which can lead to fatalities. Which event is appointed as the top event is important, because this choice dictates how the diagram will turn out. It is a very subjective choice that depends on the perspective taken. For example, an explosion can be a consequence for an organization, and a top event for a fire fighter. Sometimes a generic bowtie is created where the threats and consequences are specified in other bowties. In this case, a threat or consequence can become a new top event of another bowtie. In this manner, several bowties will be linked, creating a chain of events. This process is called bowtie chaining. Later in this paper (see 'Two visions') it is explained how this knowledge can be applied when including human factors in the bowtie model.

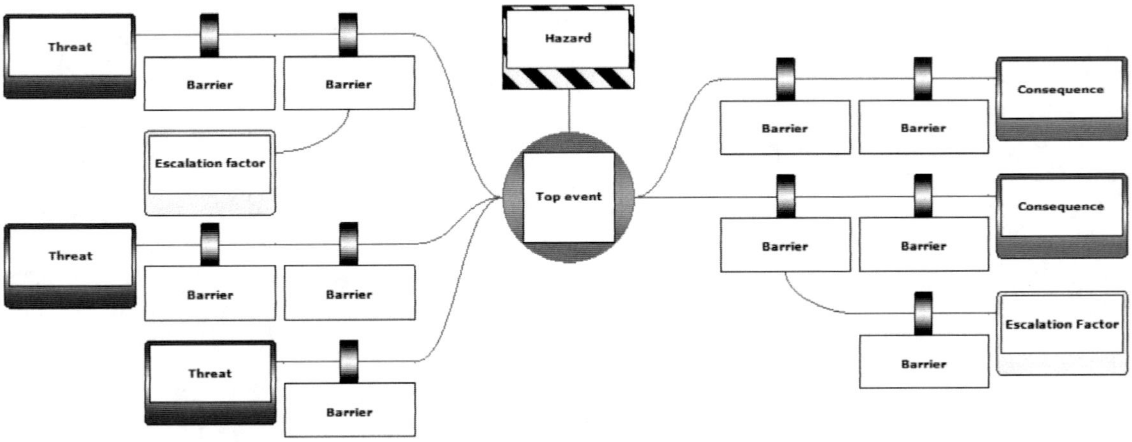

Figure 2: The skeleton of a bowtie

Barriers: risk management is about controlling risks. This is done by implementing barriers to prevent certain events from happening. A barrier (sometimes also called a control) can be any measure taken that acts against some undesirable force or intention, to maintain a desired state. Barriers are placed on both sides of the top event. Preventive barriers on the left side of the bowtie prevent the top event from happening. Recovery barriers on the right side of the bowtie either prevent the top event from resulting in unwanted consequences or mitigate the severity of the consequence. Barriers can be hardware systems, design aspects, human behavior and so on (for a more extensive analysis on barriers, see Sklet (2006).

Threats: threats are potential causes of the top event. They are located on the left-hand side of the top event. There can be multiple threats for one top event and each threat represents a single scenario that could lead to the top event directly and independently. Relevant for the human factors question is a common mistake that is often made: formulating a threat as the failure of a barrier. Barrier failures can be spotted by certain words such as 'lack of' ... or 'failure of' ..., for example 'failure of Anti-lock braking system'. Anti-lock braking systems do not cause loss of control over a vehicle; rather they are safety devices that allow the driver to remain in control over the vehicle during hard braking and slippery conditions. The described threat is in actual effect a failed barrier that, were it described positively, intends to prevent the original threat. In this case, the slippery conditions were the original threat with the anti-lock braking system as the barrier.

However, one exception to this is a primary process failure, which is a type of threat that is also described with these failure words while still being correct. If a piece of equipment that is part of the primary process fails (primary equipment such as an engine failure in a helicopter, or a pipeline integrity failure in an installation), this is an actual threat. Thus, a threat should be a force or condition that pushes an unwanted chain of events further. In the Chapter 'Two visions', it will be explained how this knowledge can be applied to include human factors in the bowtie model.

Consequences: consequences are unwanted scenarios that can be caused by the top event. Consequences are the incidents and accidents we want to prevent. Consequences are mainly unwanted because they will lead to loss or damage.

Escalation factors: once the control measures are identified, the bowtie method takes it one step further and identifies specific conditions or actions that make it more likely that a barrier will fail. An escalation factor can also be described as a failure mode of a barrier. It cannot directly cause an event but increases the chance that a certain threat or top event will lead to an event, by taking out a barrier. These are the holes in the 'Swiss Cheese Model'. Why it is agreed that human factors can be included in this bowtie element and how it can be included specifically, will be discussed later (see 'Differences').

Escalation factor barriers: similar to the barriers on the original threat line, there are barriers for escalation factors as well. These escalation factor barriers are more niche and protect the main barrier from an escalation factor.

Human error & human factors

Like the bowtie method, the topic of human factors has also become increasingly popular. More and more literature, research, books, magazines, newspaper articles, journals and conferences are available on the topic of human factors. The Ergonomics Society Conference in the UK and the Human Factors and Ergonomics Society Conference in the USA are two noteworthy institutions. Perhaps the best indicator of the rising popularity, though, is the number of human factors professionals being employed in many different industries (Jordan, 2002).

The study of human factors occupies itself with researching the interaction between the user and his/her working environment. The underlying philosophy to human factors, is that it is easier to change the working environment than it is to change the human condition (Reason, 1997). It therefore tries to design the working environment according to the physical and cognitive needs and limitations of the human mind. The goal of human factors is to optimize the performance, safety and usability of products, processes and systems.

Safety is of interest here for the purpose of this white paper. The study of human factors tries to optimize safety by designing environments in such a way that the possibility of human error is reduced as much as possible. Human error is an umbrella term that can be better analyzed by looking at the different types of errors that can be made: slips, lapses and mistakes.

Slips: This subcategory of errors are classified as visible execution failures (Reason, 1990, 1997). The plan is correct but the execution deviates from the original plan, it fails. An example is driving on the highway, but having to slow down due to traffic, thus changing gears but changing from fifth gear to third, accidentally skipping fourth gear.

Lapses: This subcategory of errors are less visible than slips as they are more 'in the mind' and are related to failures of memory (Reason, 1990, 1997). Like slips, lapses are about having the right plan, but again, the execution deviates from the original plan, it fails. Lapses involve not doing an action. An example is waiting for a red light with the handbrake on, the light turns green and you try to pull up but forgot to take the handbrake off.

Both slips and lapses involve accidentally activating the wrong routine (Reason, 1990, 1997).

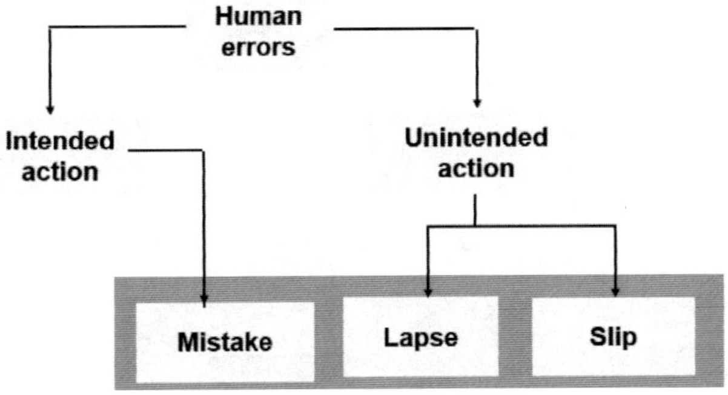

Figure 3. An abridged model of human error

Mistakes: Unlike slips and lapses, the failures of 'mistake errors' are that the plan itself is incorrect, but the execution is successful, it follows the original (wrong) plan. An example is driving on the highway and believing you need exit A and so you successfully take exit A only to find out later that you needed exit B (Reason, 1997).

More distinctions can be made in human error but are left out for the scope of this white paper. For other distinctions, please refer to James Reason, Human Error (1990 and 1997) and Rasmussen's Skill, Rules and Knowledge based model (1983) for more information.

To summarize: "Simply put, human factors can be thought of as conditions that if not managed correctly can lead to human error" (Edkins, 2016). Thus, a condition such as fatigue (human factors) could lead to inattention (human error, lapse).

Two visions

There is a clear distinction between two visions on how to include human factors into the bowtie method. Each can be related to what Hollnagel (2014) called either Safety-I or Safety- II thinking.

In short: *"A Safety-I approach presumes that things go wrong because of identifiable failures or malfunctions of specific components: technology, procedures, the human workers and the organizations in which they are embedded. Humans—acting alone or collectively—are therefore viewed predominantly as a liability or hazard, principally because they are the most variable of these components" (Hollnagel, 2015).*

Hollnagel believes safety management should move from *"ensuring that 'as few things as possible go wrong' to ensuring that 'as many things as possible go right'. We call this perspective Safety-II; it relates to the system's ability to succeed under varying conditions. A Safety-II approach assumes that everyday performance variability provides the adaptations that are needed to respond to varying conditions, and hence is the reason why things go right. Humans are consequently seen as a resource necessary for system flexibility and resilience" (Hollnagel, 2015).*

How do these visions translate to the bowtie method? Vision 1 (which is more closely related to Safety-I thinking) is in favor of the idea that human factors can be included as a threat as well as an escalation factor. Vision 2 (which is more closely related to Safety- II thinking) is in favor of the idea that human factors can only be included as an escalation factor.

Both similarities and differences between the visions will be explored in this article using the 'driving a vehicle' example. In a driving incident, imagine the driver was:

1. Speeding
2. Distracted, due to fatigue

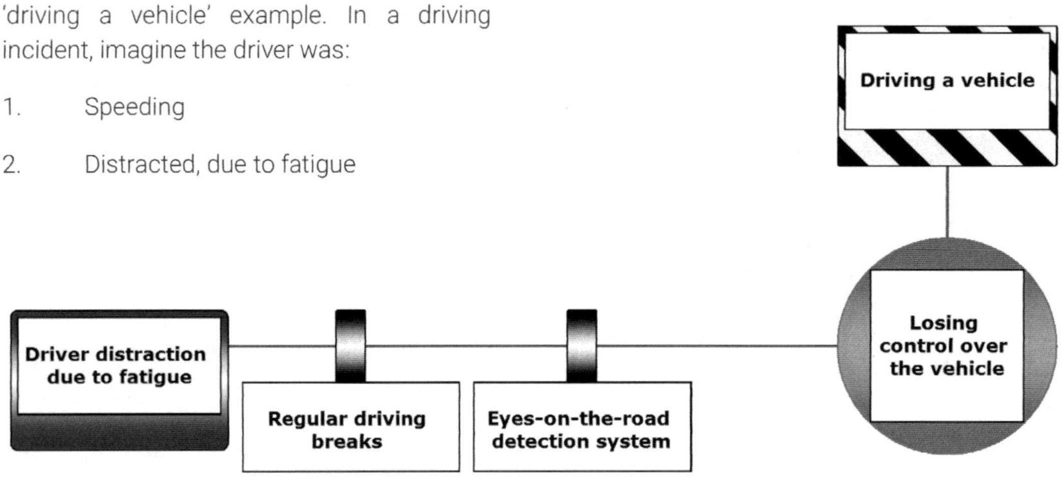

Figure 4: Including human factors in the threat

Differences in Safety-I and Safety-II

Vision 1 is in favor of the idea that 'Distracted due to fatigue' can be written down as a threat, as it can, and did, directly lead to the top event, 'losing control over the vehicle' (figure 4).

Human factors as a threat: A popular statistic is that human error represents a greater threat (80 - 90% of errors) to complex hazardous systems than technical failures do (Reason, 1997). The inclusion of human factors (or at the very least, human error) as a possible threat, therefore seems necessary to have a complete and thorough visual risk assessment. Looking back, this guideline fits well with the paradigm of thinking about safety of that time (1960s-1980s), the Safety-I kind of thinking.

An argument in favor of human factors as a threat is that human factors can sometimes be viewed as being as vital to a process as primary equipment (see section 1.1). Many, if not all, organizations need humans to operate effectively. This holds true especially for the less automatable industries such as medicine and construction. Human factors as a threat should be a force or condition that pushes an unwanted chain of events further. Therefore, human factors appear as threats in a bowtie diagram when the human act has the 'causal power' to directly influence the hazardous operation to result in a loss of control. For example, 'Driver distraction due to fatigue' directly led to the top event of 'Losing control over the vehicle'. Therefore, before deciding to include human factors, it must be considered if the human is part of a safety (barrier) system, or part of the main process. In the latter case, an error could directly lead to loss of control. In the driving example, the driver can be considered as part of the primary process as the car would not run without the driver.

Human factors as an escalation factor: Vision 2 states that human factors can never be a threat and is always a failure mode on a barrier, i.e. an escalation factor (see Figure 5). Organizations such as Environmental Resources Management (ERM) and Center for Chemical Process Safety (CCPS) argue in favor of this vision, as they believe the impact of human factors solely affects the health of barriers (McLeod et al., 2016). This is more in line with Safety-II thinking. The popular '90% of errors are human related statistic' is looked at from a different angle. Despite the errors, organizations are still growing and staying alive. Therefore, the Safety-II vision argues, it must be safe to assume that people do things right more often than they do wrong. Indeed: ***"There is growing recognition that people are often the reason operations go well despite the upsets and the everyday variability that is normal to complex activities" (Hollnagel, 2014).***

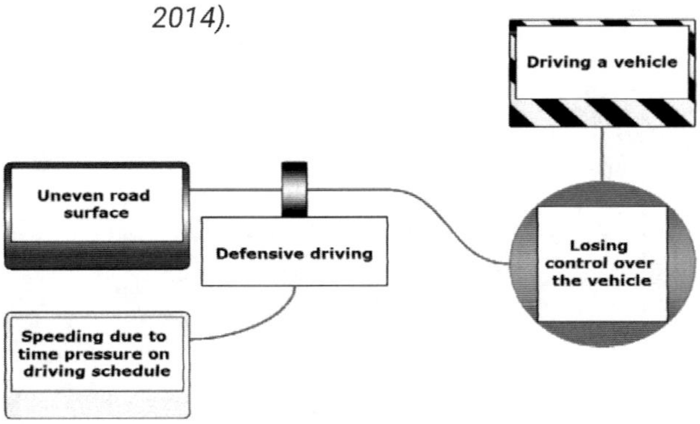

Figure 5. Human factors in an escalation factor

Furthermore, another possible negative side effect for putting in human factors as threats, as McLeod et al. (2016), argue is that, by focusing on human error as a threat, it can convey the wrong message to an organization's employees. If employees see that they are considered as threats in the organization's process, it can undermine their belief and faith in the organization and undermine the development of a strong safety culture. This runs counter to the point that human factors specialists try to make: the human is the strength of the system, not the weakness, i.e. humans are a solution to harness not a cause of failures.

Similarities

Beside the differences, there are also 4 parallels between the two visions.

Human factors must be included: both visions agree that every barrier is supported by the human element, whether it is a safety valve that needs to be maintained or an activity that requires certain competencies and/or training. Human factors are an integral part of risk management, and an organization needs to fully understand the impact these have on risks. Thus, excluding human factors from a bowtie means ignoring crucial information.

Non-generic wording to stimulate thought process: the way the threat or escalation factors should be described is not with a generic term like human error or human factor, but the specific results of those conditions, for example 'Speeding' or 'Distracted due to fatigue'. Wording the threat and escalation factor in this manner stimulates people's thought process and

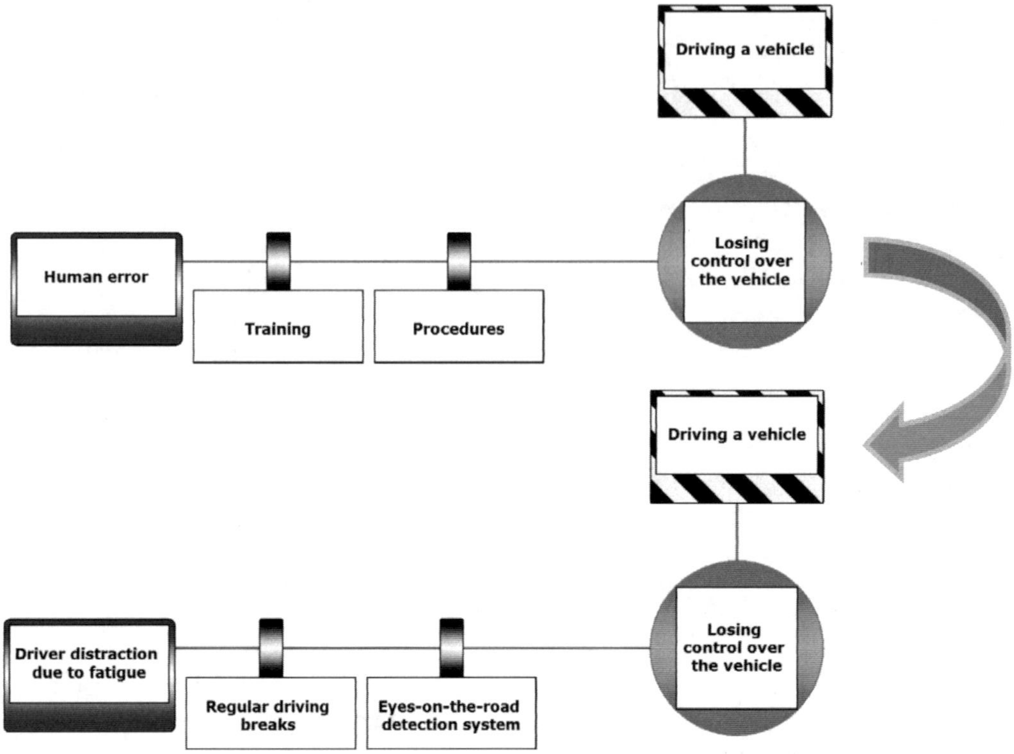

Figure 6: Improving the wording of human factors and human error

helps them complete the other steps in the bowtie method. It is much harder to come up with the barriers seen in figure 6, based on a generic wording like 'human error. 'Driver distraction due to fatigue' immediately focuses our thinking on both the result (distraction) as well as the cause (fatigue) so that we can come up with barriers that protect from either effect: regular driving breaks prevents fatigue, and an eyes-on-the-road detection system prevents distraction.

Escalation factors: Both visions agree that human factors are often part of the performance or failure modes of barriers. Failure modes are represented as escalation factors in the bowtie diagram. If the failure modes are understood, the activities a company needs to undertake to prevent these failure modes from compromising the barriers, can be known. This can be training, awareness campaigns, or it can go as far as organization development and culture transformation.

In the driving example, the driver was speeding. 'Speeding due to time pressure on driving schedule' would be an escalation factor to the barrier 'defensive driving', against the threat of the country's uneven road surfaces, see figure 5. Speeding, here, would be the human error, more specifically, a mistake (or 'violation', see James Reason, 1997). Whilst 'time pressure on driving schedule' refers to the driver's hurry to go on vacation, and therefore falls under the category of human factors. This view is in line with other organizations such as ERM and CCPS. Indeed, it can be called a standard.

A notable exception: bowtie chaining: both visions agree that it is possible to make a bowtie focused completely on human factors, with the top event usually being related to human factors by making use of bowtie chaining as described earlier in this paper (see "A' bowtie method?'). In such bowties, both visions agree that human factors can be put on the threat level to get the right level of detail and causation of the human factors.

In the driving example, 'driver distraction due to fatigue' can be chained into multiple bowties (see figure 7). In figure 7, 'driver distraction due to fatigue' has been phrased as 'Driver loss of attention (due to...)' in the bowtie with top event 'losing control over the vehicle'. Before the driver lost his attention, there were other threats that likely caused this to happen, so the threat of 'driver loss of attention' becomes the top event for the previous bowtie.

The right vision?

Perhaps the reason for the existing disagreement, is that the bowtie is, first and foremost a communicative tool. It is highly flexible and is being applied in several industries. Each industry has a different context and a different level of safety, and thus needs different guidelines for human factors and bowties. Maybe a third vision needs to be adopted, one that is as flexible as the bowtie method itself and combines the merits of both visions: For industries such as chemical, oil and gas or aviation, most processes work automatically, and their performance is protected by automatic barriers. Therefore, it is more likely that primary equipment failures involve mechanical solutions rather than human solutions. The human factor can reduce the effectiveness of the barriers but can never be a direct threat leading to the top event. For these industries, the Safety-II way of thinking supported by ERM/CCPS might be preferred, and thus human factors should only be included as escalation factors.

For the more manual labor-intensive industries such as medicine or construction, it is very difficult to automate processes and the human role is more present. In these industries, top events can directly be caused by human factor failures, as the human can be viewed as primary equipment. Therefore, as people are essential in such industries, the older style, Safety-I way of thinking might be preferred, and thus human factors can be considered both as a threat as well as an escalation factor.

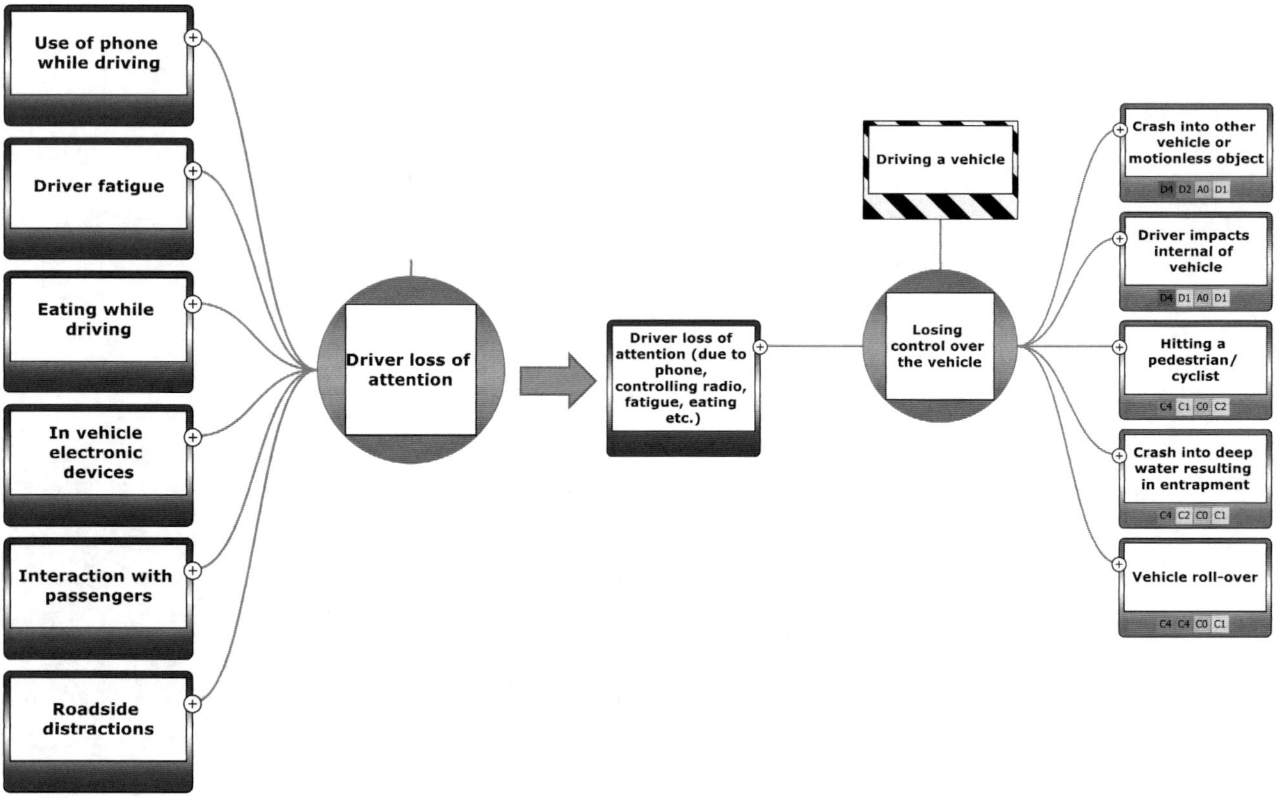

Figure 7: Bowtie chaining example of human factors

References

Edkins, G. (2016) Human Factors, Human Error & The Role of Bad Luck in Incident Investigations [Blog Post]. Retrieved from https://www.safetywise.com/single-post/2016/08/30/Human-Factors-Human-Error-The-Role-of-Bad-Luck-in-Incident-Investigations

Hollnagel, E. (2014) Safety-I and Safety-II: The past and future of safety management. Ashgate. Hopkins, A.

Hollnagel, E., Wears, R. L., & Braithwaite, J. (2015). From Safety-I to Safety-II: a white paper. The resilient health care net: published simultaneously by the University of Southern Denmark, University of Florida, USA, and Macquarie University, Australia.

Jordan, P. W. (2002). *Designing pleasurable products: An introduction to the new human factors*. CRC press.

McLeod, R., Randle, I., Miles, R., Hamilton, I., Wilkinson, J., Tomlinson, C., Thomas Jun, G., Wynn, T. (2016). *Human factors in barrier management: A White Paper by the Chartered Institute of Ergonomics & Human Factors* [White paper]. Retrieved August 17, 2018, via https://www.ergonomics.org.uk/Public/Resources/Publications/Barrier_Management.aspx

Reason, J. (1990) *Human Error*. Cambridge University Press.

Reason, J. (1997) Managing the Risks of Organizational Accidents. Ashgate
referentie in tekst: complex hazardous systems than technical failures do)

de Ruijter, A., & Guldenmund, F. (2016). The bowtie method: A review. *Safety science*, *88*, 211-218.

Sklet, S. (2006). Safety barriers: Definition, classification, and performance. *Journal of loss prevention in the process industries*, *19*(5), 494-506.

583490 Api's Process Safety Assessment Program - 8 Years Later - Learnings & Changes

Abstract:

In the past year, PSSAP has implemented two new protocols, developed PSSAP Focused (a slimmer version better suited to small sites/companies and petrochemicals) and rolled out a new pricing structure. This presentation will be an update on all these items, including learnings from the traditional 7 protocols, as well as the two new protocols (Incident Learning and Product Storage & Transfer).

20 AIChE Spring Meeting
16TH GLOBAL CONGRESS ON PROCESS SAFETY
HOUSTON • 2020

Improving Process Safety Reporting from an Insurance Perspective

Nigel Cairns CEng FIChemE
Professional Process Safety Engineer
Aon Global Risk Consulting, The Aon Centre, The Leadenhall Building,
122 Leadenhall Street, London EC3V 4AN
Nigel.cairns@aon.co.uk

Prepared for Presentation at
American Institute of Chemical Engineers
2020 Spring Meeting and 16th Global Congress on Process Safety
Houston, TX
March 30 – April 1, 2020

Improving Process Safety Reporting from an Insurance Perspective

Nigel Cairns CEng FIChemE
Professional Process Safety Engineer
Aon Global Risk Consulting, The Aon Centre, The Leadenhall Building,
122 Leadenhall Street, London EC3V 4AN
Nigel.cairns@aon.co.uk

Keywords: insurance, bow tie, process safety barrier, losses, risk rating

Abstract

Energy sites viewed and assessed by insurance risk engineers continue to have losses, seemingly independent of their reported risk quality. Aon proposes a significant change in risk quality assessment using a safety barrier model to highlight strengths, weaknesses and exposures, and give a greater focus to learning from industry losses.

1 Introduction

Energy insurance risk engineers conduct hundreds of underwriting surveys globally each year, across upstream, midstream and downstream assets. Most of these surveys are conducted by risk engineers working for (or on behalf of) insurance brokers for the benefit of insurers and insureds.

Site operators are extensively audited by stakeholders and regulatory bodies, yet we know anecdotally that insurance underwriting surveys are highly regarded for bringing independent third party and industry good practice advice to sites. However, to some operators the process can often boil down to one question: "What will this mean for my insurance premium?"

The main output from the survey process, the insurance underwriting report, is written to give evidence-based opinion of insurable risk quality to underwriters and insureds. This is, in part, used to help set insurance premiums and policy conditions, and whilst there is insurance-industry guidance on what the content of underwriting reports should be, all insurance surveying organisations have their own method of rating the risk quality of a site, against recognised good engineering and management practices.

However, the fact that the industry as a whole continues to see losses begs the question whether insurance underwriting surveys are sufficiently objective, and suitably structured and executed in order to understand and address the root causes of losses so as to prevent (or reduce) reoccurrence.

The purpose of this paper therefore is to propose a more effective process for assessing insurable risk quality within the confines of an insurance underwriting survey, using a barrier-based methodology, whilst noting that within the framework proposed, the detailed assessment of individual barrier detail is still a work in progress by the author.

2 The origins of insurance risk engineering

The origins of energy insurance go back to 1972 when the insurance mutual Oil Insurance Limited (OIL) was formed by 16 energy companies in response to two large-scale industry losses in the late 1960s. The combination of an oil spill in Santa Barbara, California and a refinery explosion in Lake Charles, Louisiana highlighted the need for specialist energy insurance, beyond that available at the time from established property insurance markets. But it was the Flixborough disaster of June 1974 which ultimately led to the formation of International Oil Insurers (IOI), and to the recruitment of risk engineers into insurance from industry in an attempt to better differentiate risks. This led directly to the insurance survey process we see today, and the associated methods of risk rating, benchmarking, establishing Estimated Maximum Losses (EMLs), and risk quality improvement through targeted risk improvement recommendations.

Early reports would review site operations, document safety culture, and attempt to give some sort of risk differentiation, but process safety as a concept was in its infancy, and there was more of a focus on developing EMLs so that underwriters and insureds could understand their maximum exposures. However, industry losses continued.

The focus started to shift more towards process safety and away from occupational safety reporting following significant losses at Phillips Pasadena (1989) and particularly after BP Texas City (2005). However, progress has been slow, as insurance has traditionally been a conservative profession, and risk engineering surveys and reports have typically lagged behind the industry-wide push for better process safety management and reporting.

The objectivity of underwriting reports varies when it comes to defining risk quality and differentiating between risks. Anecdotally, there is also a widely-held view that reports generated from some global insurance hubs still over-emphasise the importance of 'after the fact' loss mitigation, rather than scrutinising loss prevention barriers. Overall therefore, the view of much of the insurance underwriting community is that although reports generally give good risk description and information, they do not give underwriters clear guidance on which insured assets are more likely to give them a claim by policy year end.

3 Asset risk rating

To try and counter this lack of direction, report authors have devised risk rating or risk ranking tools to try to benchmark sites against Recognised And Generally Accepted Good Engineering Practices (RAGAGEP). Sites are typically scored against a number of factors, with a variety of weightings applied to distil a few dozen or few hundred individual scores down to a single indicator, within a range of (for example) 'poor' to 'excellent'.

However, this raises a number of questions, such as does the rating process:

- define the health of a site's safety barriers?
- give meaningful guidance to insureds as to where risk reduction improvements could be made?
- give a rating that merely satisfies our appetite for measurable data, regardless of how useful those data actually are?
- differentiate between risks that indicate which are the sites or the insureds most likely to have a loss? This last point is perhaps the most important to underwriters.

The reality is that the industry continues to have a significant number of losses year-on-year, and these losses are happening across all perceived risk qualities. Often the problem is that without sufficient differentiation within the risk rating methodology, ineffective risk-reduction features can get 'normalised' out, masking defective safety barriers. Too many sites which are positively rated by underwriting surveys have property-damage losses, such as fires or vapour cloud explosions (VCEs), and there appears to be no direct correlation between loss-likelihood and risk rating score: in the words of one underwriter "too many good sites are having losses".

Is loss propensity truly independent of risk quality? Or are we using the wrong tools?

Part of the problem is that the risk rating processes in common use by insurance risk engineers are often used as the 'go to' tool for a variety of conflicting purposes, for example, as a benchmarking tool for comparing individual sites or regions, or even as a means for insureds to set Senior Management bonus levels based on performance at the last insurance survey. Consequently, the number of metrics typically measured and scored has gone up rather than down, become less rather than more focused, with the result that rating quality has often been 'averaged out' or normalised into the middle of the bell curve distribution, reducing differentiation. Despite attempts by some organisations to address this by greater weighting of more critical topics, the greater number of topics being scored masks poor barrier performance, and the combining of independent loss control barriers results in them losing their individual identity as performance metrics.

For example, a low score for a poorly-resourced Inspection department could be compensated for by an adequate hazard study process, or low levels of operator competence could be offset by a high level of predictive maintenance. The evidence however is that independent barriers such as asset integrity or operator competence by themselves are key factors when looking at what causes losses and, therefore, must not be 'swapped out' for other barriers or layers of protection.

Therefore, before we can define how we can more effectively measure a site's propensity to have a loss, and thus better define its risk quality from an insurance perspective, we need to understand what actually causes losses.

4 What causes losses?

Anecdotally, insurers have known for some time what causes losses, but the first formal presentation of data was by Jarvis in September 2016 (Reference 1) on behalf of the Lloyd's Market Association (LMA). This document reviewed 100 major onshore oil, gas and petrochemical property damage losses in terms of monetary value over a 20 year period from 1996 to 2015, excluding natural catastrophe (Nat Cat) events; it was recently updated to include an additional 37 losses between 2015 and 2019.

Major losses occur because of failures of both loss prevention and mitigation barriers in response to an initiating event. The LMA data highlighted the main initiating events which led to the majority of the considered losses, as well as those loss prevention barriers which most frequently failed:

- 38% of the losses were due to mechanical integrity failure; with 71% of these due to either internal or external piping corrosion (noting the higher propensity for this in refining over other occupancies such as gas plants or terminal operations).
- 20% of the losses occurred during maintenance activity, typically due to inadequate control of work.
- Of the remaining 57 losses, 86% occurred during transient (non-routine, infrequent, abnormal or unplanned) operation.

Note that the key messages from the LMA data are supported by evidence from underwriters' own loss data and claims analysis, which typically reference a much larger industry dataset. They suggest that rating asset quality from an insurance perspective would be more effective if focussed on the health of those barriers which line up to prevent or limit losses associated with a range of initiating events.

Of course, this has the potential to significantly change the insurance underwriting survey process as we know it. Insurance surveys are currently organised by discipline; meetings are organised by department, together with a tour of the site being surveyed. Typically, this means spending a couple of hours each with the Engineering, Inspection and Operations teams, among others, with each assessed against recognised good engineering and management practices. But if we are now to focus on the effectiveness of individual barriers in response to defined initiating events, then this is drawing us down a path of site interviews by barrier rather than by discipline – of course the reality is that we'll be talking to the same people, only with greater focus on the specific barriers which stand between a site and a major property damage / business interruption (PD / BI) loss.

And if the survey and rating processes need to change, then so does the reporting process, from discipline to barrier focus.

5 The insurance underwriting report

Underwriting reports have been based on much the same format for the last 15-20 years. In recent times, some have become shorter, but only because the same information has been presented more efficiently, rather than because of a fundamental shift in what is being reported. Underwriters and their risk engineers rarely have time to read a plethora of 100-page reports, and even if they did, do the reports give a true insight into the health and status of critical safety barriers or the likelihood of a loss? The message from the insurance markets is a resounding 'no'.

Therefore, to better define and report insurable risk quality to markets and insureds, the insurance industry needs to move away from a highly descriptive report style to one which focuses more clearly on the health and quality of defined process safety barriers; one way of doing this is to adopt the look and feel of a "bow tie diagram" (see Figure 1) to give a simple, clear representation of site risk quality to underwriters to help their insurance decision-making processes.

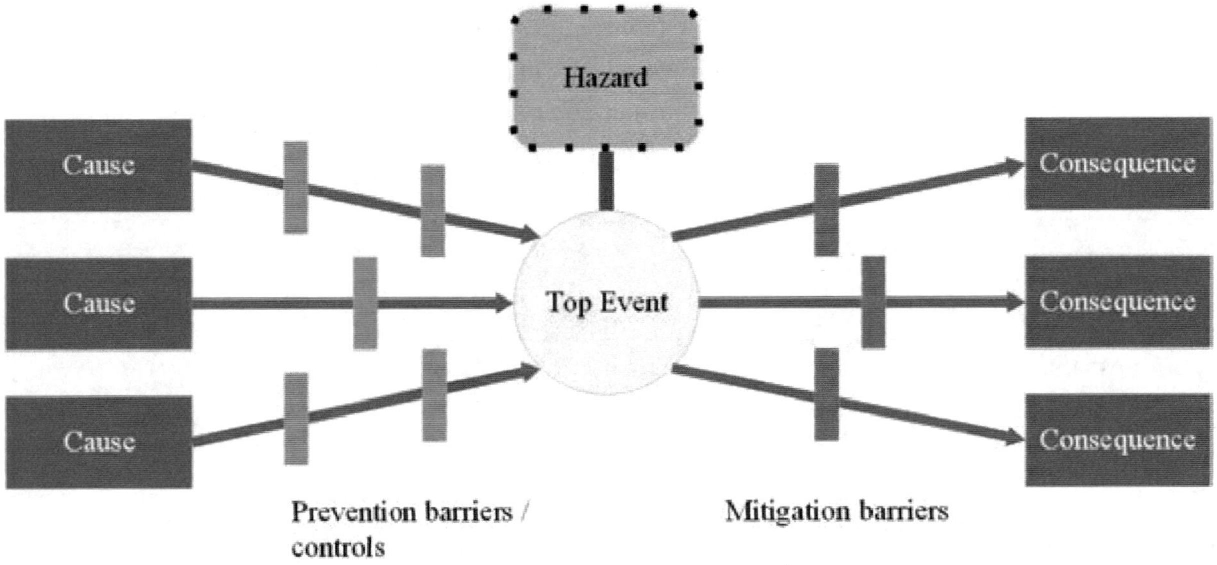

Figure 1. Typical "bow tie" diagram used for process safety reporting

The model should be simple and clear enough to effectively report on barrier health for a range of initiating events and causes, yet flexible enough to accommodate a variety of asset types and a range of review frequencies. If the model is too specific, the process becomes restrictive, cumbersome and quickly out-of-date as sites change – this is an important point for sites which will typically be surveyed by insurance risk engineers on a three or four-year cycle.

6 Adopting a barrier model

Adopting a barrier-based methodology begins with trying to build a process that focuses on the barriers which prevent what have been historically the key initiating events that lead to significant financial loss for insureds. This process can then be updated as loss initiators or barriers themselves change; for example, improved technology giving more inherently safe designs, global economics driving a shift in feedstock slates or changing demographics placing a greater strain on human resources.

Having defined the events or initiating causes which pose the greatest threats, our hazard is invariably hydrocarbons under pressure. Of course, extending the analysis to third party liability effects may require an extension of the methodology to non-hydrocarbons, for example to review the potential effects of the release of a typical refinery toxin such as hydrofluoric acid. For the purpose of the process however, the top event would not be a loss of primary containment (LOPC), but that event with which insurers are directly concerned, namely the fire or explosion which results in property damage. This is an important point, as it clearly defines those barriers which are preventing exposure and those which mitigate; it may also result in some barriers taking on a slightly different position within the model compared to a more traditional model based on an LOPC as the top event. In following this methodology through, no 'preventative' credit is taken for ignition source control on the left hand side (LHS) of our model; this is a deliberate stance from an insurance perspective to assume that a flammable cloud would invariably escalate to a VCE.

Such an analysis includes a range of barrier types and associated degradation controls (NB. passive barriers are typically stronger than active barriers) with barriers relying on human intervention typically among the weakest; refer to Table 1:

Table 1. Example of common barriers used in the process industries

Barrier type	Example
Passive hardware	Firewall
Continuous hardware	Active corrosion protection system
Active hardware	Safety Instrumented System (SIS) / automatic emergency shutdown system (ESD)
Active hardware + human	Operator-activated ESD
Active human	Visual fire detection and evacuation

The simplest and most effective way of reporting barrier health and giving a snapshot of performance (covering both the degree to which the barrier is present, and the quality of its implementation / maintenance) would be through the use of 'traffic light' colour coding as demonstrated in Table 2:

Table 2. Use of colour coding to define barrier status

Condition	Description
Effective	Barrier performing at design/intended effectiveness
Slightly degraded	Barrier performing slightly below design/intended effectiveness
Significantly degraded	Barrier performing significantly below design/intended effectiveness
Out of service	Barrier is not in place, turned-off, deactivated, fully degraded
Not applicable	Barrier not deemed applicable for that asset (for example, insurers wouldn't typically expect a remote terminal to maintain a full-time professional fire brigade, but might do so for a world-scale refinery)

So, for risk engineering reports, this might be applied to the first significant initiating event as follows.

A fire / VCE from the mechanical integrity failure of piping

For this scenario, Table 3 suggests examples of the key barriers and controls on the LHS of the model for preventing a loss caused by piping corrosion (NB. the constituent component detail of each barrier is pending continuing review by the author):

Table 3. Possible key preventative barriers and controls in a piping corrosion-initiated loss scenario

Barrier / control	Barrier / control type	Comment
Inspection competence	Degradation control	
Inspection programme philosophy	Degradation control / Preventative barrier	Identification and management of corrosion and damage mechanisms, use of appropriate NDT, etc.
Integrity Operating Windows (IOWs)	Degradation control	
Material selection	Preventative barrier	Including QA / QC

In terms of the survey process, all of these would currently be discussed with a site's Inspection department, and findings reported in the Inspection section of the underwriting report. In future, the robustness of the site's asset integrity management as a function of its defined barriers and degradation controls would be reported.

Moving from left to right along the model, further barriers and controls then play their part in preventing or mitigating a loss as a result of this threat; a selection is shown in Table 4:

Table 4. Possible key preventative and mitigating barriers in a piping corrosion-initiated loss scenario

Barrier / control	Barrier / control type	Comment
Engineering standards	Active hardware / preventative barrier	LHS of model
Gas detection	Active hardware / human	LHS – detection prompting action to prevent escalation of an LOPC to a VCE as the top event
ESD	Active hardware	LHS
Unit layout	Passive hardware / mitigating barrier	RHS of model – good layout features will reduce the impact of the top event
Fire detection	Active hardware / human	RHS – detection prompting action to mitigate the effects of the top event
Emergency response	Active human	RHS – typically the 'last line of defence' after the top event to limit escalation

This starts to build up a picture of how a new risk quality rating process might work, by identifying key barriers and degradation controls that prevent initiating events and threats becoming consequences through the top event.

A fire / VCE as a result of a transient operation

For this second initiating event, Table 5 suggests examples of those barriers and controls specific to the site's Operations team that should be in place to protect against unforeseen or transient events, such as the loss of a plant utility (e.g. power or cooling water). Again, note that the constituent component detail of each barrier is pending continuing review by the author:

Table 5. Possible key preventative barriers and controls in a transient operation-initiated loss scenario

Barrier / control	Barrier / control type	Comment
Operator competence assurance	Degradation control / preventative barrier	Includes training and recertification
Shift handover	Degradation control	
Operating procedures	Degradation control / active human	Includes Emergency Operating Procedures (EOPs), and requires action by the operator

There would also be a number of hardware controls upstream of the top event, and beyond this, the mitigation barriers on the RHS would typically be the same as in the earlier piping corrosion case.

A fire / VCE as a result of a maintenance activity

In this example, key barriers and controls on the LHS might include those in Table 6:

Table 6. Possible key preventative barriers and controls in a maintenance activity-initiated loss scenario

Barrier / control	Barrier / control type	Comment
Permit to Work (PTW) management system	Degradation control	
Training and certification of permit authorities	Degradation control	For both issuing and receiving permits
Isolation management procedures and practices	Degradation control	

Shift handover	Degradation control	

Again, the mitigation barriers on the RHS would be essentially the same as in the first initiating cause.

Aside from the specific barriers and controls postulated above for the three key loss-initiating events as identified by the LMA, incident investigations carried out by regulatory or industry bodies such as the UK's Health and Safety Executive (HSE), or the US Chemical Safety Board (CSB) have identified other key barriers whose documented absence or degradation has contributed directly to a loss.

There are many published examples of where poor or absent stewardship of key process safety elements has been the direct cause of a significant loss across a variety of initiating events and site operating modes. Examples of such elements include Process Hazard Analysis (PHA), Management of Change (MoC), Process Safety Information (PSI), process safety audits, and incident investigation.

It is worth noting that on an insurance survey, such topics don't always have a natural home for discussion, because from site to site, the processes may be owned by a variety of disciplines or functions. Not all sites have a dedicated Process Safety function managing process safety element responsibilities and ownership, so these topics may lie with any of the Engineering, HSE or even the Operations departments. Hence, they can easily be overlooked if they form part of a 'shopping list' of questions on an agenda for a particular department during a survey; or worse, the person responsible is not on site. Treating each topic as a barrier in its own right, with dedicated discussion time, could provide extra focus during a survey.

Other loss types

Adding the process safety barriers to the event-specific ones discussed earlier adds to the picture of what needs to be investigated and documented in order to determine a site's susceptibility to a large property damage (PD) loss resulting from a fire or VCE.

Of course, there are other types of losses than PD that underwriters pay out for, or that give increased exposure to insureds or clients. These may or may not have the same level of impact financially as a single catastrophic event, but could result in more frequent attritional losses that still significantly affect the bottom line. Then of course there is the Business Interruption (BI) coverage, which is triggered once the initial loss has occurred, to recompense the insured's lost income.

<u>Machinery Breakdown losses</u>

Machinery Breakdown (MB) insurance covers sudden and unforeseen losses to rotating and fixed equipment (i.e. not as a result of inappropriate levels of preventative maintenance, poor operation or corrosion over time), and is typically included on a property damage policy.

Therefore, the preventative barriers and degradation controls on the LHS of the model would be similar to those protecting against abnormal or transient plant operating conditions, as they demand a similar decisive response from plant operators. However, a degradation control covering the health of preventative and predictive maintenance schedules should also be included on the LHS to cover specific reference to this type of loss. Mitigation barriers on the RHS of the top event would be similar to those discussed earlier.

Nat Cat losses

Like MB, Nat Cat losses are typically covered by a standard property damage policy, subject to specific limits and exclusions, if the insured is operating in an area of known exposures, such as the Gulf of Mexico (windstorm). It would be unusual for a Nat Cat-related loss to be the defining maximum loss scenario for a single asset, though a Nat Cat event could be an initiator to a greater process safety loss. What tends to impact underwriters however is the aggregation of a number of insured assets in the same area. What stands out is that, aside from a decisive planned response from the site in reacting to the transient condition, the other main barrier that can be assessed is the suitability of the design and engineering codes employed during the site's construction, and the quality of the build.

Cyber event

This is only briefly mentioned here to acknowledge its existence as a threat that can lead to a fire / VCE top event, and that there are barriers that can be put in place and assessed. However, this is currently typically outwith the scope of a typical risk engineering survey and covered as part of a separate insurance programme. This could change in the future.

A note on Business Interruption

Business Interruption coverage is typically triggered by any of the above loss types. As a rule of thumb, across the Energy sector, for every £1 paid out by underwriters for a PD loss, a further £2 is paid out for lost revenues, so underwriters will look for evidence of any mitigating barriers that help keep this figure down (i.e. they will look at preventative barriers and controls to stop the loss happening in the first place). The mitigating barriers will typically be a well understood and rehearsed Business Continuity Plan (BCP), and redundancy in plant, site, or international operations. However, individual sites may not be aware of wider corporate BCPs, which suggests that a fundamental review is needed of how the insurance industry assesses those barriers that are most effective in mitigating BI exposures.

Bringing the barriers together

So far, this paper has identified the critical barriers and controls in this model that need to be examined during insurance surveys in order to try and redefine 'insurable risk quality', in response to the most common threats faced by operating sites.

For the survey process to be effective, the barriers should be grouped together to facilitate effective discussions at site level. This can then define the survey agenda based on the barriers to be assessed.

However, to define the health of each barrier, each one has to be broken down into measurable components or elements for use in site discussions.

For each element, there are aspects of recognised, good engineering practice that must be present before that element can be said to have been achieved at its designed and intended level. If it is not at the required level, then a means for scoring how near it is to the required level must be in place, not just in terms of how site procedures state it should work, but how effective its implementation has been, and what the evidence of its effectiveness is on site. As an example, in order to assess the health of the site's PHA barrier, we would need to consider at least the following elements:

- Site PHA policy and schedule
- The suitability of techniques used
- The provision for the review of process safety incidents as part of the study
- Quality of reference information such as P&IDs
- Quality of the review of abnormal operating modes
- Quality of actions generated etc.

Some barriers and controls will have fewer elements than others, but the overall health of the barrier must be no greater than its weakest element – this is a noticeable change in philosophy from current 'risk rating' systems which aggregate element or component scores, thus diluting the presence of poor performers. Scoring of each individual barrier's health would be done using the colour coding system defined earlier in this paper.

A key feature therefore of this process would be to demonstrate individual barrier health and effectiveness by a clear, colour coded status report, to define where the asset's strengths and weaknesses lie, and how likely they are to have a robust response to the main threats that cause losses.

<u>What will the barrier model look like?</u>

An example of what such a model would look like can be given using the first threat mentioned earlier, that of pipe failure due to corrosion. Figure 2 shows a number of directly applicable preventative and mitigating barriers, with those additional barriers relating to process safety shown in Figure 3, and for emergency response in Figure 4 (NB. the condition of each barrier has been randomly assigned merely to highlight the 'traffic light' nature of effectiveness reporting).

Table 7 gives the barriers under current consideration by the author relating to Figure 2:

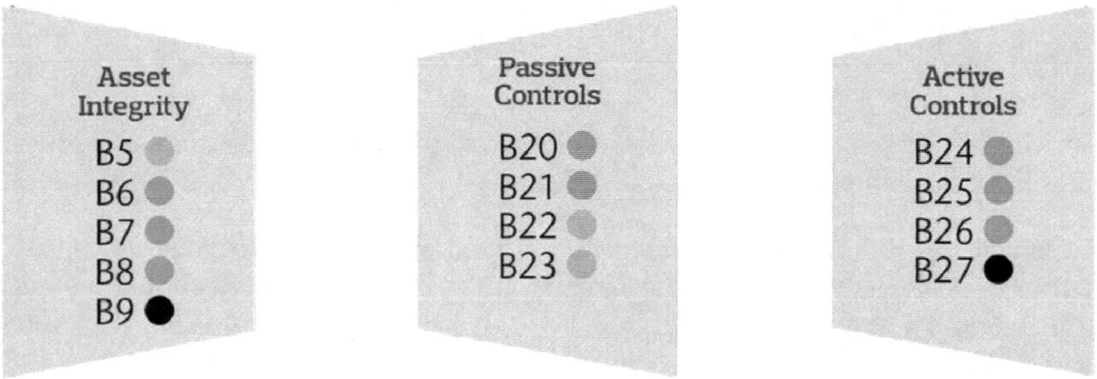

Figure 2. Asset Integrity (LHS) and loss mitigation (RHS) barriers for Passive Controls and Active Controls

Table 7. Relevant barriers under current consideration for the first threat

Barrier / control	Barrier / control	Barrier / control
B5 – Inspection competence assurance	B20 – Layout	B24 – Remote isolation
B6 – Inspection philosophy	B21 – Construction	B25 – Fire detection
B7 – IOWs	B22 – Passive protection	B26 – Emergency planning
B8 – Materials management	B23 – Inherent redundancy	B27 – Business continuity planning
B9 – Preventative maintenance		

In considering the second and third threats, the model then incorporates barriers specific to the site's Operations team; Figure 3 and Table 8 also include additional barriers from the review of learning from global process safety incidents and losses:

Figure 3. Representation of Operations and Process Safety barriers (LHS)

Table 8. Operational and process safety barriers for the second and third threats

Barrier / control	Barrier / control
B1 – Operator competence assurance	B10 – Process Safety Management (PSM)
B2 – Shift handover	B11 – Process Hazard Analysis (PHA)
B3 – Operating documentation	B12 – Management of Change (MoC)
B4 – Control of Work	B13 – Process safety auditing
	B14 – Incident investigation

Finally, Figure 4 and Table 9 demonstrate additional preventative controls and emergency response barriers which might be common across a range of initiating causes:

Figure 4. Representation of preventative (active and passive) controls (LHS) and mitigating (RHS) barriers

Table 9. Suggested further preventative and mitigating barriers across all threats

Barrier / control	**Barrier / control**
B15 – Engineering standards	B28 – Firewater system
B16 – Controls and safeguards	B29 – Active protection
B17 – Gas detection	B30 – Fire brigades and mutual aid
B18 – ESD	
B19 – Relief systems	

The ratings for the barriers above can be combined to give a summary representation of overall site risk quality, when measured against those threats known to cause energy losses (Figure 5).

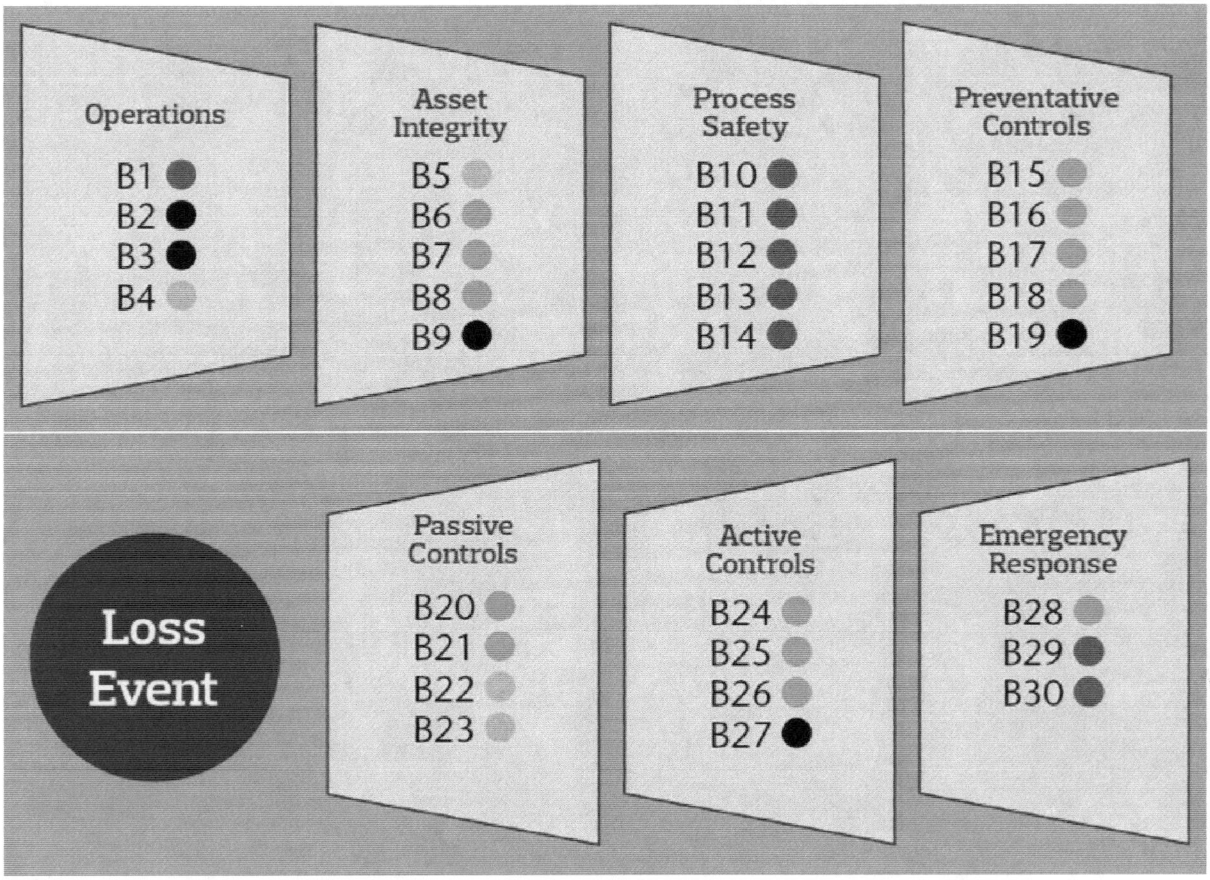

Figure 5. How a combined 'barrier diagram' might look in an Insurance Underwriting Report

Were this for an operating asset, a summary similar to Figure 5 at the front of an insurance underwriting report would quickly flag up concerns about the site's approach to process safety. Those reading the summary would be alerted immediately that the site's preventative maintenance programmes needed improvement (which would be a particular concern for underwriters were the insured purchasing Machinery Breakdown coverage). Based on the above summary, markets and the insured's senior management might also be concerned about the site's ability to respond to an abnormal or transient operation (such as loss of a utility), and depending on certain specific findings, would expect to see a number of recommendations from the survey team to address these concerns. The lack of business continuity planning (B27) might be a concern to insurance markets depending on the level of Business Interruption cover purchased by the insured.

Figure 6 shows another potential summary of findings for a site:

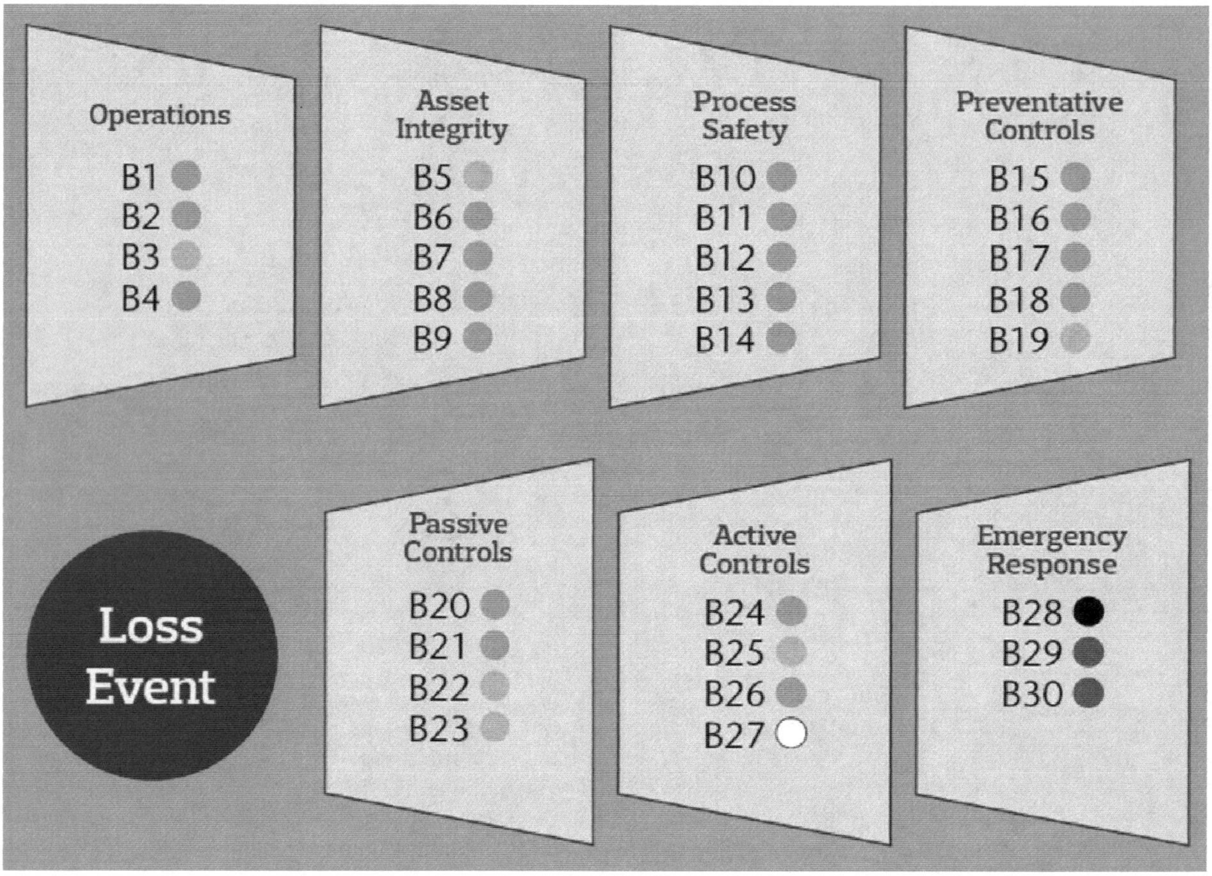

Figure 6. A further example of a possible 'barrier diagram' in an Insurance Underwriting Report

In this example, both the underwriter and insured could take comfort from the site's positive approach to installing and maintaining preventative safety barriers; if BI coverage had not been purchased in this case (note B27 is 'not applicable'), this would likely make this risk more attractive to insurance markets than that represented in Figure 5. Given the strong 'preventative' LHS of the diagram, the underwriter might not be concerned about the failings of the emergency response systems; this would depend however on underwriter 'loss appetite' and on recent loss history.

Conclusion

Aon Global Risk Consulting is proposing a significant change in the way that the energy insurance industry measures and reports risk quality in its insurance underwriting reports, giving a clear summary of the strengths, weaknesses and risks associated with each site, a greater focus on learning the lessons from industry losses, and a clear indicator of the strength of critical safety barriers at sites visited during insurance underwriting surveys for site operators, insureds and underwriters alike.

We believe that this methodology will give a more effective process to give better differentiation of risk quality across those energy assets surveyed.

7 References

[1] Jarvis, R and Goddard, A, September 2016, An analysis of common causes of major losses in the onshore oil, gas & petrochemical industries, Implications for insurance risk engineering surveys, Lloyd's Market Association (LMA).

Integration of Hydraulic Studies in Evaluating Abnormal Plant Conditions during Commissioning and Startup, Maintenance, and Intermittent Operations

Ritche Niño Li
SABIC
Petrokemya, Jubail Industrial City, Saudi Arabia
lirn@petrokemya.sabic.com
ritche.nino@gmail.com

Prepared for Presentation at
American Institute of Chemical Engineers
2020Spring Meeting and 16thGlobal Congress on Process Safety
August 17 – August 21, 2020

Integration of Hydraulic Studies in Evaluating Abnormal Plant Conditions during Commissioning and Startup, Maintenance, and Intermittent Operations

Ritche Niño Li
SABIC
Petrokemya, Jubail Industrial City, Saudi Arabia
lirn@petrokemya.sabic.com
ritche.nino@gmail.com

Keywords: hydraulic calculation, incompressible flow, transient operation, abnormal operation, management of change, pressure safety valve, preventive maintenance, commissioning, startup.

Abstract

Multiple abnormal operating circumstances are encountered in various phases of plant life cycle. During commissioning, maintenance, and irregular batch operations, challenges arise which may prevent in achieving an objective in differing modes of operation. Underestimated more often than not, a common operating problem boils down to issues of fluid transfer of process systems.

In aid of continued operation, hydraulic calculations were performed to attend on familiar dilemmas in a petrochemical plant. In the first case, simulation was utilized to determine viability of a proposed relief alternative pathway during a preventive-maintenance removal of a pressure safety valve (PSV). On the other hand, a separate calculation was also carried out to evaluate a proposal to remove a blower from a series of two units during commissioning of a project-delayed combustible dusts venting system. Lastly, a hydraulic analysis addressed an abnormal high pressure build-up and high flow in an intermittent recipe requirement of a corrosive chemical in a batch process.

The three scenarios are illustrations of process modifications in transient operating mode and abnormal operating conditions. If not decided with a sound technical judgment, process changes may induce operational and safety consequences affecting property and people. The study demonstrates the effectiveness of conducting hydraulic checks to recommend solutions or to assess proposal as part of the management of change proceedings. Based on the calculation summary, key recommendations were generated and implemented accordingly to assist requirements in varying plant conditions.

1 Introduction

The refining and petrochemical process industries have registered more than half of process safety incidents during non-routine and abnormal operations. Even though only 10% of the plant lifecycle is dedicated to transient operations, the bias of recorded process related incidents towards infrequent planned and unplanned operations is a point of concern for operators worldwide (Ostrowski& Keim, 2010; Duguid, 2008).

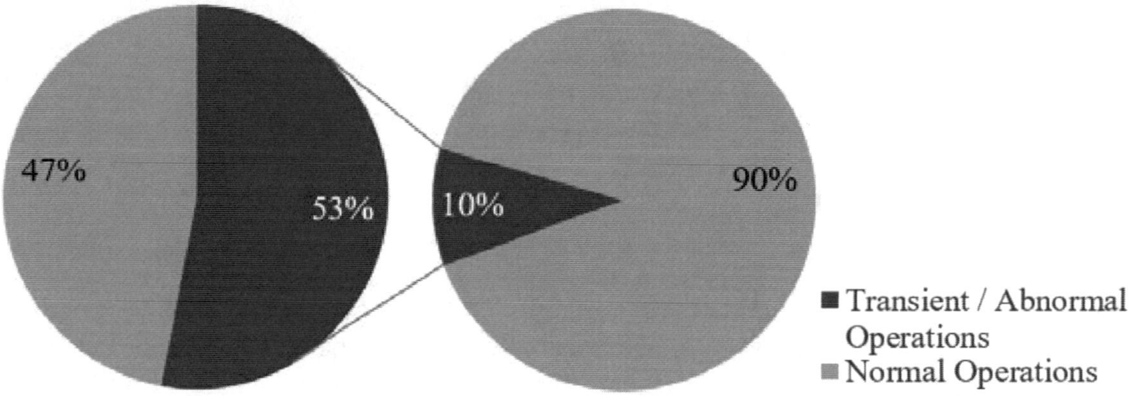

Figure 1. Percentage of Incidents Share Between Transient and Normal Operations

Figure 2. Percentage of Dedicated Time in Plant Life Cycle (Ostrowski & Keim, 2008)

A newly constructed petrochemical plant, which is the facility subject of this research, has encountered transient operations after its mechanical completion. The plant has undergone various types of operating mode from the inception of its commissioning prior to startup, then bridging to a sustainable operation, and up to halting for planned maintenance. Throughout these varying periods, a typical process plant requires to exert extra effort to overcome difficulties and to complete safely.

During the transitory phase and in some cases, changes to process units are inevitable and necessary. An example of this is providing an alternative pathway for pressure/vacuum relief to a storage vessel in the course of the temporary relief valve outage while in operation. In general, modifications in the process, if not managed properly and not critically checked, may impose hazard to the plant rather than providing a solution. One of the familiar incidents mostly associate with hazards of process modification is the Flixborough disaster. Accordingly, the introduction of uncontrolled process change to the reactor system led to the plant's demise.

With this, combining the ideas of process modifications while undergoing transient operations magnifies hazards inherent to this mode of operation. This poses a threat against the aim to preserve a safe condition for the plant, the environment, and its people.

A comprehensive evaluation of process modification is then mandatory to ensure all aspects of the change will fit the requirement of the facility. Through the process of change management, process safety information should go through a recheck.

A way of implementing confirmatory checks during management of change is by performing hydraulic studies, especially utilized to determine anomaly and solution about topics of fluid transfer. This paper promotes the use of the technique to evaluate process related inquiry. The underlying principle is to verify the compatibility of the change proposal to the system that will be modified, and to ensure potential hazards are addressed before the modification.

Moreover, hydraulic analysis can be an inexpensive alternative in answering inquiries to process systems that ranges from complex process system to hazardous materials. It is a powerful and handy tool generally used for line and pump sizing, and various piping component design.

The integration of hydraulic simulation as a tool for a safe mode of operation transition is demonstrated in this paper through the three case studies, differentiated through a define lifecycle of the plant such as commissioning and startup, maintenance, and intermittent operations. The cases are further categorized within the two types of transient operations namely non-routine planned operations, and abnormal or unplanned operations. Depending on the simulation results, a process modification (temporary or permanent) is either supported for implementation or declined to proceed.

2 Review of Literature

2.1 Transient Operations: Non-routine Planned, and Abnormal Operations

Vaughen (2020) defines transient operation as mode of operation wherein the process is undergoing a transition phase and is not at its designed normal operating mode.

Related to this operating mode, Ostrowski & Keim (2010) divided transient mode into two types: non-routine operations, and abnormal operations. Non-routine operations or planned operations are operating events that infrequently occur and have the potential for a loss of primary containment leading to a high consequence incident (Ostrowski & Keim, 2008). Some instances of non-routine operations include startup or shutdown of major units. Non-routine operations may also include continued operations in an equipment configuration beyond the intended design such as non-availability of a major pump, absence of a blower, and out of service of a boiler or a compressor (Ostrowski & Keim, 2008).

On the other hand, abnormal or unplanned operations are operating events that happen within the normal operating mode brought about by process upsets, and process parameter deviation from the normal operating limits (Vaughen, 2020). Abnormal operations have the potential for a loss of primary containment leading to a high consequence incident. Some examples of the unplanned operations involve operating instances beyond equipment's design limit or safe operating limit (SOL), responding to safety critical alarms, and loss of containment (Ostrowski& Keim, 2008).

As an illustration, the content of this paper incorporates case studies for both types of transient operations. These operating events include removal of safety equipment during preventive maintenance, blower outage during commissioning and prior to startup, and abnormal operation during an overpressure scenario in a batch process.

2.2 Commissioning

During commissioning executions, equipment is tested to check its capability to support varying operating conditions based on its design and intended function. Moreover, commissioning procedures are executed before the introduction of hazardous or process chemicals, and employs the use of safe chemicals. When executed well and along with its rigorous approach, commissioning phase is an ideal period to effectively identify design gaps or mishaps incurred during construction phase, and before startup with hazardous material.

Kletz (2009) demonstrated a near-miss incident during commissioning of a new unit. A vessel was isolated and was not pressure tested due to vessel service sensitivity to water as a test medium. The mishap led to a release of a dense vapor cloud. In spite of this, the operator was able to escape without sustaining injury (Kletz, 2009).

In general, a rightly executed commissioning plan involves the use of a disciplined, organized, and professional methodology, to bridge the newly constructed process plant into a fully integrated and operational unit. Ideally, commissioning revolves around the idea of getting it "right first time" in the most safe, efficient, cost-effective and timely manners to achieve startup and production goals (Killcross, 2012).

2.3 Startup

Process safety incidents during startups happen five times as frequent compared to events while at normal plant conditions. Interestingly in a study about a petrochemical facility, startups and other transient operations grab only a fraction of about <10% of the total operations time. The intense and stressful environment to push the plant to a start makes the startup prone to errors and incidents (CCPS, 1995; Ostrowski& Keim, 2008).

Startup may vary from initial startup, restart from a shutdown, startup after turnaround, and startup after extended outage. Initial startup aims to bring a newly constructed plant and equipment to its first operations (CCPS, 1995).

The non-conducive atmosphere of initial startup and commissioning is amplified in a case study included in this paper.

2.4 Maintenance

A huge portion of a plant's lifecycle is dedicated to operations along with maintenance. Preventive maintenance aims to ensure integrity and functionality of the assets, and is conducted on a predetermined time-based schedule (CCPS, 1995). Intuitively, about 50% of the process incidents were attributed to maintenance-related activities that took place during startups, shutdowns, maintenance, or abnormal operations (Kletz, 2009; Duguid, 2008).

Due to the significant amount of time dedicated during maintenance phase, the facility should be well prepared to determine hazards and be able to attend on it. A dilemma during online maintenance for a safety equipment is illustrated part of the research.

2.5 Batch Process

In a batch process, retention time in vessel and piping systems changes with time according to the particular processing requirement, product grade, or recipe. Time variations on reactant or monomer charging, reacting, transferring, recovering, and batch drop make it a challenging to check and troubleshoot processes (CCPS, 1999). A holistic approach integrating the operating philosophy of a batch process that is linked to a continuous process should be accounted as part of the process design. Otherwise, a mismatch of operating concept may lead to operability issues and safety concerns.

Kletz (2009) and Duguid (2008) have attributed 40% of process-related incidents to a less than adequate process design. In the report, safe design features were not applied in spite of its presence and availability industrywide.

This paper also reflects an instance when a batch process operates in interface with a conflicting design from a continuous process.

2.6 Hydraulic Calculation and Incompressible Fluid Flow

Hydraulic analysis can be an inexpensive alternative in answering inquiries to process systems that ranges from complex process system to hazardous materials. It is generally use for line sizing and determination of pump specification and equipment sizes as accurately as required for practical purposes (Moran, 2019).

Moreover, the simulation can also be used for troubleshooting issues in fluid networks. As an example, during its initial commissioning, once a pump is energized, its performance is tested to know its capability to deliver process parameters according its purpose. After individual run, its behavior will be tested with the piping systems and associated equipment. If unsatisfactory, troubleshooting onsite can be take place or a simulation can be performed if more practical. It is a powerful and handy tool in performing pipe and pump sizing, and various piping component design.

Nitrogen, air, and diluted potassium hydroxide solution, are the fluid separately involved in the three case studies. Intuitively, the aqueous solution of caustic potash was modeled through principles of incompressible flow.

During the simulation, the mach numbers, M, of nitrogen and air within their respective systems were found both below 0.3. When the mach number of a flowing fluid in any point is less than 0.3, the density variations due to pressure changes can be neglected, hence the gas density may be assumed as constant. As a general rule, gas flow with $M< 0.3$ is regarded as an incompressible flow (Pijush & Cohen, 2008).

In this paper and for calculation simplicity purposes, the flow of nitrogen and air are analyzed through incompressible flow calculations, therefore neglecting compressibility effects.

3 Hydraulic Analysis in Managing Modifications to the Process

3.1 Facility Description

Independent hydraulic analyses were carried out in response to operational requirement, troubleshooting activities, or improvement pursuit in one of the polymer plants that manufactures an engineering thermoplastic with various industrial and household applications. The facility is situated in Saudi Arabia, part of a petrochemical complex that manufactures olefins, basic petrochemicals, and other thermoplastic resins.

3.2 Problem Statements and Study Objectives

During the various plant lifecycle from commissioning and startup, operations until maintenance, the facility encountered varying challenges to run the plant in safe condition. Below case studies, describe the scenarios wherein hydraulic analyses were applied to facilitate decision-making.

3.2.1 Study 1: Maintenance

Scenario:

A conservation vent valve (PSV shown in Figure 3) was required to be removed during a preventive maintenance check. During the instance of removal, caustic storage tank was already filled-in with some level of NaOH solution.

The safety valve was installed for overpressure and vacuum protection of the NaOH solution storage tank. The governing cases of PSV by design are nitrogen blanketing control valve failure for overpressure scenario, and blocked discharge of inbreathing valve PCV_{in} for vacuum case.

Outage of the pressure safety valve for ex situ testing while operating the tank may impose lack of inbreathing and outbreathing protection for the tank. The removal of PSV may cause sustained tank damage in case of an overpressure or implosion scenario. In this paper, conservation vent valve is designated as pressure safety valve (PSV) for simplification and identification purpose.

Proposal:

During the preventive maintenance of the conservation vent valve, the breathing requirements of the tank should be fulfilled. To facilitate tank inbreathing and outbreathing conditions, continuous flow of nitrogen is introduced to the tank from PCV_{in} and released through the ¾" drain valve.

A hydraulic study was commissioned to determine the capability of the temporary arrangement to facilitate control valve failure and blocked discharge scenarios. From the PCV_{in} outlet, nitrogen supply is injected to the tank. For the blanketing outlet, nitrogen passes through antisiphon holes located at the top of the dip pipe and is ejected at the ¾" drain valve upstream of PCV_a.

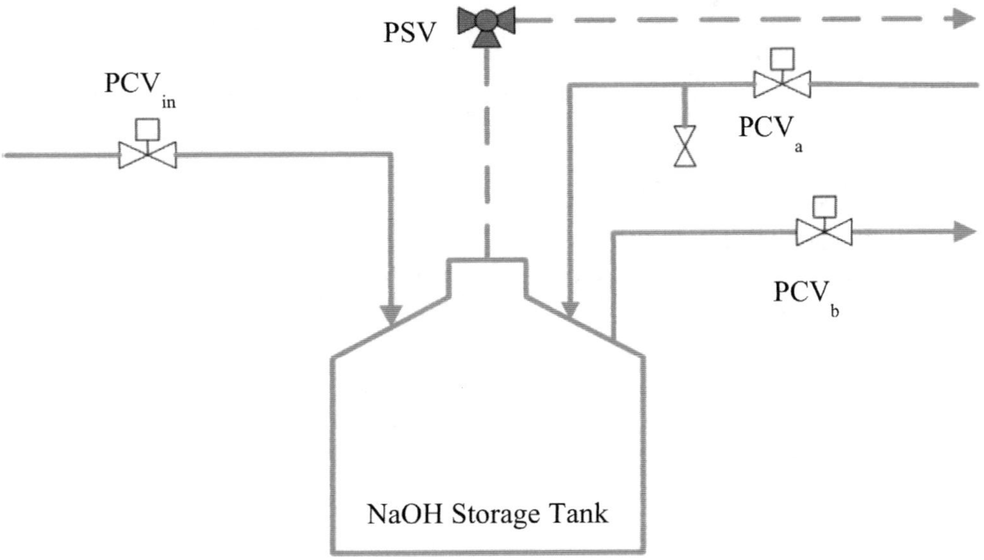

Figure 3. Conservation vent valve (PSV) will be taken for preventive maintenance.

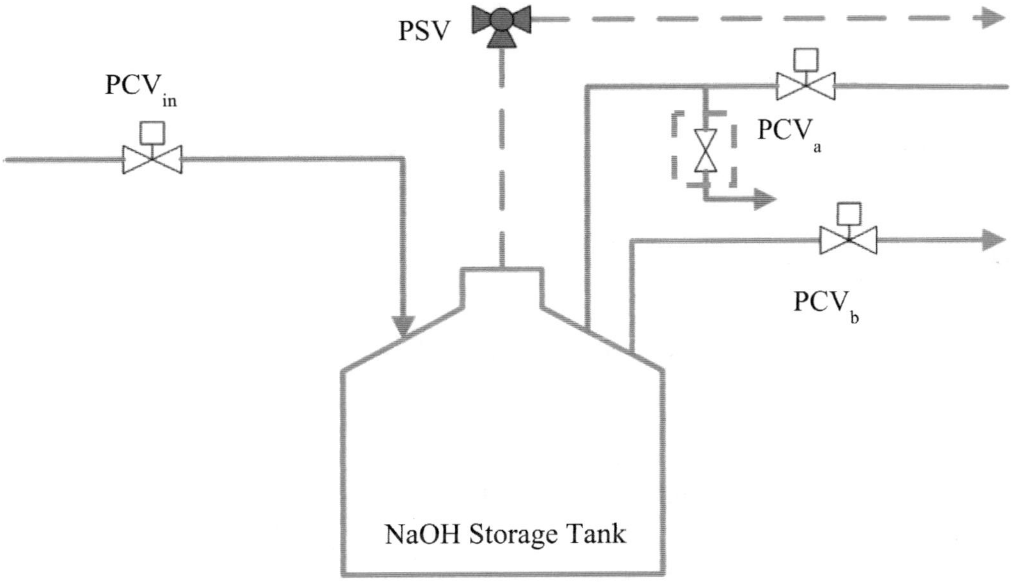

Figure 4. To facilitate tank inbreathing and outbreathing requirements, continuous flow of nitrogen will be introduced to the tank from PCV_{in} to the drain valve.

3.2.2 Study 2: Commissioning and Startup

Scenario:

During the commissioning of a vent system, Blower 1 sustained damage and required overhauling. The timeline of this event happened after the mechanical completion, part of the construction phase, bridging to commissioning phase.

Proposal:

A proposal was brought out to install bypass line across the centrifugal exhaust fan to run on a single Blower 2. If successful, the proposition will save the plant from a delayed startup.

However, the technical background of the proposal was challenged hence a confirmatory simulation was required to support the contradicting side. Since the nature of the inquiry was about fluid movement, hydraulic simulation was conducted to aid in decision-making. If not proven sufficient, the plant was required to wait until the new centrifugal exhaust fan / blower arrives.

By design, Blower 1 and Blower 2 convey combustible powder from the dryer to separator wherein wet air is also separated. Conveying air is further treated in the process.

Safety Consideration:

At low airflow, bigger particle sizes will settle in the piping vent systems. Relatively large dust particles introduce explosion risk when it adhered to surfaces and dryer internals. By design, there should be no settling of dust particles in the piping system.

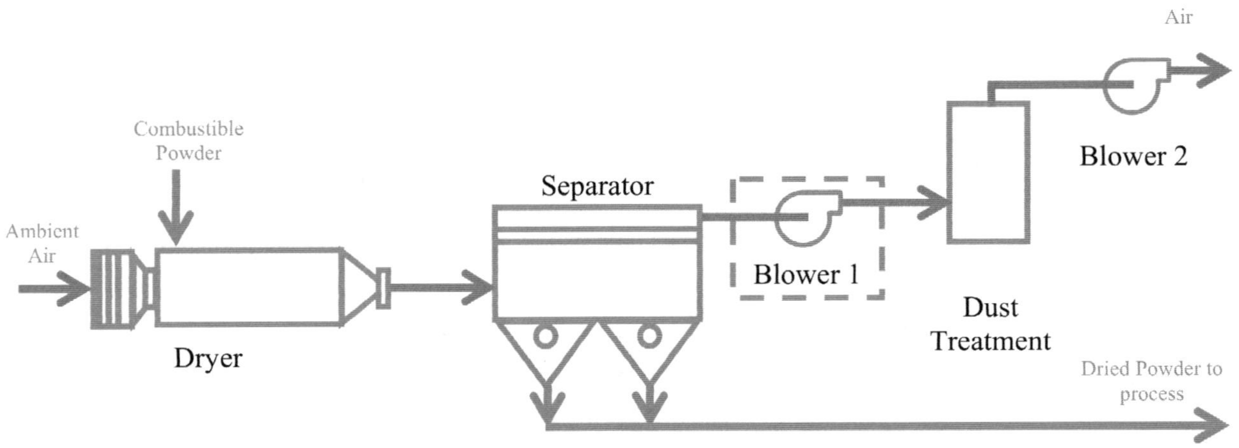

Figure 5. Schematic Diagram of Vent System by Design

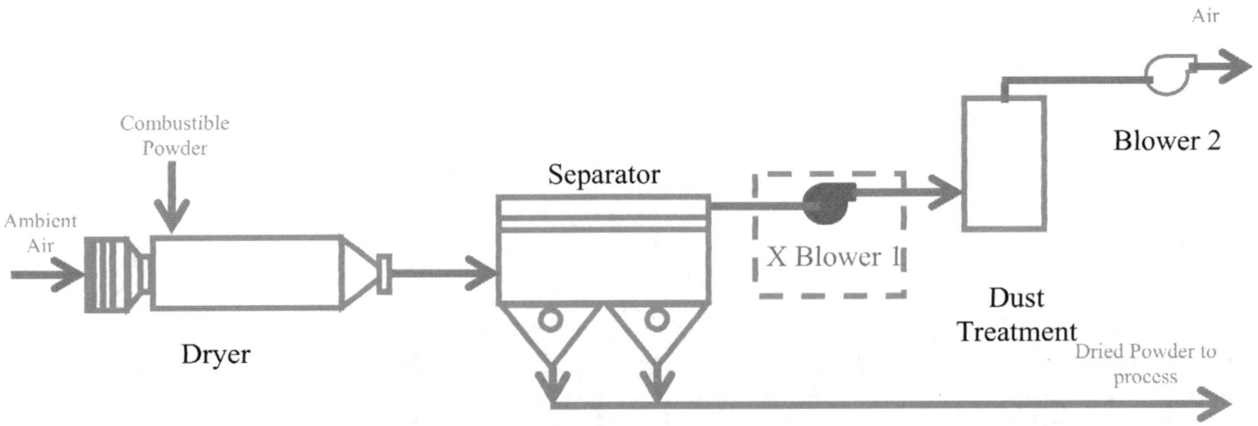

Figure 6. Proposed Schematic Diagram of Vent System wherein Blower 1 will be Bypassed

3.2.3 Study 3: Intermittent operation

Scenario:

Transfer pump delivers potassium hydroxide solution to reactors, solution make-up tank, feed tank, and storage tank. From the slightly pressurized and nitrogen blanketed storage tank, the KOH solution is delivered to the users via the pump, filters and main header. The recirculation line passes through a restriction orifice back to the KOH storage tank.

However, the system experienced excessive flow to one of the users when a batch was initially started leading to operational upset downstream. A probable flow and pressure imbalances in the system was the point of inquiry of the hydraulic simulation.

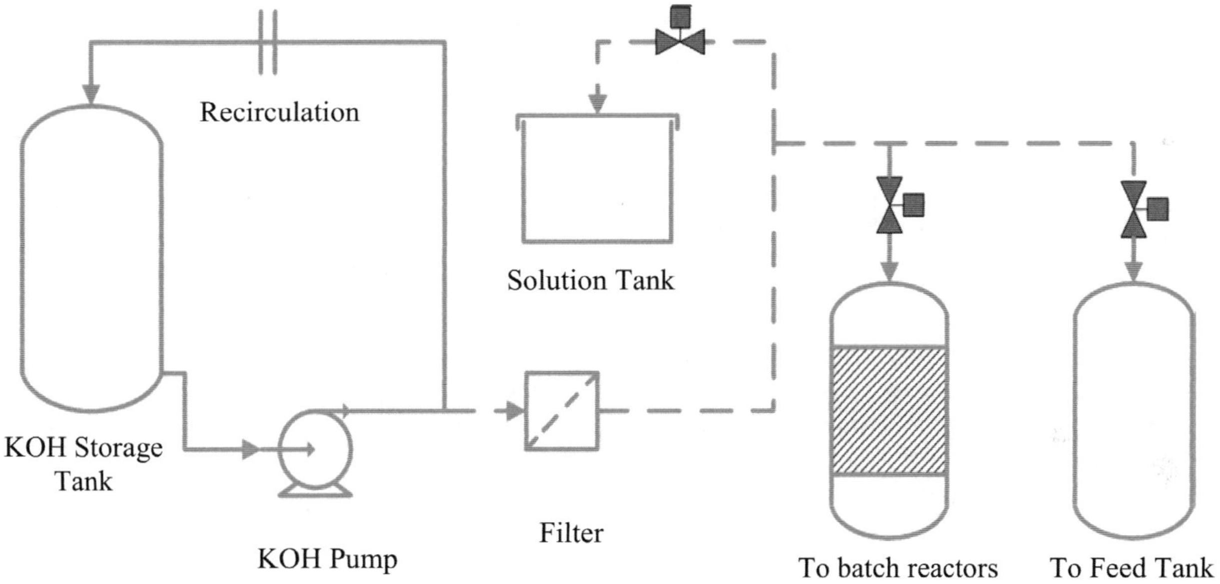

Figure 7. Schematic Diagram of KOH solution system wherein the header encounters blocked discharge when there is no demand during intermittent operation.

3.3 Problem Statements Summary

Table 1 below shows a brief summary of the background of the cases in varying plant conditions during maintenance, commissioning and startup, and batch operations including the respective problem statement or event drivers. The tandem of initiating cause (e.g., unavailability of vent system) happening at a specific operating phase of the plant (e.g., during plant initial startup) subjects the plant into an abnormal operation that requires acknowledgement to address the associated hazards.

Table 1. Problem Statement Summary

Description	Case Study 1	Case Study 2	Case Study 3
Plant Timeline / Status(A)	Online Maintenance	Commissioning and Startup	Normal Operation / Intermittent Operation
Initiating Cause(B)	Removal of conservation vent valve	Unavailability of damaged Blower 1	Closed shutoff valves to all users
Problem Statement(AxB)	Unavailability of PSV during ex situ test when the tank is in service.	Unavailability of vent system will prevent startup. Vent system is a critical path to plant initial startup.	Overpressure to KOH header due to blocked discharge, or no demand from users.
Related Asset / System	Caustic tank, inbreathing and outbreathing system	Vent system and Centrifugal Exhaust Fan 1	KOH header to users
Safety Consideration	PSV is used for tank overpressure and vacuum protection	At low airflow, bigger particle sizes will settle. Settled particles can disintegrate and initiate explosion.	The service is corrosive if exposed to personnel during relief through the PSV.
Potential Consequence	The removal of PSV may cause tank damage due to overpressure or implosion leading to potential financial impact and personnel injury.	Internal dust explosion may occur from accumulation of unstable powder due to low flow leading to potential fatality.	Overpressure in the piping may lead to release of corrosive chemical and upset to users hence potential personnel injury.

4 Hydraulic Calculations

4.1 Fulfilling Overpressure and Vacuum Requirements of NaOH Storage Tank during PSV Outage

4.1.1 Objective

Determine if ¾” drain gate valve downstream of PCV_a is enough for inbreathing and outbreathing requirement of NaOH storage tank during the temporary out of service of PSV.

4.1.2 Hydraulic Calculation Summary

Table 2 below shows the summary of the input and output of the hydraulic analysis. The hypothesis that was tested, whether a ¾” drain can be a temporary substitute for the venting requirements of the tank, was proven acceptable. The tradeoff to this arrangement is a possible dust contamination or moisture ingress into the caustic tank; hence, a quality issue may arise.

Table 2. Hydraulic Calculation Sheet – Case 1

Topic	**Fulfilling Overpressure and Vacuum Requirement of NaOH Storage Tank on PSV Outage**
Service	Nitrogen/Air Inbreathing and Outbreathing to/from NaOH Storage Tank
A. Objective	Determine if 3/4" drain gate valve downstream of PCV_a is enough for inbreathing and outbreathing requirement of NaOH Storage Tank during the temporary out of service of PSV.
B. Basis, Data and Assumptions	1. PSV Relief Calculation a. Overpressure Governing Case: Control valve failure (PCV_{in}) Maximum Relief Flowrate = approx. 100 kg/h Inlet Pressure = approx. 500 kPa_g b. Vacuum Governing Case: Blocked discharge (PCV_{in}) leading to vacuum Maximum Relief Flowrate = approx. 100 kg/h Inlet Pressure = atm 2. NaOH Storage Tank Nozzle Design Nozzle N3 Pump minimum flow nozzle has a dip tube with antisiphon holes. 3. Fluid Properties Air @ 25°C Nitrogen @ 25°C
C. Procedure	1. Determine if piping hydraulic profile including 3/4" gate valve as discharge has no significant pressure losses to be overcome for safe relieving and venting during overpressure and vacuum.
	2. Determine if equivalent area of antisiphon holes, will be sufficient as relief outlet/inlet.
D. Results and Conclusions	1. The maximum relief flow the piping can accommodate (based on the piping isometric) is about 200 kg/h during depressurization. This is greater than the maximum relief rate of both overpressure and vacuum scenario. Because of this, there is no significant piping and fittings pressure losses that can restrict safe relief venting using the drain downstream of PCV_a.

	2. Antisiphon at nozzle N3 has combined area equivalent to a 1 in. pipe. No significant impact on pressure losses brought about by antisiphon holes at dip tube that can restrict safe relief venting.
E. Recommendation	1. During this particular PSV outage, ensure 3/4" drain gate valve downstream of PCV_a is not blocked and fully open for safe relieving.
	2. Vent to safe and open area to avoid hazard of asphyxiation due to continued venting of nitrogen.
	3. Ensure antisiphon holes are not blocked.
	4. Reinstate PSV and piping as per P&ID immediately after calibration activity.

4.2 Testing the Capability of Vent System to Fulfill Duty while Running on a Single Blower

4.2.1 Objective

Determine if Blower 2 can safely handle the dust/air venting requirement when Blower 1 will be remove from the service.

4.2.2 Hydraulic Calculation Summary

Table 3. Hydraulic Calculation Sheet – Case 2

Table 3 below shows the summary of the input and output of the hydraulic simulation. The tested hypothesis, whether Blower 2 can permanently support the air movement requirement of the vent system, was not acceptable. The proposal may impose explosion hazard to the piping system causing damage to property and injury to personnel.

Topic	**Evaluating a Proposal to Remove Blower 1 from a series of two fans**
Service	Air, water vapor, and dust movement from dryer to dust treatment via Blower 1 and Blower 2
A. Objective	Determine if Blower 2 can safely handle the dust/air venting requirement when Blower 1 will be remove from the service.
B. Basis, Data and Assumptions	1. Fluid Properties Air, Water Vapor, combustible powder Normal flow @ 60°C 2. Inlet Pressure = atm 3. Elevation based on site data 4. Relatively large dust particle introduces explosion risk when it

	adheres to surfaces and dryer internals. By design there should be no settling of dust particles in the piping system.
C. Procedure	1. Determine the flowrate across the system when both Blower 1 and Blower 2 are running at normal operating condition. a. Set the inlet pressure = atm, at respective destination pressure, utilize the pump curves and piping systems.
	2. Determine the flowrate across the system when only Blower 2 is running. a. Set the inlet pressure = atm, utilize the pump curve and piping systems.
D. Results and Conclusions	1. When two blowers are running at normal operating conditions of: inlet pressure = atm to respective destination pressure the normal air flowrate is acceptable i.e., dust particles will not settle.
	2. However, when operating at a single Blower 2 (with bypassed Blower 1) at normal operating conditions of: inlet pressure = atm to respective destination pressure the normal air flowrate is not acceptable i.e., dust particles will not settle based on empirical data.
E. Recommendation	1. Neither Blower 1 nor Blower 2 can withstand total systems losses alone at normal condition. Two blowers in series must be in place to support safe movement of air, water vapor, and combustible dust.
	2. Centrifugal Exhaust Fan / Blower 1 shall be returned on service before initial startup of the vent system.

4.3 Preventing Overpressure in an Intermittent Process from KOH Header to a Batch Reactor

4.3.1 Objective

Determine the reason of buildup pressure in the KOH system. Check whether the design of recirculation line is adequate to prevent overpressure in the line.

4.3.2 Hydraulic Calculation Summary

Table 4. Hydraulic Calculation Sheet – Case 3

The following Table 4 shows the summary of the input and output of the hydraulic study. The restriction orifice located in the recirculation line back to the KOH tank was found underrated.

Topic	**Preventing KOH Header Overpressure in a Batch Process to Reactors and Makeup Tank**
Service	Transfer KOH solution from storage tank to users like batch reactors, KOH feed tank and solution tank via KOH pump.
A. Objective	Determine the reason of buildup pressure or flow imbalance in the KOH system. Check whether the design of recirculation line is adequate to prevent overpressure in the line.
B. Basis, Data and Assumptions	1. Fluid Properties KOH solution @ 25°C 2. Inlet Pressure = atm 3. Destination Pressure: KOH Feed Tank = atm Solution makeup tank = atm Storage tanks = slightly above atm Batch reactors = slightly above atm 4. Elevation based on site data
C. Procedure	1. Determine the required flowrate across the recirculation line during the normal operations. a. Set the inlet pressure, destination pressures accordingly, utilize the pump curves, piping systems, and restriction orifice size.
	2. Determine the adequacy of recirculation line to accommodate pressure buildup in the KOH header when there are no demands from the users. a. Set the inlet pressure, shutoff valves to users in closed position, utilize the pump curves, piping systems, and restriction orifice size.
D. Results and Conclusions	1. The required flowrate across the restriction orifice during normal operation is about 2 tons/h.
	2. When there is no demand from KOH header the minimum recirculation flowrate should be adjusted to approx. 5 tons/h.

E. Recommendation	Adjust the recirculation line to approx. 5 tons/h by resizing the orifice size to a bigger opening. A control valve in the recirculation line can be a long-term solution to prevent the pressure build up in the header.

5 Results and Discussions

5.1 Hydraulic Calculation Results

Table 5 exemplifies a summary of the simulation studies in each case study.

Table 5. Problem Statement and Recommendations Summary

Description	Study 1	Study 2	Study 3
Plant Lifecycle	Maintenance	Commissioning and Startup	Normal Operation / Intermittent Operation
Transient Operation	Non-routine Planned Operations	Non-routine Planned Operations	Abnormal or Unplanned Operations
Problem	Unavailability of conservation vent valve	Removal of Blower 1 from a series of two blowers	Design Inadequacy
Hydraulic Calculation Recommendation	Provide alternative overpressure / vacuum protection pathway. The use of ¾" drain valve can be used as relief outlet. **Accepted.**	Decline the proposition. The centrifugal fan must be returned on service before initial startup. **Rejected.**	Improve the process by increasing the size of restriction orifice. **Accepted.**
Type of Change	Temporary	Permanent	Permanent

5.2 Discussions: Transient Operations, Management of Change, and Hydraulic Studies

Restating the facts, almost half of the process incidents accounts for maintenance-related activities emerge during startups, shutdowns, maintenance, or abnormal operations (Kletz, 2009; Duguid, 2008). In a bigger picture, more than half of process safety incidents happen during non-routine and abnormal operations (Ostrowski & Keim, 2008; Duguid, 2008). Therefore, the statistics reiterates that a greater concern is posted by the concept of transition period operation alone.

By injecting process modification during the transient period, the risk is magnified bringing in the consequences mentioned in Table 1 Problem Statement Summary that ranges from potential operational upsets, financial impact, personnel injury, and fatality.

The recognition of hazard is critical in the course of action of the change review process. If by any chance that safety consideration was not credited in the decision-making, a catastrophic event may have happened. To demonstrate, Case Study 2 during commissioning and startup is the most critical of the three case studies because of the potential fatality consequence from an internal dust explosion in the dryer. The elements that can play around during this period include the naturally stressful commissioning environment to startup the plant, delayed commissioning schedule, inherently hazardous and combustible service, and continued operations in anon-standard equipment configuration beyond the design intention (Blower 1 removal). After the hazard was recognized, hydraulic calculation was promoted to determine whether removing Blower 1 was a safe proposal to be implemented. The result of the simulation rejected the idea (Table 5) and recommended to wait for a new centrifugal fan to arrive even in the expense of a further delayed schedule.

5.3 Effectiveness of Hydraulic Studies in Aid of Process Modifications

By performing hydraulic studies, hypothesis and suggestions were validated in solving or troubleshooting a particular process issue. This paper promotes the use of process engineering tools such as hydraulic simulation as a way to check proposals as part of management of change proceedings (MOC). The multidisciplinary team participating in the MOC must be able to recognize the underlying problem, hypothesize, and test the proposition by conducting simulation studies as deem required.

6 Conclusion

By injecting process modification during the transient period, the risk in the process is amplified bringing in the consequences that ranges from potential operational upsets, financial impact, personnel injury, and fatality.

As a resolution, hazard recognition is critical in the course of action of the change review process. If by any chance that safety consideration was not credited in the decision-making, a catastrophic event may have happened.

Moreover, conducting hydraulic simulations is a basic engineering capability that will able to support varying plant operating conditions. While this paper aims to demonstrate the efficacy of hydraulic calculations in solving various engineering questions, the study particularly focused on solving problem related to fluid transfer, a common issue across the lifecycle of the plant. The study exemplified the use of this engineering capability during transient operations from commissioning and startup, preventive maintenance, and abnormal operations from a batch process.

Finally, the criticality of conducting a thorough study before acting on a potential change cannot be overemphasized, especially during transient operation wherein an imbalance of process safety incidents reside during this period.

7 References

[1] Ostrowski, S. W. & Keim, K. K. (2010, June 23). Tame Your Transient Operations: Use a special method to identify and address potential hazards. *Chemical Processing*. Retrieved from https://www.chemicalprocessing.com/articles/2010/123/.

[2] Duguid, I. M. (1998). Analysis of past accidents in the process industries. *Loss Prevention Bulletin*. No. 142, No. 143, and No. 144. Institution of Chemical Engineers.

[3] Ostrowski, S. W. & Keim, K. K. (2008, October 28). A HAZOP methodology for transient operations. Paper Presented at the Mary Kay O'Connor Process Safety Symposium, College Station, Texas. Retrieved from http://pscmembers.tamu.edu/wp-content/uploads/15-ostrowski.pdf.

[4] Vaughen, B. K. (2020, March 14). Understanding and managing the risk during transient operations. *Process Safety Progress*. Retrieved from https://doi.org/10.1002/prs.12146.

[5] Kletz, T. (2009). *What went Wrong? - Case Histories of Process Plant Disasters and How They Could Have Been Avoided (5th Edition)*. Elsevier. Retrieved fromhttps://app.knovel.com/hotlink/toc/id:kpWWCHPPD1/what-went-wrong-case/what-went-wrong-case

[6] Killcross, M. (2012). *Chemical and Process Plant Commissioning Handbook: A Practical Guide to Plant System and Equipment Installation and Commissioning*. Oxford: Butterworth-Heinemann.

[7] CCPS. (1995). *Guidelines for Safe Process Operations and Maintenance*. Center for Chemical Process Safety, American Institute of Chemical Engineers, New York.

[8] CCPS. (1999). *Guidelines for Process Safety in Batch Reaction Systems*. Center for Chemical Process Safety, American Institute of Chemical Engineers, New York.

[9] Moran, S. (2019). *Applied Guide to Process and Plant Design* (2nd Edition). Elsevier.

[10] Pijush, P. K. & Cohen, I. M. (2008). *Fluid Mechanics* (4th Edition) - 16.1.3 Useful Thermodynamic Relations. (pp. 714-716). Elsevier.

[11] Nakayama, Y. (2018). *Introduction to Fluid Mechanics* (2nd Edition) - 13.3 Mach Number. (pp. 261). Elsevier.

[12] Shaughnessy Jr., E. J., Katz, I. M. & Schaffer, J. P. (2005). *Introduction to Fluid Mechanics* (OUP Edition) - 10.10.2 Incompressible Fluid and Incompressible Flow. (pp. 640). Oxford University Press.

586242 Occurrence of Black Powder in Gas Lines and Its Control

Abstract:

<u>Occurrence of Black Powder in Gas Lines and its Control</u>

Objective / Scope:

Black powder contamination in gas stream has been a matter of great nuisance. Its occurrence causes operational, maintenance, HSE, reliability & integrity issues and various undesirable consequences like product contamination, production loss, economic loss, and sustainability concerns (e.g. due to flaring), for both the operators and the downstream customers. The objective of this paper is to demonstrate the success of immediate / short-term measures taken to minimize its occurrence and subsequently its removal.

Methods, Procedures, Process:

Several technical reports and research works are found in the literature pertaining to black powder, still, need of more awareness has been felt by researchers / industry. In order to contribute for such cause, the present paper attempts to review the problem of occurrence of black powder and various strategies that are adopted to handle the associated issues, in general, and shares experience of minimizing the occurrence of black powder by revisiting the chemical injection dosages, frequent pigging operations and other methods at one of the oilfield areas in Kuwait, in particular classified as short-term / immediate measures.

Results, Observations, Conclusions:

Pipelines and connected systems are designed considering solid / liquid contaminations expected in the gas stream; but such provisions can become insufficient if occurrence of black powder starts, later, during operation. In view of involved complexity, problematic feature & uncertainties, operating companies may implement a black powder management system incorporating prevention, mitigation and removal strategies. Accordingly, when occurrence of black powder, along with liquid carry-over, in sour gas streams at one of the oilfield areas in Kuwait was observed, various short-term strategies in hand were reviewed such as adjusting chemical injection dosages, frequent pigging operations. This results in remarkable reduction in the occurrence of black powder and accelerating its removal from gas lines. On the basis of the results it is observed / concluded that the immediate short-term methods are effective in minimizing the occurrence of black powder in gas lines and its removal.

Novel / Additive Information:

The paper is expected to create awareness and enrich knowledge base regarding the occurrence of black powder in the middle-east region and its immediate / short-term handling solutions. This shall facilitate the Oil & Gas industry to address such problem of gas contamination and, in turn, handling its associated problems, like gas flaring, by attaining enhanced solutions based on such information.

(57ad) The Most Effective Approach to Updating PHAs: Revalidation Vs Redo (The Good, The Bad and the Ugly)

Authors:
He, E. Q., ACM Facility Safety Inc.

When a Process Hazard Analysis (PHA) needs to be updated due to a process change or regulator requirement, the two main industry options are to revalidate the most recent study or redo it from scratch.

When used correctly, PHA revalidation is an effective method to maximize the value of the PHA teamâ€™s time and keep the risk analysis consistent with previous teamsâ€™ work. However, challenges that are commonly encountered during a revalidation PHA study are:

- The tendency for revalidation sessions to either become a mind-numbing exercise in blind agreement with the previous PHA, or;
- So much needs to be changed due to insufficient description of hazards, safeguards, or risk ranking;
- Complication of including design or process modifications made since the last PHA, incidents or near-misses that occurred;
- The time and pressure caused by the expectation that a revalidation should be significantly faster than a baseline PHA, and the misconception that a good revalidation can â€œfixâ€• a bad PHA.

In many cases, revalidation can still be the best choice, as long as some effort is made before and during the session to account for these challenges and manage the expectation of the PHA team and management. However, without consideration of these challenges, many revalidation studies are set up for failure before they even begin.

On the other hand, redoing a PHA can be a more effective and efficient approach, by using the previous version of the study for reference as required. Despite the constant debate on whether PHA redo can achieve the consistent risk assessment from the original PHA findings, redo session can be proven to use the best of the PHA team time, if the following considerations are taken:

- The quality of the original PHA;
- The availability of MOC records, past incident records since the original PHA study;
- The consistency of PHA teamâ€™s expertise and experience.

This paper will discuss when a revalidation is most effective and when a redo should be considered, the key threats and opportunities for each method, and strategies to ensure the success of the PHA session whether a revalidation or redo is chosen.

586097 An Integrated Approach for Risk Management of Transportation of Hazmat: Use of Quality Function Deployment and Risk Assessment

Abstract:

An integrated approach for risk management of transportation of HAZMAT: use of quality function deployment and risk assessment

Guldana Zhigerbayeva[1], Ming Yang[1,2]

[1] Safety and Risk Engineering Group (SREG), Department of Chemical and Materials Engineering, School of Engineering, Nazarbayev University, 53 Kabanbay Batyr, Nur-Sultan, Republic of Kazakhstan 010000

[2] The Environment and Resource Efficiency Cluster (EREC), Nazarbayev University, 53 Kabanbay Batyr, Nur-Sultan, Republic of Kazakhstan 010000

Abstract

Transportation of hazardous materials (HAZMAT) is inevitable in the process industries. The statistics show a significant number of accidents has occurred during the transportation of HAZMAT. This makes risk management of HAZMAT transportation an important topic. The tree-based methods including fault-trees, event-trees and cause-consequence analysis, and Bayesian network, have been applied to risk management of HAZAMT transportation. However, there is limited work on the development of a systematic approach. The existing approaches fail to build up the linkages between the regulatory requirements and the safety measures development. The analysis of historical data from the past accidents' report databases would limit our focus on the specific incidents and their specific causes. Thus, we may overlook some essential elements in the risk management, including regulatory compliance, field expert opinions and suggestions. A systematic approach is needed to translate the regulatory requirements of HAZMAT transportation into specified safety measures (both technical and administrative) to support the risk management process. This study aims to first adapt the House of Quality (HoQ) to House of Safety (HoS) and proposes a new approach- Safety Function Deployment (SFD). The results of SFD will be used in a multi-criteria decision-support system to develop find an optimal route for HazMats transportation. The proposed approach will be demonstrated through a hypothetical transporation case in Kazakhstan.

585664 Risk Management-Resilience-Business Continuity: What, Why and How

Abstract:

Since the founding of MKOPSC 25 years ago, process safety and risk management made much progress with respect to material hazards characterization and risk methodology. Regarding the latter, fundamental issues such as parameter variability, interdependency of event probabilities and expert elicitation, all contributing to uncertainty, have been tackled. Human and organizational factors also have been studied and related with safety management system performance indicators. Static risk assessment in the design stage and of ready built plant is moving to dynamic operational risk assessment making use of data and virtual simulation-based knowledge.

However, even despite projected future major improvements in hazard identification techniques, leaving HAZOP and FMEA far behind, major hazards may escape our sight. Also, if all potential technical plant mishaps would be under control, possible external factors, e.g., human intervention, cyber-attacks, or weather will reduce certainty. Apart from not being identified, cost considerations will ignore low frequency major risks, yet, these can still materialize albeit rarely, hence coverage can never be complete.

Around 1995 in studies why certain companies are better surviving than others, the concept of "resilience" emerged. Also, the term "resilience engineering" appeared in the psychological context of organizational performance. "Engineering" meant here analyzing and improving culture. MKOPSC, though, initiated a decade ago resilience research to involve the whole plant operation. This kind of resilience based on risk assessment and indicators with early warning signals, staff prepared attitude, restrained flexible thinking, effective emergency response and recovery contingency measures, e.g., regarding supplies and customer needs, staff and funds, will lower and shorten post-disaster production output dip. Digitization and data, e.g., of various kinds of sensors and on performance indicators, will further support dynamic operational risk management, including abnormal situations, improving resilience and business continuity.

Cyber-Physical Systems Based on Plant Models, Proactive and Safe Control, Toward Full Automating Operation

Makoto Nakaya
Yokogawa Electric Co. Marketing Headquarters Innovation Center
2-9-32 Nakacho, Musashino-shi Tokyo 180-8750 Japan
Makoto.Nakaya@jp.yokogawa.com

Tadaaki Ando
Yokogawa Corporation of America
12530 West Airport Blvd, Sugar Land, TX 77478, USA
Tadaaki.andou@us.yokogawa.com
(Tadaaki.andou@jp.yokogawa.com)

Prepared for Presentation at
American Institute of Chemical Engineers
2020 Spring Meeting and 16th Global Congress on Process Safety
Houston, TX
March 30 – April 1, 2020

Cyber-Physical System Based on Plant Models, Proactive and Safe Control, Toward Full Automating Operation

Makoto Nakaya
Yokogawa Electric Co. Marketing Headquarters Innovation Center
2-9-32 Nakacho, Musashino-shi Tokyo 180-8750 Japan
Makoto.Nakaya@jp.yokogawa.com

Tadaaki Ando
Tadaaki.andou@jp.yokogawa.com

Keywords: Computing and Systems Engineering, Process Design & Development and Safe Design

Abstract

Due to the rapid development of computing, sensing and network technologies, advanced Internet of Things (IoT) technologies enable the integration of the physical and the cyber world into a cyber-physical system. Although IoT is merely a gateway into and out of the real world and the cyber world, data and information from the physical world are analyzed and learned in the cyber world, and in which a model that describes behaviors in the physical world is created. Utilizing the cyber-physical system is expected to create new value to the physical world. In the cyber-physical system, the one of the major goals must be to realize automating plant operation system.

We have developed "Mirror Plant" which is an online dynamic plant simulator. Mirror Plant receives sensor-measured data directly form the plant for model calibration. The model parameters inside Mirror Plant are automatically adjusted in the real time. As the result, Mirror Plant can perfectly simulate the dynamic plant behavior of the actual plant. By making computer run faster than reality, Mirror Plant can accurately predict the near future plant behavior. When Mirror Plant tells an alarm will occur in 30minutes, the operator can make an action to prevent predicted accident before something happened. Mirror Plant realizes a proactive plant operation yet to be seen in the conventional plant operation and contributes to safer plant operation. Mirror Plant is used in commercial LNG and chemical plants by operators who are directly involved in production and technical staffs who support production operation.

The development of IoT technology is accelerating the digital transformation in the manufacturing site, where the physical world and the cyber world will interact closely with each other. It is expected that big data utilization and AI technology will lead to more sophisticated plant model like Mirror Plant in the cyber world. By using the high-fidelity plant models, the unattended operation at night and fully automated operation of plants will be realized in future.

In this paper, we introduce Economic Model Predictive Control (EMPC), which uses nonlinear plant models created in the cyber world and delivers optimal and safe control system design. We applied EMPC to a chemical batch process with strong nonlinearity. We introduce result of EMPC shows high production with high yield and energy saving compared to the conventional PID control. If a plant model is available, anyone can use EMPC and specify any production requirements, such as maximum productivity ensuring environmental operation.

1 Introduction

The rapid development of computing and network technologies is driving forward the transformation to a smart plant using Internet of Things (IoT). With the arrival of explosionproof wireless sensors, a larger number of plants store their operation data and use them to predictively diagnose equipment failures, and today, the focus of using IoT is shifting to the prediction of product quality. There are many terms such as IoT, big data utilization, cyber physical system (CPS), and AI. Although these terms are used as synonyms, their relationships can be organized as shown in Figure 1. IoT is a gateway into and out of the real world and the cyber world, and data from the real world is analyzed and learned in the cyber world and the results are fed back to the real world in the form of operation amount and new value.

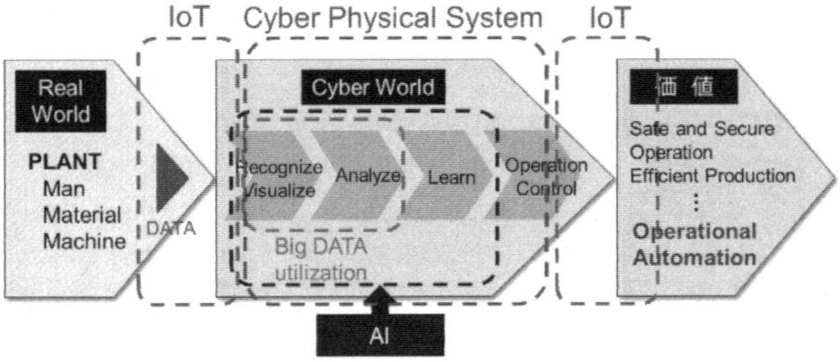

Figure 1. Relationship of IoT, Big DATA, CPS and AI

We have developed and released the Mirror Plant, which in the cyber world behaves in the same manner as a real plant. [1][2] The Mirror Plant is a dynamic simulator, which uses a physical model based on principles and rules and performs auto-tuning using the operation data supplied from the physical plant online. Plant operation is not determined by statistics and probability but determined by principles and rules. The Mirror Plant uses a physical model, since a model created by data provides good approximate accuracy only in the peripheral region where data exists.

The Mirror Plant that mirrors a real plant enables predicting plant operation in the near future by running the simulator in the cyber world faster than the real time. It can also provide a safe and secure operation environment by allowing operators to deal with alarms in advance that may occur in the future. It also allows for operators to search for the optimum operation conditions even during plant operation. The Mirror Plant, which faithfully models a plant, is widely used by operators who are directly involved in production and engineers who support the production operations. [3][4]

With the development of IoT, it is expected that a sophisticated plant model will be created in CPS as shown in Figure 1 using AI, Mirror Plant, and other technologies. For IoT, it is required to turn the results obtained in the cyber world into new value and deliver it to the real world through the Internet. Automated plant operation is one such value. There may be still a long way to go until we can use fully autonomous self-driving cars, but related technologies are being developed steadily. Meanwhile, automated plant operation requires various technical developments, such as creating a safe plant shutdown process without human intervention or measures to be taken by the operator for DCS in emergencies, as well as AI-based synthetic decision-making on operation. We are developing model predictive control using a sophisticated plant model generated by the Mirror Plant or other means in the plant control field. This paper introduces economic model predictive control (EMPC), which performs predictive control while performing an optimal computation without linearly approximating a non-linear plant model.

2 What is EMPC?

Most of the control used in plants is based on the feedback theory represented by PID control. Model predictive control (MPC) appeared in the 1980s. MPC predicts future behaviors using a mathematical plant model and calculates the operation amount while solving the optimum problem each time. Since MPC can perform advanced control for complex systems, it is used for a wide range of applications. MPC is also used in plants to increase stability against external turbulence or reduce the operation cost. EMPC described in this paper is MPC that belongs to the category of non-linear model predictive control.

Table1. Comparison between conventional MPC and EMPC

Ttem		Conventional MPC	EMPC
Prediction Model	Type	Linear	Non-linear
	Modeling	Created from operation data	Based on physical laws
	Application	only near operation point where model was created	a wide range
Optimaization	Objective Function	Minimizing the difference from target value of the prediced trajectory	Maximum yield and minimum energy consumption
	Structure	2 Layer-type (MPC+RTO)	1 Layer-type

2.1 Differences from Conventional Model Predictive Control

Table 1 summarizes the comparison between conventional MPC used in continuous plants and EMPC described herein. In model predictive control, the operation amount is obtained using a dynamic model of a plant to be controlled, so the quality of the model directly influences the control performance. In conventional MPC, the plant model is obtained by disconnecting the control loop at an operation point and conducting a step response test. Commercial MPC creates a linear model like "dead time primary delay approximation." In MPC using conventional linear models, high control performance cannot be expected in some applications such as production output adjustment, in which the plant is operated at an operation point quite different from the operation point where the model was created. Meanwhile, EMPC uses non-linear physical models based on physical and chemical laws directly for control calculations. The operating range of the model is wide and there is the advantage that significant changes in the state can also be controlled. However, calculations are very complex since a non-linear model is run, so the calculation time is significantly longer than that in conventional MPC.

In optimization calculation, conventional MPC describes the following parameters in its objective functions: difference between the trajectory predicted from the model (or reference frame) and the target value, difference in the operation amount from the previous operation, and difference between the steady-state values for the optimization target amounts and the predicted values. The steady-state values for the optimization target amounts are generally given from the outside; for example, from the real time optimizer (RTO). Therefore, conventional MPC performs the optimization control calculation in two layers (MPC+RTO). In EMPC, economic functions (EF) such as maximum product yield or minimum energy consumption are described in the objective function. Different from MPC, EMPC is a single-layer type, which obtains the operation amount while performing optimization calculation.

2.2 EMPC Alorithm

EMPC is represented by formulas (1) to (8) below. [5]

$$\min_u \int_{t0}^{tf} EF(x(t), y(t), u(t))dt \tag{1}$$

Subject to:

$$\dot{x}(t) = f(x(t), y(t), u(t)) \tag{2}$$

$$g(x(t), y(t), u(t)) = 0 \tag{3}$$

$$h(x(t), y(t), u(t)) \leq 0 \tag{4}$$

$$x(0) = x_0 \tag{5}$$

$$xl \leq x(t) \leq xh \tag{6}$$

$$yl \leq y(t) \leq yh \tag{7}$$

$$ul \leq u(t) \leq uh \tag{8}$$

Formula (1) is the objective function of EMPC as described above. EF stands for Economic Function. Formulas (2) and (3) are plant model formulas. When the model formula of the Mirror Plant where the model is updated online is used, control can be maintained even if the process changes over time. Formulas (4) to (8) are constraint conditional formulas. x represents the state amount, y the plant output, and u the operation amount.

Since EMPC calculation uses nonlinear programming (NLP), there are problems regarding the convergence of a solution, stability, and real-time property. Accordingly, attention must be paid to the influence of the scale and the number of parameters of the target plant model.

3 Application of EMPC

The following reports on the results of applying EMPC to a continuous process and a batch process. In particular, the following shows how the production is changed by the setting of EF, which is a feature of EMPC.

3.1 Application to CSTR

We applied EMPC to a continuously stirred tank reactor (CSTR) process. [6] Figure 2 shows the model formula of CSTR and the parameters.

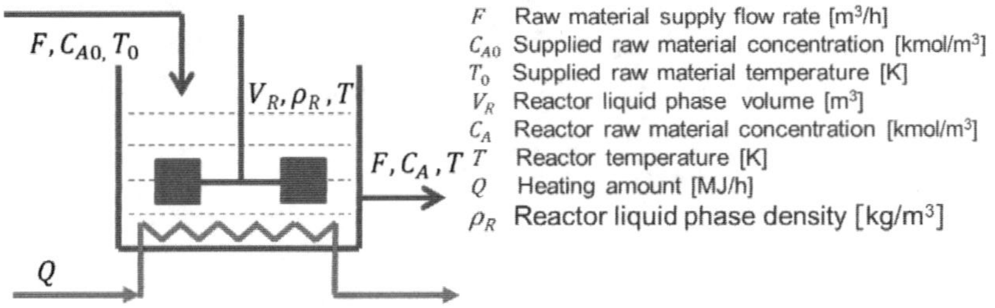

Plant model formula

$$\frac{dC_A}{dt} = \frac{F}{V_R}(C_{A0} - C_A) - k_0 e^{-E/RT} C_A^2$$

$$\frac{dT}{dt} = \frac{F}{V_R}(T_0 - T) - \frac{\Delta H k_0}{\rho_R C_p} e^{-E/RT} C_A^2 + \frac{Q}{\rho_R C_p V_R}$$

Figure 2. CSTR model

In this process, the density of the product generated in the reactor ($C_{A0} - C_A$) should be as high as possible, and heating amount Q to be input should be as small as possible in order to increase product yield. Accordingly, we set EF as shown in formula (9).

$$EF = -\alpha \cdot F(C_{A0} - C_A) + \beta \cdot Q \tag{9}$$

"α" and "β" are factors used to adjust the priority between productivity and energy saving. Figure 3 shows the results. The blue line indicates the case where priority is placed on productivity, and the red line indicates the case where priority is placed on energy saving. These shows what production can be selected by setting EF.

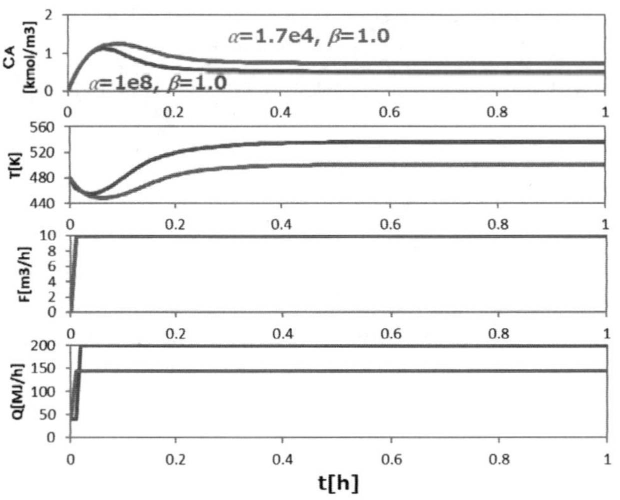

Figure 3. EMPC calculation results in CSTR

3.2 Application to a Batch Reaction Process

A batch process is used in many industries such as the fine chemical, pharmaceutical, and food industries that produce a variety of products in small quantities using the same equipment. The following describes an example of applying EMPC to a batch plant where the state keeps changing in contrast to a continuous process. [7] Figure 4 shows a batch reaction process considered in this paper.

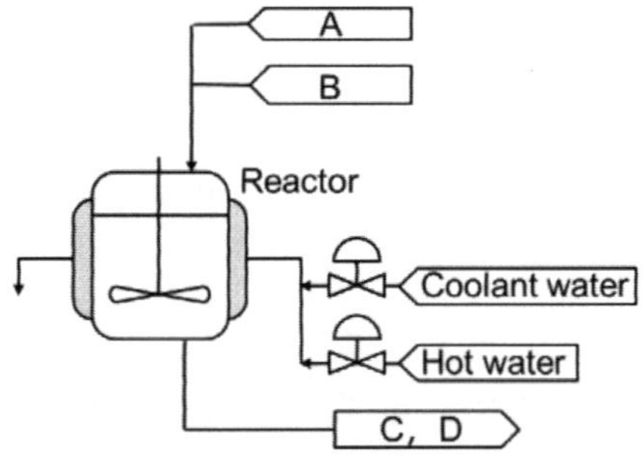

Figure 4. Batch reaction process

Raw materials A and B are input into the batch reactor, and after the reaction product C is output. A jacket is installed around the reactor to flow hot or cold water in order to regulate the temperature of the reactor. The concentration of components in the reactor is kept uniform by the agitator. Formula (10) represents a reversible reaction, and heat is generated by the forward reaction.

$$A + B \xleftrightarrow{k_1, k_2} C + D \tag{10}$$

$$C \xrightarrow{k_3} A + E \tag{11}$$

The reverse reaction in formula (10) reduces the production output of product C. There is another reaction that reduces the production output of product C, which is represented by formula (11).

The following shows model formulas that represent the balance between the materials and heat of this reaction process.

$$dC_A / dt = k_2 C_c C_D - k_1 C_A C_B + k_3 C_C^2 \quad (12)$$

$$dC_B / dt = k_2 C_c C_D - k_1 C_A C_B \quad (13)$$

$$dC_C / dt = k_1 C_A C_B - k_2 C_c C_D - k_3 C_C^2 \quad (14)$$

$$dC_D / dt = k_1 C_A C_B - k_2 C_c C_D \quad (15)$$

$$dT_r / dt = (Q_r + Q_j)/(C_r V_r C_{p_r}) \quad (16)$$

$$dT_j / dt = (T_j^{SP} - T_j)/\tau_j - Q_j /(C_j V_j C_{p_j}) \quad (17)$$

Where, C_A, C_B, C_C, and C_D represent the concentration of materials, Q_r the reaction heating amount, Q_j the heat exchange amount of the jacket, and C_{pr} and C_r are defined by formulas (20) and (21), respectively. T_j^{SP} represents the jacket entrance temperature and T_j the jacket temperature. The following shows model formulas that represent the heat quantity of the reactor and jacket.

$$Q_r = [k_1 C_A C_B * \Delta H_1 - k_2 C_C C_D * \Delta H_2 + k_3 C_C^2 * \Delta H_3] V \quad (18)$$

$$Q_j = UA(T_j - T_r) \quad (19)$$

$$C_{P_r} = (C_{P_A} * C_A + C_{P_B} * C_B + C_{P_C} * C_C + C_{P_D} * C_D) / C_r \quad (20)$$

$$C_r = C_A + C_B + C_c + C_D \quad (21)$$

EF in the batch reaction process was set as shown in formula (22) to maximize the yield.

Equation (22)

$$EF = -\alpha \cdot C_C + \beta \cdot \Delta u^2 \quad (22)$$

Figure 5 shows the results. PID control calculation results are also shown in Figure 5 for comparison with EMPC.

In PID control, the reactor temperature is kept at 50°C to prevent the influence of the reaction in formula (11). Meanwhile, EMPC sharply increases the temperature in the reactor at the start of reaction in order to give priority to the reaction speed. The temperature of the reactor is then reduced at the point when product C achieved the maximum production output in order to suppress the reverse reaction in formula (10) and the forward reaction in formula (11). As a result, product C can be produced in a shorter time compared to the PID control. The temperature of the fluid flowing in the jacket is also shown in Figure 5. Obviously, it is necessary to provide heating and cooling capabilities that quickly raise and reduce the temperature of the reactor in order to maximize the yield in the operation. EMPC uses the plant model at the plant design phase to estimate the equipment capabilities that are required to achieve target operation.

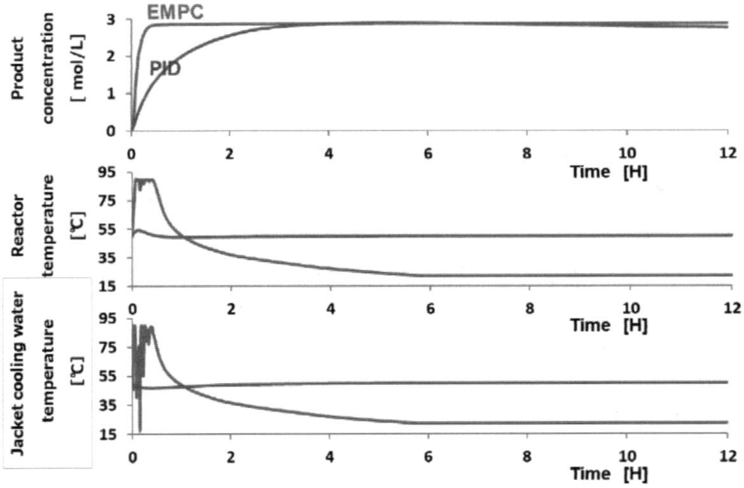

Figure 5. EMPC calculation results in batch process

4 Conclusion

The development of IoT technology is accelerating the transformation to a smart plant, where the physical world and the cyber world will interact closely with each other and efficient production and automation will be achieved. It is expected that the big data utilization and AI technology evolution will create a more sophisticated plant model in the cyber world, which will then be distributed to the real world. Use of this model opened an era in which our dreams of unmanned night operation and fully automated operation of a plant will come true.

In production at conventional plants, products were produced by controlling environmental variables in plants such as the temperature, pressure, and flow rate. In EMPC described herein, if a plant model is available, anyone can use it and specify any production requirements, such as productivity maximization and environmentally friendly operation.

With the current computing power, it is not possible to apply EMPC to the model of the whole plant and solve the overall maximization problem. Accordingly, the application of EMPC is limited to partial processes and devices. However, EMPC can also be used for designing and selecting plant equipment and devices for the optimal operation.

By leveraging the evolution of IoT which integrates the cyber world and the physical world at a high level, Yokogawa will develop and improve plant modeling and computing technologies.

5 References

[1] Makoto Nakaya, "MIRROR PLANT for Innovating Plant Operation," Yokogawa Technical Report English Edition, Vol. 56, No. 1, 2013, pp. 7–10

[2] Emiko Hatsugai, Naoto Takano, et al., "HMI of MIRROR PLANT for Facilitating Forecast of Plant Operation," Yokogawa Technical Report English Edition, Vol. 56, No. 1, 2013, pp.15–18

[3] Makoto Nakaya, Akira Yamada, et al., "Optimum Operation Supported by an On-line Simulator to Achieve Status Prediction," Instrumentation, Vol. 57, No. 6, 2014, pp.7–11

[4] K. Kato and T. Ito, "Active Use of Dynamic Plant Simulator in LNG Terminal Construction," speech draft for the 59th Automatic Control United Lecture Meeting, 2014, pp. 196–200

[5] M. Eills, H. Durand, et al., "A tutorial review of economic model predictive control methods," Journal of Process Control, Vol. 24, 2014, pp. 1156–1178

[6] X. Li, L. Zhang, et al., "Application of Economic MPC to a CSTR Process," Proceedings of 2016 IEEE Advanced Information Management, Communicates, Electronic and Automation Control Conference, 2016, pp. 685–690

[7] L. Zhang, M. Nakaya, et al., "Application of Cyber Physical Systems to Chemical Plants: Economic MPC in Batch Processes," Proceedings of the 13th IEEE Conference on Industrial Electronics and Applications, 2018, pp. 1265–1270

586511 Deployment of Process Safety Management & Risk Management for Integrated Steel Manufacturing Unit

Abstract:

This paper describes the challenges & leanings & progress made with regards to implementation of process safety management in a integrated steel manufacturing facility. The major Process risks involved are handling hot metal & toxic gases which had a risk of explosion, fire etc. A structured implementation was carried out through Center of Excellence exercise. This involved coaching leadership team , training line function on Process safety management. Apart from PSM culture development,Process Hazards were analysed using tools such as HAZOP, Consequence analysis, Bow tie , LOPA etc.Process risk was identified & communicated to management with necessary risk reduction measures. The Center of Excellence Model will cover the following activities.

Identification of High Hazard Operations

This activity focuses on the identifying process for which the system needs to be deployed, this is based on a pre-developed checklist focusing on Hazardous chemical properties such as flammability , toxicity etc.

Developing Process Design Basis

- Identification of Standard operating conditions & Safe Operating limits analyse their deviation & restoring the same to original condition.
- Identification of process safety critical equipment based on failure mechanism & developing Maintenance plan for critical equipment.

Process Hazard Analysis

- Appreciation of all process risks (Identification, Prioritization, Mitigation) using Process Hazard Analysis techniques such s *HAZOP for Batch & Continuous Processes*.
- Carrying out *consequence analysis* for toxic gas release, fires & *metal & water interaction explosions*.
- Developing *Bow ties* for the Major Accident Hazard Scenarios & identifying barriers.
- Determining the target mitigated risk through Layer of Protection Analysis.

Developing Operating Procedures & Emergency Response Plan

- Operating procedures for startup, shutdown etc were developed & operating personnel were trained.
- *Process Emergencies* were identified & response plans were developed. Control room operators were trained on handling process emergencies through *table top exercises*.

Building Process Safety Culture

PSM culture development activities such as PSM line walks focused on asset integrity, Process containment. Process safety performance Indicators were developed for monitoring the performance with a robust review mechanism.
This deployment has provided proven results in terms of achieving good operational control in the unit along with reduction in the Process Incidents.

Pressure Relief System and Safety Instrumented Systems - Common Weaknesses in Field Implementation

William Bridges
President
Process Improvement Institute, Inc.
1321 Waterside Lane,
Knoxville, TN 37922 USA
wbridges@piii.com

Stephen Bridges
Senior Process Engineer
Process Improvement Institute, Inc.
1321 Waterside Lane,
Knoxville, TN 37922 USA
wbridges@piii.com

Rashed Al-Zahrani
Manager, Process Safety
Jubail United Petrochemical Company (UNITED), a SABIC affiliate
ZahraniRA2@united.sabic.com

Prepared for Presentation at
American Institute of Chemical Engineers
2020 Spring Meeting and 16th Global Congress on Process Safety
Houston, TX
March 29 – April 2, 2019

Pressure Relief System and Safety Instrumented Systems - Common Weaknesses in Field Implementation

Presenters:

William Bridges
Process Improvement Institute, Inc. (PII)

Stephen Bridges
Process Improvement Institute, Inc. (PII)

Keywords: Human Factors, Human Error, Independent Protection Layers, IPLs, Integrity, Asset Integrity, Risk Based Process Safety, RBPS, PSV, Pressure Relief, Pressure safety Valve, Relief Valve, Safety Instrumented System, SIS, SIF, SIL, SIL Verification

Abstract

Pressure relief and safety instrumented systems are often the last line of defense before the occurrence of a catastrophic event. Their design, sizing, and reliability are hence of critical importance for the chemical process industries. Either are touted as capable of achieving a PFD = 0.01 or even 0.001. But, common weaknesses in implementation prevent this from being achieved. These include:

- Failing to take the PFD into account in the design of relief systems that require multiple relief valves to open to achieve the target capacity for the limiting case. Such systems should be designed as MooN instead of NooN.
- Failing to stagger maintenance for multiple channel systems (for multiple PSVs and for multiple sensors or multiple final element SIF). If maintenance or checks or calibrations are performed on the same day by the same person, then redundancy is lost as mistakes will be repeated on the other devices/channels.
- High error rate for control of block valves for relief valves or SIS root valves on taps and impulse lines
- Failure to account for the human error probability in the PFD estimates, resulting in expecting PFDs of 0.001 when only 0.01 is achievable long term.

Data, including from UNITED (a SABIC affiliate), are shared in this paper.

The major deficiencies found in implementation of PSVs an SIS stem from lack of understanding by design engineers of probability of failure on demand (PFD) and stems also from failure to understand and control human error probability in the inspection, test, and preventive maintenance (ITPM) programs and their practices in the field. This paper explains both issues and provides an approach to address these issues.

1 Design Deficiency in Relief Systems – PFD is not expressly considered in the design of relief systems

One major design deficiency seen across about 100 sites around the world is that the PFD of the relief system was not taken into account when designing the relief. Design engineers focus on sizing issues and setting issues such as:

- nature of the material being relieved (phase of the material, but also is it prone to cause plugging or corrosion)
- back-pressure issues in the tail pipes and discharge line/systems
- throughput necessary (and phase of the material, whether gas, vapor, two-phase, or liquid)
- design pressure (including allowable accumulation pressures)

However, very few relief system designers are trained in, or take into consideration, the PFD of the relief system. Because of this, designers specify multiple relief devices that all must open to relieve the throughput for the limiting design case. Instead of specifying a design that requires installed redundancy and instead of accounting for common cause error, they specify barely enough relief devices to me the limiting case throughput demand.

Take the case that a large petrochemical unit is required to relieve a large through of gas on the suction side of a compressor, if the compressor trips off or fails off; and further assume this release will be at relatively low pressure of about 1 atmosphere gauge pressure. For simplicity assume that the designers choosing 10 equally sized PSVs what when summed together will relieve the throughput for the limiting case (10 PSVs will be required to relieve the maximum throughput conceivable). So, if any One PSV does fails to open, then the design case throughput will not be provided; all 10 must open. If the designers do Not take into account the fault tree PFD calculation for the combined failures of the similar PSVs, then the PFD required for process safety of the relief system (all relief devices taken together as a relief system) will be too high, and the system will be under protected. This issue is summarized below:

- Target PFD for the relief system = 0.01
- PFD_1 of each relieve device (PSV) = about 0.01
- 10 PSVs required for the target throughput of the limiting case scenario, meaning if PSV_1 fails to open then we are unprotected from then limiting case, or if PSV_2 fails to open we are unprotected from the limiting case, or etc. for all 10 PSV_N. The overall system PFD is therefore calculated using fault tree as OR gates, which can be approximately reasonably well by summing the discreet PFD of each of the 10 PSVs.

- Therefore, the PFD_{SYS} of the relief system if all 10 PSVs (10oo10) must open for the limiting design case is given approximately as = $\sum$ (PFD_1 of single device) + β* PFD_1 = 10*(0.01) + 0.1*0.01= 0.101 (target PFD Not met)

  Where β is the fraction of failures due to common causes in the design of similar PSVs; this factor for PSVs is approximated here as β = 0.1 (10% fractional failure rate of a single device)

In this case, the set of NooN relief devices will be a factor of 10 less effective than the target PFD of the system (the PFD of the entire relief system is 10 times higher than the target PFD).

If the designers understand the how to estimate the PFD for a multiple device system, then they will instead include extra PSVs in the design, and achieve an MooN design, when M is the number of devices that must open to meet the capacity requirement and N is the total number of PSVs provided in the design. For such designs, then PFD of the system includes OR logic and AND logic, as the extra PSVs (those beyond what is minimally required for the throughput capacity alone) provide standby redundancy against the failure of other PSVs in the relief system. For this case, the PFD_{SYS} of the relief system of the N PSVs given MooN must open to reach the full throughput capacity of the limiting case can be approximated as:

$$PFD_{SYS} = (N!/((N-M+1)!*(M-1)!) * (2PFD_{1oo1})^{N-M+1})/(N-M+2) + \beta * PFD_{1oo1}$$

If MooN is 10oo11 (one additional PSV added to the design, and all 11 PSVs are designed to be online all of the time), then:

PFD_{SYS} (for 10oo11 case) = 0.0083 (target PFD met)

For this case the target PSV was met by adding one extra PSV and the designers simply need to account for the 11th PSV in the overall design of the hydraulics of the relief system.

2 Deficiency in the planning of ITPM activities – dependent human errors are not considered in the planning of tests or checks or PM of similar components in a system consisting of multiple key devices

Another major design deficiency seen across nearly ALL sites (hundreds) around the world is that the PFD of the relief or SIS system failed to account for dependent human errors when multiple similar devices are key to the PFD of the system. This is not a design error, per se, though designs can helped compensate for such errors. Instead this error in inherent in the programming / planning / scheduling of the ITPM activities related to a multiple channel system, as when there is more than one final element, such as multiple PSV or multiple safety block valve (XV), or more than one sensor channel in an SIS.

Probability of repeat errors (Coupled errors) – For many maintenance tasks, making a repeat error or "common cause error" or "dependent error" can lead to failures of all backup systems. Such errors have led to major process safety accidents as well as MANY airplane crashes.

Coupling represents the probability of repeating an error (or repeating success) on a second identical task, given that an error was made on the first task. The increased probability of failure on subsequent tasks given that an error has already been made is known as dependence. The list below provides some starting point guidance on values to use:

- 1/20 to1/90 – if the same tasks are separated in time and if visual cues are not present to re-enforce the mistake path. *This error rate assumes a baseline error rate of 1/100 with excellent human factors. If the baseline error is higher, then this rate will increase as well.*
- 1/2 – if the same two tasks are performed back-to-back but the worker is not in line of sight of the other component/device he/she already worked on, and if a mistake is made on the first step of two. *This error rate assumes a baseline error of 1/100 with excellent human factors. If there the baseline error is higher, then this rate will increase as well.*
- 8/10 to 10/10 – if the same two or three tasks are performed back-to-back and a strong visual cue is present (if you can clearly see the first devices you worked on), and assuming a mistake is made on the first of the two or more activities.
- Two or more people become the same as one person (with respect to counting of errors from the group), if people are working together for more than three days; this is due to the trust that can rapidly build. So, double-checking by someone you regularly work with is not a safeguard.

These factors are based on the relationships provided in NUREG-1278[1] and the related definitions of weak and strong coupling provided in the training course by Swain (1993) on the same topic [2], as shown here in Table 1. The following relationship is for errors of omission, such as failing to reopen a root valve or failing to return a safety instrumented function (SIF) to operation, after bypassing the SIF. The qualitative values in Table 1 are based jointly on Swain[1,2] (1983, 1993) and Gertman (SPAR-H, 2005 which is NUREG/CR-6883) [3].

One can readily conclude that staggering of maintenance tasks for different channels of the same SIF or for related SIFs will greatly reduce the level of dependent errors. Unfortunately, most sites PII visits do not stagger the inspection, test, or calibration of redundant channels of the same SIF or of similar SIFs; the reason they cite is the cost of staggering the staff. While there is a perceived short-term higher cost, the answer may be different when lifecycle costs are analyzed.

Simple Rule: Staggering of maintenance can prevent significant number of human errors in redundant channels. In fact, the US Federal Aviation Administration (FAA) requires staggering of maintenance for aircraft with multiple engines or multiple control systems (i.e., hydraulics) (*FAA Advisory Circular 120-42A*, as part of Extended Operations (ETOPS) approval).[4] (ETOPS is Extended-range Twin-engine Operational Performance Standards, a rule which permits twin engine aircraft to fly routes which, at some point, are more than 60 minutes flying time away from the nearest airport suitable for emergency landing.)

Table 1: Guideline for Assessing Dependence for a within-SIF Set of Identical Tasks (based partially on SPAR-H, 2005 [3] and partially on field observations by PII)

Level of Dependence	Same Person	Actions Close in time	Same Visual Frame of Reference (can see end point of prior task)	Worker Required to Write Something for Each Component
Zero (ZD)	No; the similar tasks are performed by different person/group	Either yes or no	Either yes or no	Either yes or no
Zero (ZD)	Yes	No; separated by several days	Either yes or no	Either yes or no
Low (LD)	Yes	Low; the similar tasks are performed on sequential days	No	Yes
Moderate (MD)	Yes	Moderate; the similar tasks are performed more than 4 hours apart	No	No
High (HD)	Yes	Yes; the similar tasks are performed within 2 hours	No	No
Complete (CD)	Yes	Yes; the similar tasks are performed within 2 hours	Yes	Either yes or no

The level of dependency is determined from Table 1 by assessing whether the same person is doing the tasks, the proximity of the actions in time, the proximity of the actions in space (same visual frame of reference), and whether the work is required to make a record for each component:

1. Read down the "Same Person" column and find the applicable row(s), then
2. Read down the "Actions Close in Time" column and find the applicable row,
3. Then check the two columns on the right for that row.
4. The Level of Dependence is shown in the left-most column for the applicable row.

Table 1 has two rows for Zero Dependency (ZD) because ZD can be achieved two different ways:

- Tasks done by different persons or groups (staggered people), or
- Tasks done by same persons or groups, but tasks are done several days apart (staggered times).

Once the level of dependence is known, the probability of either repeat success or repeating errors on identical tasks can be estimated. For these probabilities, we use Table 2, which is a re- typing of Table 20-17 from NUREG-1278[1] (and the similar table in SPAR-H [Gertman, 2005] [3]).

Table 2. Equations for Conditional Probabilities of Human Success or Failure on Task N, given probability of Success (x) or Failure (X) on Task N-1, for Different Levels of Dependence *Courtesy Process Improvement Institute, Inc., All Rights Reserved*

Level of Dependence	Repeating Success Equations (but shown as error probability)	Repeating Failure Equations
Zero (ZD)	$P_{Success@N} = x$	$P_{Failures@N} = X$
Low (LD)	$P_{Success@N} = (1+19x)/20$	$P_{Failures@N} = (1+19X)/20$
Moderate (MD)	$P_{Success@N} = (1+6x)/7$	$P_{Failures@N} = (1+6X)/7$
High (HD)	$P_{Success@N} = (1+x)/2$	$P_{Failures@N} = (1+X)/2$
Complete (CD)	$P_{Success@N} = 1.0$	$P_{Failures@N} = 1.0$

How to Avoid these Repeat Errors (dependent human error) – The straightforward solution is to follow the rules implemented by the US FAA for aircraft maintenance. This requires staggering the maintenance for each similar device in a set of devices that comprise the IPL.

- So, if there are 11 PSVs for the earlier case where 10oo11 PSVs must open for the design case, then set up the ITPM program such that either (1) there is 3 days between the servicing of each PSV in the set of 11 PSVs (if serviced by the same crew) or (2) have a different crew do the blocking, removing, checking, re-installing of each PSV that is checked on the same day.

- Similarly, if there are 3 instrument channels on a SIL 2 SIF, then either (1) ensure there are 3 days between the servicing of each channel in the SIF if performed by the same technician or (2) have a different technician do the testing, calibration, or checking of each channel that is checked on the same day.

To implement staggering, the site will need to track the ID number of each worker assigned to inspect or test or maintain devices that make up an IPL so that (1) staggering can be ensured up front and (2) staggering can be verified by auditing.

For more details on the major effect dependent errors can have on SIFs, read the earlier papers on the topic [5,6].

One means to improve the reliability and independence of the instrumented system is to use a smart sensor/transmitter for the LSH which will detect null movement of the sensor reading, indicating the valve is closed on the tap is plugged. Another possibility is to implement a limit switch (or captive key system) on the root valve. There is a probability that these safeguards against human error will also fail or be bypassed by the staff, but assuming the probability of that failure is the same as other human errors for this example, 0.02, then the systemic human error drops to about zero as the probability of leaving the root valve closed is now ANDed with the probability of smart sensor/transmitter or limit switch failing, [6].

3 Failing to Account for the Human Error Inherent in the positioning of Root Valves (block valves)

Another major deficiency in high integrity systems is failing to account for the relative probability of human error for such simple tasks like re-opening the block valves on either side of a PSV before the PSV is returned from servicing or similarly failing to re-open the root valve between the process and sensor of SIFs after servicing/checking. One common approach that used by industry is require a lock, or car seal (put in place by a human) on the block valve or root valve, and then administratively ensure the valve is indeed open by an checking the valve position and the placement of the Open Tag using someone other than the person who changed the position of the valve back to open when the maintenance activity was completed. The most common term for such administrative systems is a Car Sealed Open (CSO) or Locked Open (LO) or Chain Locked Open (CLO) system.

The problem observed with such systems, is when the checker is from the same site /local organization, then checking is not truly independent. During the writing of *Guidelines for Initiating Events and Independent Protection Layers*, 2015, CCPS/AICHE [7], the committee members stated that checkers from a site would report that all CSO or LO valves were found in the as-desired position (typically Open) but when an independent checker (from outside of the site) was used to do an unannounced check, they found 4/100 block valves in the wrong position. So, for a PSV, the impact on the system PFD is large:

- Target PFD for the relief system = 0.01
- PFD_1 of each relieve device (PSV) = about 0.01
- Probability of leaving at least one block valve upstream or downstream of a PSV closed is about: P_{HUM_ERR} = about 0.04
- Therefore, the PFD_{SYS} of each relief system with a block valve upstream or downstream is approximately = $PFD_1 + P_{HUM\_ERR} = 0.01 + 0.04 = 0.05$

In this case, the relief system for even a single PSV with block valves will be a factor of 5 less effective than the target PFD of the system (the PFD of the system is 5 times higher than the target PFD).

SIL 2 and SIL 3 SIFs are impacted to the same extent.

How to Avoid these Increases in the Overall PFD of the IPL – The straightforward solution is to not have block valves upstream or downstream of PSVs and not have root valves for SIL 2 or SIL 3 SIFs. But this solution is not always possible due to shutdown schedules for the plant (since if there are no isolation valves, maintenance can only be done when the process line is shutdown).

Another solution is having a truly independent check by an outside auditor about once per year, to find any valves in the incorrect position and to continually encourage accurate internal auditing of CSO, CSC, LO, LC block valves and any SIF root valve. This solution will cost about 1 to 2 staff-days of labor per 100 PSVs or 100 SIFs.

CASE STUDY on Third Party Inspection of CSO Valves for 448 PSV at UNITED (a SABIC Affiliate) in Jubail, Saudi Arabia:

PII was contracted by UNITED to check enough PSVs to determine the actual error rate at the site of PSV block valves being in the wrong position (and also CSO tags missing, though the tag does not affect the functioning of the PSV, obviously, although a missing tag implies that it is more likely in the future for the). There are approximately 2000 PSVs at the site and about 1423 in the 4 units that PII inspected. In these 4 units, 890 of these PSVs were inspected, but only 448 of the PSVs had a block-valve upstream, downstream, or both. (Since the likely error rate was expected to be between 0.005 to 0.05, PII chose to check about 500 PSVs with block valves.) One process safety engineer with 10+ years' experience was assigned to perform this check. Of the 448 PSVs (with block valves) that were checked, zero (0) were found to have a block valve upstream or downstream or both closed instead of open, without there being a valid reason (such as out-of-service for a valid reason); one (1) was found in the wrong position but there was a maintenance work-order for the related vessel. In addition, of the 840 block valves checked (counting both upstream and downstream), 79 were missing car seals, but the valve was in the proper position. The data from the inspection is summarized on the next page.

In this case, the positioning of the block valves were excellent, meaning that the PFD of the relief system is essentially the same as the PFD of the PSV alone, so the human error rate of mispositioning of the block valves are low enough to not effect the estimated PFD of the relief system at this site. But roughly 9% of the car seals were missing (mostly in one unit) which needs to be corrected. The site 50% converted to a more robust car seal mechanism, using uniquely printed metal seals, with plastic encases metal cables. They believe this system will greatly reduce the number of missing seals.

One last caveat on the data reported above: The inspection by a third party (PII) of the car seals was announced in October 2019, but the actual inspection by PII slipped until February 2020. It is possible, therefore, that the staff in the units took extra care in the 4 months to ensure the block valves were in the proper position. But there is no way to know for sure. But if special care was taken, then why were so many car seals missing?

RESULTS of THIRD PARTY INSPECTION of CSO Block Valves for PSVs at UNITED

			Upstream						Downstream					
Unit	PSV Total in Unit	PSV Total Sample	# BVs	# BVs wrong	# CSs on BV	# CS Missing	Fraction (Probability) BVs wrong	Fraction (Probability) Missing CS	# BVs	# BVs wrong	# CSs on BV	# CS Missing	Fraction (Probability) BVs wrong	Fraction (Probability) Missing CS
A	534	363	238	0	237	1	0	0.0042	154	0	153	1	0	0.007
B	65	49	35	0	35	0	0	0.0000	22	0	22	0	0	0.000
C	276	211	43	0	40	3	0	0.0698	154	0	7	0	0	0.000
D	548	267	132	0	89	43	0	0.3258	62	0	31	31	0	0.500
TOTAL	1423	**890**	**448**	**0**	**401**	**47**	**0**	**0.1000**	392	**0**	**213**	**32**	**0**	**0.127**

4 Failing to Account for the Human Error in the Baseline Estimation of the PFD.

The last major deficiency in high integrity systems to be covered in this paper is failing to account for the specific probability of human error in the PFD of the IPL, including for a single PSV installation, a multiple PSV set IPL, and SIL 2 or SIL 3 SIFs. A few of the specific human errors were discussed in sections 1-3 of this paper. For a SIF, below are the typical human errors of interest:

- Probability of leaving the root valve for the level switch (sensor/transmitter) closed
- Probability of leaving the entire SIF in BYPASS after maintenance or after some other human intervention (such as an inadvertent error or as a necessity during startup)
- Probability of miscalibration of the level transmitter/switch.

The new issue here applies to SIL 2 and SIL 3 SIFs more than to PSV or multiple PSV or single or multiple checkvalve IPLs. The reason is that for a SIF the user is required to a complete a SIL Verification calculation that states that the target PFD was achieved by the design. But, as explained in two earlier papers [5,6], the guidelines and standards for SIS from IEC and from ANSI/ISA do not require inclusion of the probability of human error (P_{HUM_ERR}) in SIL Verification. If the human error is not required to be in the estimate of the overall PFD of the SIF, then why try to control the human error? Further, leaving such terms out may give false hope that a high integrity IPL was actually achieved, whereas it nearly impossible to actually achieve a PFD of 0.001 for a SIL 3, due to the human error probability that cannot be lowered enough. A SIL 3 with a PFD of 0.001 is as rare as a Unicorn (they very likely do not exist) [6].

In this case, for any IPL with a target PFD of 0.001, the system will typically have a PFD that is 10 times greater than predicted, because the human error contribution was omitted from the PFD estimate. This can lead to the risk of a scenario being too high, and can lead to wasting resources on an IPL that the company believes is much more reliable than it actually is [6].

5 CLOSING

High integrity IPLs, such as PSVs, multiple PSVs with extra capacity, and SIL 2 and SIL 3, are highly vulnerable to human error and to mis-design. These errors have been observed across hundreds of facilities that use or make highly hazardous chemicals. But once these vulnerabilities are known, an organization can take measures to reduce the vulnerability. Not knowing about these vulnerabilities or not addressing them can lead to much higher risk of major accidents, or misallocation of funds and efforts on the wrong IPLs.

6 Acronyms Used

10oo10	ten out of ten voting architecture; all 10 devices must work for the safe state
10oo10	ten out of ten voting architecture; all 10 devices must work for the safe state
ANSI	American National Standards Institute
CD	Complete Dependence
CSC	Car Sealed Closed
CSO	Car Sealed Open
D	Dangerous
ETOPS	Extended-range Twin-engine Operational Performance Standards
FAA	US Federal Aviation Administration
IPL	Independent Protection Layer
ISA	International Society of Automation
LC	Locked Closed
LD	Low Dependence
LO	Locked Open
LOPA	Layer of Protection Analysis
MooN	M devices out of N total devices
NooN	N devices out of N total devices
P	Probability
PFD	Probability of Failure (dangerous) on Demand
PII	Process Improvement Institute, Inc.
PSV	Pressure Safety Valve
SIF	Safety Instrumented Function
SIL	Safety Integrity Level
SIS	Safety Instrumented System
SYS	Systematic failure contributions to overall PFD
HUM_ERR	Specific human errors and failures generated by human error
ZD	Zero Dependence
CSO, CSC, LO, LC	

7 References

References Cited

[1] A. Swain and H. Guttmann, *Handbook of Human Reliability Analysis with Emphasis on Nuclear Power Plant Applications, Sandia National Laboratories*, 1983 [this document became NUREG/CR-1278– The Human Reliability Handbook, guidelines from the US NRC on Human Reliability Analysis].

[2] A. Swain, *Human Reliability Analysis*, Training Course, ABS Consulting (formerly JBF Associates), 1993.

[3] D. Gertman, H. Blackman, J. Marble, J. Byers, and C. Smith, *The SPAR-H Human Reliability Analysis Method*, NUREG/CR-6883, U.S. Nuclear Regulatory

Commission, Office of Nuclear Regulatory Research, Washington, DC, August 2005.

[4] US Federal Aviation Administration, "AC 120-42B - Extended Operations (ETOPS and Polar Operations)", Washington, DC, June 13, 2008

[5] W.G. Bridges and H.W. Thomas, "Accounting for Human Error Probability in SIL Verification Calculations", 8th Global Congress on Process Safety, Houston, TX, April 1-4, 2012, American Institute of Chemical Engineers.

[6] A. Dowell III, W.G. Bridges, H.W. Thomas, and M. Massello, "SIL-3, SIL-2, and Unicorns", 15th Global Congress on Process Safety, New Orleans, LA, March 31 - April 3, 2019, American Institute of Chemical Engineers.

[7] CCPS, *Guidelines for Initiating Events and Independent Protection Layers in Layer of Protection Analysis*, Center for Chemical Process Safety, American Institute of Chemical Engineers, New York, NY, 2015.

(57au) Heat of Cracking Determination for Heavy Petroleum Stocks

Authors:
Self, F., Bechtel OG&C
Sielegar, D., Bechtel Oil, Gas & Chemicals, Inc.
Arora, G., Bechtel Oil & Gas
Quiroga, V., VMG-Schlumberger

Heat of Cracking Determination for Heavy Petroleum Stocks

Heavy petroleum resids are processed in refinery atmospheric columns, vacuum columns and other equipment. If a fire occurs on the columns, the column may experience temperatures exceeding the design temperature with large relief rates.

In actuality the resid may experience mild thermal cracking into lighter components thus lowering the vaporization temperature. Further, the heat utilized to crack the resid will not be used for vaporization and thus reduces the fire relief rate.

Determination of the heat of the cracking reaction is key but has been historically difficult due to the large amount of computations. Indeed, seventy years ago, Professor W.L. Nelson, in the 3rd edition (1949) of his classic book "Petroleum Refinery Engineering", remarked that "Although the method outlined [for heat of cracking] …is correct from the standpoint of theory, it is hardly satisfactory for practical work unless great care is used." However, recently simulators have been developed that can handle the computations.

This paper will outline a method to characterize the resid and products and use them to determine the heat of cracking. Heats of cracking at different cracking temperatures will be presented and compared against experimental data. The method is practical yet theoretically rigorous, utilizes publicly available information, and is sufficiently accurate for pressure relief determination.

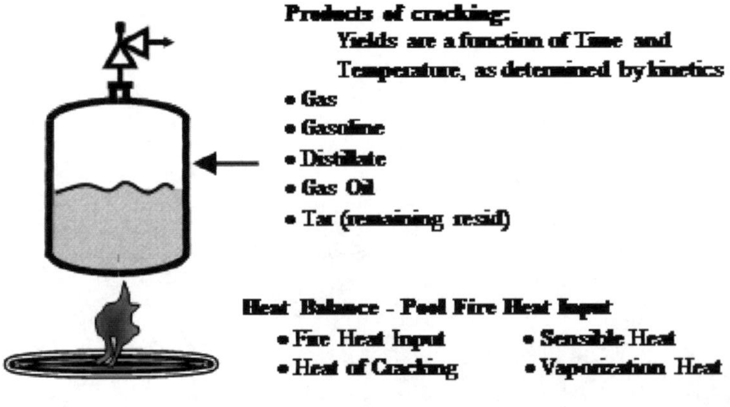

584034 Methodology for Conducting Flare System Assessment in Petroleum Refinery

Abstract:

Presenter Information

Presenter Name:

Bejoy Bharatiya

Presenter Biography: (50 words Maximum)

Bejoy Bharatiya a Chartered Professional Engineer (CPEng), is Senior Process Safety Engineer at Saudi Aramco Shell Refinery (SASREF), and has over 30 years' of professional experience in the upstream oil & gas, refining, petrochemical, and mineral processing industry. Prior to joining SASREF, he was the Process Safety Consultant at SANTOS (South Australian Oil Search) Pty Ltd. Australia, Senior Operations Engineer with Saudi Aramco, and the Principal Process Engineer of Brown & Root Energy Services (currently, Kellogg, Brown & Root), Australia.

Contact Information

Contact Name: Bejoy Bharatiya

Telephone No.:

+61 8 9440 0440

M: +966 553 543 149

Fax No.: +61 413 098 008

E-mail Address (Required):

pinkubejoy@iinet.net.au

Paper/Presentation Title: Flare System Assessment-A Basic Approach

Abstract: (1500 words Maximum):

Methodology for Conducting Flare System Assessment in Petroleum Refinery

Bejoy K. Bharatiya

Saudi Aramco Shell Refinery

The petroleum industry has long since recognized certain risks are involved in the flare and relief system of a hydrocarbon processing facility. Over the years, the oil and gas industry has learned lessons from a number of process safety events and various hazards to flare systems such as loss of control and loss of containment top events that could result in: fires, vapour cloud explosions (VCE), toxic and asphyxiant exposure, which occurred in various parts of the world and are not limited to the country of origin. The design and operation of flare system s have been the subject of concern to the oil industry.

It has been the continuous effort of the industry to update the design and operating philosophies to improve the inherent safety of the flare and relief systems. The flare system is the last line of defense for various emergency scenarios. The present paper is concerned with defining the major safety considerations of international oil companies including Saudi Aramco and Shell, and International Industry Standards in designing and operating the flare systems. The flare system scope is defined in this paper as the piping and equipment downstream of the relief devices, control valves and equipment vents tied into the flare system all the way to the flare tip including knock out vessels and seal vessels.

These design considerations include basic design philosophies, and safety instrumentations in relief valve back pressure limitations and sizing of the discharge piping. Major design considerations and hazard assessment for flare system to be discussed in this paper are described below. Some oil companies use required relief valve capacity for sizing the closed discharge systems, rather than rated capacity which is mentioned in API 520 Part II. The following relief load reduction credits are used for designing the flare system:

1. Flare system design philosophies - Generally, there are three permitted methods typically adopted by some of the leading oil companies for flare load mitigation: Use of 100-50-50 method; Flare IPF (Instrumented Protective Functions) method; and Flare Bundle method.

a) Use of 100-50-50 methodology

· The impact of the common discharge system design produced by relief flows from the connected process units as a result of global utility failure is calculated by taking 100% of the quantities established for the relief and de-pressuring load from each unit in turn, together with 50% thereof for the other units.

· If it is determined that multiple trains will discharge simultaneously, the relief loads from those trains shall be taken at 100% (e.g. parallel trains that would simultaneously relieve or de-pressure during a utility failure).

· The individual relief valve backpressure shall be taken as the highest value from any of the cases mentioned above.

· 100-50-50 method shall not be combined with the other methods described below.

This methodology considers beneficial effects of dynamics, operator intervention, control and safeguarding systems without the need to specify these. The resulting design is ALARP based on the long operational experience demonstrated by the company.

b) Flare IPF method - This considers beneficial effects of all safeguarding systems on the flare loads. Additional inputs required are: mitigated relief loads (relief loads when the safeguarding is effective) and the SIL classification of the safeguarding systems.

• The flare IPF method shall not be combined with the 100-50-50 methodology to analyze the same common mode scenario.

• The system shall meet the tolerability requirements with any single IPF failing (not working as designed).

• The combination of up to three concurrent individual or partial IPF failures resulting in total flare loads exceeding the acceptance criteria be designed to the appropriate SIL level to meet the acceptance criteria.

• It results in an enumerated probability of exceeding design requirements such as backpressure, flare radiation etc. The resulting design is ALARP based on the application of SIL classification to manage the risk.

c) Flare Bundle method - This considers beneficial effects of safeguarding systems as well as basic process control and operator intervention. Required inputs are following parameters:

I. common mode relief loads for all utility failures

II. as I above. but mitigated relief loads (relief loads when the basic process control or operator is response is effective)

III. the probability of the mitigation of effective operator response (to be established and documented in a team review)

It results in a design with probability of exceeding design requirements being managed both by systems and by operator competencies. The resulting design is ALARP.

This method requires a commitment to establish systems capable to maintain each of the flare load management instrumentation barriers as completely independent of each other. This method requires lifetime commitment to maintain and assure the ongoing effectiveness of the barriers.

The advantage of this approach is that additional instrumentation (basic process control or IPF) barriers can be installed on a system to reclassify loads from the Likely Load, or Somewhat Likely categories, where statistically a fewer number of loads are required for the analysis.

The above methods (a, b, c) rely on the hydraulic model for the flare system network based on the accurate input data (relief loads and piping data). Generally, software/applications such as Aspen Flare System Analyzer or, Honeywell UniSim Flare is used for the network analysis. There are also other software available in the market for performing the hydraulic calculations.

1. **Hazard assessment** – Hazard assessment is based upon the hazard of the materials present in the processing facility. Some of the highest ranking included in the risk assessment (taken from facility Hazard Register) are:

Hazard	Top Event	Consequence
All Liquid Hydrocarbons (HC) including condensed flare gas	Loss of Containment	Fire
	Loss of Control or	

Toxics / H2S / N2	Containment	Personnel Exposure
All HC, including H2 enriched HC	Loss of Containment	Generation of vapour cloud leading to VCE or fire
Heavier than air HC	Loss of Containment	Vapour cloud and/or Fire resulting from heaving than air HC

1. Some of the major **process threats** to the flare systems will be discussed are following:

I. O_2 Ingress into the Flare System

II. Loss of Pilots / Flame

III. Flare Line Entry / Live Flare Opening

IV.Overpressure due to Blocked Flow

V.Flare Header Liquid Hold-up / Slug Flow

VI.Flare Stack Structural Failure

VII.Hydrocarbon / H2S Breakthrough to Sewer / Atmosphere

VIII.Backpressure Exceeding Flare System Design leading to Equipment Overpressure

IX.Flow induced vibration

X.Acoustic Fatigue

XI.Low-temperature effects/brittle fracture

XII.Radiation

Etc.

4. Flare Network Analysis – In the industry, there are many software that are being used to carry out the network analysis, viz. UniSim Flare and Aspen Flare System Analyzer.

Overview of our network analysis as an example will be presented to describe the effect of global power failure case (100%) and the results of high backpressure and/or high noise from certain relief valves in the process unit.

This paper discusses industry experience and summarizes the salient design considerations and risk assessment to improve safety integrity of the flare system. An effort has been made to highlight the major design approaches/ methodologies adopted by oil companies using company design standards, and industry standards like API. Emphasis has been made to increase the awareness among the process design engineers to develop safer designs of the flare systems.

586906 Develop a Hazard Index Using Machine Learning Approach for the Hazard Identification of Chemical Logistic Warehouses

Abstract:

With the rapid development of chemical process plants, the safe storage of hazardous chemicals become an essential topic. Several chemical warehouse incidents related to fire and explosion have been reported recently. Therefore, an accurate hazard identification method for the logistic warehouse is needed not only for the facility to develop a proper emergency response plan but also for the residents who live near the facility to have an effective hazard communication. Furthermore, the government can better allocate the resources for first responders to make fire protection strategies, and the stakeholders can lead to improved risk management. Hazard index is a helpful tool to identify and quantify the hazard in a facility or a process unit. The challenge for this research is to improve the current method with the novel technique to implement our purpose.

The first objective of this research is to develop a "Storage Hazard Factor" (SHF) to evaluate and rank the inherent hazards of chemicals stored in logistic warehouses. In the factor calculation, the inherent hazard of chemicals is determined by various parameters (*e.g.*, the NFPA rating, the flammability limit, and the protective action criteria values, *etc.*) and validated by the comparison with other indices. The current criteria for flammable hazard ratings are based on flash point, which is proved to be insufficient. Two machine learning based methods will be used for the classification of liquid flammability considering aerosolization based on DIPPR 801 database. Subsequently, SHF and other warehouse safety penalty factors (*e.g.*, the quantity of the chemicals, the distance to the nearest fire department, *etc.*) are utilized to identify the Logistic Warehouse Hazard Index (LWHI) of the facilities. Last but not least, this method is applied to real-time data from Houston Chronicle, and several statistical analyses are used to prove the hazard index is helpful for hazard identification to emergency responders and hazard communication to the public.

Use of Historical Data to Provide Process Leak Frequency Estimates

Brian Bain
DNV GL
Cromarty House, 67-72 Regent Quay, Aberdeen, AB11 5AR
brian.bain@dnvgl.com

Zoë Wattis
DNV GL
Ashby Road Holywell Park, Loughborough, LE11 3GR
zoe.wattis@dnvgl.com

Carol Humphreys
DNV GL
Ashby Road Holywell Park, Loughborough, LE11 3GR
Carol.humphreys@dnvgl.com

Prepared for Presentation at
American Institute of Chemical Engineers
2020 Spring Meeting and 16th Global Congress on Process Safety
Houston, TX
March 30 – April 1, 2020

Use of Historical Data to Provide Process Leak Frequency Estimates

Brian Bain
DNV GL
Cromarty House, 67-72 Regent Quay, Aberdeen, AB11 5AR
brian.bain@dnvgl.com

Zoë Wattis
DNV GL
Ashby Road Holywell Park, Loughborough, LE11 3GR
zoe.wattis@dnvgl.com

Carol Humphreys
DNV GL
Ashby Road Holywell Park, Loughborough, LE11 3GR
carol.humphreys@dnvgl.com

Keywords: Hydrocarbon, Leaks, Releases, Frequency

Abstract

A key driver in the risk assessment of an offshore installation or process plant is the leak frequency as this is combined with the consequence information to determine the process safety risks. Frequencies are normally determined by processing historic information on incidents.

The UK's Hydrocarbon Release Database (HCRD) captures the details of almost 5000 leaks for UK North Sea offshore installations from 1992 to the present day. It is the most commonly used source of leak frequency data both in the UK and the US. It is also applied as many other parts of the world as well as being used for onshore process plants and refineries.

Revised correlations for leak frequency and hole size distribution have been developed by DNV GL on behalf of the International Association of Oil and Gas Producers (IOGP). This analysis indicates a significant reduction in leak frequency which has implications for risk-based design and operational decision making. Alternative correlations produced by organizations in the Norwegian offshore industry show a similar significant reduction.

With such a high reliance on the use of the HCRD as a source, it is important that the reporting, recording, collation and interpretation of the data is thorough and technically sound. There are initiatives in place or planned to further improve the data collection and thereby the accuracy of the estimates produced.

This paper discusses how the IOGP leak frequency correlations were derived from the HCRD and the measures taken to ensure robustness of the conclusions. It also explores the impact of using the updated frequency correlations in quantitative risk assessments and future work investigating the technical quality of the HCRD.

1 Introduction

The oil and gas industry make considerable investments to reduce the likelihood and consequence of accidents. It is a complex task to assess the relative values of these investments to achieve an appropriate balance of financial expenditure and risk reduction.

While all hazards need to be considered in a full risk assessment, the focus of attention is often the hydrocarbon loss of containment events. Hence, one of the key inputs to a risk assessment is the estimate of frequency of accidental releases of different sizes. The leak frequency estimate is based on analysis of historical incidents. Of the various sources of data available, the UK Health and Safety Executive's (HSE) Hydrocarbon Release Database (HCRD) [1,2,3] has become established as the prime source of data for the development of leak correlations for used in risk assessments [4]. Correlations based on the HCRD are often used for onshore facilities as well as offshore even though the environmental and operational conditions may be rather different from the North Sea. Due to its maturity and comprehensiveness, the HCRD is the most significant source of information for use in the majority of risk assessment of petrochemical plants and offshore installations throughout the world. The risk is directly proportional to the leak frequency, hence the estimate of the leak frequency will directly affect risk based decisions on the provision of safety barriers and the costs of constructing and operating facilities.

In recent years, hydrocarbon leak events in the Norwegian Continental Shelf have been analyzed to provide an alternative set of correlations known as the Process Leak for Offshore Installations Frequency Assessment Model (PLOFAM) [5]. Both this and the most recent analysis of the HCRD indicate significantly lower frequencies than may have been used in the past and this will reduce the calculated risks when used in the analysis of a given installation.

Recent work by DNV GL identified some areas for improvement in the way in which information in the HCRD is collected, collated and analyzed. Some of these issues were addressed in the information provided on the HSE's web site [1,2,3] which covered the period from October 1992 until December 2015. A model for calculating suitable estimates for process equipment leak frequencies has been developed using this dataset and is discussed in this paper.

In September 2019, the International Association of Oil and Gas Producers (IOGP) revised their guidance on the calculation of process leak frequencies and used the analysis of the HCRD carried out by DNV GL as the primary source [6]. Details and parameters for the PLOFAM model are also included in the IOGP report for use in risk assessment of Norwegian offshore installations.

2 The UK HSE Hydrocarbon Release Database

2.1 General

The UK offshore industry has been required to provide data on loss of containment incidents ever since the Piper Alpha disaster and the subsequent enquiry led by Lord Cullen [7] In his report, Cullen stated; *"I am convinced that learning from accidents and incidents is an important way of improving safety performance"*. One way of achieving this was to systematically record incidents that occurred on installations in UK waters. This was intended to provide a structured set of data that would identify trends and could be used as a source of information for risk analyses.

Thus, the UK HSE set up a reporting system to gather data on loss of containment incidents that occur on installations operating in UK waters alongside the population of equipment in which these incidents were occurring so that frequencies could be derived.

2.2 Incident Data

The latest publicly available information from the HCRD gives details of 4656 incidents until 31st December 2015. Provisional data for 2016 and 2017 has also been made available to a joint industry project for research purposes.

The data collected up until 2015 has 118 fields of information for each record. The most important of these are;

- Date of incident
- Installation Type
- Process System Type
- Equipment Type
- Quantity of Hydrocarbon Released
- Equipment Size
- Equivalent Hole Diameter
- System Pressure at the time of the release

Other information provides more details on the installation, the area where the release took place, the cause of the leak, how it was detected, information on ignition and any emergency action which followed as a result. While these are important and merit detailed analysis, they are not required to establish frequency distributions for various types of equipment.

Figure 1 shows the variation in the number of incidents reported on a year by year basis. This shows a downward trend indicating that appropriate process leak frequencies used in risk assessments may also have decreased; however, the population of process equipment should also be taken into consideration.

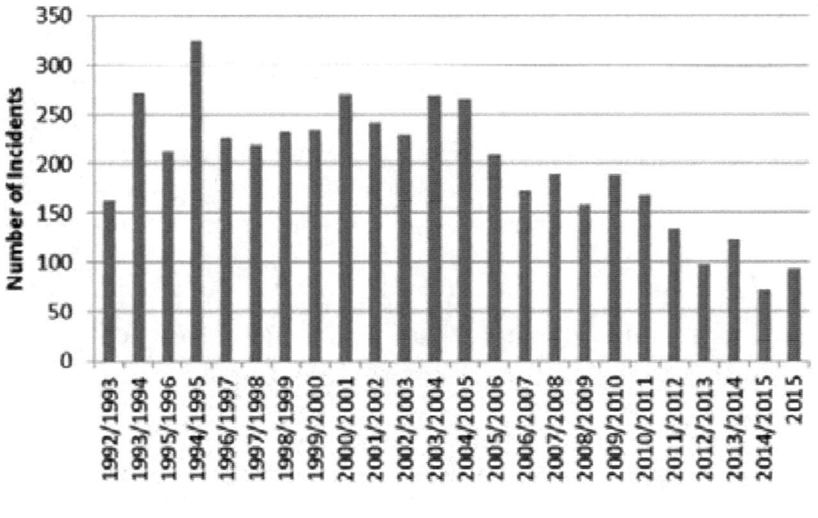

Figure 1: Number of Reported Incident per Year in the HCRD

2.3 Population Data

The successful operation of the HCRD depends not only on the correct submission of reportable hydrocarbon releases, but also on the retention of an accurate record of all duty holders (operators/owners) together with full details of their systems and equipment population data.

Between 1992 and 1994, the HSE collected systems and equipment population data for UKCS installations. This provided data on the total population of 120 categories of process equipment in the UKCS. In 2002 an on-line system was introduced to enable operators to submit and update their equipment population data.

3 UK HSE Dataset Update

3.1 Population Data

Some anomalies were observed in the population data set and in 2016 this prompted members of the industry body Oil and Gas UK (OGUK) to work in collaboration to analyze the available data and estimate appropriate values. This included an exercise to provide estimates of missing data through comparison with equivalent process systems and installations. The dataset produced by this exercise was made available on the HSE web site [3] and is one of the main inputs to the derivation of the IOGP correlations.

The UK industry has continued to work to improve the estimates of the equipment population. In 2020 a system will be implemented which requires operators to review the equipment counts for their installations when updating their Safety Case documents, a 5-yearly process. This process should ensure an improved estimate of the equipment count and enable refinements to the parameters used in the leak frequency correlations.

3.2 Incident Data

The incident data collected is extensive with approximate 100 data fields applied to close to 5000 records.

As a result of legislation introduced by the European Union in 2014 [8], the requirements for reporting changed. The new ROGI form [9] had a mixture of compulsory and voluntary information. However, some data considered necessary by risk analysts, such as the equipment type and the equivalent hole diameter fell into the voluntary category reducing the completeness and quality of the data.

4 Norwegian Database

4.1 General

Previously, risk assessment of offshore installations in Norwegian waters used the HCRD as the leak frequency data source. However, a recent study involving a number of operators and consulting organizations in Norway adopted an alternative approach for deriving frequency estimates; the Process Leak for Offshore Installations Frequency Assessment Model (PLOFAM2) [5]. Although, the model is mainly validated against available Norwegian Continental Shelf (NCS) leak data, it also takes account of data from the HCRD. Leak frequencies are based on the period from 2006 until 2017 while the whole period (2001-2017) is used for the relative leak rate distribution.

4.2 Incident Data

Information was gathered on 260 incidents recorded at installations located on the Norwegian Continental Shelf (NCS) in the period 2001 to 2017 inclusive.

4.3 Population Data

Data was obtained from equipment parts counts compiled for use in quantitative risk analyses for 79 installations which were in operation in the NCS at some point during the period 2001-2017. For a further 25 installations, where equipment counts were not available, estimates were made by comparing them with similar installations.

5 Derivation of Leak Frequencies

5.1 Selection of Relevant Incidents

Not all incidents reported are considered relevant to a QRA study. In addition, the population data needs to match with the incident data. Therefore, incidents relating to installations not included in the population data set are excluded as are systems such as water treatment, flaring/venting systems or utilities. Some holes were very small and those less than 1 mm in diameter were excluded along with incidents occurring in depressurized process systems.

Taking all of these criteria into account, the number of incidents from the HCRD data set was reduced to 1964 from the original 4563.

Similar but not identical criteria were applied to the NCS dataset resulting in 217 relevant incidents for use in the PLOFAM model.

Both studies based the frequencies on experience from 2006 onwards.

The hole size distribution is also important. It is assumed that the underlying distribution has been constant over the whole period. The hole size model thus uses the selected data over the whole period; 1992 Q3 to 2015 Q4 for the UK data and Q1 2001 to 2017 Q4 for the PLOFAM study.

5.2 Frequency Calculation

The frequency of leaks greater than or equal to 1 mm can generally be determined by dividing the estimated number of leaks in that period by the population to which it applies measured in equipment years. For cases where the equipment type was not recorded or there are no recorded incidents further statistical analysis was conducted.

5.3 Hole Size Distribution

The available data for a given type of equipment can be arranged in order of hole size and a plot of the frequency of exceeding that hole size produced. Typically, this produces a plot similar to the blue line shown in Figure 2.

5.4 Mathematical Correlation

The general equation used in the correlations reported by IOGP [6] is of the form

$$F = Cd^m + B \qquad \text{for } d = 1 \text{ mm to } D$$

$$F = 0 \qquad \text{for } d > D$$

Where F is the frequency of a hole greater than or equal to d

d is the hole size

D is the equipment size

and C, m and B are coefficients for each equipment type.

For some equipment types there is sufficient data to determine a variation in the value of these parameters as a function of the equipment size. For other types, constant values are assumed.

A least squares curve fitting process was used to determine the values of the C, m and B parameters which gave the best fit to the historic data. This gives a function represented by the black line in Figure 2.

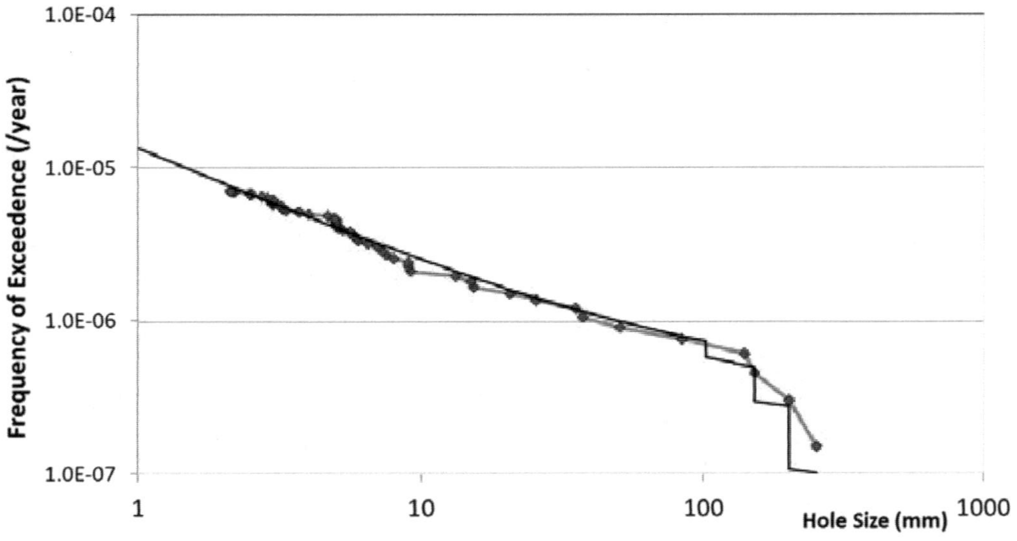

Figure 2: Mathematical Correlation Fitted to Historic Data

The IOGP Guidance [6] provides details of the parameters along with graphical and tabulated representations for 24 equipment types. As an example, Figure 3 shows the graph for flange joints which indicates the effect of equipment size.

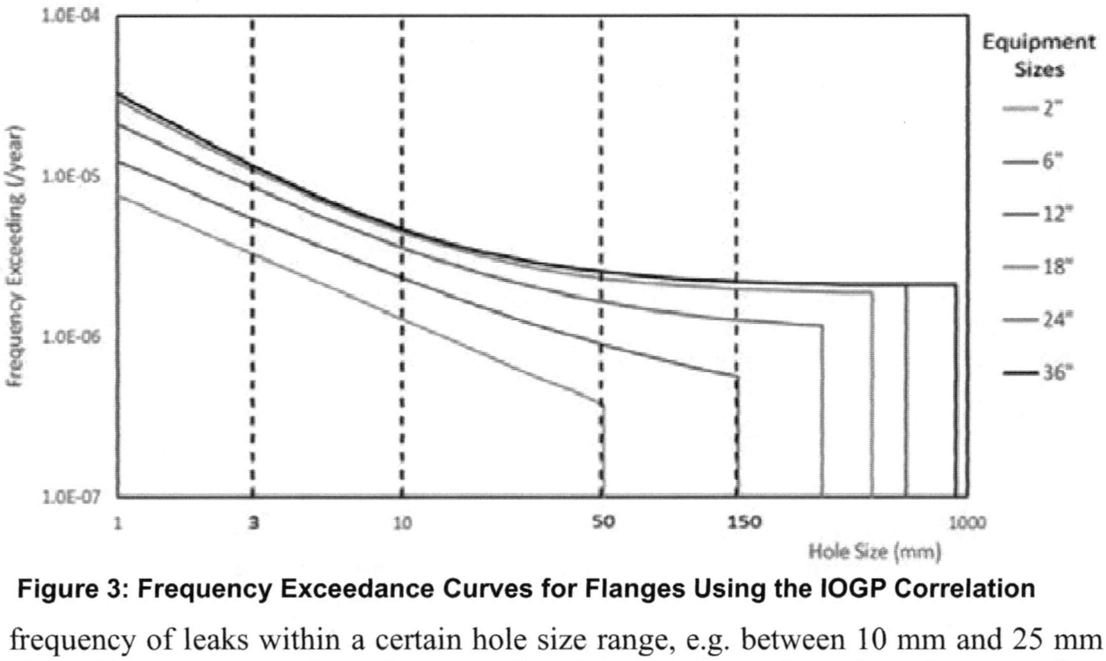

Figure 3: Frequency Exceedance Curves for Flanges Using the IOGP Correlation

The frequency of leaks within a certain hole size range, e.g. between 10 mm and 25 mm is evaluated by subtracting the value exceeding the upper range from the value exceeding the lower range.

5.5 PLOFAM Model

The PLOFAM study proposed a more complex formulation;

$$F(d,D) = [F_0(D) - \alpha F_D(D)]\, d^{m(D)} + \alpha F_D(D), \qquad 1\,mm < d \le D$$

$$F(d,D) = 0, \qquad d > D$$

Where, $F_0(D) = F_{hist}(A_0 D^{M_0})$

$F_D(D) = F_O(D)(A_D D^{M_D} + B_D)$

and $m(D) = \frac{log(F_D - \alpha F_D) - log(F_0 - \alpha F_D)}{log(D)}$

where $F(d, D)$ is the hole size frequency distribution per equipment-year.

F_0 is the total leak frequency per equipment-year. The subscript 0 indicates the total leak frequency for an equipment and hence the "starting point" on the y-axis.
$F_0 = F(d = 1, D)$.

F_D is the total full bore hole frequency per equipment-year. The subscript D indicates the frequency for getting a hole equal to the equipment diameter D.
$F_D = F(d = D, D)$.

d is the hole diameter in millimetres

D is the equipment diameter in millimetres

m Slope parameter

F_{hist} is the average leak frequency (independent of equipment diameter) for the relevant equipment type per equipment-year

A_0 is a parameter in equation for total leak frequency, F_0

M_0 is a parameter in equation for total leak frequency, F_0

A_D is a parameter in equation for full bore hole frequency, F_D

M_D is a parameter in equation for full bore hole frequency, F_D

B_D is a parameter in equation for full bore hole frequency, F_D

α is a dimensionless parameter, independent of equipment diameter D, $\alpha \in [0, 1\rangle$.

As an example, Figure 4 shows the exceedance curves for a standard flange using the PLOFAM model.

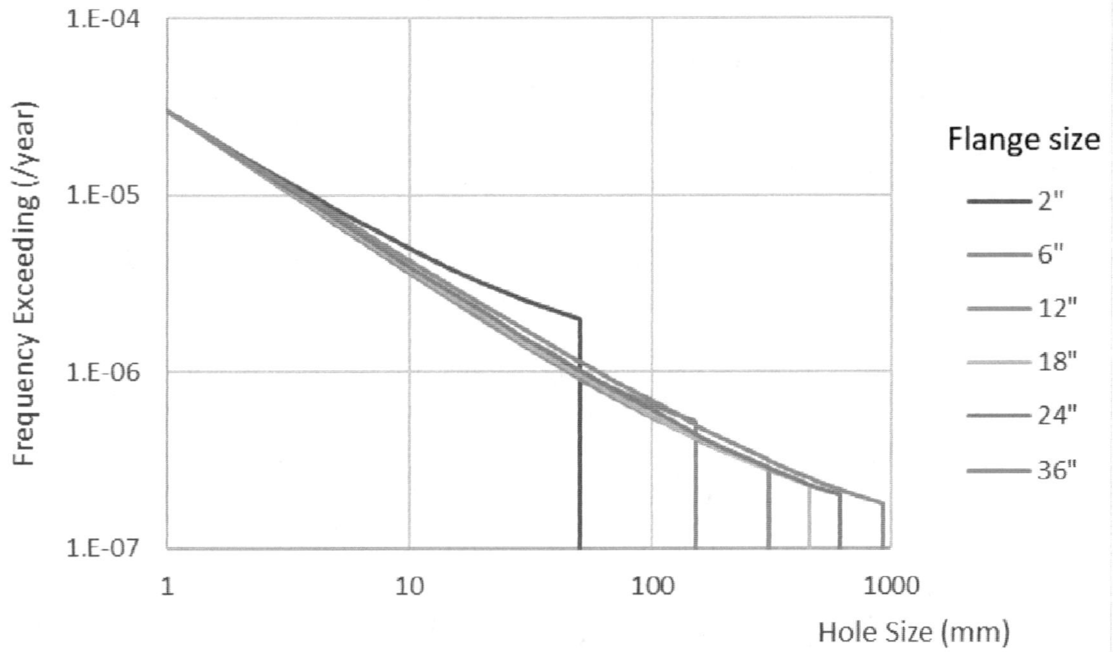

Figure 4: Frequency Exceedance Curves for Standard Flanges Using the PLOFAM model

6 Effect on Using the New Leak Frequency Models

Both the IOGP and PLOFAM models give estimates of significantly lower frequencies than previously used. In the case of the IOGP formulation the values are approximately 60% lower. Given that the risk is directly proportional to the leak frequency, the calculated risk will fall by a similar amount. As such, remedial measures which were previously judged to be required may now be considered unnecessary.

The main factors for the reduction in the IOGP estimates are;

- The reduction in the annual number of incidents reported
- The decision to base frequencies on the last 10 years of available data
- A reassessment of the incidents to be included in the analysis.
- A reassessment of the equipment population data.

The PLOFAM model produces estimates for leak frequencies which are lower than the IOGP correlations and this is particularly the case for larger hole sizes. This implies a bigger reduction relative to previous estimates compared with those obtained from the IOGP model.

7 Future Work – Towards Improved Accuracy and Derivation of More Value

The collection of data in UK and Norway should continue in order to gain more statistical significance for the various equipment types. It is also necessary to confirm or otherwise that the downward trend continues. Some observers believe that the current level of reporting marks a low point in the frequencies of events and that a combination of aging and the effects of the downturn in the oil industry on maintenance levels will lead to an increase in coming years.

A review of the population data in the UK is required to ensure a robust database is available to accurately assess the leak frequencies.

The quality of the incident data collected needs to be actively managed with improvements in the UK reporting form and encouragement to complete all data fields as accurately as practical.

These two initiatives will improve and maintain the quality of the data gathered in the UK in future years and thereby make the complete dataset more valuable.

More information can be obtained from the existing data. Relatively few of the data fields are used in the calculation of the leak frequency functions described in this paper. The data can be used to address some important questions being asked by the industry such as;

- Is there an increase in the frequency during start-up operations compared with normal operating conditions?
- Do leaks on normally manned installations occur more frequently during the period that a workforce is present?
- What proportion of leaks are due to design/maintenance issues compared with procedural/operational failures?
- What effect does the age of the equipment have on its likelihood of failure to contain the hydrocarbon?
- Does the trend of leak frequencies vary between types of equipment and can these changes be related to improved design or operation?

8 References

[1] HSE, Spreadsheet entitled "Offshore Hydrocarbon Releases 1992 – 2016" http://www.hse.gov.uk/offshore/statistics.htm (accessed Dec 31, 2019).

[2] HSE Spreadsheet entitled "Offshore Hydrocarbon Releases 2015 – 2016" available from http://www.hse.gov.uk/offshore/statistics.htm (accessed Dec 31, 2019).

[3] HSE Spreadsheet entitled "Offshore Hydrocarbon Population Data 2015 – 2016" available from http://www.hse.gov.uk/offshore/statistics.htm (accessed Dec 31, 2019).

[4] Bain, B, "Updated Leak Frequency Modelling Based on the UK Hydrocarbon Release Database, Hazards 27 Symposium, 2017.

[5] Lloyd's Register Consulting, Process Leak for Offshore Installations Frequency Assessment model – PLOFAM(2). Report No. 1007566/R1 Rev: Final, December 2018.

[6] IOGP, Risk Assessment Data Directory, "Process Release Frequencies", Report 434-01, September 2019.

[7] Cullen, The Hon. Lord W. Douglas 1990. The public inquiry into the Piper Alpha disaster. London: H.M. Stationery.

[8] European Union, 2014, "Commission Implementing Regulation (EU) No. 1112/2014.

[9] Offshore Safety Directive Regulator, Report of an Oil and Gas Incident (ROGI) Form, https://www.hse.gov.uk/osdr/assets/docs/rogi.pdf (accessed 31st December 2019).

Application of Management of Change (MoC) in Refinery

Seckin Gokce, Process Safety Chief Engineer & Gozde Turk Eski, Process Safety Superintendent, Tupras Izmit Refinery, Kocaeli/TURKEY

Many companies are forced to make new investments / innovations in various fields to make them stronger due to changing and developing conditions or to sustain current operational availability. Facilities need to consider the changes to be implemented in terms of safety, environment, health and reliability. If the planned changes are applied on site without appropriate hazard analysis the risk of a process safety accident could increase significantly. This paper covers management of change and risk assessment studies with the examples of refinery applications.

INTRODUCTION

In all chemical process plants especially in refineries, it is critical to achieve high operating plants operational and mechanical utility. In such continuously working systems, there is maintenance activities done within a schedule but there may be unplanned maintenance. Mostly maintenance activities are repair and replace in kind. Besides these in order to improve operational availability, operational safety or keep up with the technological improvements there are many changes done in the units. Change means here any kind of process, hardware or technology differentiation in the process plant related with the process. All these changes come up with a need but also they may cause new risks or create new needs. Any kind of change inside a process plant should be reviewed and discussed with all the related disciplines (operations, technology, maintenance, design) in order to define all the possible outcomes of the change and assure all hazards are adequately addressed to meet company and regulatory requirements. Poor change management mostly causes new Process Safety related incidents which may lead to deaths, huge asset damages, irreversible environmental damages and also reputational effects. In chemical industry around the world there are so many examples of poor change management (for example Bhopal, Flixbroug).

Management of Change, as a Process Safety Management System task, should be rigidly addressing the scope of change, process of change management and closure and approval of the change. All the people in a company can come up with the change request that is why it is important to clearly convey the message from top to bottom that all kind of changes must be following up the MOC process regarding to the company standards.

MANAGEMENT OF CHANGE APPLICATIONS

Equipment Changes

In the refinery, most typical equipment changes we come across are pump renewals, new pump implementation, instrument additions, pipeline changes and rarely new process equipment like drum or column additions. Any not-in kind changes/additions, including emergency repairs and replacement, are in the scope of MOC if they are in the same kind. For example, think about the renewal of a pump if there is any difference from the current pump like material, shut off pressure, temperature specification, mechanical seal plan etc. that is under the scope of MOC. As defined above this change should follow the MOC path. During this MOC process, all design parameters and current process conditions should be compared to decide the change is safe to continue or not. In current application, we mostly go over a checklist besides what if or HAZOP risk assessment in order to be sure that we covered all parts. These checklists are the attachment of MOC standard, which is specific to equipment type.

These kind of changes are the ones which can be detected by people and easier to do risk assessments. There are other important tasks. This includes training, maintenance records and project documents revisions. Frequently these tasks are poorly done.

When we first initiated the MOC standard the hardest part was making people finish the change process completely. They think that if the approval is taken and equipment is implemented on site change is completed. Actually no, change is continuing until you finish all the documentation and trainings. That is why there is a closure approval in MOC system in order to make both parts, user of the change (which is mostly operations) and doer of the change (which is mostly maintenance or project departments), efficiently handover the change.

Process Changes

Another type of change in the refinery is process changes, which covers the operating temperature/pressure changes, chemical changes, process flow changes or by-pass of equipment or systems etc. These changes are mostly requested and done by the same team, which are operations and process development team. That is why most of the time they try to by-pass the MOC process intentionally or unintentionally. This kind of changes might be seen as innocent test runs but industry experience shows that these are the ones that have the highest potential to cause the incidents.

Here the critical thing is all related disciplines, to be affected by the change, gathering around the table and discussing the change and potential outcomes of the change. Defining all related disciplines is important and needs a good knowledge about the process, company standards and experience. In some cases you may need to gather the team again and again, and ask for third parties like process designer company etc. Experience level of the risk assessment team is important for all MOCs but that is more critical for process changes compared to other types. It is important that during the hazards review any new risks are identified including discussion of potential failure modes that cold be introduced.

Service Changes

So many plants, so many products and so many pipelines… What you see when you glance over the refinery! The term operational availability sometimes requires to use the same pipeline or equipment for different purposes. That is the reality inside the refinery, we have a good tool to manage the need of this flexibility which is MOC. These changes also easily captured under the scope of MOC since it is unusual path or service to be used. Here the critical thing is risk assessment teams are more crowded since that is mostly affecting most of the plants and disciplines. There may be meeting scheduling problems but beware of that all the team should be together for a better brainstorming. Especially, in MOC related risk assessments brainstorming and welcoming all levels of questions during the risk assessments is important.

Organizational Changes

At first glance, organizational changes seems like an HR issue but it is more than that. Since as chemical industries we have been dealing with risky processes, competency in the job is critical. Especially for HSE critical roles, any kind of organizational changes should be assessed in terms of competency.

In order to have a systematic MOC approach for organizational changes the first step is having HSE critical roles and their competency level requirements already defined.

Temporary Changes

Any kind of temporary technical, facility or organizational changes sometimes assumed as a minor issue and not classified as an MOC. However, experience and critical chemical industry accidents show that there are so many examples of major accidents caused by these temporary changes. Since they are seen as minor, most of the time these are not very well assessed. Any changes should follow the same risk assessment procedure the only change in temporary changes is to follow up the change in order to re change it. According to our standard any changes that will last less than 6 months is considered as temporary changes.

Management of Change Steps

Any change that affects process chemicals, technology, equipment, procedures, or facilities, as well as any organizational, personnel, and policy changes that could impact these elements. For applying on facilities successfully, MOC system needs to be well organized to make good risk assessment with relevant steps. Facility has own MOC software and employees use it when they have any demanding changes which is related with scope of MOC.

Step 1: Determine if an MOC is required or not.

Facility has own standard about MOC and it gives definition and scope of changes. The process safety department decide it. If it is appropriate change demanding the MOC steps are started or rejected.

Step 2: Process Hazard Analysis (PHA) and Coordination

After determining of the desired change is required MOC, PHA coordinator start to pre-assessment. Firstly, the change package need to include such kind of detailed information; description of change, type of change (temporary/permanent), related site, drawings, technical calculations & data etc. It is very critical and important that all related documents are provided and studied well by the proposing the change. If all needed items are prepared properly, PHA coordinator decides the risk assessment tool or tools (HAZOP, What if, HAZID, Fault Tree and Event Tree Analysis, Scenario Modeling (Toxic Gas Dispersion, Fire and Explosion Analyses), Facility Siting Studies and organize meeting with all related departments to evaluate the risks against corporate risk criteria (usually a risk matrix) and define the necessary additional barriers to risk.

Process Safety Management System starts with managing the process related risks and from the first step of establishing a new process unit that is needed to evaluate the risks and define the necessary measurements. Process Hazard Analysis methodologies are the systematic tools that help to evaluate the risks in a process. For different needs there are several methodologies like DOW FETI, HAZOP, What if, HAZID, Fault Tree and Event Tree Analysis, Scenario Modeling, Facility Siting etc. Some of these methodologies are semi quantitative some are quantitative methodologies. In order to define and prevent major industrial accidents in our refinery we use DOW FETI, HAZOP, FTA and ETA and Scenario Modeling. Moreover for some kind of projects we may also apply What if and HAZID analysis instead of HAZOP depending on the complexity. Facility Siting is another tool for risk related studies that is used to evaluate the risks for buildings in order to protect people inside the buildings.

The workflow for a full unit PHA study starts with DOW FETI calculation that is used for risk ranking of al static equipment inside the unit. DOW FETI calculation is done in accordance with the equipment temperature, pressure, hold-up amounts, components and also reaction types if there is any. Then risk assessment process continues with HAZOP studies that are for scenario identification. In HAZOP, first step is chopping the unit into small parts to be evaluated that is called nodes. After that all nodes are evaluated using P&IDs and PFDs by keywords like temperature, pressure, flow, level etc. HAZOP helps to understand what kind of scenarios are possible inside the unit and these critical scenarios are taken through FTA and ETA studies. HAZOP is a semi quantitative risk assessment method where FTA and ETA are quantitative methods. FTA is applied for critical scenarios in order to find out the probability of top event then ETA is then applied in order to find out the probability of scenarios like fire, explosion, toxic dispersion etc. caused by this top event. FTA and ETA analysis all use the statistical PFD and failure rate values of all initiating events and barriers defined in HAZOP for the related scenario.

Scenario Modeling is the following step that is used for understanding the effect of the probable scenario. The most probable scenarios coming from ETA is modeled to see the effect zones. These effect zones are used to plan Emergency Response Studies for all these potential major accidents.

Facility Siting is another check point in a unit PHA study. All potential scenario modeling results are used to evaluate the risks for the buildings and also if there are any in the effect zones of fire, explosion or toxic release then there should be actions defined.

Also in this step, PHA coordinator and facility superintendent give the decision of application of Prestart-up Safety Review (PSSR) before the operate it.

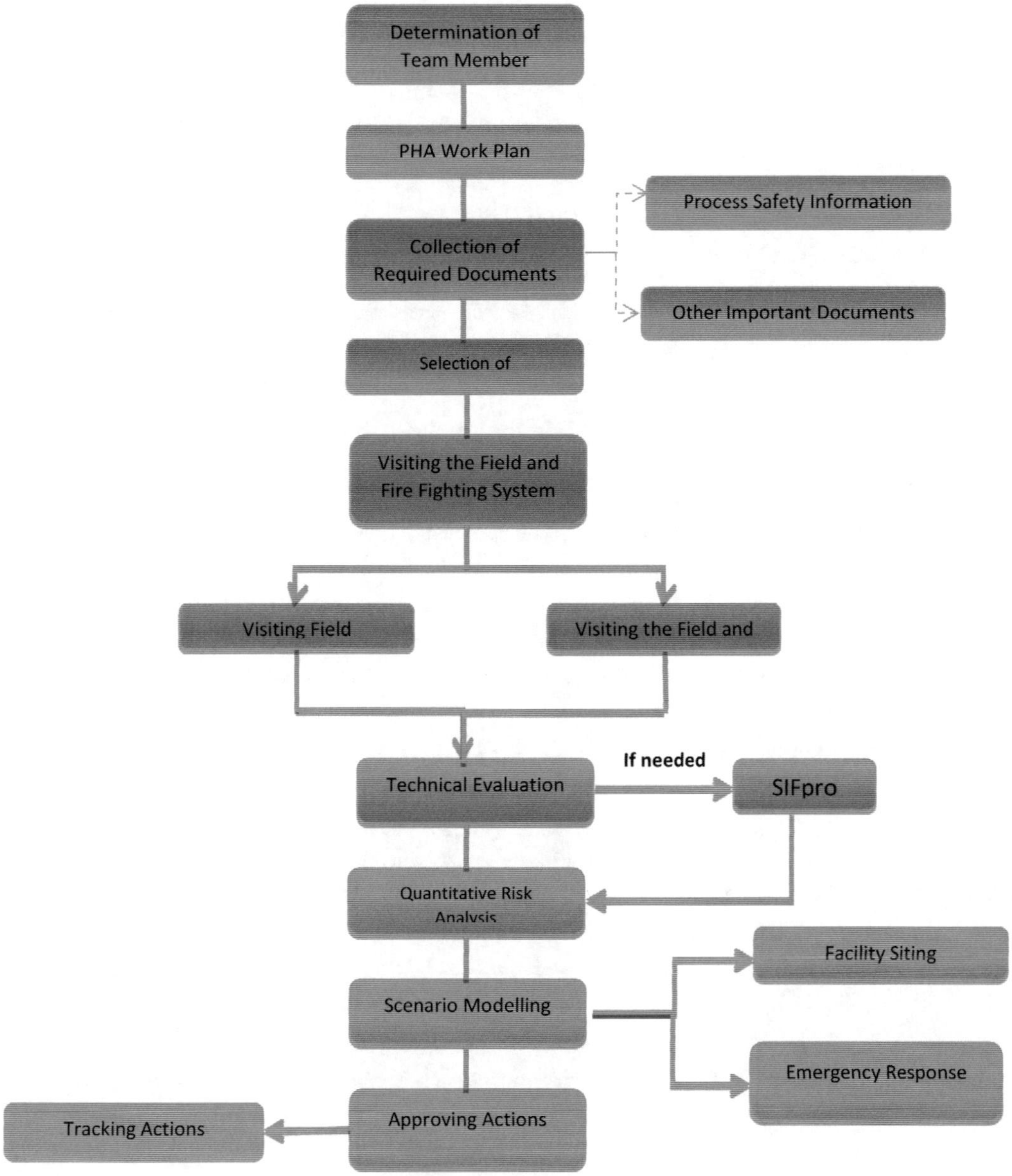

Figure 1: General Process Hazard Analysis Step

Emergency cases;

Sometimes, in facilities have emergency cases that change needing immediate implementation, the process should allow for administrative steps to be done quick. If such cases occur, facility's responsible person need to get verbal confirmation from site manager then should apply the desired change. In 48 hours, the proposer or implementer have to create change package on the MOC system then detailed risk assessment will be done and documented by PHA team.

Step 3: Technical Approval-PHA members

After the PHA studies, team analyze the request and evaluate the risk if it is acceptable risk level or not. If it is not, the team can make suggestions to reduce risk or rejected the request due to the safety issues. If the change request's details are proper and covered all safety and technical issues, the request is approved by team members.

Step 4: Technical Approval-Site Managers

The change request will be sent as a package that includes all risk assessments and technical drawings to the all related site managers for reviewing and approving. If the approval steps could be done then the change request will be started applying on site.

Step 4: Implementation of Change

Following the approval process, the implementer can start his desired change on the field. During the PHA studies if there is an additional suggestions/actions are advised due to the mitigation of risk, the proposal change have to cover it.

During that period the revision of the relevant documents should be done, employees and affected sites to be provided training about the changes by their site leaders. It's very important that to share new situation's risks and its barriers/procedures, which are new for related, sites. All steps of the risk assessment studies and updating of documents are applied and tracked by MOC system.

Step 5: PreStart-up Safety Review (PPSR)-If needed

PSSR ensure that all necessary checks have been made before commissioning a plant, and to ensure that all issues that can be checked for commissioning so far are safe. PSSR is the last checkpoint to inquire whether all equipment has been installed or maintained according to the standards and all necessary checks have been performed in accordance with the Process Safety Management System. PSSR team uses check lists during the evaluation. There are two possible results from the PSSR. First are necessary items that must be completed before start-up, and secondly, items that must be completed after start-up. For items to be completed after start-up, completion dates need to be describe, along with the person responsible for completion. The additional actions are tracked by MOC software too.

Step 6: Implementation of Action Items (Pre-Startup & Post-Startup Actions) and Tracking Actions

Actions items that were deemed "post-start-up" in the evaluation phase are completed at this step. While these action items were not required for cutover to the new state, they should be completed before the MOC is closed to ensure the sustainability and on-going safety of operations. Overdue actions are tracked on MOC software as KPI's.

Step 7: Closing Approval and Commissioning

In fact this step is final control step. All issues relevant with the desired changes are implemented on site, technical documents are updated and uploaded on system, training & instructions are provided related employees and additional recommendations are done. The implementer and site manager review the all steps and agree that MOC package satisfy all requirements. They both approve it close the MOC package.

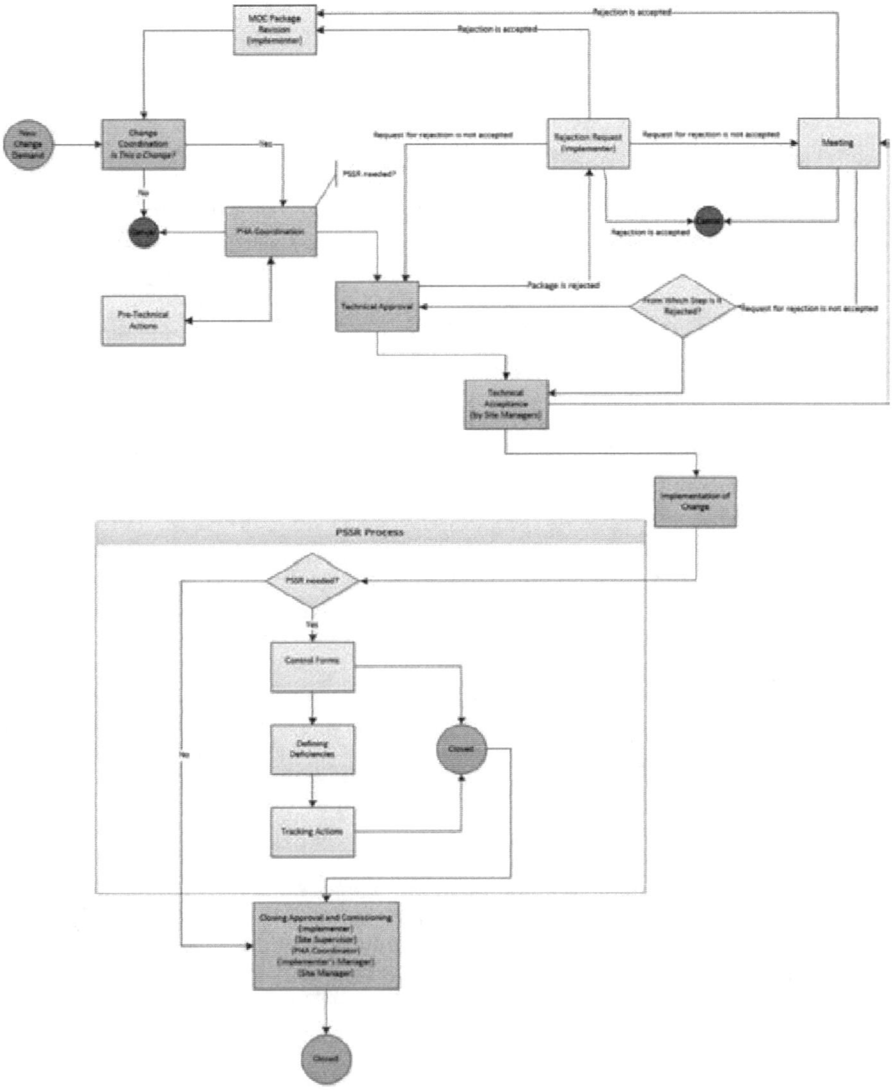

Figure 2: MOC and PSSR Steps

Auditing

Management Systems are audited in accordance with the Independent Auditing Standard for Internal Auditing Standards in order to ensure that the MOC is operated as specified in this standard, to find and remove the deficiencies in the system.

In order to identify, classify and resolve the deficiencies that occur in the Management of Change, the Change Coordinator informs the refinery management at the Annual Review of Management. In accordance with the decisions taken in this meeting, the problems that prevent the operation of the system are eliminated.

Example of MOC Studies

Case 1: Capacity Increasing Demand on Hydrocracker Unit

Case Description

Faculty has Hydrocracker Unit, which has been operated as 8800m3/day capacity. Due to the some profit issues, management would like to increase capacity. To define acceptable capacity value, process simulation team has started the studies if the desired change is possible or not.

What Happened?

Process team has created a new MOC package to discuss their findings. After the studies, they found it 9200 m3/day for new capacity value. PHA coordinator started the change organization with related departments. It was very important that the process hazard analysis-meeting schedule fit to everyone. Before the meeting, all data/reports/documents/graphs were prepared by process team and shared with the coordinator to send it all PHA team members.

During the meeting, all information was discussed by team members. In such cases, What If analyze form has used to describe all new potential risks, their results and existing barriers. All items were examined carefully and checked if the risk of item was ALARP level or not. If not, additional recommendation was given by team and it was tracked on MOC system. According to the result, team has signed What If analyze form, and then it was uploaded on system which was available for all employees. Then, Operation Manager and Refinery Manager also gave technical approval to final form.

Since the unit is very critical, its all parameters need to be checked and evaluated carefully.

For example;

- Checking unit vendor's report – operational parameters/design parameters
- Evaluation of new speed values and related erosion effects in critical lines that was determined by Reliability Department according to simulation results.
- Evaluation of relief valve capacity considering with blocked outlet scenarios were done by Process Department.
- Evaluations of pumps and compressors capacities were done by Mechanical Department.
- Evaluations of control valves capacities were done by Instrument Department.
- Evaluation of furnace load and static equipment's capacities were done by Process Equipment and Development Department.
- Evaluation of process parameters (reactor temperature profiles, residence time etc. effect on system.) were done by Process Department.
- Evaluation of supports due to new speed and load change in lines were done by Mechanical Department and Reliability Department.
- Reviewing of refinery flue gas/recycle gas flow effect on DEA system due to increasing of those flows were done by Process Department.
- Evaluation of the sufficiency of sulfur, hydrogen and contaminated water stripping units according to the simulation results were done by Process Department.
- Evaluation of consumption of washing water and corrosion inhibitors according to the simulation results were done by Process Department.
- Evaluation and suitability of process instruments (set value, critical alarm, range wide etc.) according to the simulation results were done by Project Instrument and Process Department.
- Evaluation of electrical motors, reviewing power consumption were done by Project Electrical Department.
- Checking and evaluating the steel structures from the relevant documents according to the new situation were done by Civil Engineering Department.
- Evaluation of fan capacity according to the simulation results was done by Process Department.
- Evaluation of catalyst activity at new situation was done by Process Department.

What Happened If MOC hadn't been Done?

Hydrocracker facilities are operated under very high temperature and pressure. Other danger is toxicity due to H2S content. Any type of change that would like to implement on facility's any process item need to be evaluated. A project that is intended to be applied to the site without a MOC coordination may result in larger industrial accidents, leading to a loss of life and property.

In that case, the desired would have been implemented to site without any considering safe design & operation parameters/risk assessment and then possibly facility had been faced serious process safety incidents.

Case 2 : Tank Service Change-Waste Water Treatment Unit (WWTU)

Case Description

WWTU unit had decided to make tank service change due to some operations causes. In the previously, storage tank had been used to hydrogen peroxide (H2O2) tank then services was changed to iron 3 chloride (FeCl3) storage tank. The unit made this change on its own.

What Happened?

After a short time, during the routine field check, operator had saw the corrosion in tank's transmitter connection cables which was used as FeCl3 storage tank.

Picture 1: Deformation of connection cables

The unit had bypassed the MOC system so such type of near misses occurred. WWTU had stopped chemical operations and inform the Process Safety Department. MOC package was created to make risk assessment study. Same change was made on other WWTU too. After this near misses, other WWTU checked their equipments and connection cables. They realized that tank feed pump had corrosion too. Same mistake, similar near misseses…

What is the Correct Approach?

Sometimes, the units want to make quick changes and apply it on site without any risk assessments. It is very important to be able to work safely in the field. But how? Operational disciplines are key parameters for every employee. If WWTU has such kind of intentions, firstly they had to create MOC package to make comprehensive risk assessment study.

What Would Happen If MOC Package Created?

PHA coordinator organized meeting with all related disciplines. The team considered desired change conditions if it was fit or not. For example, material of tank, connection cables, instruments materials and services, impact of process, pump materials etc. were evaluated according to new service.

Also, one of other critical point is unit's chemical interaction matrix. In the meeting, this matrix was checked by members. If the MOC was done, it was certain that the tank service change could not be done due to chemical interaction matrix. As it is seen below, this two chemicals has redox reaction.

$$2FeCl_3 + 3H_2O_2(l) = 6HCl(a) + Fe_2O_3 + 1.5O_2(g) \quad \Delta H(20C) = -520.0kJ$$

WWTU	Sodyum Karbonat	Fosforik Asit	Karbon giderici	KOI giderici	H202	Sülfürik Asit	Kostik	Cationic Polielektrolit (Sıvı)	FERROFOS 8411	Sodyum Hipoklorit	Hidroklorik asit	FECl3	Ascorbik Asit
Sodyum Karbonat		Heat+CO2+ Pressure			Heat	Heat+CO2 + Pressure					Heat+CO2+ Pressure+HCl		Heat+CO2+ Pressure
Fosforik Asit					Heat		Heat			Heat+Cl2		Heat	
Karbon giderici													
KOI giderici													
H202						Heat+O2+ Pressure	Heat			Heat+O2+ Pressure	Heat+O2 Pressure	Heat+O2+ Pressure	Heat
Sülfürik Asit							Heat			Heat+Cl2		Heat	

Figure 3 : Part of WWTU Chemical Interaction Matrix

Also, there was tank service change checklist, it's also examining chemical compatibility.

What is the Key Lessons?

In fact, even it was a near misses this event has many messages to employees. Applying and dominating the standards, importance of communications and using MOC to make comprehensive risk assessment study are main messages to everyone. Sometimes the change is not as simple as it seems. Every desired changes should be evaluated carefully to work and implement safely.

References

1-CCPS, Guidelines for the Management of Change for Process Safety

2-CCPS, Guidelines for Process Safety Metrics, 2010, 4.5.1 Choose Metrics Appropriate for the Operation, 48p

20AIChE Spring Meeting

16TH GLOBAL CONGRESS ON PROCESS SAFETY

HOUSTON • 2020

Explosion mitigation using aqueous solutions of flame inhibitors

Kees van Wingerden
Gexcon AS
Fantoftvegen 38, 5072 Bergen, Norway
kees@gexcon.com

Dirk Roosendans
Total SA
Tour Michelet, La Défense, A0819, Paris, France
dirk.roosendans@total.com

Pol Hoorelbeke
Total SA
Tour Michelet, La Défense, A0819, Paris, France
pol.hoorelbeke@total.com

Prepared for Presentation at
American Institute of Chemical Engineers
2020 Spring Meeting and 16th Global Congress on Process Safety
Houston, TX
March 30 – April 1, 2020

Explosion mitigation using aqueous solutions of flame inhibitors

Kees van Wingerden
Gexcon AS
Fantoftvegen 38, 5072 Bergen, Norway
kees@gexcon.com

Dirk Roosendans
Total SA
Tour Michelet, La Défense, A0819, Paris, France
dirk.roosendans@total.com

Pol Hoorelbeke
Total SA
Tour Michelet, La Défense, A0819, Paris, France
pol.hoorelbeke@total.com

Keywords: explosion mitigation, flame inhibition, water mist, superheated water, water deluge

Abstract

The efficiency of mists of aqueous solutions of flame inhibitors as an explosion mitigation means was investigated on small-scale showing that by increasing the amount of flame inhibitor (potassium carbonate) in water, an increase of the degree of mitigation (reduction of flame velocity and explosion overpressure) is observed up to full quenching.

To generate water-mists directly at a larger-scale in sufficiently high concentrations dispersion of superheated aqueous solutions of potassium carbonate was chosen. Without inhibitors dispersion of superheated water resulted in a clear explosion mitigation in a 50 m^3 obstructed vented volume. Introducing superheated aqueous solutions of potassium carbonate however resulted in a reduction of the mitigation effect.

As an alternative an industrial water deluge system was applied. Again, both pure water and inhibitor-solutions were applied. In this case the explosion itself generates fine droplets due to water droplet break-up. At high water loads explosion peak pressures were reduced considerably by the water deluge but a clear effect of adding inhibitor to the water could not be differentiated.

The paper discusses the possible reasons for the difference of behaviour on small and large scale.

1 Introduction

Vapour cloud explosions are among those events causing the highest losses in the petrochemical industry. Hence, preventive measures are given highest attention: correct design of the facilities (incl. material choice), its maintenance and inspection and choice of equipment approved for use in potentially explosive atmospheres. Research in the past has especially been devoted to predicting the consequences of these events [1, 2, 3]. Less attention has been given to mitigation of vapour cloud explosions, i.e. limitation of consequences.

Considering research that has been performed water deluge has been shown to be effective as a means of explosion mitigation. The water deluge system shall be activated before ignition. The positive impact of water deluge has been documented extensively [4, 5]. Only droplets smaller than 10-20 μm will affect flame propagation due to evaporation in the flame (in methane-air mixtures). Larger droplets need to break-up before having an impact. Flow accelerations occurring during explosions in congested areas are able to break-up droplets larger than typically 200 μm [5]. Hydrodynamic forces acting on droplets in an accelerating flow will cause droplets to break up if these droplets are not able to adapt to the flow accelerations.

In [6] it was suggested to use solid flame inhibitors dispersed into the vapour cloud upon its release as an alternative mitigation technology. Relatively low concentrations (50-100 g/m^3) of solid flame inhibitors such as sodium bicarbonate, sodium chloride, potassium carbonate and potassium bicarbonate are able to reduce flame speeds and overpressures in hydrocarbon-air mixture explosions considerably [7]. A system to disperse the solid flame inhibitors in practice has been developed and its properties to disperse the inhibitor [8] and to suppress explosions [9] has been tested on full scale.

In [10] it was suggested to investigate the possibility of combining the two: applying aqueous solutions of flame inhibitors to mitigate vapour cloud explosion. Experimental investigations were performed on small-scale using water mist-potassium carbonate solutions. Whereas water mist alone already had an impact and caused a reduction of flame speeds and overpressure introducing potassium carbonate in the water resulted in an additional reduction.

The current paper briefly describes the results of the small-scale experiments performed in a 135 l channel but especially addresses the results of tests performed with aqueous solutions of flame inhibitors at a larger-scale. Two sets of large-scale experiments were performed: experiments using superheated aqueous solutions and experiments where aqueous solutions were dispersed using a water deluge system. The main aim of these experiments was to verify the effects seen on small-scale on a larger scale and the possibility of using this mitigation technique on industrial scale. Both sets of large scale experiments were performed in a congested $50m^3$ chamber provided with explosion venting. In all experiments potassium carbonate was used as a flame inhibitor.

2 Description of the test set-ups

The small-scale experiments were performed in a horizontal 135 l test channel with a length of 1.5 m and a cross section of 0.3 m x 0.3 m. The large-scale experiments were performed in a 50 m^3 vessel with a length of 8 m and a cross section of 2.5 m x 2.5 m. Both facilities are described below.

2.1 Small-scale 135 l channel

The small-scale vessel is only open at one of the short ends. Ignition is effected at the other closed short end of the channel. One of the long walls is transparent allowing for high speed filming. Baffles can be inserted in the channel at three different locations.

- at the ignition end (31.0 cm from ignition end)
- in the middle (67.5 cm from ignition end)
- at the open end (148.5 cm from ignition end)

Each baffle consists of five 1.5 cm strips evenly distributed over the cross section and blocking 25 % in total. During the tests none, 1 or 2 baffles were used.

As a flammable gas propane was used which was added to a gas recirculation system, gradually, and mixed with the air until a homogeneous set concentration is reached. During filling the open end is closed with a thin plastic sheet. To inject aqueous solutions a total of 10 fogging nozzles (type BETE® PJ6) are used (5 in the roof, 5 in the back wall). The nozzles generate droplets with 90% < 25 µm at a pressure of 70 bar g which give a flow rate of 0.114 l/min per nozzle (assuming pure water).

Figure 1 shows the 135-l channel.

Figure 1 135-l channel looking towards transparent long side wall. Five (5) of the water fog nozzles are directly connected to the water hoses seen on top of the channel. The recirculation system is visible on the left side behind the channel.

The gas concentration is measured using a Servomex XA4200C1 gas analyser. The explosion pressure is measured at 3 locations in the channel. A high-speed video camera (3000 fps) was sued to record flame propagation in the channel. The water/aqueous solution mist concentration was estimated on the basis of the total water/aqueous solution injected and collected at several locations in the channel and the injection time. In the majority of the tests the injection of the mist lasted 10 s resulting in average water concentrations of 30 – 80 g/m^3 (4 – 11 mol %). Concentrations were generally higher close to the open end of the channel.

2.2 Large-scale 50 m^3 vessel

The two series of large-scale tests were performed in a congested chamber 8.0 m long, 2.5 m high and 2.5 m wide, thus having a volume of 50.0 m^3. The chamber has a coarse steel grating held by beams forming a mezzanine deck approximately 1.25 m above the floor of the vessel. During the current tests one or both of the short ends were used as vent openings. As Figure 2 shows the chambers contains several objects (obstructions) which cause explosion flames to accelerate due to turbulence generated at these obstructions.

Figure 2 50 m^3 explosion chamber used to investigate effect of aqueous solutions of flame inhibitors

Also during the large-scale experiments propane was used as a flammable gas. As on small-scale the propane-air mixtures were prepared by adding the propane to a gas recirculation system, gradually, and mixed with the air until a homogeneous set concentration was reached. During filling the open ends of the chamber were closed with a thin plastic sheet. The gas concentration of the mixtures was monitored using an infrared gas analyser (type Servomex 4900).

To generate mist-air mixtures as obtained during the small-scale experiments would imply the use of a very large number of nozzles (3700) all to be provided with hoses and pressurised water/aqueous solution. This is not a feasible solution for industrial applications. To generate water-mists directly at a larger-scale in sufficiently high concentrations a dispersion of superheated water/aqueous solution seemed to be the only viable solution. To investigate the possibility of using superheated water in the 50 m^3 vessel three 4 litre suppression bottles supplied by IEP Technologies UK were applied. Each of the bottles were modified and provided with a 1500W heating jacket and a controller from IsoHeat, a thermocouple and a pressure transducer. The pressure inside the bottles was controlled by adjusting the temperature of the heating jackets. The pressure inside the bottles was also monitored during the dispersion of the water/aqueous solution. The temperature of the water/aqueous solution was monitored using thermocouples. The bottles were either filled with 2 or 3.5 litres of water or aqueous solution. Each bottle was connected to a nozzle and mounted along the centre line of the roof of the explosion chamber, regularly divided over its length.

Ignition was effected 6 s after the moment of dispersion to limit the effect of turbulence caused by the injection. The ignition source was an exploding wire located close to one of the short walls of the chamber (1 m from the end in the centre).

As a second part of the experimental work, the superheated water/inhibitor-solution dispersion system was substituted by a regular deluge system consisting of 7-11 nozzles of type MV12 125° with K-factor of 18.7. The nozzles were mounted regularly divided against the roof of the chamber. The droplet size delivered by these nozzles was not measured but on the basis of results presented for other MV nozzles it is expected to be between 200 and 600 μm with an median value of about 400 μm [11].

For the experiments using water deluge the configuration of the 50m^3 explosion chamber was changed from two open ends to one open end and one closed end. Ignition was effected close to the closed end to provoke higher flow velocities promoting droplet break-up. The total water/aqueous solution application rate was varied between 9.5 $l/min/m^2$ to 22 $l/min/m^2$.

3 Results

3.1 Small-scale tests

The parameters varied during the small-scale experiments in the 135-l channel were especially the propane-air concentration, the degree of congestion and the composition of the aqueous solution. All small-scale experiments were performed using potassium carbonate (K_2CO_3).

Generally speaking, the experimental results show that water mist in itself already causes a minor reduction of both flame speeds and overpressures (for the water concentrations used of 30 -80 g/m^3). Introducing potassium carbonate into the water the degree of mitigation is observed to increase up to full quenching of the explosion flame (for potassium carbonate concentrations of 60-70 g/l in water. This was seen for all obstacle configurations investigated including a channel without obstacles. An example of typical results obtained is summarised in Figure 3.

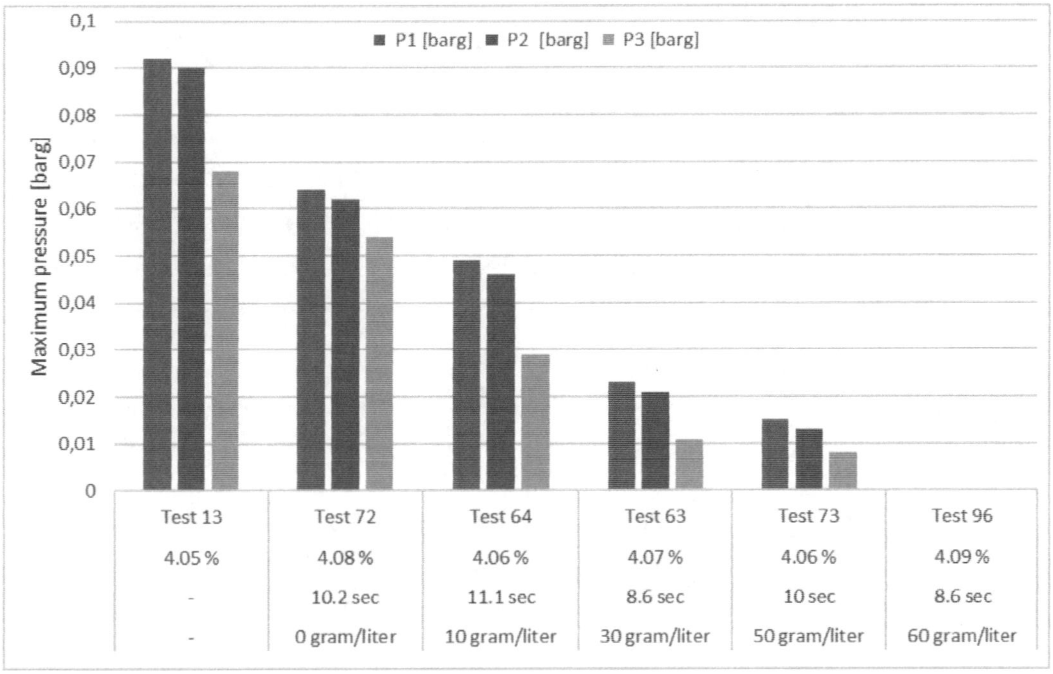

Figure 3 Effect of aqueous solutions of potassium carbonate on explosion overpressures in 4.05-4.09 % propane-air mixtures in a 0.135 m^3 channel with two baffles (at 31.0 cm and 148.5 cm from ignition end). Overpressures are measured at 3 locations (P1-P3) for increasing amounts of the flame inhibitor dissolved in water (increasing from 0 to 60 g/litre). The water mist concentration is determined by the injection duration varying from 8.6 to 11.1 s.

The overall results show that water droplets of a size of 90% < 25 μm in a concentration of 30-80 g/m^3 appear able to reduce the overpressures in the channel by 44-83% depending on the obstacle configuration. Adding potassium carbonate to the water causes this reduction of overpressures to increase. The more potassium carbonate is added to the water the stronger the mitigation of the propane explosions is up to full quenching (i.e. no ignition) when the potassium carbonate concentration in the water is 60 g/l or more.

3.2 Large-scale tests: Superheated aqueous solutions

The principle of superheated water is using stored thermal energy. The bottles are heated up to temperatures of 180 °C or higher resulting in a saturated water pressure in the bottles of 10 bar or higher. The thermodynamic condition of the water is now such that upon a sudden removal of the confinement (opening of the bottle) the water flashes immediately resulting in a water vapor cloud. This cloud cools off rapidly upon contact with the surrounding air and condensates giving rise to a cloud of very fine droplets. A part of the superheated water does not flash resulting in larger droplets. As a result, discharge of a superheated water system as described above results in a bimodal droplet diameter distribution. For the system used in the current investigation the droplet distribution was not measured but in [12] a similar system was reported to have a bimodal droplet distribution with modes located at 10 μm and 120 μm. The relative composition by weight was reported to be 25 % of 20 μm droplets and 75 % of 180 mm droplets.

In the current tests the distribution may be different due to generally higher temperatures used (on average 206 °C). It is expected that this thermodynamical condition leads to a higher fraction of fine droplets.

Nevertheless, assuming that approximately 25 % is present as small droplets one can assume that after 6 s delay time this is the only fraction left in the gas-air mixture. The terminal velocity of droplets of 180 μm is about 0.7 m/s and since the initial jet is pointing downwards this seems a valid assumption. If 25 % of the initial dispersed water is present in the mixture at the moment of ignition this implies a water fraction of 4 mol % if 6 l is dispersed and a water fraction of 7 mol % if 10.5 l is dispersed.

Initially reference tests were performed to determine the explosion overpressures in the obstructed 50 m^3 vessel without any mitigation except from the vent openings. Next tests were performed with superheated water without the introduction of inhibitor into the water. Results obtained for 4.15 % propane-air mixtures are summarised in Figure 4 showing a clear reduction of overpressures when increasing the amount of superheated water released from 0 to 10.5 l. The effect of the water can also be recognised from the delay the maximum overpressure is generated.

In [12] the design efficiency $D_E$of the superheated water mitigation system was expressed as:

$$D_E = \frac{Measured\ value\ of\ C_E}{Theoretical\ maximum\ value\ of\ C_E\ for\ the\ mitigation\ method\ used}$$

Where $C_E = (P_u - P_S)/P_u$

With P_u = the maximum pressure without the superheated water system activated

P_s = the maximum pressure with superheated water system activated

Using this as a way of presenting the results of the experiments performed the effect of the superheated water system using water without adding any inhibitor would be as shown in Figure 5 presenting the results for both lean (3.5 vol % propane-air), optimal (4.1 vol % propane-air) and rich propane-air mixtures (5.0%vol. propane-air). The theoretical maximum value of C_E has been assumed to be 1 implying full quenching of the flame as seen for the small-scale experiments. The Figure shows that the mitigation is similar for lean and rich mixtures showing an increase of the design efficiency when the amount of water released increases. For rich mixtures the effect of the water is however marginal: hardly a reduction of the overpressure is seen even for a 7 mol % water fraction.

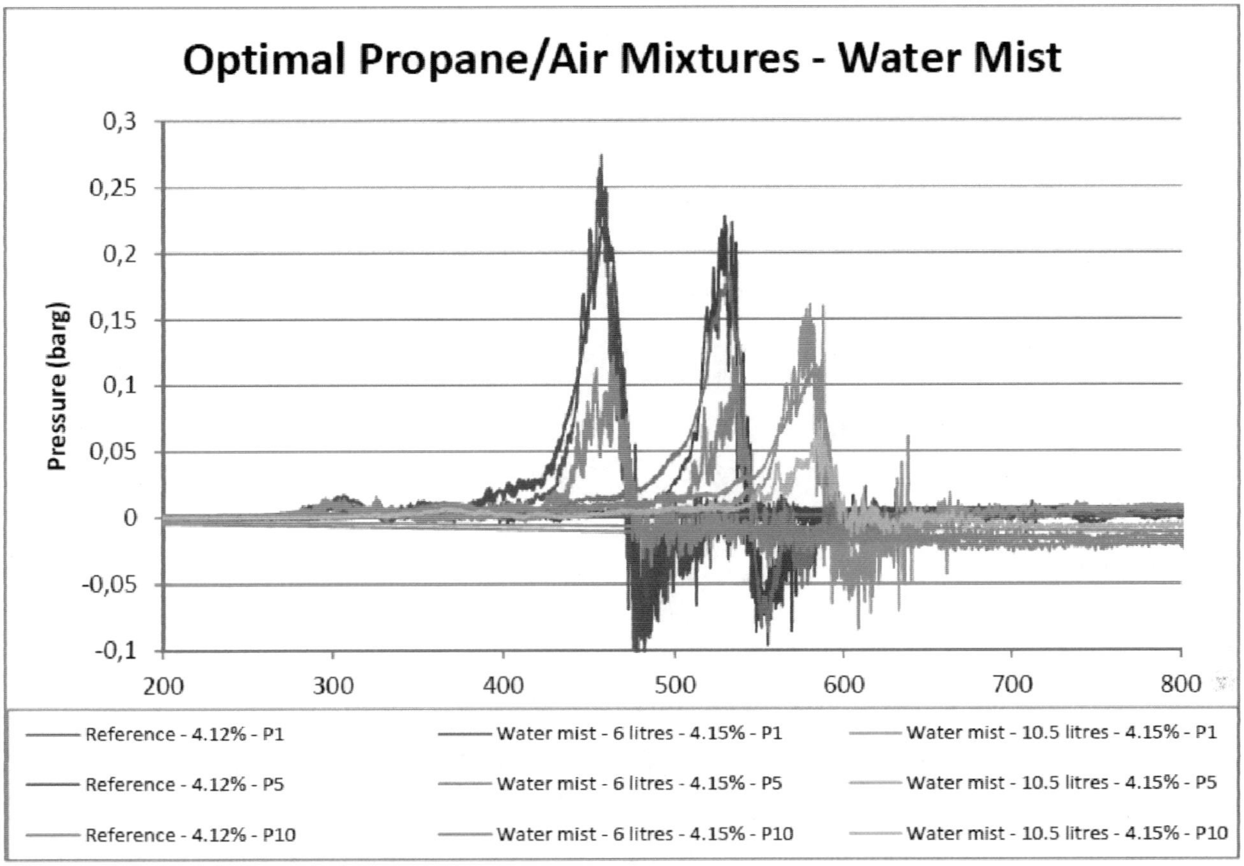

Figure 4 Effect of superheated water on overpressure generated by optimal propane-air explosions in a vented 50 m^3 obstructed chamber. The amount of water released was increased from 0 to 10.5 l.

As a comparison: the small-scale experiments (for optimal propane-air mixtures) resulted in a design efficiency of 0.32 considering pure water mists (for a water fraction of 4-11 mol %) which is very comparable.

Introducing potassium carbonate into the water in relative amounts up to exceeding those used in the small-scale experiments did not result in a considerable additional reduction of the overpressures seen. In fact, pressures seem to increase slightly the more potassium carbonate was added. An example of results is shown in Figure 6. The tests were performed injecting 10.5 l water, a water fraction of 7 mol % (assuming that 25 % of the initial water injected remains air bound as small droplets (< 20 μm)). The potassium carbonate concentration in the water was gradually increased from 30-210 g/l (the latter is far more than used in the small-scale experiments where full quenching was observed for concentrations of 60 g/l). Figure 6 shows that flame propagation seems to be affected in the initial phase causing the moment of occurrence of the maximum overpressure to be delayed for increased amounts of potassium carbonate. This could indicate a reduction of combustion rates at least during the initial phases. The maximum pressure is however increasing. The reason for this behavior is not clear. [13]

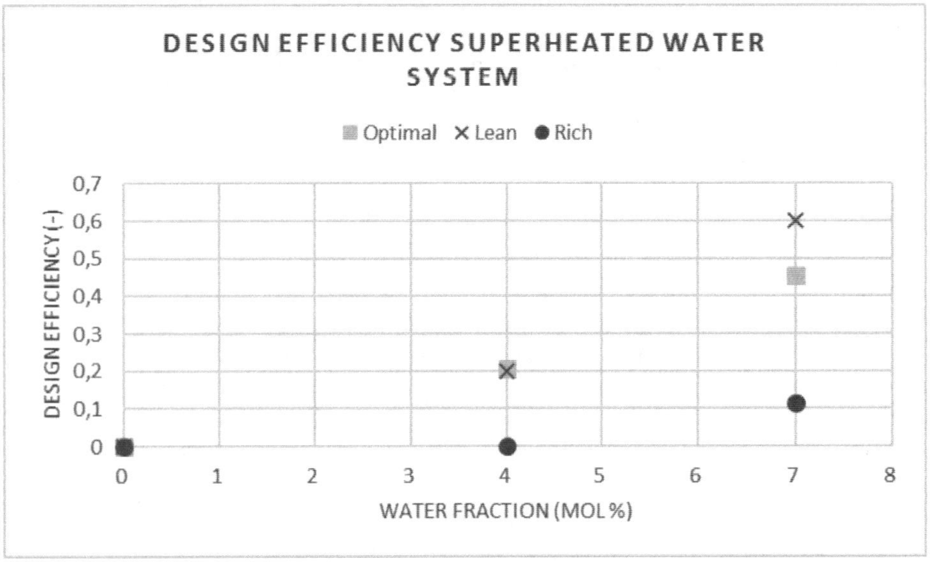

Figure 5 **Design efficiency of superheated water system for lean, rich and optimal propane-air mixtures as a function of the water fraction.**

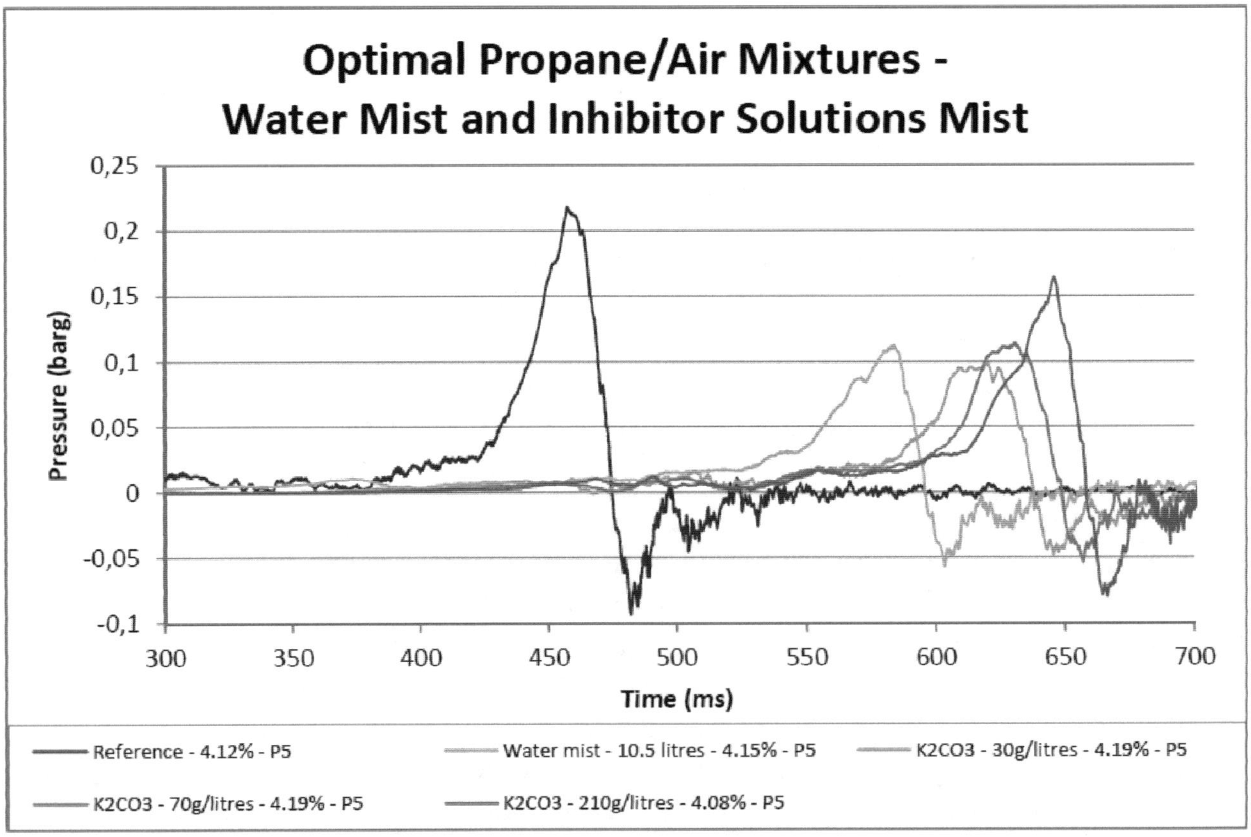

Figure 6 **Effect of superheated aqueous solutions of potassium carbonate on overpressure generated by optimal propane-air explosions in a vented 50 m^3 obstructed chamber. The amount of aqueous solution released was 10.5 l. The potassium carbonate concentration was increased from 0 to 210 g/l water.**

points in the direction of the higher viscosity of the solution as a possible explanation. A higher viscosity would however result in a higher concentration of fine droplets.

A better explanation is to be found in [14] where it was concluded that dissolved salts tend to displace superheated conditions to higher temperatures. The higher the potassium carbonate concentration the higher the temperatures necessary for obtaining superheat conditions and hence the lower the efficiency of the mitigation system. Using superheated aqueous solutions of flame inhibitors as a technology to generate fine mists to mitigate explosions is therefore not a feasible technology.

3.3 Large-scale tests: water deluge system

Since the superheated water/aqueous solution system did not result in the mitigation seen on small-scale it was decided to apply an industrial water deluge system instead. In this case the explosion itself generates fine droplets due to water droplet break-up. This implies that the aqueous solution mitigation system, if it works, only can be applied in congested areas where flame accelerations, due to the interaction between combustion generated turbulence and the combustion, occur. Droplet break-up occurs when a droplet is exposed to a flow resulting in hydrodynamic forces stronger than the forces keeping the droplet together, viz. surface tension. This can be expressed by the Weber number:

$$We = \rho v^2 d/\sigma$$

where

We = Weber number

ρ = density of the gas mixture stream (kg/m^3)

v = velocity of the gas mixture stream relative to the velocity of the droplet (m/s)

d = diameter of droplet (m)

σ = surface tension of liquid (N/m)

The critical Weber number where droplet start to break up is We = 10-12 [5].

The experiments were performed in the 50 m^3 vessel with a similar obstruction layout. The only difference was that one of the open ends was closed resulting in stronger flame acceleration and thereby higher pressures. Ignition was effected close to the closed end of the vessel. involved both reference tests, tests with water and tests with aqueous solutions. Figure 7 shows an example of results obtained. The maximum pressure in case of no mitigation was 1.7 bar. Using a relatively high water application rate of 21.9 $l/min/m^2$ results in a maximum pressure of 1.2 bar. Using the same application rate for an aqueous solution of potassium carbonate of 100 g/l results in a maximum pressure of 1.05 bar, implying some effect of the inhibitor but not as much as one would expect from the small-scale experiments. The application of water deluge as a mitigation system implies that the explosion is allowed to develop initially up to the moment droplet break-up occurs and where mitigation sets in. Turbulence generated by the water deluge system causes the flame to propagate faster during the initial phase than the non-mitigated test. The maximum pressure obtained is therefore a combination of the effect of turbulence during the initial phases and the degree of mitigation afterwards. Figure 8 shows the effect of varying the water application rate on the maximum explosion overpressure. A lower water application rate implies less turbulence generation initially but also lower water fractions after droplet break-up, i.e. less mitigation.

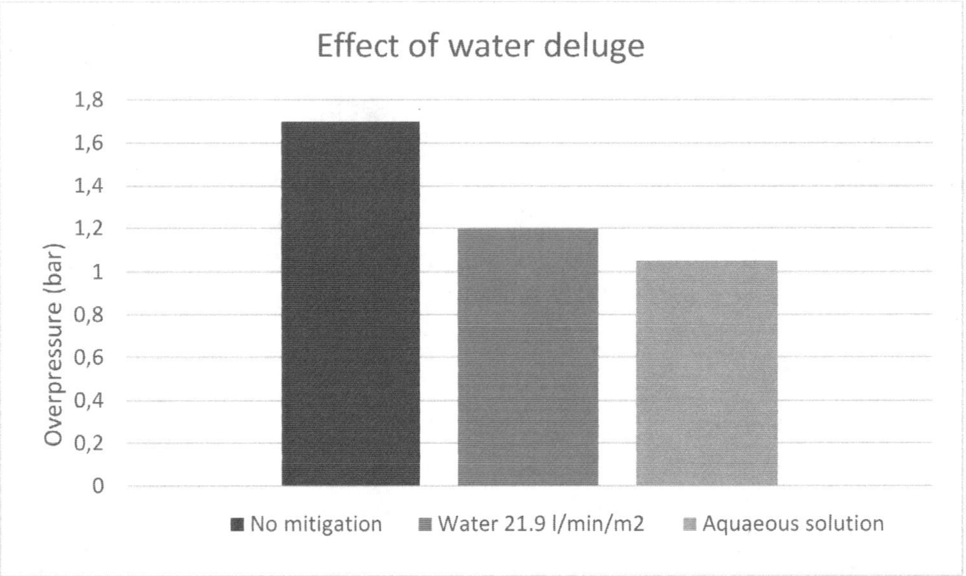

Figure 7 Maximum overpressure obtained for 4.1 vol % propane/air explosions with a water loading of approximately 21.9 l/min/m^2 with and without a potassium carbonate solution of 100 g/l, compared to the dry reference test.

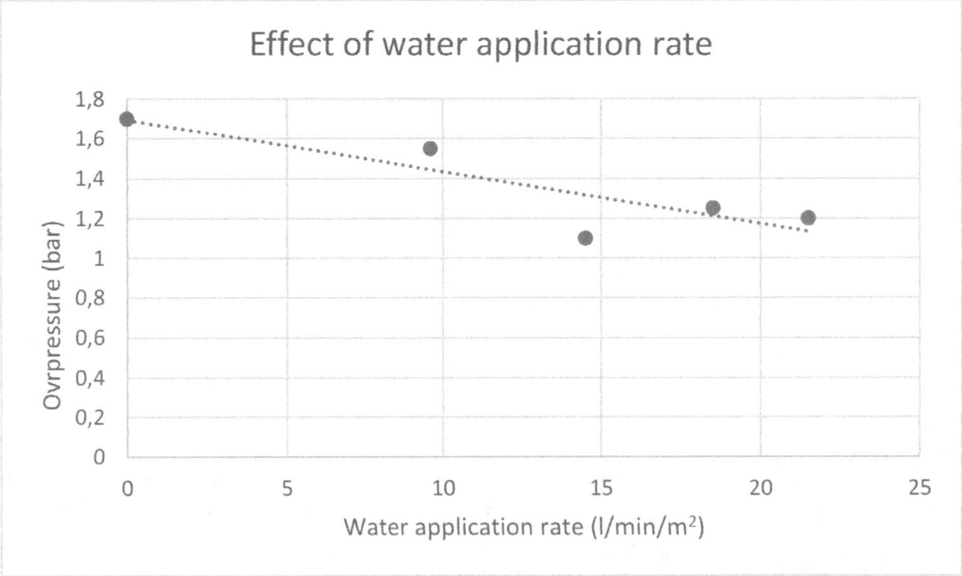

Figure 8 Maximum overpressure obtained for 4.1 vol % propane/air explosions varying the water application rate.

In an attempt to distinguish the effect of initial turbulence from the mitigating effect of the aqueous solution tests were performed with low application rates reducing the initial turbulence whereas the mitigating effect was still expected to be considerable. This, however, appeared to be disappointing as seen in Figure 9.

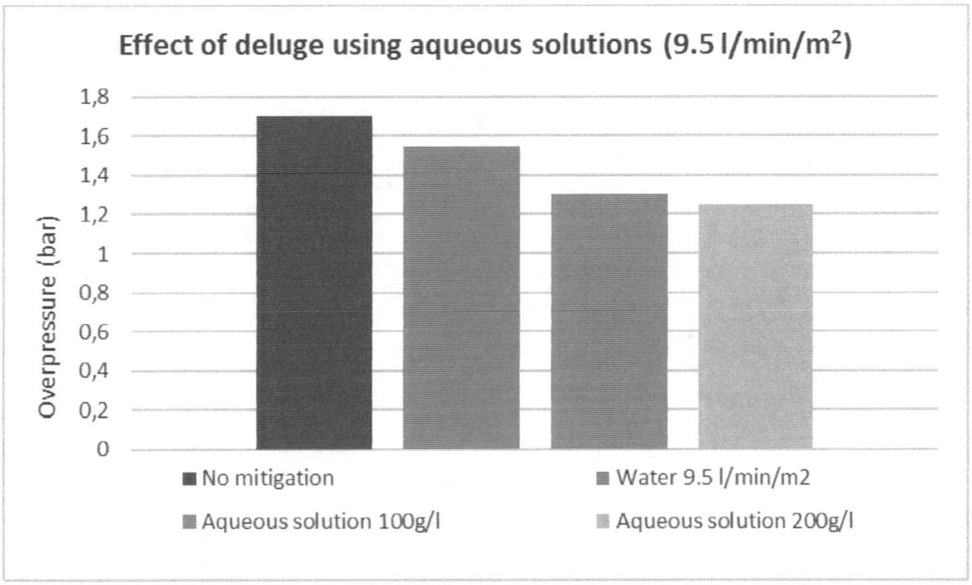

Figure 9 Maximum overpressure obtained for 4.1 vol % propane/air explosions for a low application rate (9.5 l/min/m^2) with water and aqueous solutions of 100 g/l and 200 g/l of potassium carbonate.

Although aqueous solutions do result in some improved mitigation compared to pure water the effects are limited. A possible reason may be that some salts (such as NaCl) cause an increase of the surface tension of water making it more difficult for droplets to break up [15] by flow as the Weber number suggests. The effect of an increased water surface tension and increased viscosity due to the introduction of potassium carbonate into the water has hardly any influence on the droplet characteristics of the water nozzle used [16].

4 Conclusions

Whereas direct injection of mists of aqueous solutions of water and potassium had considerable improved mitigation effects on explosions in a small-scale congested channel compared to injection of water mist only a similar effect could not be reproduced on a larger scale.

Generation of water mist using a superheated water injection system resulted in a considerable reduction of explosions overpressures in a congested 50 m^3 chamber when using pure water. This effect was however not improving any further when applying aqueous solutions of water and potassium carbonate. An explanation for this is that the adding of potassium carbonate causes superheated conditions to be displaced to higher temperatures. The higher the potassium carbonate concentration the higher the temperatures necessary for obtaining superheat conditions and hence the lower the efficiency of the mitigation system. Using superheated aqueous solutions of flame inhibitors as a technology to generate fine mists to mitigate explosions is therefore not a feasible technology.

Using a water deluge system instead, aqueous solutions were shown to have a small improved mitigation effect on explosions compared to that of pure water only. A possible reason may be that introduction of potassium carbonate in the water causes an increase of the surface tension of water making it more difficult for droplets to break up in an accelerating flow.

As such there is a challenge to apply aqueous solutions as a measure for mitigation of large-scale vapour cloud explosions since generation of sufficiently fine mists of aqueous solutions on an industrial scale is not possible with current technologies.

5 Acknowledgments

The authors acknowledge the financial support obtained from Total SA for performing these experiments.

6 References

[1] Van den Berg, A. C., The multi-energy method: A framework for vapour cloud explosion blast prediction, Journal of Hazardous Materials, 12 (1), (1985), 1-10

[2] Tang, M.J. and Baker, Q. A., A new set of blast curves from vapor cloud explosion, Process Safety Progress, 18 (4), (1999), 235-240

[3] Puttock, J.S., Developments in the Congestion Assessment Method for the prediction of vapour-cloud explosions, Loss Prevention and Safety Promotion in the Process Industries, Proceedings of the 10th International Symposium (2001), 1107-1133

[4] Catlin, C.A, Gregory, C.A.J., Johnson, D.M. and Walker, D.G., Explosion mitigation in offshore modules by general area deluge. Trans. I.Chem.E. 71B, (1993), 101-111

[5] van Wingerden, K., Mitigation of gas explosions using water deluge. Process Safety Progress, 19 (3), (2000), 173-178

[6] Hoorelbeke, L., Numerical and Experimental Study of Vapor Cloud Explosions, PhD Thesis. Univ. of Brussels, 2011

[7] van Wingerden, K. and Hoorelbeke, P., On the potential of mitigating vapour cloud explosions using flame inhibitors, American Institute of Chemical Engineers, Seventh Global Congress on Process Safety (GCPS), Chicago, Illinois, USA, 2011

[8] Roosendans, D., Hoorelbeke, P., and van Wingerden, K., Mitigation of vapour cloud explosions by chemical inhibition, IChemE Symposium Series 158, (2012), 313-326

[9] Davis, S.G., Merilo, E., Engel, D., Ziemba, A., Pinto, M. and van Wingerden, K. Large scale detonation testing: New findings in the prediction of DDTs at large scales, Journal of Loss Prevention in the Process Industries 48, (2017), 345-357

[10] Roosendans, D., van Wingerden, K., Holme, M.N. and Hoorelbeke, P., Experimental investigation of explosion mitigating properties of aqueous potassium carbonate solutions, Journal of Loss Prevention in the Process Industries, 46, (2017), 209-226

[11] Wighus, R., Droplet size, velocity and area distribution – Deluge nozzles, SINTEF NBL report no. A12114, 2012

[12] Tam, V.H.Y., O'Connell, M., Pedersen, G. and Renwick, P., Testing of the Micromist device: an active soft barrier for explosion control, Journal of Loss Prevention in the Process Industries, 16(1), (2003), 81-88

[13] Roosendans, D., Mitigation of Vapor Cloud Explosions by Chemical Inhibition using Alkali Metal Compounds, PhD Thesis Vrije Universiteit Brussel, 2018

[14] Thiéry, R., Loock, S. and Mercury, L., Explosive properties of superheated aqueous solutions in volcanic and hydrothermal systems, Metastable Systems under Pressure (NATO Science for Peace and Security Series A: Chemistry and Biology), 2010th Edition (ed. Rzoska, S. et al), 293-310

[15] Okura, H.I., Chena, Y., Wilkins, D.M. and Rokea, S., The Jones-Ray effect reinterpreted: Surface tension minima of low ionic strength electrolyte solutions are caused by electric field induced water-water correlations, Chemical Physics Letters, 684, (2017), 433-442

[16] Kooij, S., Sijs, R., Denn, M. M., Villermaux, E. and Bonn, D., What Determines the Drop Size in Sprays? Physical Review X 8, 031019 (2018)

585375 Contain the Tiger: Accounting for All Energy from Reactive Chemicals in Process Design

Abstract:

Chemical manufacturing processes rely on intended chemical reactions to create desired products. During this process, energy from both intended and unintended sources can be generated, and needs to be safely accommodated through process design and safety systems. This paper will identify test methods for characterizing the energy produced by chemical reactions and review process design strategies including inherently safer design options. Several types of chemistries will be reviewed using methods contained in the Center for Chemical Process Safety (CCPS) book Guidelines for Chemical Reactivity Evaluation and Application to Process Design.

A case study will be reviewed to illustrate the concept and application of these critical process design elements for reactive chemical systems, focusing on the proper use of kinetic parameters, thermal stability data, and heat transfer requirements to develop an appropriate Basis of Safety for the reactive system design.

The following learning objectives will be achieved:

1. Introduce a practical approach for Reactive Energy Hazard Evaluation.
2. Overview a disciplined approach for reviewing test data to identify potential energetic scenarios.
3. Understand critical elements of process design for reactive chemical systems.

Chemical Process Safety and Fire Protection – the Role Fire Protection Plays in Your Process Hazard Analysis

Ron Melucci, Technical Manager – Fire Protection
Janet Persechino, Technical Manager – Chemical Process Safety
Engineering Planning and Management, Inc.
959 Concord Street
Framingham, MA 01701
rbm@epm-inc.com
jap@epm-inc.com

Abstract

In any chemical process the fundamental principles of process safety should be applied to decrease the likelihood of a release. Similarly, the fundamental principles of fire protection should be applied for mitigation of the consequences of a release. Focusing on process safety guidance from the Occupational Safety and Health Administration (OSHA), the Environmental Protection Agency (EPA) and other relevant industry practices such as the Center for Chemical Process Safety (CCPS) Risk Based Process Safety and FM Global Data Sheets will decrease the likelihood of releases and catastrophic losses. Fire protection practices outlined in the National Fire Protection Association (NFPA) Fire Code and the International Fire Code are also essential for loss mitigation. These process safety and fire protection methodologies work in tandem and should not be considered as part of the overall risk assessment process.

Chemical plants require more specialized fire protection features and systems than most industrial facilities. The extreme hazards associated with the chemical processing industry (CPI) can quickly lead to a cascading sequence of events that can result in a catastrophic loss. An unwanted release of flammable liquids, if not properly controlled, can overtax sprinkler systems with several minutes and compromise the integrity of vessels, piping, safety interlocks and structural steel elements. This can lead to a collapse of process equipment/structures, loss of process control, devasting off-site consequences and significant business interruptions.

The CPI faces unique challenges when it comes to the automatic sprinkler protection placement and installation and other fire mitigation techniques. This paper will outline some of the different protection schemes and how to decide which protection option is best suited for a particular application.

The overall conclusion is that process safety must be applied for loss prevention and this should be combined with active and passive fire protection features to mitigate the consequences of a release should lapses in process safety occur. These topics should not be considered separately but viewed as working in partnership. It is important that these concepts are implemented and reviewed by experts in the fire protection and process safety fields to ensure adequate protection schemes are employed in a proper manner and the information is correctly covered and credited in the Process Hazard Analysis (PHA).

Introduction

In any chemical process the fundamental principles of process safety should be applied to decrease the likelihood of a release. Similarly, the fundamental principles of fire protection should be applied for mitigation of consequences. Focusing on process safety guidance from the Occupational Safety and Health Administration (OSHA), Environmental Protection Agency (EPA) and other relevant industry practices geared specifically for chemical process safety management will decrease the likelihood of releases and catastrophic loss. Fire protection practices are outlined in the National Fire Protection Association (NFPA) Fire Code NFPA 1 and the International Fire Code (IFC). These process safety and fire protection methodologies work in tandem to achieve loss prevention and mitigation.

The applicable codes (mentioned above) require Process Hazard Analyses (PHAs) and Fire Hazard Analyses (FHAs) for most new chemical plants and those undergoing changes. The codes state that the analyses are to be performed by someone with expertise in chemical process safety and fire protection. Additional building and system safeguards are often implemented as a result of these structured and systematic analyses. This can include improved exothermic reactor vessel cooling systems, isolation valve installation, pipe rack protection, etc.

FM Global Datasheet 7-14, *Fire Protection for Chemical Plants,* indicates that over 60% of the surveyed chemical plant losses between 2003 and 2014 were attributed to equipment design problems and operator error. These items are evaluated when the correct hazard analysis methodology is employed. A PHA will include an evaluation of adequate system safeguards and human reliability aspects. This type of comprehensive review should decrease the chance of having a loss due to a design or human element issue.

Chemical plants require more specialized fire protection features and systems than most industrial facilities. The extreme hazards associated with the chemical processing industry (CPI) can quickly lead to a cascading sequence of events that can result in a catastrophic loss. An unwanted release of flammable liquid, if not properly controlled, can result in a fire that can overtax sprinkler systems within several minutes. Such uncontrolled fires can compromise the integrity of vessels, piping, process safety systems or structural steel elements. This can lead to a collapse of process equipment/structures and a loss of process control which can result in:

- ✓ Fatalities
- ✓ Serious injuries
- ✓ Large fires/explosions
- ✓ Significant property damage
- ✓ Environmental issues
- ✓ Extensive Downtime
- ✓ Loss of market share
- ✓ Damage to company reputation
- ✓ Difficulty in obtaining insurance coverage
- ✓ Fines/penalties

A sprinkler system may control a chemical plant fire by cooling and wetting surrounding areas but may not extinguish fires involving flammable gases, liquefied flammable gases and low flash point flammable liquids. Additional consideration may therefore be needed for special fire protection systems that protect critical equipment from fire damage, particularly where loss of equipment/components can result in catastrophic losses as described above.

Fire Protection and the PHA

The fire protection systems should be designed in conjunction with, or in consideration of, the PHA. The PHA will include identification of critical equipment that is needed to prevent catastrophic loss and will estimate the likelihood and severity of unwanted releases.

Fire protection system design for chemical plants with flammable liquid hazards requires careful consideration of the fire code requirements as well as other requirements such as NFPA 30, *Flammable and Combustible Liquids Code,* and the guidelines in FM Global Datasheet 7-14, *Fire Protection for Chemical Plants*. Installation guidelines such as NFPA 15, *Standard for Water Spray Fixed Systems for Fire Protection* and NFPA 13, *Standard for Installation of Sprinkler Systems,* are also applicable.

To meet the fire protection objectives and thus loss mitigation, both passive and active (fire) protection features and systems should be implemented. Common and effective features include:

Table 1 Accidental Release Consequence Mitigation Methods

PASSIVE PROTECTION	ACTIVE PROTECTION
Spatial or Fire Barrier Separation	Area-type Automatic Sprinkler or Deluge Protection
Fire Proofing of Structural Steel/Equipment Supports	Fixed Directional Deluge Water Spray Protection
Containment and Emergency Drainage	Process Safety Features such as Safety Shut Off Valves

Active fire protection should cover all areas where a flammable liquid release may flow. Containment systems may include emergency drainage that drains to a secondary containment area to help remove the released liquid and discharged water to an impoundment area that is located far from equipment and buildings.

Fire Protection Installation Challenges for the CPI

Installation of sprinklers close to the ceiling may not be effective in many CPI applications. Sprinklers are often installed far below ceilings, or below open metal grating, which can significantly delay sprinkler activation. There are two common reasons for this:

- Stratification - There may be insufficient fire plume buoyancy and temperature at high ceiling levels. It is imperative that sprinklers activate in the first several minutes of a fire and even this type of small delay in sprinkler activation, especially given the high hazard materials involved in the CPI, can mean complete loss of fire control which can result in a catastrophic event

- There may be no ceiling jet formation to cause sprinkler activation as in the case of a fire occurring below open metal grating. Improper sprinkler system design can lead to complete loss of fire control and a catastrophic loss scenario

Due to the installation challenges, sprinklers with heat collectors installed far below ceilings and below metal grating, have become standard practice. For this type of design, sprinkler activation is only achieved if the sprinkler(s) are located within the fire plume geometry (i.e., close to the vertical axis of the fire plume), to intercept the buoyant hot gases needed to actuate the sprinkler fusible element that allows the application of water.

The industry guidelines for such sprinkler design are not specific, prescriptive requirements. Therefore, a performance-based analysis is required of the fire protection engineer. NFPA 13 *Standard for the Installation of Sprinkler Systems*, does not specifically address the spacing requirements for sprinklers installed under open metal grating. However, section 8.5.5.3.4 of the 2016 edition of NFPA 13 contains the following verbiage: "*Sprinklers installed under open gratings shall be of the intermediate level/rack storage type or otherwise shielded from the discharge of overhead sprinklers*." The annex material elaborates that shields (not heat collectors), should be installed on sprinklers under open grating. The shields are intended to protect sprinklers from discharge of overhead sprinklers, but the intent is often misunderstood as heat collection.

NFPA 13 Section 8.5.4.1.4 states: "*Heat collectors shall not be used as a means to assist the activation of a sprinkler*." Studies have shown that "heat collectors" can actually delay the activation of the sprinkler by forming a dead air pocket. For sprinklers installed far below solid ceilings or below metal grates, sprinkler activation is only achieved as stated above, if the sprinklers are within the fire plume geometry.

Sprinklers are sometimes installed far below the ceiling or below metal grate platforms for protection of critical equipment. This may include directional sprinklers or nozzles to provide exposure protection to the vessel or equipment, in lieu of fire extinguishment or control. This can often be the preferred fire protection approach to mitigate fire damage. The exposure protection objective described in NFPA 15, *Standard for Water Spray Fixed Systems for Fire Protection*, notes that the protection objective is to keep the vessels and other critical equipment sufficiently wet and cool to protect these assets from fire exposure.

Conclusion

The CPI faces unique challenges when it comes to automatic sprinkler protection. This type of protection can be used for different purposes in the CPI. An assessment of the risk, plant layout and overall loss mitigation goals will help decide which protection option is suited to the application. Due to the subjectivity and lack of definitive industry guidance for every situation, a qualified fire protection engineer who is very familiar with these requirements and guidelines, should make the fire protection design decisions described herein.

The overall conclusion is that process safety must be applied for loss prevention to decrease the likelihood of a loss occurrence and this should be combined with active and passive fire protection for mitigation of consequences. These topics should not be considered separately, but viewed as working in partnership as proactive and reactive systems. It is important that these subjects are administered and reviewed by experts in fire protection and process safety fields to ensure adequate protection schemes are employed.

583240 Experimental Study on the Characterisation of Flammability of Methane Enriched in Hydrogen in the Framework of Risk Assessment of Injection of Hydrogen in Natural Gas Network (concept "POWER TO GAS")

Abstract:

In the context of planning a decarburization of the European Union energy sector, in order to reach a reduction of carbon emissions up to 95 % by 2050, several studies on the compatibility of the existing gas networks with blends of natural gas enriched with hydrogen up to 20% of hydrogen have been tested in industrial scale projects.

The explosion process of multi-component gas mixture is extremely complex and may cause serious disaster effects. The safety issue concerning explosion of multi-component gas mixture is urgent to be investigated on account of its wide range of applications. In current work, series of experiments to characterize severals parameters (Lower Explosive Limit - Upper Explosive Limit - Lower Oxygen Concentration - Maximum Pression - Gas explosion constant - Minimum Ignition Current - Maximum Experimental Safe Gap - Ignition Sensitivity) were performed in a 20 L spherical explosion vessel at initial conditions of 1 atm and 293 K, involving methane-hydrogen/air mixtures.

In relationship to the European Union ATEX Directive 2014/34/EU and protective systems intended for the use in potentially explosive atmospheres, gases and dusts are classified according to their ignition properties. The explosion group for gases and vapors is determined by the maximum experimental safe gap (MESG) and/or minimum ignition current (MIC) ratio values, both defined in the ISO/IEC 80079 (Explosive Atmosphere—Part 20-1: Material characteristics for gas and vapour classification—Test methods and data) standard.

This paper presents the experimental results produced by Ineris highlighting the influence of the addition of hydrogen on the flammability and explosive properties of the natural gas-hydrogen mixture (0-20% Vol.).

The presence of hydrogen has a negligible influence on certain properties such as:

o the lower explosive limit,

o the oxygen limit concentration,

o the maximum explosion pressure,

o the self-ignition temperature;

The following properties are more sensitive to hydrogen content:

o the upper limit of explosiveness increases linearly with the hydrogen content in methane;

o Industrial methane, like natural gas, is classified in Group IIA on the grounds that it contains no more than 25% Vol. hydrogen. A mixture of methane with other compounds of group IIA of any proportion is classified in Group IIA;

o the maximum pressure-mounting speed remains constant and equal to that of methane up to about 15 % hydrogen, but increases rapidly for higher hydrogen levels;

o the minimum ignition energy decreases rapidly as the hydrogen content increases, suggesting an increase in the probability of inflammation with the hydrogen content (which, however, has yet to be confirmed by specific tests). However, for low hydrogen levels (10% max), the probability of ignition, although greater than for natural gas alone, would remain in the same orders of magnitude.

587043 Maintenance Error Management for Improving Mechanical Integrity

Abstract:

Weak management systems lead to Human Errors that lead to accidents. The paper covers significance of maintenance errors (ME) in process industry, what and why of ME, principles of ME, extent of ME, and how to deal with ME. It proposes to have a structured program in place and what must be considered while formulating the program at a company. Limited deployment of the program at some intense maintenance environment at process facilities has shown promising results.

Maintenance errors have been among the principal causes of several major accidents in a wide range of technologies, for instance Piper Alpha disaster. For process industry, maintenance, particularly maintenance of safety critical equipment, has always been addressed as important means to prevent major accidents. Maintenance is indissolubly linked with safety in a number of ways - errors in planning, executing or checking maintenance could cause system faults; or inadequate or erroneous maintenance may also mean that existing failure or degradation is not discovered and corrected – and thereby contributes to production shut-downs, work accidents and/or major incidents. Maintenance errors not only endanger lives and assets, they are also extremely bad for business.

Process plant maintenance is a particularly challenging task due to several factors such as frequent removal and replacement of a large numbers of components, cramped, arid, poorly lit spaces and different groups work on the same item of equipment. Turnaround maintenance poses further heightened risk.

The remedy is to have a structured Maintenance Error Management Program. A wide variety of error management (EM) measures are resorted to targeting person and team, workplace and task, and organizational factors. Different types of error require different counter-measures; engineering solutions, collective mindfulness and organizational culture that is just, willing to report its unsafe acts and able to learn from them. It is not enough to deploy some tools but need to be watched, adjusted and managed on regular basis.

Aviation industry has recognized this and has mature programs in place. These could serve as examples for chemical process industry. For instance Maintenance Error Management System by Civil Aviation Authority, UK and Boeing's human factors tool called the Maintenance Error Decision Aid (MEDA) for investigating contributing factors to maintenance errors.

Human error in process plant maintenance in the past has not been given the attention it deserves. It is a pressing Human Factors issue. Practitioners of Mechanical Integrity element of PSM need to factor its impact and develop actions to mitigate its effect. This paper would help them present the case to management for recognizing human errors in process plant maintenance as a major risk factor that ought to be controlled by having a structured program in place.

A Machine Learning-based Approach to the Identification of Key Material Failure Modes of the Process Facilities

Rustam Abubakirov
Safety and Risk Engineering Group,
Dept. of Chemical and Materials Engineering, Nazarbayev University
Nur-Sultan city, Kazakhstan
rustam.abubakirov@nu.edu.kz

Ming Yang
Safety and Security Science Section,
Faculty of Technology, Policy, and Management, Delft University of Technology
Delft, The Netherlands
m.yang-1@tudelft.nl

Prepared for Presentation at
American Institute of Chemical Engineers
2020 Spring Meeting and 16th Global Congress on Process Safety
August 17 – August 21, 2020

A Machine Learning-based Approach to the Identification of Key Material Failure Modes of the Process Facilities

Rustam Abubakirov
Safety and Risk Engineering Group,
Dept. of Chemical and Materials Engineering, Nazarbayev University
Nur-Sultan city, Kazakhstan
rustam.abubakirov@nu.edu.kz

Ming Yang
Safety and Security Science Section,
Faculty of Technology, Policy, and Management, Delft University of Technology

Keywords: Asset Integrity, Reliability, Damage Mechanisms, Process Safety.

Abstract

Identification of credible failure modes and associated damage mechanisms is a critical part of asset integrity management program for the process units in the oil and gas industry. The Failure Mode and Effect Analysis (FMEA) has been widely used for these purposes. At the stages of units' design and early operation, subject matter expert's opinions (SMEs) with qualitative or semi-quantitative approaches for the damage risk assessment are often used. However, the quality of the assessment can be heavily dependent on the experts' knowledge and information available for the studied system. The present study aims to improve the reliability of the existing credible failure mode identification by using machine learning techniques of regression and feature engineering in the framework. The developed model assesses the contribution of individual damage mechanisms to the overall component's risk level given the available operational, design, and inspection metrics. By means of feature engineering, the limited input information is expanded for damage factor/applicable mechanism analysis by API RP 581 and API RP 571, used as a reference. Meanwhile, the several regression techniques will be used for risk analysis of individual damage mechanisms, in terms of their relative probabilities of occurrence and consequences. Based on which, the damage mechanisms are prioritized. The performance of the model will be tested on the example of several process units, including sour gas treatment, thermal cracking, separators, and several simulated flowlines.

1. Introduction

Material failures and their degradation or corrosion present a significant challenge to the present-day process industries. Wood et al. (2013) note that corrosion-related phenomena were the primary contributing factor of 20% of major accidents across the EU refineries since 1984. The range of the affected equipment included almost all process units, including trays, drums, towers, with the piping being the primary group (71%). With this regard, early identification of credible damage mechanisms, their characterization, and development of mitigation strategy is the critical task for maintaining the components' integrity.

There are several practiced methodologies for damage mode identification and its cause and effect characterization with FMEA (Failure Mode and Effect Analysis) among the widely-used ones. FMEA implies a systematic quantitative framework for probabilistic risk assessment to describe the state of all credible failing events of the considered component. Applied in asset integrity management (AIM), the framework deals with the characterization of all potential damage mechanisms throughout the component's lifecycle, during both normal and transient operations where the equipment is subjected to the interaction with the aggressive environment/ conditions. The framework is generally reliant on the subject matter experts' (SME) opinion in estimations of its target metrics as risks or risk priority numbers. The holistic quantitative methodologies as API RP 581 (2016) might be applied for these tasks as well. In API RP 581, Probability of Failure (PoF) assessment is based on the concept of the statistically derived generic failure frequency for different types of components and damage factors quantifying the severity of material degradation. The framework provides extensive calculation protocols covering a wide range of damage mechanisms inherent to the petrochemical industries. However, the massive data inputs and its pre-processing time is often not entirely available for quantitative assessment; thus, making qualitative and semi-quantitative approaches more commonly used for the risk assessments (Bai and Bai, 2014). This explainins active SME involvement in the analysis. The involvement of SMEs is not something undesirable in risk assessment. Nevertheless, it can be subjective and dependent on the surveyor's experience and the information available for the assessment.

There have been several attempts to introduce alternative ways of quantitative judgment into the risk assessments. The works might complement the existing standard frameworks or might offer its own probabilistic or reliability-based theory for the risk assessment. For example, Rachman and Ratnayake (2019) proposed a machine learning-based approach for the component screening assessment as per API RP 581. Paltrinieri et al. (2019) have developed a neural network for the dynamic risk assessment in drilling operations. Similar applications of machine learning are available in adjacent fields as process control and optimization (Min et al., 2019; Wu et al., 2020). In general, the main advantages of these approaches lie in their efficiency in solving high-dimensional problems as the discussed cases are and availability of computational environments and tools for it. Another important competitive aspect of machine learning approaches, which is less discussed in the area of chemical process safety, is their usefulness in cases with missing information inputs. There is a significant advancement in missing information handling in the area of bioinformatics (Lopes and Ribeiro, 2012; Thomas et al., 2020). Recently Martinez-Luengo et al., (2019) illustrated missing values imputation through employing neural networks in wind turbine fatigue life assessments. In the process component integrity assessments, the issue of missing information is not less critical: some commonly reliant metrics as thicknesses or corrosion rates are scarcely available in real operations.

The aim of the study is the development of a flexible and transparent approach supporting quantitative identification of critical failure modes in the cases with incomplete information available for conducting conventional quantitative assessments. The emphasis of missing parameters handling simulations will be put on cases with inspection metrics missing for the considered component. The approach will use the methodologies on damage mechanisms evaluation of API RP 571 (2011) and API RP 581 (2016) to develop a model allowing risk assessment of each damage mechanism through machine learning techniques. Performance and applicability of two regression algorithms of Deep Neural Networks and Random Forest will be illustrated throughout the study. To deal with the cases of incomplete input information the approaches of data imputation and integrated model development with the support of missing features will be introduced. Performance of the proposed method will be evaluated through the case study on several chemical process units, including sour gas treatment, thermal cracking, separators, and several simulated flowlines operating in a corrosive environment.

2. The proposed methodology and its application

2.1 The basic principles

The main challenge for operators to perform a valid quantitative risk assessment is their limited access to some types of data prescribed in the standard frameworks. Moreover, some critical metrics in the assessments, such as measured thickness or actual corrosion rates, are scarcely available due to their relatively high costs. It is rarely prioritized to obtain these data during the component's early operational history. To assess the component's integrity state quantitatively in these cases, we may adopt the power of statistical and machine learning approaches. Figure 1 presents a schematic flowchart of the proposed approach.

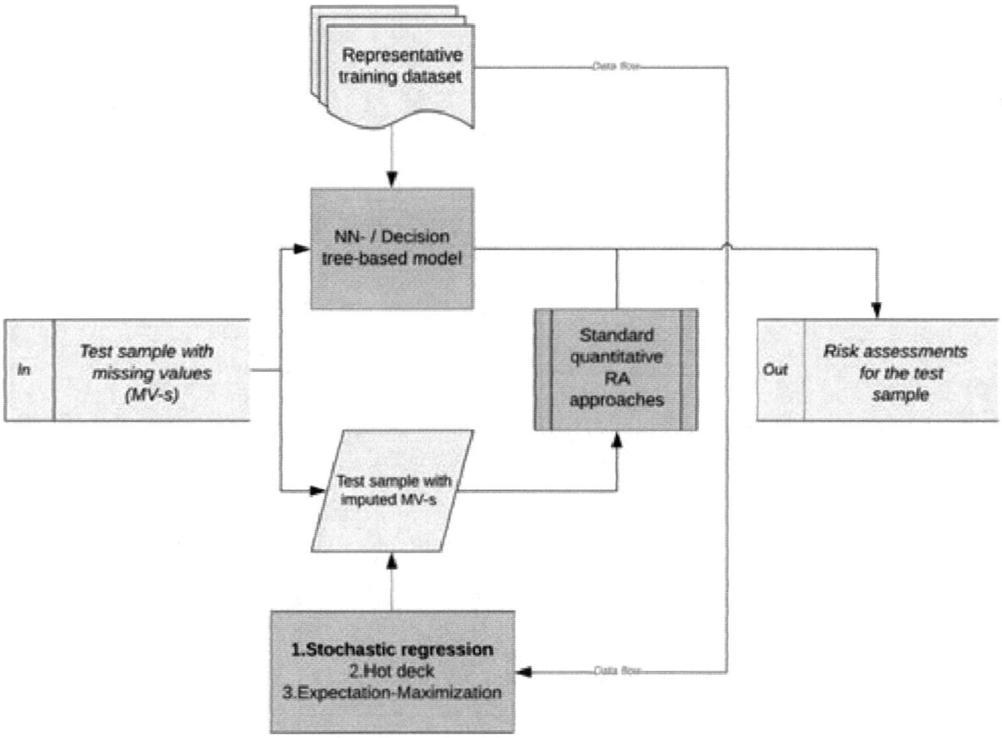

Figure 1. Flowchart of the proposed approach

In the identification of critical damage mechanisms and failure modes, we may encounter some sample components with possibly missing input data on the design, operational, or other inspection parameters. In these cases, these input data need to be pre-processed with all missing values imputed. This step will help to prepare the sample ready for a standard risk assessment. The multiple imputation algorithms like expectation–maximization, hot deck, or stochastic regression can be used for that purpose. Alternatively, the models like Neural Networks or Decision Tree-based algorithms with built-in support of missing values can be powerful as well (Lopes and Ribeiro, 2012). The latter ones require less time to implement them but at the cost of higher required computational power. However, the successful implementation of both approaches is dependent on the quality of the training datasets used. An ideal training dataset is believed to cover a full range of credible damage mechanisms inherent to the potentially assessed components (i.e., being representative); meanwhile they should be consistent in the data entries over the relationships that they carry.

2.2 Data inputs and their pre-processing

A training dataset comprised of 100,000 samples with labeled 62 features and 13 target values has been created for model training and validation. The features represent the standard design, operating, and inspection metrics used in API RP 581(2016) procedures with 13 target variables, 12 of which were the damage factors, and 1 being the consequences of the failure (CoF) all calculated by the framework above. The damage factors have been accounted for damage mechanisms as general thinning, both internal and external, stress corrosion cracking, mechanical fatigue, high-temperature hydrogen attack, and brittle fracture. The detailed list of the considered mechanisms is presented in Table 1.

Table 1. Damage mechanisms considered within the present study

#	Governing damage mechanism	Individual damage mechanism	Nomenclature
1	Internal thinning	General internal thinning	Int Thin
2	External thinning	General external thinning	Ext Thin
3		Corrosion under insulation	
4	Stress corrosion cracking	Amine cracking	SCC – AC
5		Caustic cracking	SCC – CC
6		Chloride stress corrosion cracking	SCC – ClSCC
7		Hydrogen induced cracking	SCC – HIC
8		Sulfide stress cracking	SCC – SSC
9	High temperature hydrogen attack	High-temperature hydrogen attack	HTHA
10	Brittle fracture	Brittle fracture	BF
11		885° embrittlement	885F
12	Mechanical fatigue	Mechanical fatigue	MF

The samples' feature distribution was based on several operational service classes (e.g., sour hydrocarbons or lean amine) with corresponding fluid, material, control and corrosion parameters interrelatedly distributed within regular operational ranges. The majority of the remaining features were not apparently dependent on any other one. They were set to be randomly distributed within their characteristic ranges. The entries in the training dataset do not need to be real-world operational data. Even on the contrary, the authors believe that the sufficient presentation of instances with deviations from the normal operational conditions is crucial while training. It would allow the model to learn from a full spectrum of potential failure conditions and their effects. This property is called representativeness of a dataset (Schaafsma and Vihinen, 2018).

Data consistency and bias are other two important aspects to consider before the model training stage. Apart from its general usability, consistency accounts for the absence of contradicting entries within the dataset. Especially it is relevant to the integrated datasets comprised of smaller datasets coming from differing origins, i.e., simulations to the real-world entries, etc. For example, in completely dry and clean hydrocarbons, internal thinning is not expected; while even a small fraction of contradicting samples in the dataset (noise) could significantly disturb the learning. Bias in the data is something more challenging to trace. It is related to the representativeness and is the state when the variables capturing the desired phenomenon are misrepresented in the dataset due to the lack of actual observations or their prejudice. The case in the area of AIM can be misrepresentation in training of rarely captured metrics as for corrosion under insulation or, in contrary, including only heavily corroded well-observed samples to the process.

Once the quality of data is ensured, it needs to be pre-processed for the use of ML algorithms. Feature selection has been performed because of relatively high dimensionality of the data. The procedure aims to eliminate redundant and irrelevant features compromising the accuracy and speed of model learning (Maldonado and Weber, 2009). Several methods can be performed, which include knowledge-based selection, where domain knowledge is used for well-understood cases like RBI. The alternatives can be the filter methods checking for excessive correlation between the features or the wrapper methods utilizing the ML algorithms themselves to set the best list of features (Rachman and Ratnayake, 2019). API RP 581 clearly defines the set of required metrics required for the assessments with API RP 571 providing detailed information on applicable damage mechanisms. Thus, the subsets of features for each DF and CoF were manually defined for the reference model (no missing values). Even though a few features had both Pearson and Spearman correlation above 0.95, suggesting the repetition of information. These features, along with the ones lacking any variability within itself, were excluded from the dataset. These and further procedures have been performed in the Google Colaboratory environment with Python 3.7 and standard packages as Pandas, NumPy, Scikit-learn and Keras.

2.3 Machine learning

Two types of regression algorithms have been used for predicting the target variables (damage factors and CoF-s) based on the feature set provided for each sample component. These were Random Forest Regressor and Deep Neural Network. Generally, the choice of the algorithm is primarily dependent on the data to which it is applied, its type, quality, and the available computational power. Therefore, no universal algorithm can be applicable and effective for a wide range of cases, even though some general performance trends might be drawn. In some studies, such as Rachman and Ratnayake (2019) and Paltrinieri et al. (2019), they have discussed the application of machine learning to process risk estimation. According to the works, Neural Networks and the ensemble algorithms as Random Forest were expected to perform effectively on presumably similar in structure highly non-linear data. Olson et al. (2018) in its widely referred comparative study of the most common algorithms also notes about the powerfulness of the ensemble algorithms. However, the main accounted advantage of the chosen methods was their ability to handle missing inputs. The standard algorithms and especially the networks are not able to handle the missing values on their own; it generally requires extensive modifications. Nevertheless, given the successful model performance on the complete dataset, the model's re-training on a dataset with reduced dimensionality, which fits a specific problem, makes it possible to deal with missing values. Application of such integrated model is believed to require significantly less computational time compared to imputation techniques with the prediction efficiency retained.

Deep Neural Networks (DNN) uses multiple layers to progressively extract the features in increasing complexity from the input information (Kelleher et al., 2015). Within its structure, the first layer will represent the input set of unnormalized features, the last layer defined as the output with the layer in between being a set of model-derived layers. The overall length of the model chain there defines its deepness, referring to the term "deep" in its name (Goodfellow et al., 2016). Within the present study, a standard Neural Network Sequential functional in Keras open source library has been used for the DNN model.

Random Forest Regressor (RFR) is an ensemble algorithm which comprises a set of decision trees built independently and with a different structure. Building several decision trees and taking their weighted vote, rather than relying on a single decision tree, generally prevents the overfitting, which is inherent to the decision tree-based algorithms (Pal, 2005). Overfitting is an undesired condition when the obtained model fits extensively to the data at which it was trained, including noise in data. Random Forest algorithm can also be used to calculate feature importance distribution based on Gini impurity reduction. For the practical application, to some extent, it allows tracking the structure of the model and its decision criteria, which is challenging otherwise with DNN. In the present study, a standard Random Forest Regressor estimator in sklearn.ensemble library has been used.

There are several metrics used for the model performance assessment in regression problems. The list includes but not limited to the Mean Absolute Error (MAE), Mean Squared Error (MSE), Mean Absolute Percentage Error (MAPE), with each having its application specifications dependent on the type of model, data structure, the fraction of outliers, etc. For illustration purposes, MAPE was used due to its intuitive simplicity and independence from the scale of the assessed parameters. MAPE is given by:

$$MAPE = \frac{1}{n} \sum_{t=1}^{n} \left| \frac{A_t - P_t}{A_t} \right| \quad (1)$$

Where A_t is the actual value; P_t is the predicted value of the target variable.

In addition to the standard metrics, the results with the following discussion will be illustrated through the examples of four chemical process units: sour gas treatment, thermal cracking, and two separators for which the identification of critical failure modes has been performed. Another two representative simulated flowlines have been picked up from the testing dataset and added to the analysis to assess the overfitting potential of the resultant model. A short specification of the components is presented in Table 2.

Table 2. General information on the process components used for model benchmarking

Sample #	1	2	3	4	5	6
Component type	Vessel	Vessel	Vessel	Vessel	Pipe (simulated)	Pipe (simulated)
Process medium	Sour HC gas, amine	High temperature sweet HC gas	Sour HC liquids, saline water	Sour HC liquids, saline water	Sour HC liquids, lean amine	Sour HC liquids, saline water

3. Results and discussions

Benchmarking of the model has been performed on the six process components in Table 2. Figures 2-3 illustrate the results. Figure 2 presents the probabilities of the failure caused by the governing damage mechanisms. They are predicted by the DNN-based model based on different data availability scenarios and compared to the reference predictions as per API RP 581. Concerning inspection metrics, three cases have been considered:

(1) All data is available. The model's accuracy is tested.
(2) Measured remaining thickness with missing data.
(3) Measured remaining thickness and both internal and external corrosion rates with missing data.

Figure 3 presents a hierarchical risk value chart for each damage mechanism - failure modes, based on which the prioritization and identification of critical failure modes could be performed. The data used to plot the figure, including all the damage factors (DF), were taken from the same predicted outcomes as those in Figure 2 was constructed; thus, not accounting for the natural randomness in the model. The consequence of failure (CoF) for each corresponding case was predicted by the same DNN-based model, input data for which is based on Level 1 CoF Analysis as per API RP 581.

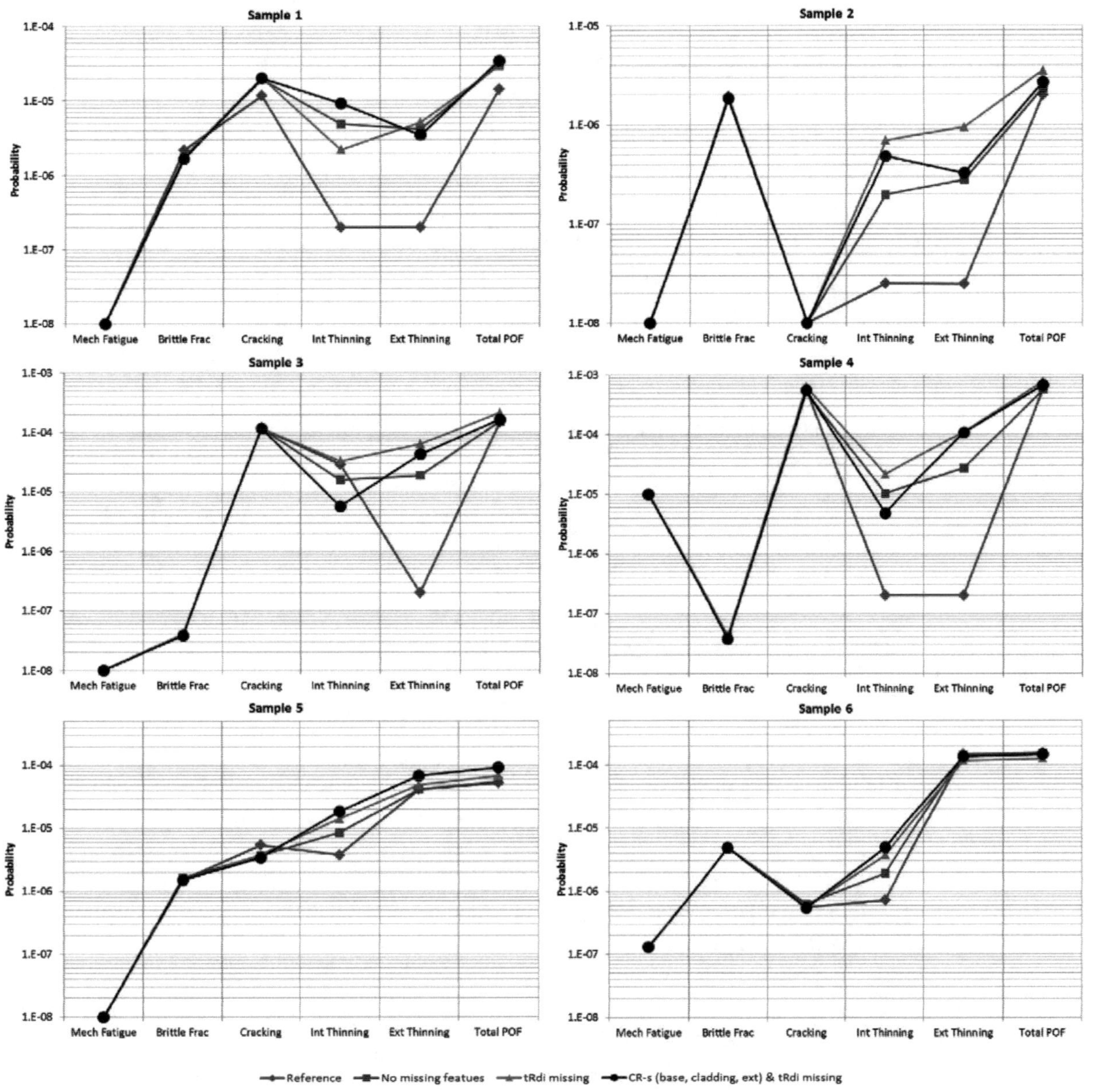

Note: Red line indicates the reference predictions as per API RP 581.

Figure 2. PoF-s by the major damage type groups at different data availability scenarios obtained by DNN model (blue lines).

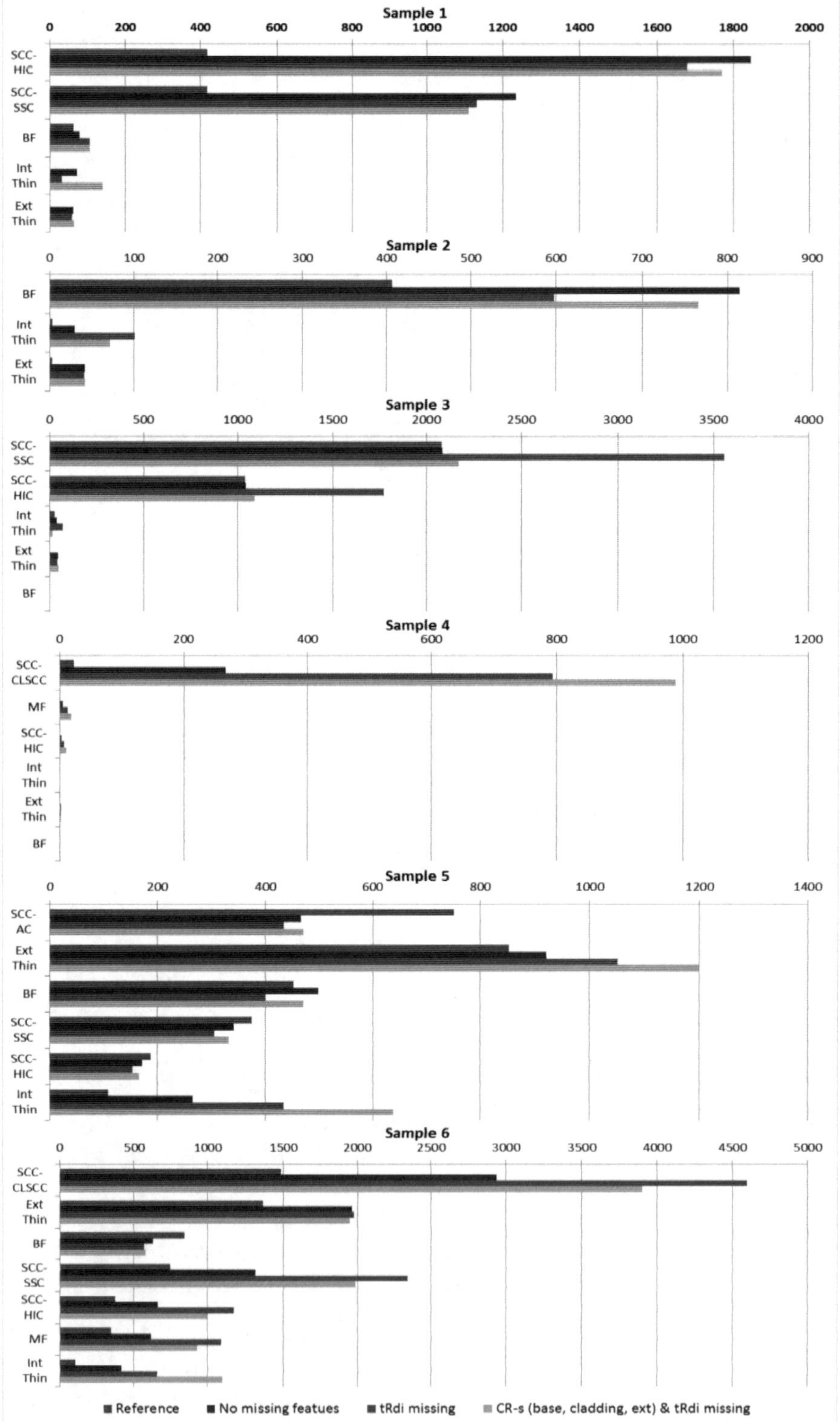

Note: Predictions produced by DNN model at different data availability scenarios are marked by blue, reference calculations as per API RP 581 by red.

Figure 3. Estimated risk values ($/year) for each individual failure mode inherent to the corresponding process component.

3.1 Integrated models

As indicated by Figures 2-3, the central question of interest is concerned about the quality of the obtained predictions and overall reliability of the proposed approach as a whole. The problem statement required identification of critical failure modes and their risk ranking; and, as Figure 3 illustrates, the hierarchical order of the failure mode risks was successfully predicted even in the cases with limited input data. However, the significant discrepancies were inherent to the magnitudes of the obtained predictions. In some cases, as for Samples 2, 3 and 4 from Figure 2, the difference between the reference values and baseline predictions with no missing values for PoF values accounted for two orders of magnitude, i.e., between 2E-7 and 2E-5. There is ample space for discussion on these two numbers. In the domain of statistics, various assessment metrics, e.g., MAE, MSE or MAPE, being applied would eventually indicate tremendous discrepancies. On the other hand, looking from the probabilistic perspective, the difference between "practically impossible" and "barely possible" as the two numbers imply would not seem as significant. To the authors' point of view, the interpretation of the results may vary depending on the user's objectives of the performed analysis. However, the authors also want to emphasize the importance of the signs of two phenomena causing the obtained discrepancies: data bias and model overfitting.

The primary assumption of the machine learning theory lies in the premise that the distribution of the test samples would be identical to the distribution of training ones (Mitchell, 1997). Once either of the sample sets is not comparable to each other and does not represent the overall actual population, the condition is called bias. The phenomenon can be classified into several types depending on the quality of the samples or prejudice of their measurements, and the distribution of the results in Figures 2-3 shows the signs of sample bias in the training data. For example, according to the review of the training samples, the absolute majority (93%) of them had a moderate or severe level of internal thinning (DF >10); meanwhile only 2% of the entries having the damage factor of around 1; thus, explaining the general conservatism of the predictions. A similar trend was inherent to the CoF data, where high-consequence entries were dominant as well. Despite the seeming simplicity of the phenomenon, managing bias in the data might not be as easy: eventually, the properties distribution is dependent on the user's expectations on the test samples. Ideally, the training dataset represents evenly-distributed data with the full set of relevant features. However, even with the ideal data model, precision is critically dependent on the used ML model itself, one of the critical properties of which is bias–variance tradeoff.

Underfitting and overfitting are the two opposing states in the model's bias-variance balance. Overfitting occurs when the model includes excessive features than was needed and fits the training data replicating its structure. Overfitting is undesired in machine learning as it not only complicates the model wasting resources, but also leading to incorrect predictions (Hawkins, 2004). From Figure 2, the significant difference in the accuracy of predictions was observed between Samples 1-4 and Samples 5-6, i.e., between the component sample groups being slightly different in the property's distribution. The model performance was significantly better in the predictions on simulated samples, which were firmly related to the model's training samples. This indicates overfitting. Considering the conservative values of predictions, the DNN model, in analogy with the discussed bias phenomenon, possibly prioritized the widely represented biased-high corrosion rates related features over the actual thickness measurements when it was available in the reference model. Later, when the thickness and corrosion rates were not available, the extended set of fluid property and material specifications features were used to support the conservative trend. The general ways to combat the overfitting are improving procedures of the model cross-validation by providing more test data to it and tightening the feature selection process.

Figure 4 presents the model's feature selection process and also overfitting. It represents the feature importance distribution charts for Internal Thinning DF predictions. Random Forest models were developed at different data availability scenarios. As it can be observed for the reference model, the features which were actually used in the Thinning DF occupy only the upper 0.8 fractions of the overall domain. Meanwhile, features like "MatS" – material sulfur content with the rest 47 features having no straight relationship to the thinning estimations occupied the remaining domain; thus, complicating and worsening the model. This is the sign of overfitting, which could be minimized by proper manual feature selection. However, such flexibility allows the model to perform prediction even when some of the features were missing. When "tRdi" (remaining measured thickness) was not present, the model placed "tOv" (design thickness) in the first place with the corrosion rates and time features re-distributed as well. In the third case, when no inspection metrics were available, the model placed fluid chemistry and material properties features like "pH" or "MCrC" (material chromium content) to the higher orders of importance trying to relate them to the missing corrosion rates. In this case, the missing features of measured thickness and corrosion rates could be called latent or 'hidden', as the information on their values might be extracted from a more extensive set of known parameters with some degree of certainty. Thus, the proper feature selection is highly important and it is always a balance between the model's simplicity and its flexibility. A great number of irrelevant features lead to the model's complication, overfitting, and unnecessary computational resources. On the other hand, some indefinite set of features might carry additional information which might be unrepresented otherwise in the models.

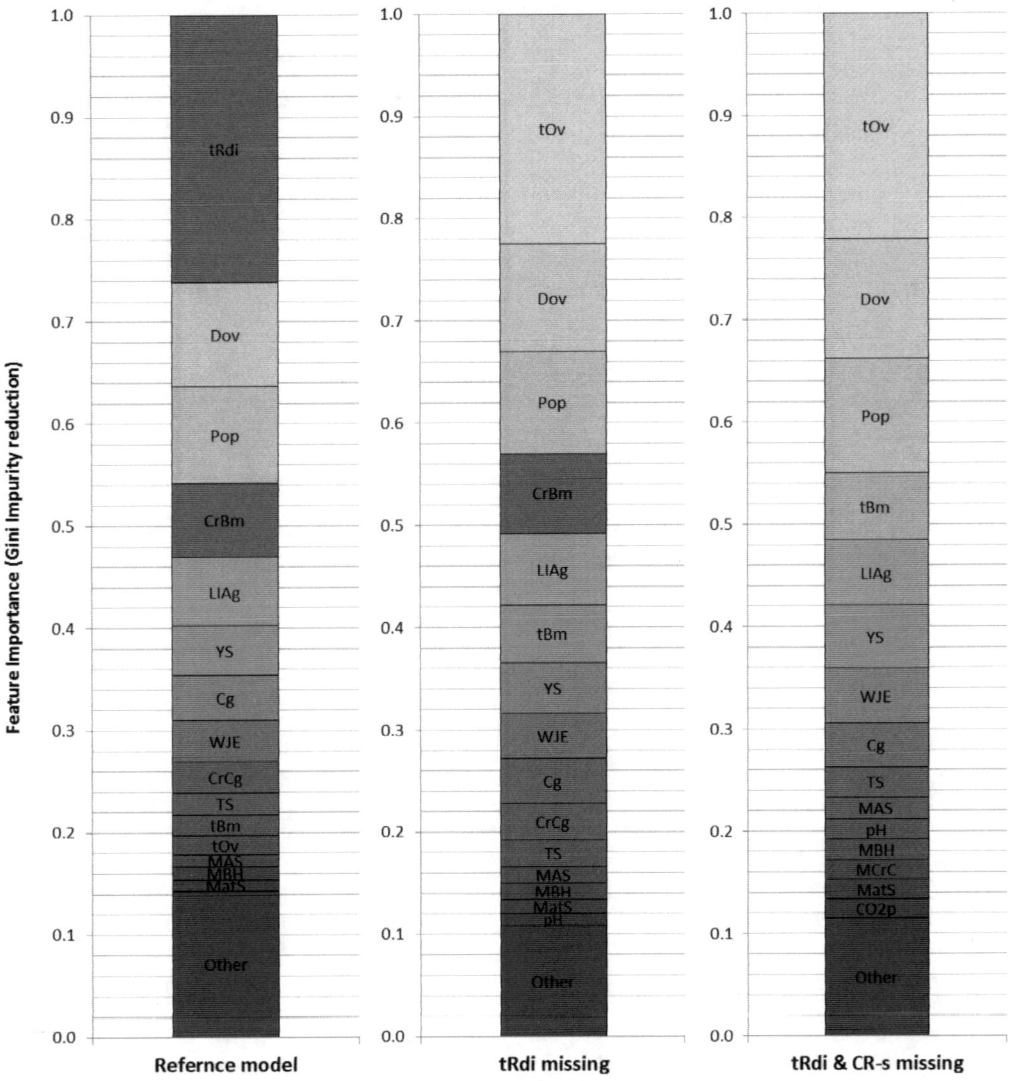

Figure 4. Feature importance distributions for Internal Thinning DF prediction model produced by Random Forest algorithm at different data availability scenarios.

3.2 Imputation models

Another proposed option for performing the risk assessment with limited data concerned imputation of the missing features employing stochastic regression. Unlike the integrated models, which predicted the required risk metrics directly, the approach implies a prediction of the missing metrics first with them to be used in the standard risk assessment calculations next. Theoretically, prediction of simpler metrics, e.g., corrosion rates or thickness instead of the damage factor, allows better control over the input set of features; thus, reducing the chances of overfitting or bias. Similar to the integrated models, two DNN models have been trained to predict Internal and External Thinning DFs at different data availability scenarios. To account for natural randomness of the models, the predictions have been made repeatedly for multiple times, allowing obtaining the whole distribution of the predicted metrics. Models' assessment has been carried on the test dataset to diminish the data bias effect with MAPE chosen as the assessment metrics. Figure 5 presents the obtained results in comparison with the integrated models set to predict the DFs directly.

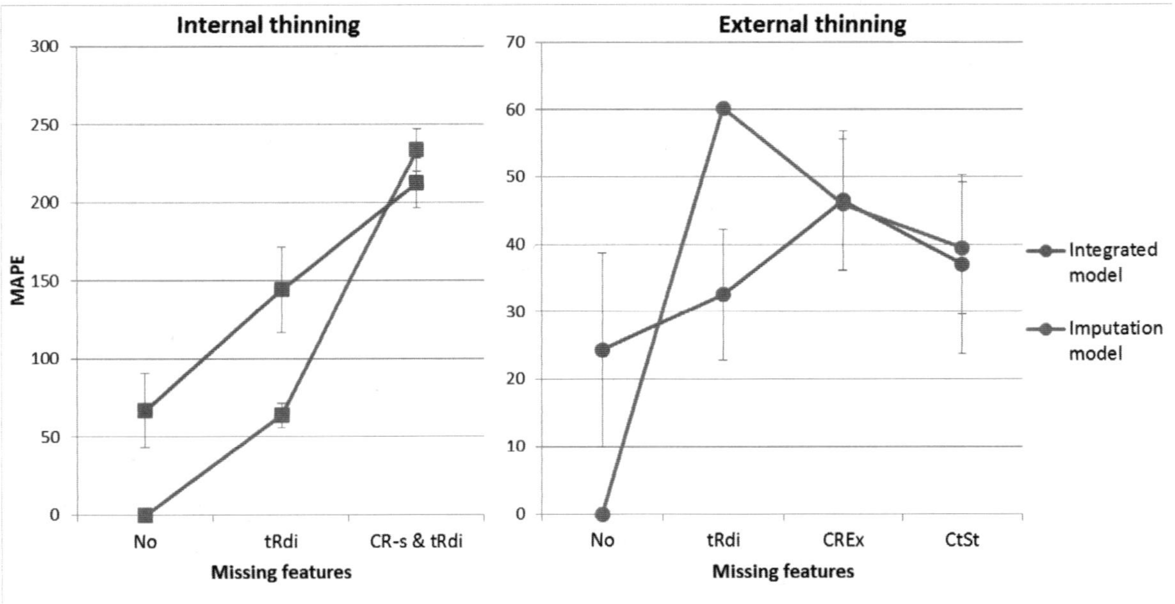

Note: the error bars on the chart indicate standard deviation of the results with dot indicating their mean.

Figure 5. Imputation-based and integrated model performance on the Internal and External DF predictions with denoted features missing in the analysis.

Despite the seemed competitiveness of the imputation approach, the results of the simulations did not reveal any significant advantages. In cases with the measured thickness, corrosion rates or coating integrity state missing, the approaches' performance was approximately equal with imputation results being generally less variable. Meanwhile, there was no practical meaning to assess imputation with no missing features. In general, the obtained results might indicate that the main complexity associated with predictions of the safety metrics with limited data lies in recovering of thickness, corrosion rate information or any other metrics missing. Thus, regardless of the applied approach of the chosen, the models' performance is critically dependent on the training data quality and the set of model features chosen to predict the optimal parameters.

4. Conclusions

Identification of credible damage mechanisms and associated failure modes is a critical task in asset integrity management for the chemical process components. FMEA is a framework widely used for that purpose. It relies on SMEs-led quantitative estimations of the risk associated with individual failure modes. Quantitative frameworks for the risk assessments as API RP 581 might be applied as well. The major challenges for their direct application are the amount and accessibility of data it requires. Some of the data entries, especially inspection metrics as measured thickness or actual corrosion rates, might be difficult to obtain to the operators. The present paper demonstrated a structured approach to deal with the limited data inputs by incorporating existent knowledge on the missing parameters into the calculations using machine learning techniques; thus, providing a quantitative base for decision-making in the analysis with missing parameters present. The methodology has been presented through a case study on several process components for which the failure mode risk assessments have been conducted. Two general ways of doing it have been illustrated: through data imputation and integrated models, with Deep Neural Network and Random Forest algorithm applied.

Results of the study shew that the approach's performance in risk metrics prediction is critically dependent on models' training data quality and bias-variance characteristics of the used model. Concerning the data quality, its representativeness, consistency, and bias level are essential. The data bias phenomenon, which significantly compromised the model's prediction accuracy for the real cases, has been traced throughout the results. Therefore, ensuring that the training samples' feature distribution is statistically close enough to the real population of samples is critical. The proper model tuning across bias-variance properties is not of less significance as well. From one side, the model variance allows it to optimize its feature set in cases with limited inputs; from the other, it might be prone to unnecessarily complicate the model leading to the loss of accuracy. Finding an optimal model for the risk assessment with limited input data includes a balance between the properties mentioned above in ratios that need to be further investigated.

Acknowledgements

This study was supported by Young Researchers Alliance and Nazarbayev University Social Development Fund through the Fostering Research and Innovation Potential (FRIP) program.

References

1. API RP 571. *API Recommended Practice 571, Damage mechanisms affecting fixed equipment in the refining industry.* USA: American Petroleum Institute, 2011.
2. API RP 581. *API Recommended Practice 581, Risk-based Inspection Methodology, Third Edition.* USA: American Petroleum Institute, 2016.
3. Bai, Yong, and Qiang Bai. *Subsea Pipeline Integrity and Risk Management.* New York: Gulf Professional Publishing, 2014.
4. Goodfellow, I. J., Y. Bengio, and A. Courville. *Deep Learning*. Citeseer, Cambridge, MA, USA: The MIT Press, 2016.
5. Hawkins, Douglas M. "The Problem of Overfitting." *J. Chem. Inf. Comput. Sci.*44, 2004: 1-12.

6. Kelleher, JD, Mac Namee B, and D'Arcy A. *Fundamentals of machine learning for predictive data analytics: algorithms, worked examples, and case studies*. Cambridge, 2015.
7. Lopes, Noel, and Bernardete Ribeiro. "Handling Missing Values via a Neural Selective Input Model." *Neural Network World*, 2012: 357-370.
8. Maldonado, Sebastián and Richard Weber. “A wrapper method for feature selection using Support Vector Machines.” *Information Sciences*, 2009: 2208-2217.
9. Min, Qingfei, Yangguang Lu, Zhiyong Liu, Chao Su, and Bo Wang. “Machine Learning based Digital Twin Framework for Production Optimization in Petrochemical Industry.” *International Journal of Information Management*, 2019: 502-519.
10. Mitchell, Thomas. *Machine Learning*. McGraw-Hill Education, 1997.
11. Olson, Randal S., William La Cava, Zairah Mustahsan, Akshay Varik, and Jason H. Moore. “Data-driven advice for applying machine learning to bioinformatics problems.” Pacific Symposium on Biocomputing, 2018. 192-203.
12. Pal, M. “Random Forest Classifier for Remote Sensing Classification.” *International Journal for Remote Sensing*, 2005: 217-222.
13. Paltrinieri, Nicola, and Louise Comfo, Genserik Reniers. "Learning about risk: Machine learning for risk assessment." *Safety Science*, 2019: 475–486.
14. Rachman, Andika, and R.M. Chandima Ratnayake. "Machine learning approach for risk-based inspection screening assessment." *Reliability Engineering and System Safety*, 2019: 518–532.
15. Schaafsma, Gerard C. P., and Mauno Vihinen. “Representativeness of Variation Benchmark Datasets.” *BMC Bioinformatics*, 2018.
16. Thomas, Rajat M., Willem Bruin, Paul Zhutovsky, and Guidovan Wingen. “Dealing with missing data, small sample sizes, and heterogeneity in machine learning studies of brain disorders.” In *Machine Learning. Methods and Applications to Brain Disorders*, edited by Sandra Vieira Andrea Mechelli, 249-266. Academic Press, 2020.
17. Wood, Maureen Heraty, Ana Lisa Vetere Arellano, and Lorenzo Van Wijk. *Corrosion-Related Accidents in Petroleum Refineries*. Luxembourg: European Commission, Institute for the Protection and Security of the Citizen, 2013.

Ecolab Process Safety Qualification Process: Executive Summary

Ecolab, a global petro/chemical manufacturer, had recently seen their Process Safety performance begin to plateau, and not see the improvements of past years. They were in danger of not hitting their targets across their global plant sites for Process Safety Event Reduction.
They had also recently developed an Engineering Career Ladder within Human Resources. With help from the Global Process Safety Team, and Global Engineering, the career ladder created consistent job titles and roles for all engineering roles with criteria that employees must achieve to move through the ladder. This created a path for engineers in the company to follow when developing their own careers.
With these two key data points in mind, the Process Safety Training Program, and senior leadership, recommended developing a consistent *Process Safety Qualification Process* to drive performance improvement. There were two goals for developing /deploying this process:

1. Ensure engineering employees at all global plant sites have the knowledge and expertise to complete key Process Safety tasks at the plant level. These tasks are crucial to Process Safety Event Reduction, and driving a strong consistent program across all global plant sites.
2. Provide tools, support, and resources for engineering employees to develop their careers, and complete the requirements listed in the Engineering Career Ladder

Outline of Qualification Process:

1.Employee completes self-assessment	30% Training, Testing, Evaluation
2.Employee completes training and testing	70% Examples of experience and skill application
a.Passes tests in all 12 core elements	
4.Employee shows skill application with examples of their work and coaching	
a.Submits project to Review Panel to show skill application	
6.Review Panel confirms qualification through consensus	

Core Elements of Qualification:

Compliance with Global Standards	RBPS/PSM Program Knowledge	Managing/Using Process Safety Information	Regulatory Compliance/ RAGAGEP
Chemical Compatibility/ Red List	Management of Change	HIRA/ PHA Tools	Plant Design
Measurement and Metrics	Root Cause Analysis	Critical Equipment	Process Safety Culture

The process is currently in the pilot phase, and will be ready for global deployment by the end of 2019. This sets the template for future evolution of the global program, and will be incorporated into other disciplines within the organization. This project was selected by the 5th CCPS Global Summit on Process Safety, Singapore, 2019. Nathan Thompson was also selected as co-leader of the *CCPS Early Career Professionals* project, now launching in North America.
Nathan Thompson: Global Process Safety Training Program Manager, St Paul, USA

585087 Right Sizing Combustible Particulate Management for R&D Processes

Abstract:

Combustible dust hazards are routinely recognized in a manufacturing setting but are commonly overlooked in laboratories. Most companies use research and development (R&D) departments to create new products but fail to recognize all new potential hazards associated with their development. Part of the struggle is the recognition of a combustible dust hazard and completing the right size review for an R&D setting.

This session will guide participants through educating non-safety personnel on combustible dust hazards, how 3M is screening combustible dust processes to drive the appropriate level of hazard analysis, and how to right size the R&D Dust Hazard Analysis (DHA) based on equipment in an R&D setting.

584646 Right Sizing Combustible Particulate Management While Meeting NFPA 652 Requirements

Abstract:

Working in the manufacturing industry at a Fortune 500 company with multiple commodity classes and size of manufacturing locations presented an interesting challenge when trying to control combustible particulate solid hazards and implement requirements from NFPA 652. Each location dealt with different combustible solid particulates, different processing methods and unique equipment. To handle this challenge, it was important to develop a combustible solids particulate management program that would meet the needs of every facility.

The global Combustible Solid Particulates management program that 3M has developed and is in the process of implementing, involves a three-step process which includes creating an inventory of solid particulates, analyzing the hazards of each process utilizing combustible solid particulates, and managing the hazards identified.

586586 Engineering Design for Ergonomics - the Missing Link

Abstract:

Process safety addresses many issues related to the safety of employees in the workplace and human performance. Human factors is included in the safety culture element, however ergonomics is often neglected. It has been said that human factors addresses worker safety from the neck up (cognitive issues) and ergonomics addresses safety from the neck down (anthropometric aspects). Admittedly, this is a gross analogy, however it is useful for discussion purposes.

During the design phase, it is rare that ergonomics is included as a priority in the decision process. How many times have we seen valves controlling processes placed eight feet off the ground where operations personnel are required to place chains to articulate them. This is often the cause of injuries we see surface in PHAs. These hazards are often disregarded as unavoidable because it was part of the original design, however if ergonomics would have been considered in the original design, then the valve would have been properly placed where operations personnel could easily access the valve without causing injury.

This paper will discuss the importance of incorporating ergonomics principles into the design of process equipment with the goal of reducing the potential for employee injury.

585298 New Wine in New Cups: Measuring Operator Performance for Safety-II Perspective

Abstract:

Standard operating procedures (SOPs) play a critical role in achieving safety and productivity of day-to-day operations in the chemical industries. Investigations of previous incidents also indicate that a majority of adverse events during routine and non-routine operations are attributed to issues associated with SOPs. For example, a recent investigation of fatalities that occurred at DuPont La Porte facility points out the absence of formal procedure required to remove plugging (e.g., hydrate) from a waste gas piping system and inadequate emergency procedures for hazard notification and evacuation as major causes of the incident [1]. To ensure the safety of a complex system such as chemical plants, there have been two viewpoints towards human operators' use of SOPs: *Safety-I* and *Safety-II*. First, Safety-I perspective defines safety as the absence of undesired events [2]. Thus, it looks for things that went wrong (e.g., slip, mistake, violation) and seeks to minimize the deviations from prescribed tasks, which is assumed to be a cause of adverse incidents . Second, Safety-II views safety as the presence of desirable and successful outcomes. Therefore, this standpoint highlights things that went right (e.g., adaptive behavior, workaround), and considers performance variability of human operators from SOPs to be inevitable and even necessary to maintain safe operations of the system [3]. In favor of Safety-I approach, conventional measures of SOP performance have long been established and utilized to indicate the degree of safety in relations to procedural systems. Nonetheless, no measures of human operators' SOP performance that adopt or support Safety-II framework exist to date. To address such a gap, this paper identifies limitations of traditional and dominant measures of SOP performance and then proposes a novel idea that harmonizes the two views towards safety (i.e., Safety-I and Safety-II). The new measurement concept embraces not only the conventional measures regarding the implementation of SOPs (i.e., following or not following a procedure), but also incorporates human adaptive behaviors. To instantiate the new measures of SOP performance, case studies using real-world examples in the chemical industries are presented. Following this, implications for the proposed measure of SOP performance based on Safety-II viewpoint and future research proposals to support the benefits of the measure are provided.

References

1. U.S. Chemical Safety Board, *Investigation Report: Toxic Chemical Release at the DuPont La Porte Chemical Facility*. 2019.
2. Hollnagel, E., *Safety-I and Safety–II: The Past and Future of Safety Management*. 2014, Surrey, UK: Ashgate Publishing.
3. Sujan, M.A., H. Huang, and J. Braithwaite, *Learning from incidents in health care: Critique from a Safety-II perspective.* Safety Science, 2017. **99**: p. 115-121.

586223 Incorporating Abnormal Modes of Operation into Pressure Relief System Design and Documentation

Abstract:

Pressure relief systems are commonly referred to as the last line of defense in oil refineries and chemical facilities; however, the design of pressure relief systems does not typically consider abnormal modes of operation such as startup, shutdown, and other temporary operations. Past studies have shown that a majority of incidents occur during abnormal modes of operation including several significant incidents that helped shape the state of process safety as it is today. This paper will review potential sources of overpressure during various abnormal modes of operation and analyze how these overpressure scenarios can be incorporated into pressure relief system design and documentation.

2020 AIChE Spring Meeting
16th Global Congress on Process Safety
HOUSTON • 2020

Facilitating Barrier Integrity Management During PHA Revalidation

Edward M. Marszal
Kenexis
3366 Riverside Drive, Columbus, OH 43221, USA
edward.marszal@kenexis.com

Austin Bryan
Kenexis
3366 Riverside Drive, Columbus, OH 43221, USA
austin.bryan@kenexis.com

Prepared for Presentation at
American Institute of Chemical Engineers
2020 Spring Meeting and 16th Global Congress on Process Safety
Houston, TX
March 30 – April 1, 2020

Facilitating Barrier Integrity Management During PHA Revalidation

Edward M. Marszal
Kenexis
3366 Riverside Drive, Columbus, OH 43221, USA
edward.marszal@kenexis.com

Austin Bryan
Kenexis
3366 Riverside Drive, Columbus, OH 43221, USA
austin.bryan@kenexis.com

Keywords: bowtie, LOPA, integrity management

Abstract

Currently, risk in process industries is managed through analysis of process hazard and identification of barriers to prevent or mitigate hazards (typically through a PHA), a design team which takes the specification for barriers to perform detailed design and ultimately construction of barriers, and operations which monitors and tests barriers. However, current practices leave many potential gaps in the workflow which may allow for inadequate risk mitigation and intolerable risk within a facility. These can include incomplete or inadequate specification of barriers, incorrect design, or inability to validate designs once in operation. The use of numerous software applications with differing database structures leads to teams and individuals operating in isolation, often with little visibility on how information given to them was generated and with no impetus to follow up and ensure that the final design fulfills their recommendations / specifications in actual operation. Using enterprise software tools with a standardized database structure that allow personnel and applications that are used in the different phases of the plant lifecycle to communicate and synchronize data, and can organize and present data in a more easily understood visual representation, such as bow-tie diagrams. Such tools can allow for both feed forward and feedback of information, ensuring that pertinent information is available to people when needed, and validation of the barriers through the lifecycle is streamlined.

1 Introduction

Process Hazards Analysis (PHA) considers the risk of accident scenarios based on the consequences of realization of the scenario, the likelihood at which the scenarios can occur, and the safeguards (a.k.a., barriers) that prevent or mitigate the scenarios. Risk analysis teams rely on barriers to reduce risk to tolerable levels, but during the analysis process can only set performance targets. These performance targets and conceptual specifications are passed down to design teams as recommended specifications with the expectation that that they will be faithfully implemented as the hazard analysis team intended. Constraints on budget, timing, equipment availability, and plant logistics can prevent recommendations from being executed, or in some cases modifications to the recommendations are made to make implementation practicable. Furthermore, it is also expected that once those well-designed barriers are put into place, they will continue to operate in conformance with their performance targets throughout their entire lifecycle. This requires that they are tested, and preventive and corrective maintenance is performed as specified. Even when recommendations are implemented, if maintenance and testing expectations are not properly communicated or implemented, barrier performance may not match the design targets and can prevent tolerable levels of risk from being achieved.

Without some sort of system of follow-up, plants might operate for long periods of time with intolerable levels of risk due to improperly implemented barriers. In order to ensure that barriers achieve the desired risk reduction throughout their entire lifecycle, a feedback mechanism must be used. Initially, feedback to the PHA team from the design team can be performed, prior to plant construction, to ensure that all recommended barriers are in place as specified. If gaps between specifications and actual design are identified they can then be addressed prior to start up. During plant operation, continued achievement of barrier performance targets, considering actual testing, inspection, maintenance, and bypass, can be confirmed by having operational phase tools. This includes management of bypass systems, event documentation systems, and computerized maintenance management systems, that communicate and synchronize with analysis phase tools to allow the analysis phase team an ongoing view of the status of barriers, in real time, throughout the lifetime of the facility. This synchronization is then particularly important during PHA Revalidation where the automated tools and bowtie visualization allow for easy verification that design assumptions match actual performance, and if they do not, allow for modifications to barrier design or process operating parameters to allow risk to be moved back into the tolerable region.

Use of enterprise tools with a standardized database structure begins with the hazards analysis documentation tool being able to automatically exports lists of barriers to designers along with their performance requirements, automatically, without having to enter data twice. The design tools should be able to link backward and view the hazard scenarios that the barriers are intended to protect against to improve designer decision making because they have a better idea of the purpose of the barriers. After completion of design, the design phase performance calculations can be fed back to the analysis phase tool to allow comparison of analysis phase expectations with design phase specifications. The design phase information should then be fed forward into operation and maintenance phase tools such as computerized maintenance management software (CMMS) for scheduling and tracking of testing and maintenance, management of change software for the tracking of bypass, and event management systems to record barrier activation events. These tools should be able to feedback, throughout the lifecycle to design phase tools, and subsequently the analysis phase tools. In this way, all tools can have a complete view and barrier requirements and actual performance throughout the entire lifecycle. All along the way, bow tie diagrams provide an easily understood visualization mechanism, across all tools, for visualization and communication of barrier status and thus achievement of tolerable risk.

Defining Barriers and Setting Their Performance Targets

PHA is initially performed during the analysis phase of the safety lifecycle of a facility. In this phase, the design of the process equipment is analyzed by a multidisciplinary team and performance targets for barriers defined. PHA provides discussion prompts, usually deviations from design intent, that lead to hazard scenarios that are fully developed by the PHA team. These scenarios will include a cause, such as an equipment failure or improper operator interaction with the process. This cause, and potentially other causes can then result in a loss of containment of hazardous materials or energy from the process. The loss of containment can then result in one or more consequences, which represent harm being caused to assets, environment or people. The team will also discuss the safeguards or barriers that are in place to prevent or mitigate the consequences. Barriers are either preventive or mitigative. In general, preventive barriers prevent the loss of containment from occurring, preventing any related consequences. Mitigative barriers do not prevent the loss of containment. Instead, they function by reducing the magnitude of the consequences that result from the loss of containment. The PHA team then considers the scenario's likelihood and consequence in order to determine if the overall risk is tolerable or if barrier addition or improvement is required in order to make risk tolerable.

At this time, the work of a PHA team is generally performed using computer-based worksheets of information. Rows of data present scenarios, causes, consequences, and safeguards. In order to better allow teams and other stakeholders of the PHA to visualize the information contained in a PHA scenario bow-tie diagrams are being used more frequently. Furthermore, the development of Unified Hazard Assessment[1] is allowing for the interchangeability of data in HAZOP, LOPA, Hazard Register, and Bow-tie diagram formats – all automatically in a single tool through a unified data structure that serves all those purposes. Consider a typical hazard scenario, as shown in the figure below. The LOPA includes threats (green), preventative barriers (purple), the top level event (blue), mitigative barriers (yellow), and consequences (red). The rows of a LOPA worksheet can be easily represented in the associated bowtie, providing an easier to understand picture of process risk.

[1] *Applying Data Science to Hazard Analysis*, Edward M. Marszal, InTech, January/February 2020, ISA, Research Triangle Park; NC.

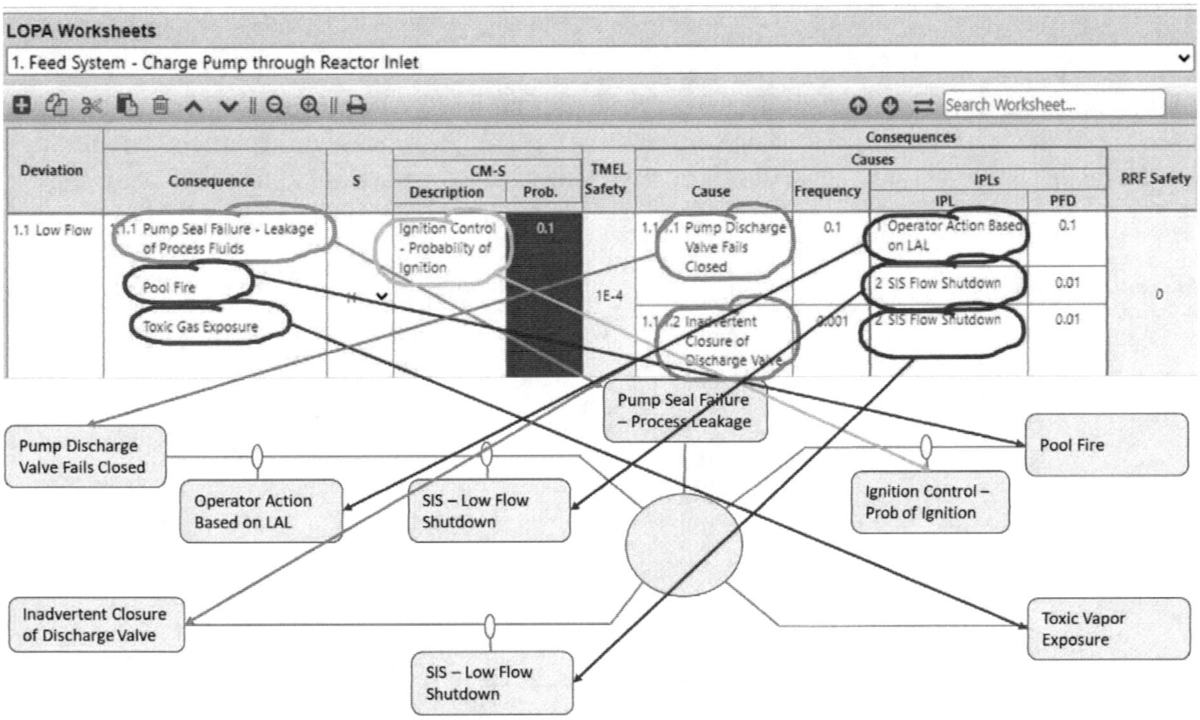

Figure 1 – Converting a LOPA Scenario into a Bowtie Diagram

A shortcoming of many PHA studies in this phase is that they do not develop the barriers in sufficient detail for persons responsible for the barrier design to effectively perform their task. This lack of data can manifest itself in poor choices by the design team that will ultimately prevent the expectations of the analysis phase team from being realized. This failure, at a minimum, could result in costly re-work late in the lifecycle of the plant and at worst could result in failure of tolerable risk goals to be achieved by the facility. Consider the following LOPA data sheet. While it appears to have a comprehensive listing of the information that is required by the design team, in fact, there are a few parameters that are required by the design team that are left out.

IPLs				
IPL	IPL Tag	PFD	IPL Type	Selected SIL
1 Operator Action Based on LAL-102	LAL-102	0.1	ALARM	
2 Charge Pump (P-101) Discharge Low-Low Flow Stops Pump	FZL-P101	0.01	SIF	SIL 2
2 Charge Pump (P-101) Discharge Low-Low Flow Stops Pump	FZL-P101	0.01	SIF	SIL 2

Figure 2 – Typical LOPA Safeguard Listing

Furthermore, some of these parameters are only appropriately specified by the analysis team and if left to the design team could be ignored or inappropriate choices made. Some of the common parameters that are often left out of analysis phase documentation include:

- Process Safety Time
- Shut off Valve Leak Tightness Requirements

The analysis phase is where the expertise to properly define these parameters resides. Furthermore, if the analysis team does not define them, these parameters are frequently ignored, potentially resulting in designs that will not allow a safe state to be achieved upon activation of the barrier. While the figure above shows the listing of barrier attributes in a worksheet format, there is no reason that the same information cannot be presented in an enhanced bow-tie diagram format as shown in Figure 3.

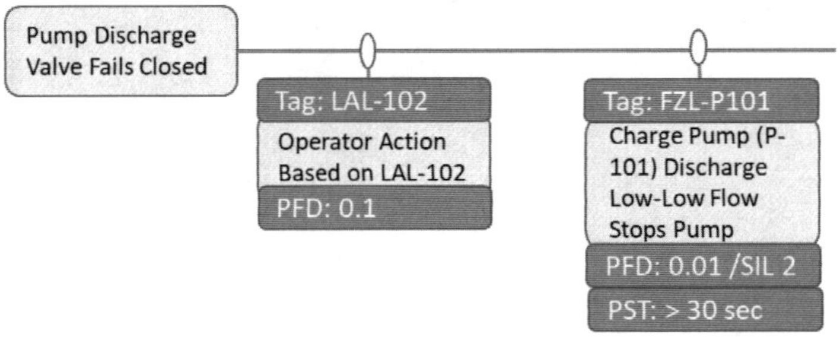

Figure 3 – Barrier Details View in Bowtie Diagram

It is even reasonable to expect that a comprehensive tool could allow the user to click on the barrier in the bow-tie diagram and have the data sheet(s) for the associated barrier be presented for a more comprehensive view. This is presented in Figure 4.

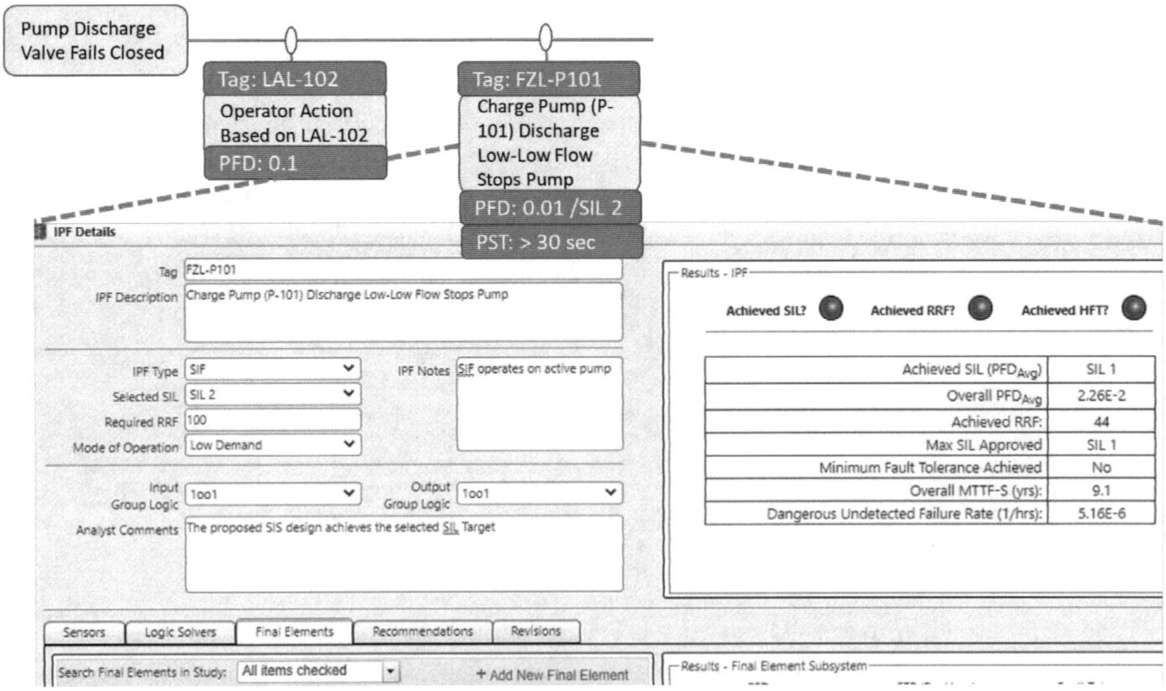

Figure 4 – Bowtie Linked to Barrier Specifications

Using advanced tools, and implementing administrative controls and training, will ensure a more comprehensive specification of barriers by the analysis team. This information is automatically moved, copied, or synchronized with the design phase tool. The synchronization needs to apply regardless of the type of barrier being employed, whether this is a SIF, a relief valve, or critical alarm. Automatic synchronization allows for complete and comprehensive transfer of requirements from the analysis phase to the design phase with a minimum of effort and no chance of error transposing information from one tool to another. The use of standard and open data structures for different portions of the plant safety lifecycle will greatly improve the workflow of barrier design.

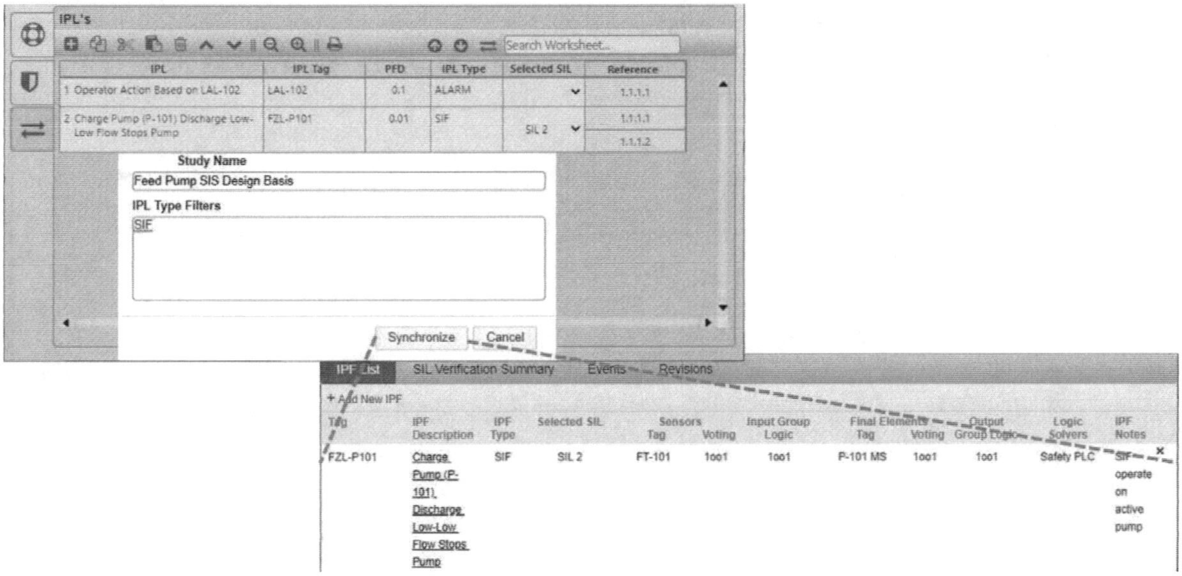

Figure 5 – Synchronization of Barrier Specifications Between Analysis Phase and Design Phase Tools

Ensuring Barrier Integrity During the Design Phase

Once barrier requirements are defined in the analysis phase and then transferred into design tools, conceptual design and detailed design can then begin. A common barrier that might be designed for a process facility is a safety instrumented function (SIF). Analysis phase requirements that might be defined for a SIF include a description of the measurements and actions taken by the SIF, process safety time, and a safety integrity level (SIL). Designers of SIF will make many design decisions about the SIF including the specific instruments used, the amount of redundancy employed, and the frequency and type of testing of the SIF components. All these factors are considered during the design process which culminates in a quantitative calculation of the actual achieved SIL in comparison with the performance targets selected by the analysis phase team.

As designers perform their tasks, there is great benefit in their understanding the hazard scenario when making decisions about the appropriateness of various options. For instance, a SIF designer may choose to implement a "double block" arrangement of shutoff valves (i.e., two valves piped in series) to lower the probability of system failure and allow the achievement of higher SIL targets. While this design choice might seem appropriate to the designer, there is a possibility that trapping of chemicals that are prone to self-accelerating decomposition (e.g., cumene hydroperoxide) in between the two closed valves could introduce a new serious unintended hazard.

For designers to make the best choices, the ability to directly view the hazard scenario for the equipment they are designing is often essential. While designers are working with options for their equipment, it should be possible to present the hazard scenario(s) that employ those barriers. Optimally, a single click in the design phase tool should be able to present a list of related PHA scenarios and their bow-tie diagrams. This allows the design team to better understand the scenarios developed by the analysis phase team, the rationale behind their performance target selection, and the ramifications of the decisions that they are making regarding the options for achievement of the selected targets. This concept is presented in the figure below.

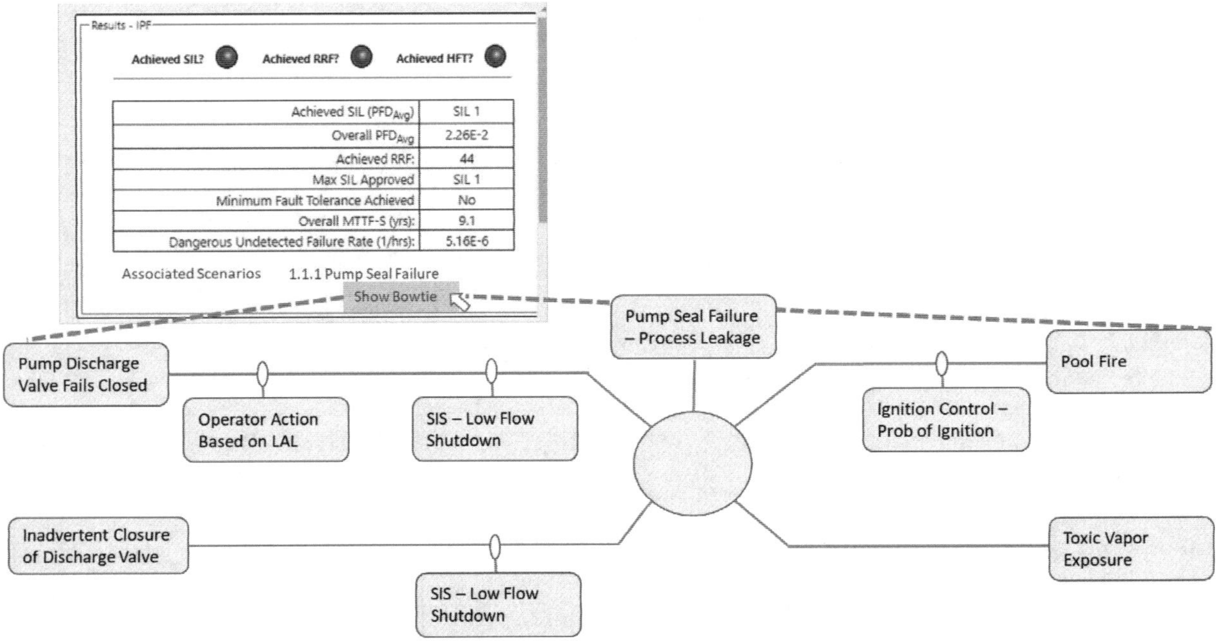

Figure 6 – Accessing Hazard Scenario from Design Phase Tool

After the design team has completed its work, or at least the first pass at its work, data developed during the design needs to be synchronized with other phases in the safety lifecycle of the facility. This requires both feed forward and feedback. The feed forward portion requires that design data be transferred to the operational phase. Operational phase data includes testing frequencies and procedures along with procedures for response to failures of components and subsystems of barriers. The feedback includes parameters that are achieved by the barriers as developed by the design team so that a comparison against targets can be made and confirmed. At this stage, it is critical for the analysis phase team to view the design team's output vis-à-vis their expectations that were developed in the analysis phase to ensure that the barriers, as designed, achieve the risk reduction the analysis team identified and do not create any additional hazard scenarios. Feedback of the design to the analysis tools (Figure 7) allows for rapid review of designs by the analysis team and identification of any gaps which will need to be addressed prior to installation and start up. Again, the bowtie diagram provides an easily understood visual for both the analysis team and management.

IPL	IPL Tag	PFD Target	PFD Actual	IPL Type	Selected SIL
Operator Action Based on LAL-102	LAL-102	0.1		ALARM	
Charge Pump (P-101) Discharge Low-Low Flow Stops Pump	FZL-P101	0.01	0.022	SIF	SIL 2

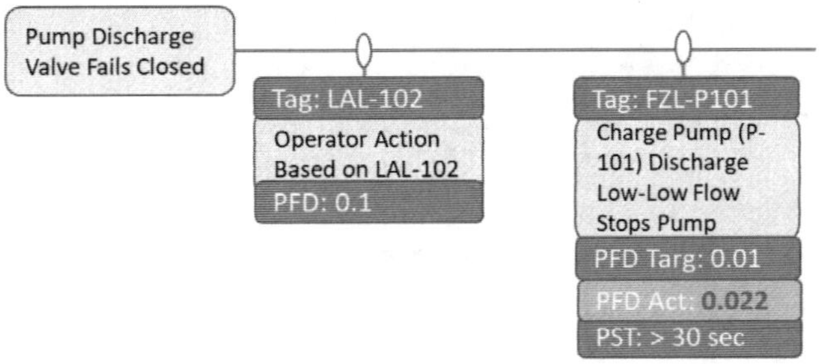

Figure 7 – Display of Design Result Feedback to Analysis Phase Tools – Worksheet and Bowtie

Ensuring Barrier Integrity During the Operational Phase

Interactions between the plant personnel and the barriers, including proof testing and inspection, are required during the operational phase. These activities are required to occur at a frequency and in accordance with the procedures developed during the design phase. The frequency and comprehensiveness of testing has a significant impact on the achievement of performance targets, as delayed or incomplete testing may impact the capability of the barrier to achieve its performance targets. Typically, during the design phase, a certain frequency and level of comprehensiveness of testing is determined to be required.

It is ideal if the design phase tool allows for automatic population of the operation phase tool, commonly a computerized maintenance management system (CMMS) (Figure 8). This population will need to include what devices need to be tested and inspected, at what frequency this testing should be performed and what procedures should be utilized for the testing. Seamless integration of software ensures the analysis and design teams' requirements are faithfully implemented and eliminate the potential for transcription errors.

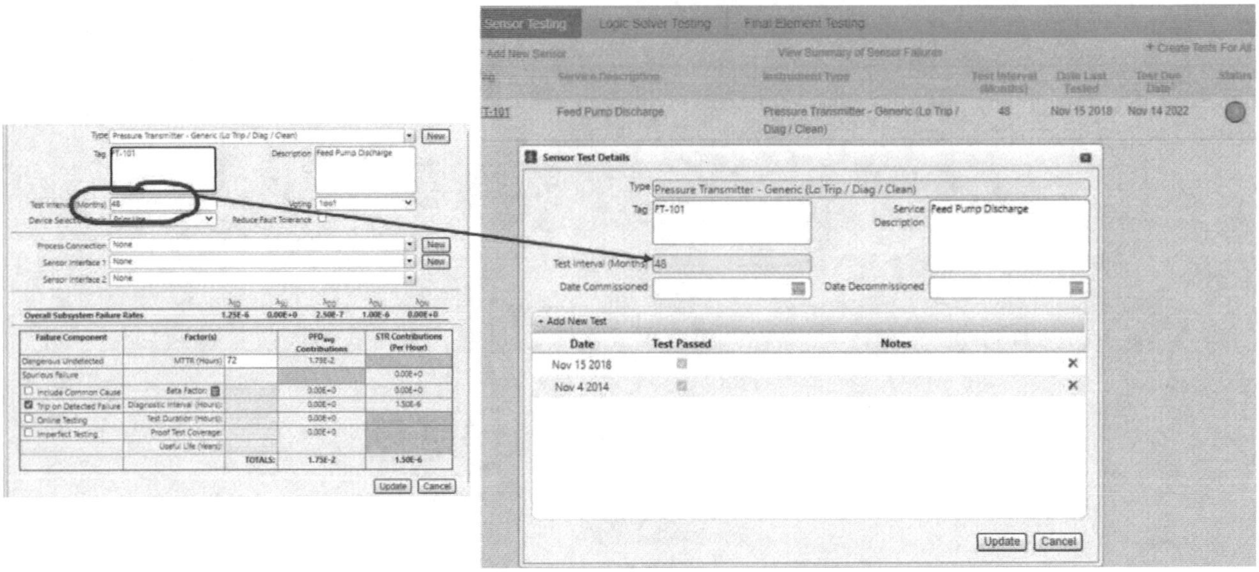

Figure 8 – Automatic Population of CMMS Database from Design Tool

Again, it is useful for the operational phase tool to be able to display information from the design phase, especially if a failure occurs during testing and repairs must be made, in some cases where replacement in kind might not be available. The data associated with the failures of barrier components is essential, as it will be compared with design phase assumptions during revalidation. As such, tracking them in the operational phase tool and feeding back summary information to the design phase tools becomes critical.

In addition, barriers might be fully or partially bypassed while a process plant is still in operation to allow maintenance and testing without adversely affecting the plant. These bypasses must be tracked and authorized carefully, often using a management of change software tool. During the bypass of a barrier, several items of information are critical to the team applying the bypass. First off, during the bypass authorization process, an analysis of the hazard scenarios impacted by the bypass of the barrier is essential. This analysis needs to be documented in the bypass authorization tool, and is greatly facilitated by having the barriers listed in the bypass authorization tool's data base and allowing selection of the barrier in the tool to provide access to a display of the hazard scenario in a bow tie diagram (Figure 9). Providing the bowtie diagram of the relevant scenarios allows for better understanding of the risk being mitigated and an informed decision on the duration of the bypass and potential additional mitigation that may need to be implemented while the bypass is in place.

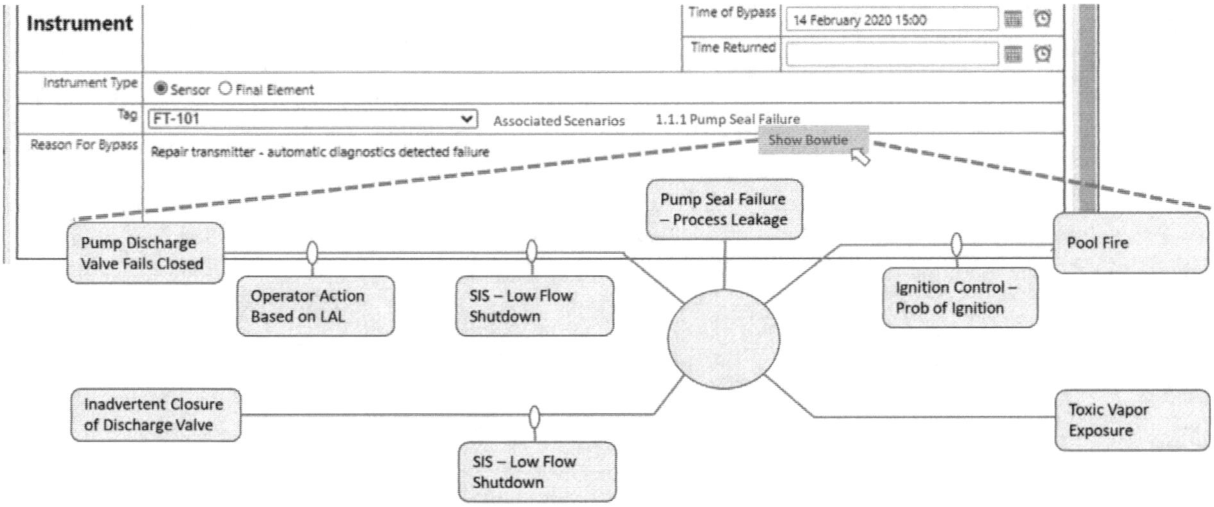

Figure 9 – Display Bowtie of Hazard Scenario from Bypass Authorization Tool

Also, during bypass of a barrier it may be important to have compensating measures, also known as an alternate protection plan, that will serve as a replacement for the bypassed barrier if the level of risk is determined to be beyond tolerable levels while the barrier is bypassed. Alternate protection plan information can easily be developed for each barrier during either the analysis phase or design phase of the plant safety lifecycle. Once developed, this information can be automatically included in a bypass authorization worksheet by selecting the barrier, and automatically communicated to operation staff through emergency SOP databases that can be linked to the design tools.

Instrument		Time of Bypass	14 February 2020 15:00
		Time Returned	
Instrument Type	◉ Sensor ○ Final Element		
Tag	FT-101		
Reason For Bypass	Repair transmitter - automatic diagnostics detected failure		

	Type	Description	Additional Action Required	
			Alternate Protection Plan	Bypass Risk Assessment
○	Type 1	Bypass an instrument for repair or maintenance; instrument is part of fault tolerance system where SIF will still activate upon process demand; repair completed in less than MTTR	No	No
○	Type 2	Bypass an instrument for repair or maintenance; instrument is part of fault tolerance system where SIF will still activate upon process demand; repair requires more than MTTR	No*	Yes
◉	Type 3	Bypass an instrument for repair or maintenance; instrument is NOT part of fault tolerance system; repair completed in less than MTTR	Yes	No
○	Type 4	Bypass an instrument for repair or maintenance; instrument is NOT part of fault tolerance system; repair requires more than MTTR	Yes	Yes
○	Type 5	Bypass instrument for any reason other than instrument repair or maintenance	Per Bypass Risk Assessment *	Yes

* May be required if the Bypass Risk Assessment indicates that it is necessary

Alternate Protection Plan

Item	Value / Description
What process variable or variables must be monitored?	Local Flow Indicator at Pump P-101 Discharge
What are the manual trigger points for the monitored variables?	Flow of less than 10% of linearized scale
Who is responsible for performing the process variable monitoring?	Dedicated outside operator
Who is responsible for performing the manual shutdown action?	Dedicated outside operator, same as monitoring variable
What specific actions must be taken to manually shutdown?	Set Pump P-101 Hand / Off / Auto switch to the OFF position
Can a manual shutdown be performed within the process safety time?	Yes
Is there sufficient independence between the normal operating staff and the alternate protection?	Yes, a dedicated separate operator is used
A Bypass Risk Assessment has been performed and is acceptable (if required)	No bypass risk assessment is required.

Figure 10 – Automatic Completion of Alternate Protection Plan from Design Phase Tool Data

The operational phase of the plant lifecycle is where actual performance metrics about the plant evidence themselves. Tracking actual performance and comparing it against design phase and analysis phase assumptions is critical to ensuring achievement of tolerable risk. Critical events that must be tracked include the occurrence of initiating events and activation of barriers. The key to tracking is to associate each initiating event occurrence and barrier activation with an alarm (audible or journal) that will be documented in the plant control systems alarm log. Then, on a periodic basis the alarm log can be parsed to identify the occurrence of the alarms that indicate initiating events or barrier activation. This information can then be fed back to the analysis phase for revalidation.

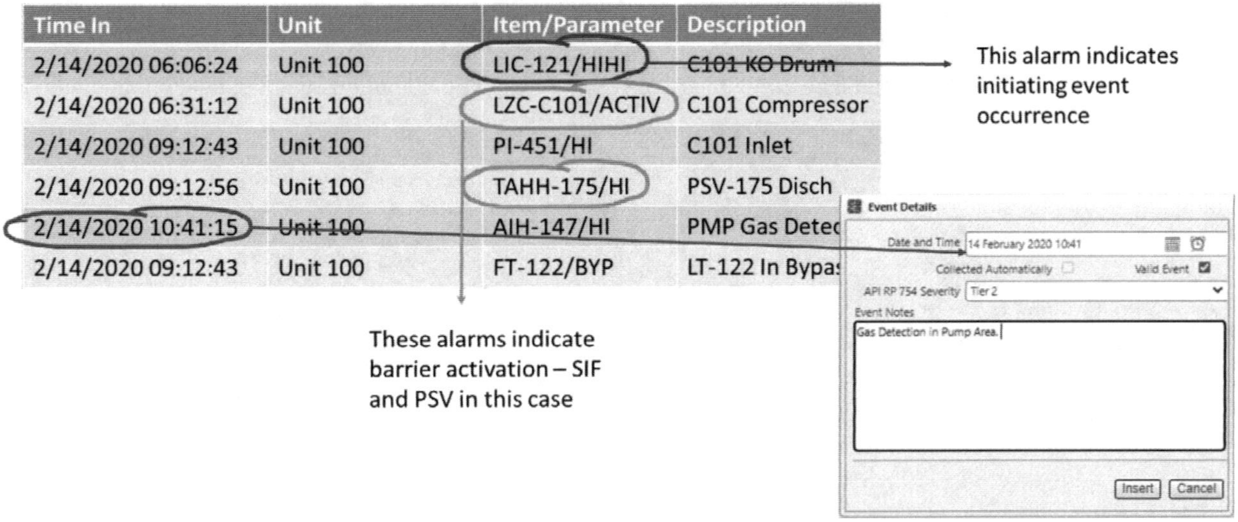

Figure 11 – Alarm Log Parsing for Barrier Activation and Other HSE Relevant Events

Revalidation Workflow for Barrier Consideration

PHA, the analysis phase of the process safety lifecycle, typically needs to be revalidated on a regular basis to ensure that design assumptions were appropriate, actual performance matches expectations, and any changes to the plant operating conditions or equipment modifications were appropriately handled. While revalidation is often considered a simple task of relooking at the PHA document it is the starting point for revalidating all facets of the plant process safety lifecycle including not only the analysis phase, but also the design and operational phases. Some of the items that need to be fed back and considered during revalidation include the following.

- Initiating event frequency
- Barrier activation rate
- Barrier failure rate (by component)

These actual numbers must be compared against design assumptions to determine if there are gaps between the risk of the plant as designed, as opposed the reality of an actual operating plant. The analysis phase calculations, such as LOPA, can then be reviewed to identify gaps. In some cases, if the actual plant conditions resulted in higher demand rates and barrier failure rates than the design assumed, which results in a higher intolerable level of risk, the difference between design and actual could result in the need to change the original design, possibly by adding redundancy, increasing testing and maintenance, selecting different equipment, or even changing the operating conditions of the plant. This task can be greatly facilitated using connected enterprise software tools that allow for communication from operational phase tools that gather actual operating data back to the analysis phase tools. Once again, communication of these results is facilitated by presented in a bowtie diagram.

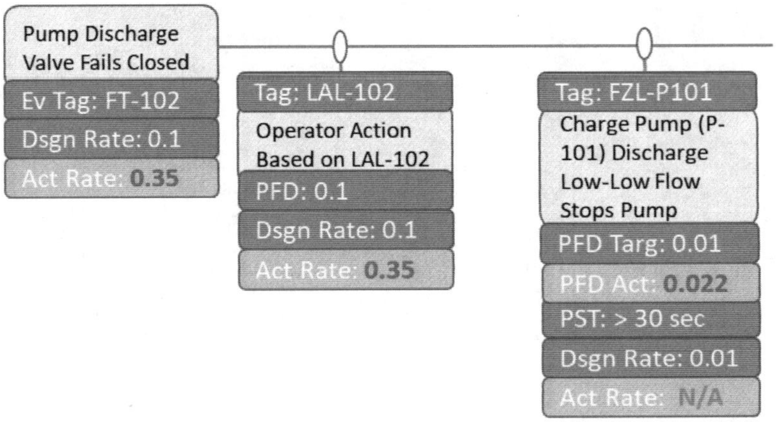

Figure 12 – Comparing Design and Actual Performance in Bowtie

Conclusions

Integrity management of process plant safety barriers is a critical activity that is suffering from lack of communication and lack of data. The process plant safety lifecycle contains three basic phases – analysis, design, and operation. During each phase, critical data is generated that must be passed back and forth from the different phases to ensure that designs are appropriate and tolerable levels of risk can be achieved. Enterprise software tools that utilizes standardized and open data structures facilitate this type of communication between applications and between phases using bowtie diagrams as the mechanism to communicate all this information in an easily understood format. Industry needs to work together to standardize the structure of the data that is passed between the phases and demand that software vendors adhere to these open standards. In this way, critical data can be passed from phase to phase facilitating the integrity management of barriers throughout the whole lifecycle, including the analysis phase where discrepancies between desired performance and actual performance can be assessed and, if necessary, design modifications can be introduced.

Conclusions

Marszal, E.M., *Unified Hazard Assessment – Bringing Together HAZOP, LOPA, Hazard Registers, and Bowtie in a Unified Structure*, Proceedings – Mary Kay O'Conner Process Safety Center 22nd Annual International Symposium, October 2019, College Station, Texas.

Marszal, E.M., *Applying Data Science to Hazard Analysis*, InTech Magazine, Jan-Feb 2020, International Society for Automation, Research Triangle Park, North Carolina.

586239 A Case Study on KOC Approach for PHA Revalidation of Existing Gathering Centre

Abstract:

Introduction:

A hazard is an inherent physical or chemical characteristic that has the potential to do harm (to people, property, or the environment). A hazard evaluation study is an organized effort to identify and analyze the significance of hazardous situations associated with a process or activity.

There are number of reasons such as changes in Process or Equipment, in on-site or off-site occupancy patterns and Gaps and Deficiencies in Prior PHA. In addition, new knowledge is now available to better understand the hazards potential and control, safeguard previously credited in the PHA have been removed / compromised or discredited, and Actual incidents have revealed scenarios not previously identified in the PHA. Positively, process changes have actually removed hazards from the process, prior PHA did not take proper credit for some risk reduction measures and new risks reduction measures have been added, which clearly reduced the original risk.

Objective:

The primary objective of a PHA revalidation is to produce an updated PHA that adequately identifies, evaluates, and proposes controls for the hazards of the process, as they are currently understood.

Methodology:

KOC has established KOC.PS.017 - Process Hazard Analysis (PHA) Revalidation Procedure to determine an appropriate approach for the revalidation (Redoing vs. Update and Revalidate), conducting revalidation in timely manner, establish responsibilities and confirming that the PHA is accurately reflects the hazards of the current process.

The revalidation process build upon the still pertinent portions of the prior PHA. Corrections are made and new content is added to the PHA as required (i.e., the PHA is Updated), and the results are documented to serve as the new PHA (i.e., the PHA is Revalidated).

KOC appointed well known third party consultant for onsite PHA revalidation and below are stages for PHA revalidation.

Team Selection

Selecting the Appropriate PHA Revalidation Approach

- Update and Revalidate.
- Retrofit, Update, and Revalidate
- Redo Review and Analyze Information Review of Operating Experience
- Evaluating of PHA Quality and Completeness
- PHA Revalidation Information Requirements

- Changes Review

- Review Previous Incidents
- Review the Status / Resolution of Previous PHA Recommendations
- Update Human Factors
- Update the Facility Siting Analysis
- Address Hazards Associated with Non-Routine Operating Modes
- Address Compliance of PHA with Regulatory Requirements Documentation of PHA Revalidation Recommendation Follow - up
- PHA Report Contents
- Wrap-Up Discussions

Results

Type of studies carried out and number of new recommendations will be discussed for Gathering Center 24.

Conclusion:

Continuous improvement in the control of newly introduced process safety risks.

Ensured the effective close out of pending actions raised during previous process hazard analysis.

Addresses all four pillars of Risk Based Process Safety.

Considerable time saving in comparison to Redo PHA.

Improvement in employee engagement.

Visual Dust Hazard Analysis – Understanding Threats and Assuring Controls

Michelle R. Murphy
Mica, LLC
Atkinson, NH
murphy@micadhm.com

David Hatch
Process Safety Integrity
Derby, UK
dhatch@psintegrity.com

Prepared for Presentation at
American Institute of Chemical Engineers
2020 Spring Meeting and 16th Global Congress on Process Safety
Houston, TX
March 30 – April 1, 2020

Visual Dust Hazard Analysis – Understanding Threats and Assuring Controls

Michelle R. Murphy
Mica, LLC
Atkinson, NH
murphy@micadhm.com

David Hatch
Process Safety Integrity
Derby, UK
dhatch@psintegrity.com

Keywords: Dust Explosions, Bowties, Flammable Atmospheres, Risk Management. Ignition Controls, Dust Hazard Analysis

Abstract

The potential harm and asset damage from a dust explosion have been known for many centuries. A wide variety of materials that are explosible in dust form exist in many industries, however knowing about an issue is not the same as properly understanding and adequately addressing the challenges.

The authors draw on their complementary experience in Dust Hazard Analysis (DHA) application (1) and BowTie implementation (2) to present a novel approach to scenario visualisation and asset management using bowties.

Just as a conventional HAZOP or Process Hazard Analysis (PHA) subdivides the process into nodes, the DHA breaks the plant into sections which are systematically challenged to determine if explosion or fire hazards exist. For dusts, this includes identification of competent ignition threats. Subsequent evaluation is conducted to determine which are prevented and/or mitigated by safeguards specific to each threat and consequence.

Risks are assigned and actions issued as per a normal hazard analysis, however the graphical presentation of the ignition scenarios and safeguards and the illustration of potential ignited particles which can travel and ignite within downstream equipment offers significant communication benefits. This can be particularly helpful in facilities or countries where English and/or Risk are not the primary language or familiar terminology. This paper will show how BowTies are an effective tool to engage less technical stakeholders and offer a robust framework to assure the health (presence and performance) of the assumed or planned human and hardware controls which must be sustained to provide the necessary risk reduction.

In summary, this novel approach evolves analysis into an operational framework to ensure that the ignition risks are better understood, and the controls are properly implemented, operated and maintained.

1 Combustible Dust Hazard Analysis

The potential harm and asset damage from a dust explosion have been known for centuries. A wide variety of materials that are explosible in dust form exist in many industries, however knowing about an issue is not the same as properly understanding and adequately addressing the challenges.

Figure 1 shows the progression from fire to flash fire to explosion for combustible dusts. These hazards exist in dust processing equipment and in building spaces where dust is escaping equipment. To prevent the realization of these unwanted events, one of the elements of the original fire triangle needs to be removed. While explosions are the most destructive, flash fires often result in fatalities, and having one of these events can trigger secondary events.

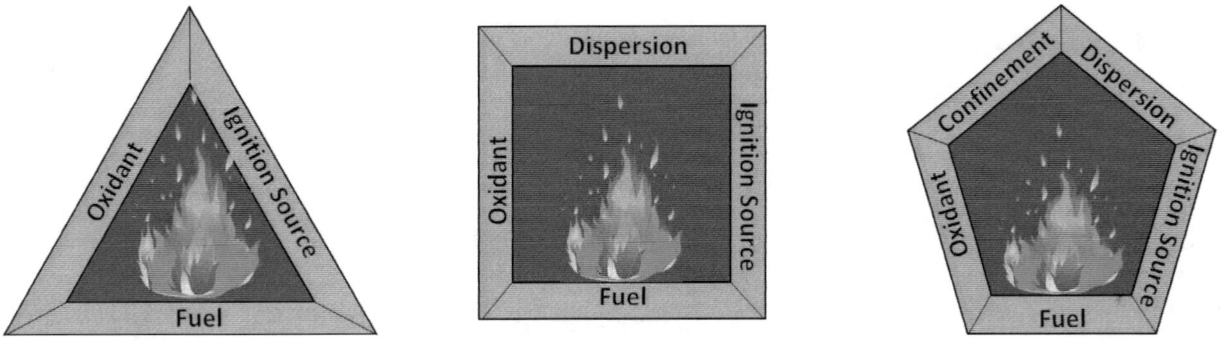

Figure 1 – Progression of combustible dust hazards from fire to flash fire to explosion

Dust hazard analyses (DHAs) are conducted to identify and assess dust flash-fire and explosion hazards in both process equipment and building spaces. Like PHAs, many methodologies are available for completing hazard identification and assessment of dust hazards. Most involve recording of the discussion in a spreadsheet format. Visual DHAs (bowties) are a novel approach to scenario visualization that can simplify communication of DHA results, elevate appreciation of safeguards (barriers), and improve the management of change (MOC) process.

At present, most applications involve development of bowties for higher hazard scenarios after a more traditional hazard analysis has been completed. This paper will present bowties as developed from traditional DHA records.

1.1 Regulations

In the US, the Occupational Health and Safety Administration (OSHA) does not have a comprehensive combustible dust standard. It relies on existing related regulations and recognized and generally accepted good engineering practices (RAGAGEPs). The most widely applied of these standards are from the National Fire Protection Association (NFPA®). Both the fundamentals standard NFPA 652 (3) and the industry-specific standards (4; 5; 6; 7), require dust hazards analysis (DHA) of dust handling operations. The NFPA does not specify a DHA technique to be adopted however one example is provided and several commonly applied hazard analysis techniques are listed in the appendix. In their recent paper, Murphy and Borene (8) detail the NFPA requirements for conducting DHAs and present a typical DHA process.

In Europe, ATEX (ATmosphères EXplosives) is the name commonly given to the two European Directives for controlling explosive atmospheres:

- Directive 1999/92/EC (9) (also known as 'ATEX 137' or the 'ATEX Workplace Directive') on minimum requirements for improving the health and safety protection of workers potentially at risk from explosive atmospheres.
- Directive 2014/34/EC (10) (also known as 'ATEX 95' or 'the ATEX Equipment Directive') on the approximation of the laws of Members States concerning equipment and protective systems intended for use in potentially explosive atmospheres.

Article 8 of the ATEX Workplace Directive requires the employer to draw up (and maintain) an Explosion Protection Document (EPD) which includes:

- Identification of hazards
- Evaluation of risks
- Definition of specific measures to safeguard the health & safety of workers at risk from explosive atmospheres

In the UK, the Dangerous Substances and Explosive Atmospheres Regulations 2002 (DSEAR) regulations (11) put into effect the requirements of ATEX 137 and require employers (duty-holders) to identify dangerous substances (including flammable gases, vapours, mists and combustible dusts), evaluate the associated risks, implement controls to remove/reduce flammable atmospheres and/or ignition sources, prepare emergency plans and ensure employees are properly informed about and trained to control the risks from dangerous substances. Although DSEAR does not require an EPD (since its contents are covered by Regulation 5 *Risk Assessment* and Regulation 7 *Places where explosive atmospheres may occur*) these are often prepared by duty-holders as a clear, concise, basis of safety that can be communicated with employees and other interested parties. This approach is intended to complement and/or supplement such documents.

1.2 Regulatory Deadlines

In the US the deadline for most facilities to complete a dust hazard analysis (DHA) is 7th September 2020 (later for agricultural and food manufacturers) in accordance with NFPA Standard 652 on the Fundamentals of Combustible Dust and the associated industry-specific standards.

In the UK and Europe, the ATEX Workplace Directive and DSEAR regulations are already effective and the obligations on duty-holders are long established. This approach aims to improve the awareness and understanding of those responsible for and/or affected by flammable atmospheres.

1.3 DHA Elements

NFPA states that a dust hazard analysis is intended to identify and evaluate areas of the process or building spaces where fire, flash fire, and explosion hazards exist. Once identified, they are to be evaluated for applicable fire and deflagration scenarios, including:

1. Safe operating ranges
2. Existing safeguards
3. Additional safeguards, as needed

Like all hazard analyses, an identification method is employed to identify hazards. Once identified, causes and consequences are postulated to define scenarios. In the case of combustible dust, these would focus on fire, flash fire and explosion scenarios. Existing safeguards are identified and evaluated for their ability to manage these events and remove or limit the consequences. Where existing safeguards are deemed inadequate, additional safeguards are recommended. Finally, a plan for implementation is put together. The process is depicted in Figure 2. An implementation plan is typically completed after the hazard analysis step and is not shown in the hazard analysis flow diagram.

Figure 2 – Hazard analysis flow

1.4 Hazard Identification Methods

Many methods exist to identify dust hazards. Typical identification methods include:

- What-if/Checklist
- HAZOP
- Event Tree/Fault Tree
- Failure Modes and Effects Analysis

Due to the straightforward operation of combustible dust handling equipment, a checklist approach is often employed to identify hazards. Such a checklist can be based on specific requirements and recommendations found in several reference documents (e.g., NFPA codes and standards, CCPS Guidelines for Safe Handling of Powders and Bulk Solids (12) , CCPS Guidelines for Combustible Dust Hazard Analysis (13) and Lee's Loss Prevention in the Process Industries (14)).

CCPS Guidelines for Risk Based Process Safety (15) states “A company that uses its risk and understanding is better able to deal with the resultant risk, and subsequently, sustain long-term, accident-free, and profitable operations.” DHAs are the first step in the risk assessment process.

2 Application of Bowties

2.1 Purpose of Bowtie

A 'bowtie' is a diagram that visualizes the risk you are dealing with in just one, easy to understand picture. The diagram is shaped like a bowtie, creating a clear differentiation between proactive and reactive risk management. The power of this type of diagram is that it gives you an overview of multiple plausible scenarios, in a single picture. In short, it provides a simple, visual explanation of a risk that would be much more difficult to explain otherwise.

Bowtie diagrams are now commonplace in understanding and managing process safety risks, particularly associated with major accidents. Historically they have been applied most frequently in the chemical, petrochemical and oil & gas industries but are equally applicable in all industries, e.g., transport, mining, energy, finance or healthcare and can be used to manage all risks (effect of uncertainty on objectives) including safety, environmental impact, asset damage or loss of reputation.

Bowties demonstrate how hazards are controlled and illustrate the links between controls (barriers) and the relevant components of the safety or risk management system.

2.2 Methodology

The bowtie diagram is based on the concept of the “Swiss Cheese Model” developed by James Reason (16). This model illustrates system failures which can be addressed by controls (barriers) represented by swiss cheese slices which are intended to provide ‘defence in depth’.

Barriers have weaknesses which are either inherent i.e., always present or are active failures created during the scenario which, when aligned, allows the ‘accident trajectory’ to pass through the holes, as depicted in Figure 3.This model has evolved into the familiar and increasingly popular bowtie diagram which allows multiple trajectories (Cause-Consequence relationships) to be visualised on the same diagram.

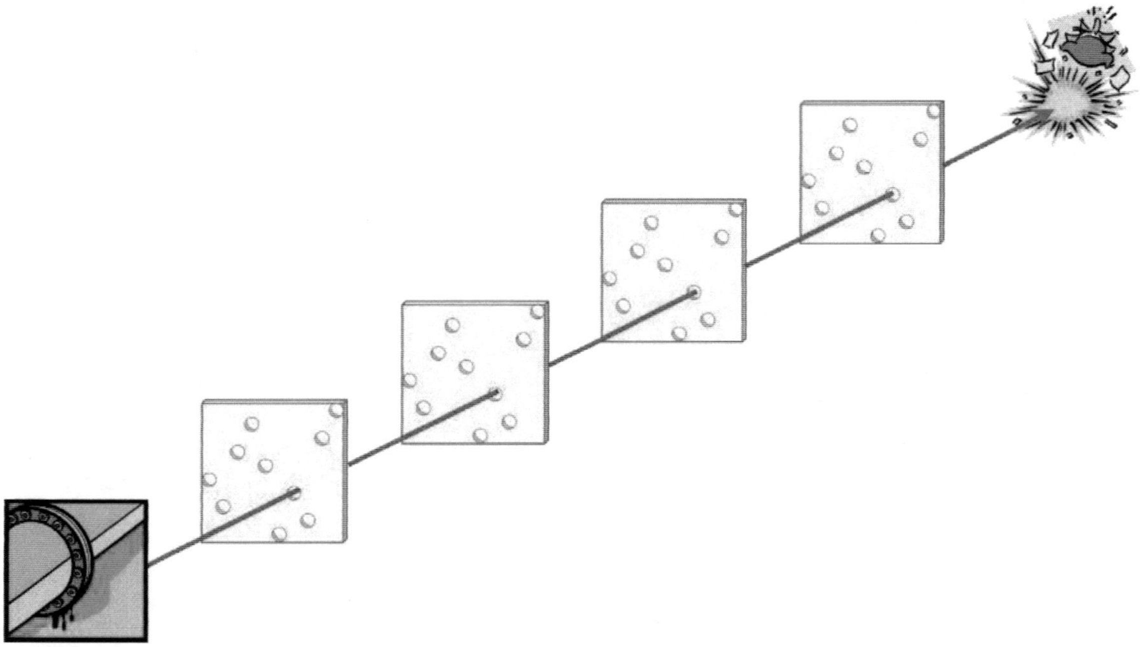

Figure 3 – Swiss Cheese Model

The components of a bowtie diagram are summarised as follows:

- Hazard = Operation, activity or material with potential to cause harm
- Top Event = Initial loss of control of Hazard
- Consequences = Negative effects (harm or damage) that could result from Top Event
- Threats = Potential reasons for loss of control of the Hazard leading to the Top Event
- Barrier = Engineering or Administrative risk reduction measure
    - Prevention
    - On its own can prevent a Threat from developing into a Top Event
        - Eliminate the Threat
        - Prevent the Top Event
    - Mitigation
    - On its own can mitigate the Consequences of the Top Event once it has occurred
        - Prevent consequence (likelihood)
        - Reduce impact (severity)

Barriers can either be hardware, human, or combinations of both which are implemented and operate as follows:

- Passive hardware
    - Always present e.g. structural or containment
- Active hardware

    - Automated response to detection e.g. control action or trip
- Active hardware + human
    - Human response prompted by hardware detection e.g. alarms
    - Hardware response prompted by human detection e.g. manual call point
- Active human
    - Human response prompted by observation e.g. walk around & intervention
- Continuous hardware
    - Always operating e.g. ventilation

These components are assembled into scenarios which are read from left to right as shown in the overview in Figure 4 below (with good practice guidance from the CCPS Bowtie book (17)):

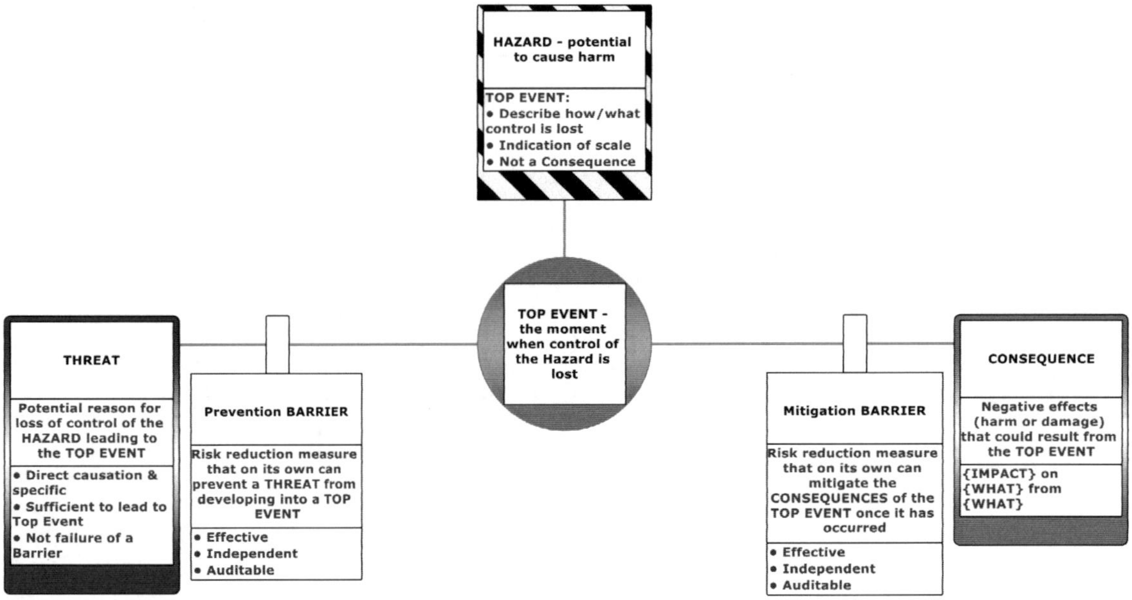

Figure 4 – Bowtie Model

A Top Event may have several Consequences (safety, environmental, asset, etc.) that result from it and several Threats (e.g., equipment failures, environmental influences or operational issues) that could cause it.

2.3 Application to Dust Hazard Analysis

Just as a conventional HAZOP or PHA subdivides the process into nodes, the DHA breaks the plant into systems which are systematically challenged to determine if explosion or fire hazards exist. For dusts, this includes identification of competent ignition threats. Subsequent evaluation is conducted to determine which are prevented and/or mitigated by safeguards specific to each threat and consequence.

Annex B of NFPA 652 (3) provides an example DHA of a wood milling process. Wood chips are brought in via rail car and trailer truck, unloaded and pneumatically conveyed into a storage silo, transported through a screw conveyer to a size reduction mill, and further processed for product delivery. The DHA is applied to each process component and is documented in paragraph format. Figure 5 shows how the results of the evaluation of the first node, offload duct to offload fan, would look in a simple bowtie representation.

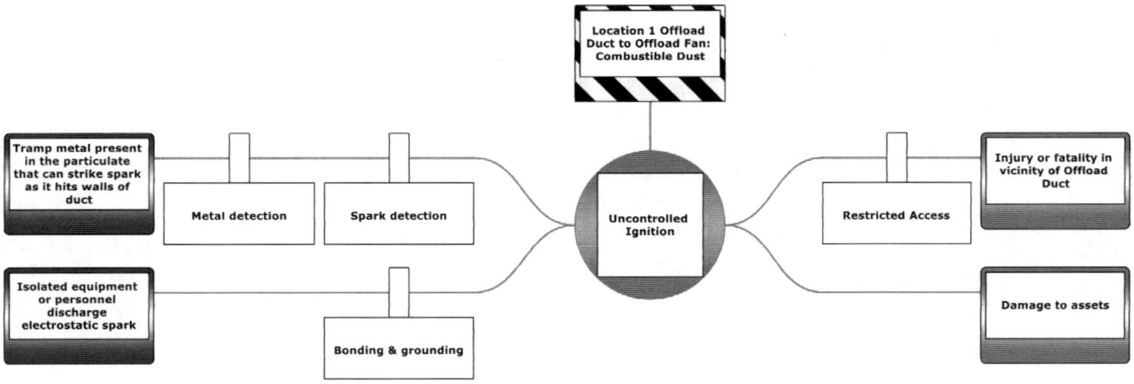

Figure 5 – Basic DHA of a process unit

A more detailed example is discussed later in this paper.

In the manner presented in Figure 5, each DHA node (major asset or sub-system) is represented by a separate bowtie with the credible/competent ignition sources represented as threats e.g. from EN 1127-1 (18) . To simplify the documentation of ignition sources, these can be coded to make the diagrams concise:

- AC Adiabatic compression and shock waves
- EA Electrical apparatus
- EC Stray electric currents, cathodic corrosion protection
- EW Electromagnetic waves
- EX Exothermic reactions, including self-ignition of dusts
- FL Flames and hot gases (including hot particles)
- HS Hot surfaces
- IR Ionizing radiation
- LT Lightning
- MS Mechanically generated sparks
- RF Radio Frequency (RF) Electromagnetic waves
- SE Static electricity

- US Ultrasonics

Hazard data can be added to the diagram including:

- Material and properties
- Safe Operating Range
- Deflagrable (Yes/No)
- Suspended in Air (Yes/No)
- Above Minimum Explosive Concentration (Yes/No)

The potential effects of an uncontrolled ignition (some processes involve controlled ignition e.g., biomass fuel) include fire, flash fire, or explosion that result in fatality, injury, and/or property damage (including business interruption). These can occur in the node/asset/system under review or in connected nodes/assets/systems e.g., ignited or smouldering particles passing downstream. Figure 6 shows the connection of consequences (upstream) that become threats (downstream). In the figure, uncontrolled ignition in the offload duct propagates to the offload fan. These have become threats to the downstream system.

NFPA specifically calls out the evaluation of potential deflagration propagation between parts of the process. The bowtie simplifies the application and visualization of this critical part of the analysis.

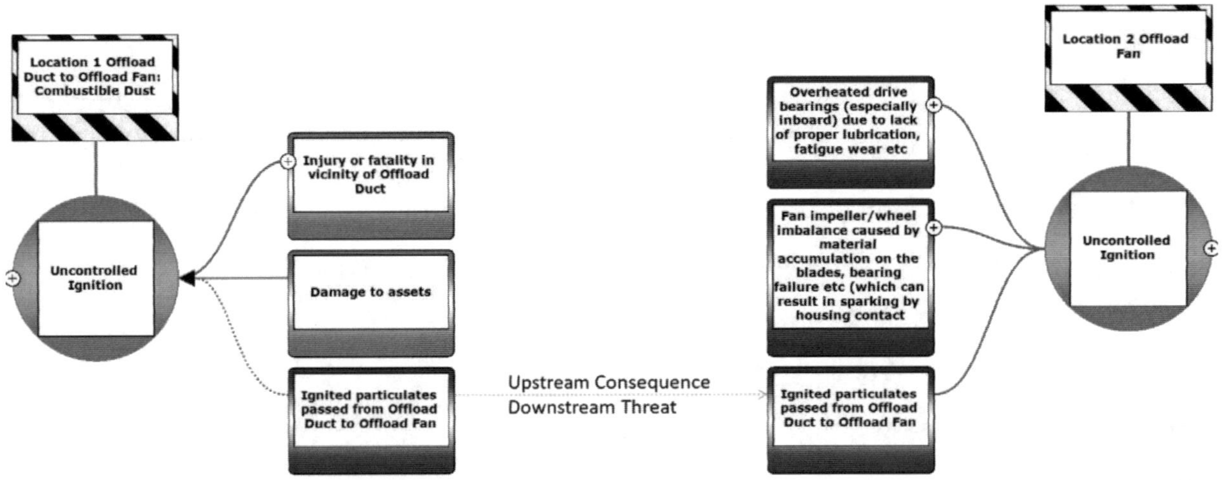

Figure 6 – Indirect Consequences (chained bowties)

Risks are evaluated based on the likelihood of a coincident flammable dust atmosphere (in ATEX, this is defined in hazardous area classification) and an ignition source (based on threat frequency or potential) and the severity of the fire or explosion impact. These are assigned and actions issued as per a normal hazard analysis. The graphical presentation of the ignition scenarios and safeguards and the illustration of potential ignited particles which can travel and ignite within downstream equipment offers significant communication benefits. This can be particularly helpful in facilities or countries where English and/or Risk are not the primary language or familiar terminology.

The visual nature of the bowtie helps confirm existing control measures (prevention & mitigation barriers) e.g. explosion vents, temperature sensors, etc. and identify where additional measures to meet risk target are needed.

2.4 Risk Management

Effective risk management requires an understanding of hazard scenarios and an appreciation (and respect) of the measures to prevent loss of control and mitigate the consequences. Bowtie diagrams provide clear, consistent information to a wide audience compared to detailed analyses which are limited to specialist interpretation.

Once barriers have been identified, a barrier-management program is necessary to ensure that controls are implemented and continue to operate and perform as required to provide ongoing risk reduction. This program involves monitoring the presence and measuring the effectiveness of hardware (technical) and human (organisational) barriers and repairing, replacing or refreshing them as and when appropriate in advance of any demands to act, i.e., before the scenarios occur. The visual nature of bowties can simplify the identification of barriers for program development.

3 Example

The approach of converting from a traditional spreadsheet format is presented in an example of a filter receiver (Figure 7). Filter receivers are one type of air material separator and are often found in dust handling operations as part of the pneumatic transport system. In this example, material is transferred from a silo to a day bin, through a filter receiver. The receiver has a rotary air lock before discharging by gravity into the bin. The material properties and safe operating ranges would also be documented for the DHA. The example employs the risk-ranking presented in the paper by Murphy & Borene (8), wherein C, L, and R represent a ranking of consequence, likelihood, and risk. The rank is determined from the consequence and likelihood assigned by the DHA team. The resultant rank is color-coded to depict a prioritization of risk, moving from green (acceptable) to red (highest priority).

Table 1 shows the results of an analysis of ignition sources, as part of an overall checklist review of the filter receiver. Note that the example is provided for demonstration purposes only and is not intended to represent a complete DHA.

The same record is presented in a bowtie format in Figure . The first impression is one of simplicity. The information is presented clearly and concisely, allowing a quick review of the barriers that are imperative to preventing the fatality. This allows for further discussion on the availability/reliability of the barriers. For example:

1. Equipment material is fixed so very effective barrier
2. Overload trip needs to be regularly tested

3. Ignition energy is a robust barrier as long as material doesn't change i.e., analysis only fit for specific dust
4. Explosion vent needs to be regularly inspected
5. Explosion vent size/capacity needs to match dust (explosion) properties i.e., if material changes vent may not be adequate.

The visual design can be used as a communication tool for design, operation, and maintenance to understand the importance of specific activities and the role they play in preventing and mitigating the event of an explosion in the receiver resulting in a fatality. Additionally, during the MOC process, it will be straightforward to see what changes will impact the barriers (here change of material, change of rotary valve speed, removal of explosion vent, for example).

In this example, the following bowtie features are exploited:

1. Threat types categorised from EN 1127-1
2. Barrier (safeguard) types
3. Consequence risk for 4 types of receptors:
    a. Safety = D
    b. Environment = not determined (grey box)
    c. Asset = not determined (grey box)
    d. Reputation = not determined (grey box)

It is also possible to show individual Threat likelihoods (with or without safeguards or Barriers) and the individual Barrier effectiveness to appreciate the risk reduction contribution that each makes.

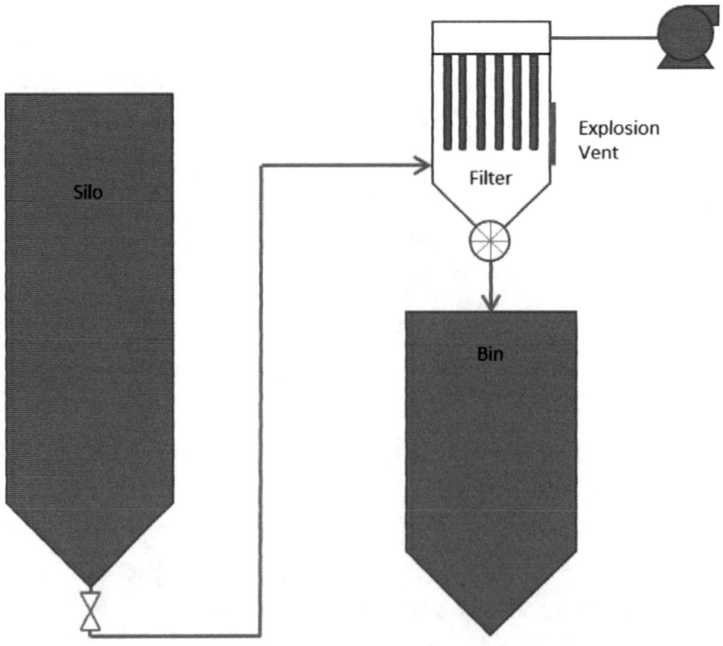

Figure 7 – Example receiving system

Table 1 – Excerpt of a traditional recording of a checklist DHA on a filter receiver

Item No.	Question	Answer	Cause	Consequence	Safeguards	C	L	R	Recommendation
1.1	Are competent sources of ignition present?	Bearings	Bearings cause ignition - N/A as bearings are outboard						
		Electrical spark	Electrical spark causes ignition - N/A as no electrical equipment in receiver						
		Impact spark	Impact spark caused by tramp material	Explosion in receiver, one fatality	Material of construction is stainless steel, explosion venting on receiver	1	1	D	
		Frictional spark	Frictional spark caused by tramp material getting caught in rotary valve	Explosion in receiver, one fatality	Rotary valve is low rpm, valve has an overload trip, explosion venting on receiver	1	1	D	
		Electrostatic spark	Electrostatic spark from unbonded flex boot causes ignition	Explosion in receiver, one fatality	MIE 100-300 mJ, explosion venting on receiver to safe location	1	1	D	1.1.1 Update grounding/bonding program to assure bonding is maintained across flex boots.
1.2	Do AMS with explosion hazards have isolation devices?	Yes, rotary valve on outlet of receiver	Electrostatic spark causes ignition in filter receiver and propagates forward to day bin	Explosion in day bin, one fatality	Rotary valve designed for material blocking deflagration isolation downstream	1	1	D	1.2.1 See Recommendation [1.1.1] to update grounding/bonding program.
		No isolation back to the silo	Electrostatic spark causes ignition in filter receiver and propagates back to the silo	Explosion in silo, one fatality		4	3	B	1.2.2 Add an isolation device (in accordance with NFPA 69) upstream of the filter receiver to prevent propagation back to the silo.
									1.2.3 See Recommendation [1.1.1] to update grounding/bonding program.

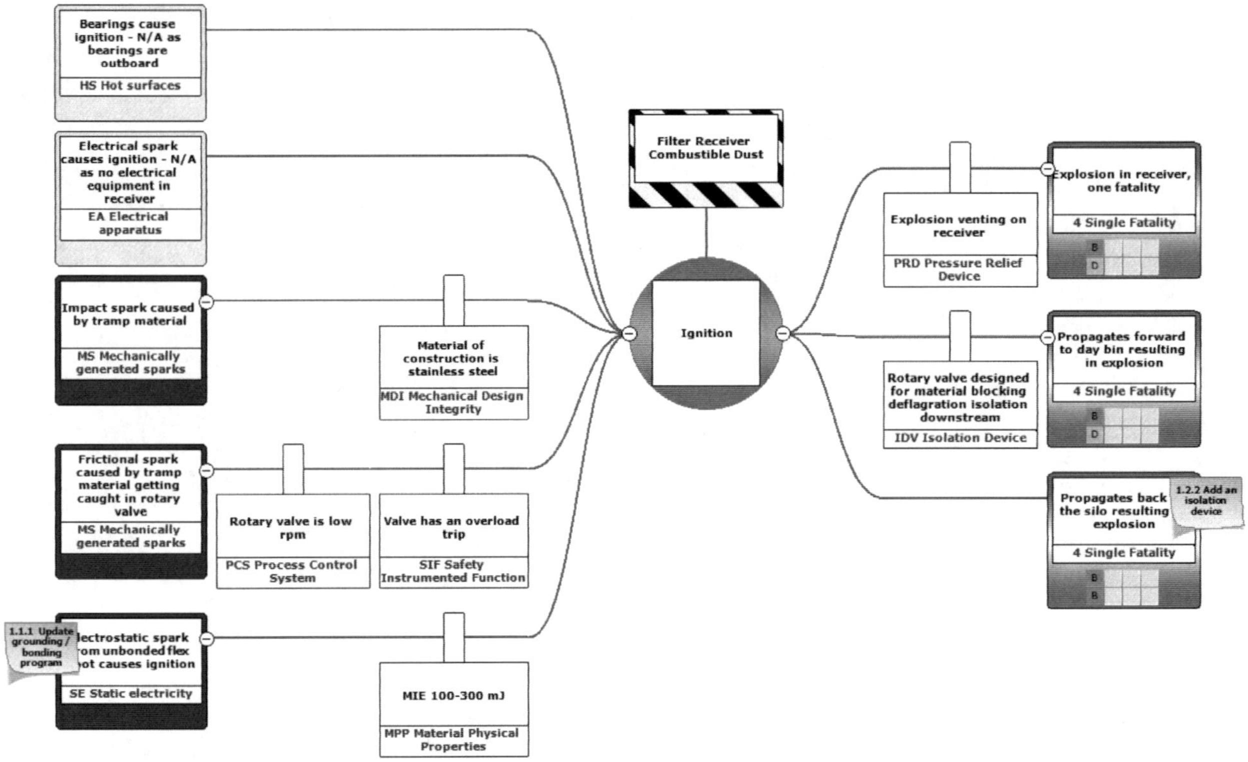

Figure 8 - Traditional DHA in bowtie format with ignition as the top event

4 Conclusion

4.1 Challenges of traditional hazard identification methods

The consequences and risks of a dust explosion have been known for many years and assessments have been conducted to address these. A traditional hazard analysis format (e.g., Checklist, HAZOP) is often applied which may have the following weaknesses:

1. Controls wrongly assigned to causes/threats/ignition sources
2. Controls missing from appropriate causes/threats/ignition sources
3. Failure to distinguish between prevention and mitigation/recovery measures
4. Failure to identify and evaluate connections between nodes/equipment where smouldering particles can ignite downstream dust

Furthermore, the presence (availability) and performance (reliability) of barriers is not part of the assessment and therefore a static snapshot of assumed or planned controls.

4.2 Benefits over traditional hazard identification methods

There are a number of identification methods ranging from simple checklists to more rigorous, systematic techniques which are well publicised (19) and employed. Bowties provide context to the explosion scenarios in a simple but effective visualisation of the relationships between:

- Causes of uncontrolled ignition;
- Consequences of those events, and;
- Barriers that prevent the event and/or mitigate the effects

A graphical representation offers a language that facilitates a common understanding of hazards, unwanted events, their risks, and controls to help direct or deploy resources appropriately and proportionately.

Classification of barriers, either by type or owner, allows dependencies or common mode vulnerabilities to be evaluated and addressed. A common mode failure occurs when a single event causes multiple barriers to fail (either within the same bowtie or across multiple bowties) and therefore barriers should be independent of the threat that they prevent and of other barriers on the same threat to top event or top event to consequence pathway.

4.3 Communication

The aim is not just to make existing assessments more engaging i.e. to involve all stakeholders but also to provide a life-long dynamic framework where threats (competent ignition sources) and barriers (control measures) are monitored, evaluated and actions taken to ensure that protection is sustained and risk targets are maintained. Visualisation enables duty holders not only to analyse their dust hazards but also to communicate the analyses to front line personnel to ensure that they understand the risks they are responsible for managing and sustain the protection measures for which they are accountable.

The current aim is to enhance not replace traditional methods.

4.4 Operational vs design focus

Bowties are not a static snapshot of assumed/planned controls but a live asset/risk management platform that can be updated (manually or automatically) to show the current state of health (presence and performance) of controls and the current risk exposure.

Barriers degrade over time and their performance must be monitored, measured and sustained at the required level to achieve the necessary risk reduction.

4.5 Better management of change implementation

Identification of barrier criticality and ownership makes management of change (MOC) more robust since the potential impact of defeating, degrading or deleting barriers within a single or across multiple scenarios (bowties) is more apparent. Suitable bowtie software with a barrier database can be filtered/sorted to focus on the deployment of barriers. This allows proper addressing in all scenarios where an affected barrier or barriers is/are implemented. Because barriers can be hardware, human or a combination of both, technical and organisational changes can be evaluated to ensure that the risk reduction is not unduly compromised.

The authors believe this approach highlights and addresses existing deficiencies whilst providing the same risk assessment functionality expected or required. For combustible dust this means consideration of probability of a flammable atmosphere (using ATEX (20), NFPA classification of dust flash-fire or explosion hazard area (3), or NFPA classification as a hazardous location for electrical installations (21)), probability of competent ignition source, and severity of fire, flash fire, or explosion consequences.

The goal is to operationalise scenarios and demonstrate that duty-holders are and remain in control through ongoing barrier maintenance and robust change management.

An initial step for duty-holders would be to migrate their existing tabular (worksheet) assessments into bowties to expose and address weaknesses. Visual DHA is designed to be different (evolution rather than revolution) with a format where existing information is not lost but knowledge is gained.

References

1. **Murphy, M.** Making Sense of Combustible-Dust Hazard Analysis. *Chemical Engineering Progress.* 2016, April 2016.

2. **Hatch, D., McCulloch, P., Travers, I.** Enhancing PHAs: The Power of Bowties. *Chemical Engineering Progress.* 2019, February 2019.

3. **NFPA.** *NFPA 652 Standard on the Fundamentals of Combustible Dust 2019 Edition.* May 2018.

4. **NFPA**. *NFPA 654, Prevention of fire and Dust Explosions from the Manufacturing, Processing, and Handling of Combustible Pariculate Solids.* Augst 2019.

5. **NFPA .** *NFPA 61 Standard for the Prevention of Fires and Dust Explosions in Agricultural and Food Processing Facilities.* November 2019.

6. **NFPA.** *NFPA 484 Standard for Combustible Metals.* May 2018.

7. **NFPA**. *NFPA 664 Standard on the Prevention of Fires and Explosions in Wood Processing and Woodworking Facilities.* June 2016.

8. **Murphy, M. and Borene, M.** How to Conduct a Dust Hazards Analysis. *Process Safety Progress.* 2018, October 2018.

9. **European Commission.** *ATEX 137 workplace directive 1999/92/EC.* 1999.

10. **European Commission .** *ATEX 95 equipment directive 2014/34/EC.* 2014.

11. **UK Government.** *Dangerous Substances and Explosive Atmospheres Regulations (DSEAR).* 2002.

12. **CCPS.** *Guidelines for Safe Handling of Powders and Bulk Solids.* November 2004.

13. **CCPS .** *Guidelines for Combustible Dust Hazard Analysis.* May 2017.

14. **Elsevier.** *Lees' Loss Prevention in the Process Industries 4th Edition.* 2012.

15. **CCPS.** *Guidelines for Risk Based Process Safety.* March 2007.

16. **Reason, J.** *Managing the risks of organizational accidents.* 1997.

17. **CCPS.** *Bow Ties in Risk Management: A Concept Book for Process Safety.* September 2018.

18. **British Standards Institute.** *BS EN 1127-1:2011, Explosive atmospheres. Explosion prevention and protection. Basic concepts and methodology.* 2011.

19. **CCPS.** *Guidelines for Hazard Evaluation Procedures 3rd Edition.* April 2008.

20. **British Standards Institute.** *BS EN 60079-10-2: 2015, Explosive atmospheres. Classification of areas. Explosive dust atmospheres.* 2015.

21. **NFPA.** *NFPA 70 National Electrical Code.* January 2020.

585578 Vinyl-Siloxane Cleavage: Applying Thermo-Analytical and Hybrid Calorimetry Techniques to Inform Process Safety and Design

Abstract:

The use of homogeneous and heterogeneous acid catalysts is widespread in the processing of siloxanes as they protonate Si-O bonds, enabling a vast array of rearrangement reactions. Therefore, developing a fundamental understanding how acid catalysts interact with various siloxane moieties is crucial to safe and sustainable operation. In this study, we present our findings on the interaction of acid catalysts with vinyl pendant siloxanes, specifically focusing on the exothermic acid-catalyzed cleavage of the C-Si bond with gas evolution. Thermo-analytical techniques including Differential Scanning Calorimetry (DSC) and Thermogravimetric Analysis (TGA) were used to screen various moieties to evaluate gas and heat generation. Adiabatic Accelerating Rate Calorimetry (ARC) measurements were performed to quantify the kinetics of C-Si cleavage and evaluate effect of operational conditions on the kinetics of the cleavage reaction. Then, a hybrid technique, ARCxGC/MS was utilized to identify the various reaction products and the correlate them with the detected onset temperatures. Using molecular modelling and additivity techniques like secondary Benson Group estimation, the theoretical Heat of Reaction for the proposed reaction was determined and compared to predictions from the afore-mentioned calorimetry techniques. From the above effort, it was concluded that cleavage of the vinyl-silicon bond due to activation by the acid catalyst is an exothermic and gas-generating reaction. The findings were used to generate, for the first time, a detailed quantitative thermo-kinetic model of the vinyl-Si cleavage reaction pathway useful for design safety. Our findings were supported by a cursory review of available literature that qualitatively suggest that scission on the C-Si bond is activated in the presence of electrophiles and protic acids.

586618 Safety Critical Elements (SCE): Rationalizing Our Barriers

Abstract:

Safety Critical Elements (SCE) can be described as part of an installation and such of its structure, equipment and systems or any part thereof, whereby failure of which could cause or contribute substantially to the release of a hazard with major risks. In other words, SCEs are the barriers which are in place to prevent or limit the effect of a major accident.

In early 2013, the organization has embarked on the adaptation of SCE as part of Hazard and Effects Management Process (HEMP) and subsequently, business units within the organization has adopted the concept of SCE within their respective processes. Offshore business unit, particularly, had incorporated SCE identification process as part of Safety Case and subsequently, SCE management process within their maintenance system. The SCE identification and management protocols for offshore business units were established based on available standards and practises, from both within the organization as well as from external sources.

As part of the Safety Cases revalidation, which needs to be done every 5 years, the review of the SCEs that were previously identified for operating facilities under offshore business units was also triggered. These reviews highlighted the real challenges or pain points encountered by operating facilities in managing SCEs as part of their day-to-day operation and maintenance. The challenges or pain points can be summarized as follows:

1. The number of items / equipment / system identified and included as SCE for a particular operating facility is high. On an average, it approximately constitutes about 30-40% of the overall asset register.
2. The high number of items / equipment / system tagged as SCEs has resulted in high number of SCE related work orders i.e. inspection, testing and preventive maintenance (ITPM) task being generated through the maintenance management system, which overwhelms the maintenance personnel at site.
3. This causes overdue and subsequently leads to backlog of the SCE specific ITPM task to be carried out, which is reflected in the Facility Status Monitoring (FSM) system.
4. FSM is being directly monitored by higher management and it raises a serious concern on the healthiness status of the SCEs at operating units in relation to its specified function.

Based on the feedback gathered from operating facilities, it is agreed that there is a need for rationalization of the SCEs currently listed in the respective facilities asset register. This paper details the approach for SCE rationalization which was established and implemented for operating facilities, in view of streamlining the list of SCEs to reflect only those items / equipment / system which are critical in-line with the purpose and definition of SCE.

586170 Advances in Layers of Protection for Combustible Dust Hazards

Abstract:

Often considered as two separate scenarios, combustible dust fires and explosions can often be related. Layers of protection are a benefit in the mitigation of both of the potential hazards presented by combustible dust. New solutions in the last few years for hazard assessment, prevention systems, and mitigation systems have helped create new layered approaches to combustible dust. Dust hazard analysis are the starting point with more elaborate alternatives to prescriptive approaches like risk-based analysis. New test protocols for quick water mist suppression systems have provided an improved layer to the traditional fire protection solution for combustible hazards in process equipment. Intelligent and connected mitigation equipment now provides improved reaction times, the ability to remotely monitor, and more reliable communication networks between components.

This presentation will examine the layers of mitigation protection for both fire and explosion hazards. Specifically, it will discuss how these layers relate and when designed together can provide a higher level of safety. Factors such as reliability (SIL), hazard assessment methodology, intelligent "smart" mitigation solutions, the impact of human error, and active vs passive safeguards will be discussed. A case study of a process area will be walked through during the presentation to show how layers of mitigation protection for both combustible dust fire and explosion hazards can improve safety.

587094 Leveraging the Success of Accelerative Learning and PHAs

Abstract:

The concept of Accelerative Learning has been known and practiced for several decades. This technique has proved to be a very welcome tool for adult learning, time and time again. This concept has been widely applied in the industries in lieu of the conventional classroom training methods, and its results have benefitted in achieving success for various corporations. The success has meant that the audience has improved their engagement in the class, embraced the challenging topics, increased the awareness and helping in bringing about a mind shift in some cases, through this learning exercise

On the other hand, Process hazard analysis (PHA), have also been in existence for the past few decades. Some of its methodologies are preferred over the other for level of details, and are preferred over the other for its brevity and time saving feature. These are always conducted in a workshop environment, and its success is dependent on the level of engagement and preparation. Meanwhile, the sessions can also become a victim on team that has been reluctant to participate for various reasons.

The last decade 2010-2019 has come with its set of challenges. Industries cut back on various workshops due to the financial crunch, due to their ever shrinking budgets and also the economic downturn in the oil & gas industry, which led to a cascading effect in other industries. This led to changes in industry approaches being altered to focus on "More with Less". One emergent outcome of this need, was the philosophy of Smart PHAs or Smart HAZOPs

The author has applied the concepts of AL to PHAs, and has changed the way the PHAs are facilitated in a workshop environment. This has resulted in higher level of personnel engagement, that has fostered the brain storming needs of a PHAs, and has left the audience highly energized after the PHA sessions, such that they want to return for me. Through this paper, the author will share the success stories while highlighting the benefits of this hybrid technique.

586044 Initiative for Advanced Risk Assessment to Prevent Accidents in Our Complex

Abstract:

Against recent accidents at chemical plants in japan, we have implemented improvements on tangible and intangible elements in our company. It is based on Guidelines for Risk Based Process Safety（RBPS） for the purpose of properly managing risk sources than before.

On the intangible elements, we made a new risk assessment (RA) system for quantitative evaluation in order to manage properly risk source. In addition, to operate the RA system, we created qualification system to raise the knowledge and skills of employees who conduct RA.

With regard to the tangible elements, we are considering the practical use of advanced technologies in order to grasp the quantitatively risk which causes the leakage of hazardous material.I will introduce about details next.

592114 Process and adsorbent for separating ethanol and associated oxygenates from a biofermentation system

Abstract:

Abstract for poster presentation or talk:

Abstract Title: Process and adsorbent for separating ethanol and associated oxygenates from a biofermentation system

Presenter Information: Deepak Sharma; Sr. Process Engineer, Engineering and Technology, Bayer AG

Patent number: US8658845B2

Patent Authors:

Deepak Sharma - Sr. Process Engineer, Bayer Ag, St. Louis; (Presenting Author)

Anil Oroskar – CTO, Orochem Technologies, Napervielle, IL;

David W. House – Scientist, Orochem Technologies, Napervielle, IL;

Alice M. Havill – Scientist, Lanzatech, New Zealand

The principles of process intensification if used on existing processes can lead to development and redesign of existing processes to become more sustainable. This presentation will discuss a process and an adsorbent for the separation of ethanol associated oxygenates from a dilute mixture of ethanol and associated oxygenates in water in the presence of organic compounds derived from a biofermentation process. After pretreatment, the separation is carried out in a simulated moving bed adsorption system employing a stationary phase adsorbent comprising fluorinated carbon or modified C18 silica gel selective for the adsorption of ethanol and associated oxygenates, such as 2,3-butanediol, with a mobile phase desorbent selected from the group consisting of methanol, ethanol, propanol, and methyl tertiary butyl ether. The process is useful for removing water from dilute aqueous mixtures of organic compounds comprising ethanol in dilute concentration in water and produced by fermentation, biomass extraction, biocatalytic, and enzymatic processes which are not economically recoverable by conventional distillation methods.

Keywords: Separation; Adsorption; SMB; Purification

Patent Assignee: Orochem Technologies LLC

592111 Pump safety, the undiscussed topic in chemical industry leading to several fatal incidents

Abstract:

Abstract Title: Pump safety, the undiscussed topic in chemical industry leading to several fatal incidents

Presenter Information: Deepak Sharma; Sr. Process Engineer, Engineering and Technology, Bayer AG

Pumps are one of the equipment's in the process industry that require a thorough review during HAZOP studies. The constant source of energy given to the pump, if not converted to kinetic energy to flow the fluid, can lead to heating of the pumped fluid. If the fluid keeps accumulating the energy provided by the motor, in a blocked-in scenario, it can lead to very high temperatures eventually leading to loss of containment of the pump and the pump can get damaged. There have been instances in the industry where fragments of damaged pumps have flown several feet away from the point of loss of containment leading to injuries. The poster will outline a scenario where a pump is run blocked in on suction and discharge for water and a chemical which has a low thermal decomposition temperature. Arc data is used to evaluate the onset temperature of thermal decomposition. The poster will outline safeguards that can be employed to avoid catastrophic loss of containment of the pump. Outlined below are safeguards used to avoid loss of containment of a pump in case of a blocked-in scenario:

- Install PSV on suction of pump
- Employ a low amp/power interlock on the pump to turn off the pump in case of blocked in situation
- Have a high-pressure interlock on the discharge line of the pump to direct flow back to the tank in case of pump deadhead
- Install an Interlock to turn off the pump in case the feed forward valves (automatic valves) close inadvertently

Pump deadhead scenarios are not only catastrophic when in flammable service there are several salts that have low thermal decomposition temperatures. Pumps in water service left blocked in could also blow catastrophically leading to reportable incidents. I feel this is an important topic to bring to the attention of students and professionals in the chemical industry. Poster will include pictures of actual installed pumps along with required safeguards in the field along with supporting P&ID depictions.

589440 New Kind of Linear Nanocomposite Applied in Improving Rheological Behavior of Crude Oil and Its Mechanism Studied through Molecular Simulation Method

Abstract:

In this work, a new kind of linear nanocomposite produced by carbon nanotube(CNT) as pour point depressant (C2-100) was developed by blending with MCNT nanoparticle and copolymer named Ethylene-vinyl acetate (EVA). Its effect of reducing yield stress and modifying the morphology of wax agglomerate were also evaluated. Besides, the synergetic effect of C2-100 nanocomposite with the magnetic field was also conducted. The mechanism of modifying rheological behavior of waxy crude oil was studied through molecular simulation method, which aims at understanding the interaction between this nanocomposite and wax molecule. The results indicated that C2-100 nanocomposite could reduce the yield stress of waxy crude oil more effectively than the copolymer (EVA) even at low concentration, as well as changing the morphology of wax crystals and agglomerate. Moreover, its effect of reducing yield stress becomes more obvious when the alternate magnetic field was applied. In addition, the results of molecular simulation also presented that the effect of improving rheological behavior of C2-100 nanocomposite is related to the adsorption behavior of wax molecules on the surface of MCNT nanoparticle, which is adsorbed with EVA molecules. In addition, the adsorption behavior of different fragments of EVA molecules can also influence the adsorption behavior of wax molecules on the surface of EVA-attached MCNT nanoparticle, which may influence the effect of C2-100 to improve the rheological behavior of waxy crude oil.

589683 Solar Energy Conversion and Noise Characterization Sensing in Building Integrated Solar Photovoltaic Devices with Ventilation

Abstract:

A study on solar energy conversion and noise characterization sensing in solar photovoltaic devices with ventilation is conducted. A building integrated parallel plate solar photovoltaic (PV) device was built with a pair of PV modules, a ventilated air cavity and an insulating back panel of plywood board filled with polystyrene installed in an outdoor test-room. Photovoltaic amplification was attained with power output from a potentiometer through rotation of its circular knob. A parallel plate PV device was studied for its
electrical parameters as Resistance-Capacitance (RC) electrical analog circuit. The effect of inductive and capacitive heating losses was considered in evaluating electrical characteristics of a PV device exposed to solar radiation. The characterization of noise interference, due to power difference of two intensities in a wave on PV Device is presented. Noise characterization of a wave is obtained depending on type of wave. The difference of two power intensities is due to transmission of light, sound, heat, electricity, fluid, and fire into a parallel plate PV device. Based on analysis of a solar PV device, a definition with measurement expression and unit for an electrical noise were coined. Standard definitions of noise sources, their measurement equations, their units and their origins under limiting reference conditions were also devised. A review for sensing noise of physical agents is conducted for sensing noise of physical agents (sun, light, sound, heat, fluid, electricity and fire). With integrated environmental control, these environmental parameters are to be characterized through various environmental monitoring sensors.

589945 Structural Characterization and Model Representation of Vitrinite-Rich India Coal By Using Advanced Analytical Techniques

Abstract:

The detailed investigation of the chemical structure of Karharbari coal (vitritine rich subbituminous coal from India) by utilizing advanced instrumental analytical techniques such as solid state ^{13}C nuclear magnetic resonance (^{13}C-NMR) spectroscopy, Fourier transform infrared (FT-IR) spectroscopy, X-ray diffraction (XRD), Flash-pyrolysis gas chromatography mass spectroscopy (Py-GC/MS), Laser desorption ionization time-of-flight mass spectroscopy (LD-TOF-MS), X-ray photoelectron spectroscopy (XPS), and high-resolution transmission electron microscopy (HRTEM). The Karharbari coal was an amorphous loosely organized lower rank coal with aromaticity 0.79. This was indicated by XRD results and also confirmed with ^{13}C-NMR results (aromaticity =0.84) and FTIR results (aromaticity = 0.82). The average molecular weight of the aromatic cluster is 507 amu by NMR result which is closer to the result obtained by HRTEM (MW = 530.46 amu). The presence of different heteroatoms like carbon, oxygen, nitrogen and sulphur with their functionalities, were determined by using a XPS technique. Different carbon/oxygen forms present in the Karharbari coal are (C-C) and (C-H) 68.53%, (C-O) 23.46%, (COO^-) 1.96%, (C=O) 6.05% and different nitrogen forms such as pyridine, pyrrolic, quaternary, oxidized and their distribution (mol %) are 19.36%, 45.68%, 31.25%, 3.71 respectively, and different sulphur forms were present i.e. corresponding to thiophenes, sulphones, sulphuric acid and sulphates with the molar content of 16.47%, 41.61%, 21.37%, 20.55% respectively. The construction of the individual molecules will be discussed in detail in the rest of this paper. The individual molecules were geometry-optimized and Materials Studio Amorphous Builder module was used to construct a three dimensional associated structure.

589479 Life Management of High Energy Piping with Highlights from ASME B31.1

Abstract:

High energy piping (HEP) systems – or as defined by the ASME B31.1 Power Piping Code, Covered Piping Systems (CPS) – are critical assets with respect to production facilities. A failure can have catastrophic consequences from a safety perspective and in terms of equipment loss and downtime. In fact, loss of high-pressure steam due to a failure in a main run of steam supply piping could shut an entire production unit down. Because of this and the requirements of the Code, HEP maintenance and inspection programs are established to ensure that the integrity of the systems is maintained in a safe and reliable condition throughout their lifecycle. There is no one-size-fits-all for the life management of these critical assets because there is a great variety in age, design, materials, and owner objectives. Life management of these assets is beneficial in that it provides a structured, systematic approach to creating and implementing a program including evaluating the current program (whether formal or informal), what the objectives of the program are (short and long term), what equipment is in place, and its history.

A HEP program put in place for overall life management is not synonymous with an inspection program, although inspections are an important part of an overall program. Inspections should be used in conjunction with analytical tools, real time monitoring, and laboratory examinations. The overall approach is life management – not inspection management – of the equipment.

This paper describes a HEP life management approach which includes identifying locations of interest, risk-ranking those locations, implementing a plan for determining the health of the highest risk locations, and taking actions based on the system health that has been determined. This paper also presents and discusses some of the requirements of the B31.1 code, which at the highest level states that each operating company shall develop operation and maintenance procedures for piping systems deemed necessary to ensure safe facility operations based on the provisions of the code, relevant industry experience, the operating company's experience (and knowledge of its facility), and the conditions under which the piping systems are operated.

590131 Optimization of C3MR LNG Liquefaction Process Using a Parallelized Simulator

Abstract:

Controlling the cost of energy-intensive processes such as C3MR (propane/mixed refrigerant) LNG liquefaction is a prime opportunity for optimization. Our work combines different C3MR models to develop an optimization that uses multiple independent variables, relaxed constraints, and flowsheet alterations to achieve an optimal operating cost. This work uses a rigorous model of the process in a process simulator rather than a reduced model of the process response. Use of a first-principles process simulator increases the accuracy of the model and therefore allows a more realistic optimization, at the cost of increased computing time. Parallelization (using multiple CPU cores) in the process simulator greatly increases the speed of the optimization calculations. CHEMCAD's optimization tools and multi-core technology make it possible to easily define the optimizations, run the optimizations efficiently, and generate global optima with a high degree of confidence.

590196 Modifying Fischer-Tropsch Reaction Sequence with Addition of Zeolite to Iron-Based Activated-Carbon Supported Catalyst

Abstract:

In addition to previously presented scientific work, modified reaction sequence of Fischer-Tropsch (FT) process was studied by addition of Zeolite ZSM-5 to an iron-based activated-carbon-supported catalyst. Main reactions involved due to FT catalyst are Fischer Tropsch Synthesis (FTS) and Water Gas Shift (WGS) reactions. Main reactions due to ZSM-5 are cracking, oligomerization, isomerization, cyclization, aromatization, conversion of alcohols to hydrocarbons, and coke formation. The waxy and linear structure FT products are converted to gasoline range, lighter hydrocarbons, iso-paraffins, and aromatics using ZSM-5. The catalyst ratio of 1:2, FT catalyst: ZSM-5, is used at 300 PSIG and 300°F. Separate-bed configuration, i.e. FT and ZSM-5 are separated in a reactor, arrangement is used for the reaction sequence analysis. Alcohols with carbon-chain from 3-5 produced by FTS are converted into hydrocarbons over ZSM-5. Waxy hydrocarbons with carbon-chain greater than 23 produced by FTS are cracked and isomerized to small n-paraffins and iso-paraffins using ZSM-5. The hydrocarbon with carbon chain from 7-10 produced by ZSM-5 cracking reactions are converted to aromatic products over ZSM-5. Produced gasoline to kerosene range n-paraffins are further cracked into smaller chain (1-5) iso- and n-paraffins. Some of the iso-paraffins are oligomerized and aromatized over ZSM-5 to form more aromatics. Further, multiple aromatic rings combine to undergo cyclization and dehydrogenation reactions over ZSM-5 to form coke. Water produced as a byproduct of FTS is combined with some of the carbon monoxide, which is the WGS, to form carbon dioxide and hydrogen.

590427 Lateral Retention Force Required for Sliding Saline Droplets over Silianized Silicon Surface

Abstract:

The lateral retention force required to slide a liquid drop on a surface is expected to increase with the time that the drop rests on the surface prior to sliding till it reaches a plateau. The molecular reorientation of the molecules at the periphery of the solid surface subjected to the unresolved vertical component of surface tension force causes this phenomenon. In this system, the surface tension of deionized water is increased by adding NaCl and is expected that the lateral retention force is increased proportionally with concentration. The retention force increased with increase in concentration reaching a maximum only to drop to a minimum with concentrations closer to saturation. That is, the lowest sliding force is required for both lower and higher concentrations of NaCl. This sudden decrease in observed lateral force is not a function of surface tension and the surface on which the drop slides. This phenomenon is associated with adhesion of salt to the surface and is supported by the drop sliding motion and depinning of the liquid from the solid.

590433 Study of Droplet Motion across Polymeric Substrates By Centrifugal Adhesion Balance (CAB)

Abstract:

On a given surface, the contour of the drop changes accordingly based on factors of the force applied, chemical structure of the fluid, drop size, and surface topography. With increasing force that is exerted on the drop, we determine the comprehensive trend for drop motion with respect to the controlled factors mentioned above. Drop motion is categorized into four phases and the point at where the initial displacement of receding line is observed is defined as the onset of motion. In this study, the onset of motion is affected by the factors mentioned above and aids in the determination for another value correlation known as Gs, known as Interfacial Modulus. The resistance to interaction towards the droplet as observed is quantified with respect to change in surface topography. This study therefore characterized drop motion and specifically onset motion based on controlled variables (factors). Work of adhesion is also quantified across different substrates to characterize the detachment of liquid from its substrate. All these parameters such as Gs, onset of motion, contact angles were analyzed with different polymeric substrates, drop sizes, and various fluids used.

590665 Population Balance Modeling and Spatially-Oriented Control for Crystal Size and Shape in Crystallization Processes

Abstract:

Crystallization is an important separation and purification process with relevant applications in the food, pharmaceutical, fine chemical and fertilizing industries. In these processes, modeling and process control for crystallization are focused on crystal size distribution (CSD) and shape of the crystals in order to make it possible to achieve the desired quality and purity of the product of interest. The control of crystal size and shape is of considerable importance for industrial crystallization processes, as crystal morphology has great influence on downstream steps and also from the point of view of product specification criteria. The consideration of crystal shape in modeling of crystallization processes is also important due to its impact on product effectiveness, such as the bioavailability and tablet stability in pharmaceuticals. Control of the crystallization process is can be facilitated with online measurements, which provide information to be used in the modeling of the process. However, if online measurements are not available, the strongly nonlinear dynamics of these processes makes difficult to establish relations between some inputs (e.g. temperature) and the final desirable characteristics of the crystal product. In this context, image analysis has been shown to be a suitable technique to monitoring the process [1].

In our present work, a population balance model-based scheme for controlling the size and shape of crystals is presented. The modeling was performed using a two-dimensional population balance (2D-PBM) in terms of the two characteristic dimensions of the crystals, describing the evolution of crystal size and shape distributions along the batch time. To identify the nucleation, growth and dissolution kinetics parameters using the 2D-PBM approach along the length and width directions, QICPIC-LIXELL's high speed dynamic image analysis equipment was used for the online measurement of the CSD, shape parameters and aspect ratio distribution of crystals during the experiments. In this way, the developed model allowed for the prediction of the crystal size and shape distributions during growth-dissolution cycles.

The developed framework was employed to obtain optimal temperature policies for controlling the crystal size and shape distributions that meet target criteria for size, morphology and productivity using the dynamic image analysis along the runs. Based on the concept of using measurements along batch crystallization processes as a trajectory in a phase space [2], a trajectory-endpoint control problem is presented with an oriented control scheme that is applied to the shape of the crystals. The scheme was used to control the batch cooling crystallization of potassium dihydrogen phosphate. The reported experimental data demonstrate the application of the control policies for temperature to produce crystals of the desired average shape and size in different batch times. Thus, using the first principle modeling coupled to the developed trajectory tracking control, the proposed framework has the potential to be applied to industrial crystallization processes for shape control and to be extended to polymorph control investigations.

References:

[1] Eisenschmidt, H.; Bajcinca, N.; Sundmacher, K. (2016) Optimal Control of Crystal Shapes in Batch Crystallization Experiments by Growth-Dissolution Cycles. *Crystal Growth & Design*, v. 16, p. 3297-2803.

[2] Griffin, D. J., Grover, M. A., Kawajiri, Y., Rousseau, R. W. (2015) Mass-count plots for crystal size control. *Chemical Engineering Science*, v. 137, p. 338-351.

590450 Auditing the Process Hazard Analysis - Key Characteristics to Scrutinize in a Good-Quality HAZOP Study and Supporting Effective HAZOP Study Updates

Abstract:

Although Process Safety Management (PSM), requiring good-quality Process Hazard Analyses, was promulgated in the United States in 1992, and HAZOP Study techniques date back to the mid-1970's, the HAZOP Study technique is still not uniformly applied in a high quality fashion. The objective of this paper is to provide experience-based tips on Process Hazard Analysis auditing, and some key characteristics that the auditor can scrutinize to assure quality implementation. Analogously, a similar set of characteristics that can facilitate future HAZOP Study updates can also contribute to a quality HAZOP Study that supports the spirit of Safety Management System objectives for the identification of hazards potentially impacting plant personnel, the community, and the environment. This paper will also review tips for structuring the HAZOP Study to facilitate later updates and revalidation.

590342 Development of High Performance Organic-Based Cathodes

Abstract:

For lithium-based batteries, structure/property relations in terms of cell voltage are desired for new organic-based cathode active materials. In this initial pursuit of structure/property relations, ab-initio calculations were performed on a few organic-based molecules to show the effect on the theoretical cell voltage by inclusion of nitrogen and/or oxygen. Preliminary test results are presented on a newly developed molecule and methods to potentially improve on the cell performance are given.

590705 Electrochemical Kinetics of Degradation of Graphite Anodes in Lithium Ion Batteries

Abstract:

Growth of the solid electrolyte interphase (SEI) is a major driver of capacity fade in LIBs. Despite its importance, the fundamental mechanisms remain unclear, primarily because of the complicated reaction pathways involved[1–3]. SEI growth can be both electrochemical and chemical in nature[4], and thus, it is a strong function of the potential and degree of lithiation of the electrode. We model the early-stage and long-term growth of SEI by accurately capturing the potential dependence of its formation kinetics as well as long term rate limiting steps, and validating it against the world's largest open source battery cycling data, generated in-house[5]. This is done using the Multiphase Porous Electrode Theory (MPET) framework[6] on graphite (phase separating) and carbon black (non phase separating) particles.

Lithium plating is another key degradation phenomenon that has been elusive, and it becomes important while trying to fast-charge batteries, i.e., 0% - 80% state-of-charge in 30 mins. We show that lithium plating is a key function of electrode morphology, phase-separation dynamics and potential. Phase-separation in graphite is modeled in the electrode using the Cahn-Hilliard Reaction framework described by Bazant[7]. We understand the electrochemistry of the onset of lithium plating with in-situ measurements connected to real time cell potential in a phase-separating electrode for the first time[8].

Results indicate that the peak SEI-forming currents are higher for higher driving currents. Also, we find that SEI only grows during electrode lithiation, i.e. the battery only degrades while being charged. We also find that onset of lithium plating is correctly captured only when phase separation in active material is accounted for. Further, the onset of plating is delayed on electrodes with a thick SEI layer – understanding SEI/plating coupling is integral to predicting fast charging manufacturing protocols for LIBs. This work holds promise for the predictive design of procedures[9] for manufacture and formation of LIBs.

[1] Cohen, Y. S.; Cohen, Y.; Aurbach, D. Micromorphological Studies of Lithium Electrodes in Alkyl Carbonate Solutions Using in Situ Atomic Force Microscopy. J. Phys. Chem. B 2000, 104, 12282–12291. https://doi.org/10.1021/jp002526b.

[2] Horstmann, B.; Single, F.; Latz, A. Review on Multi-Scale Models of Solid-Electrolyte Interphase Formation, Current Opinion in Electrochemistry 13, 62-69 2019.

[3] Nie, M.; Abraham, D. P.; Seo, D. M.; Chen, Y.; Bose, A.; Lucht, B. L. Role of Solution Structure in Solid Electrolyte Interphase Formation on Graphite with LiPF6 in Propylene Carbonate. J. Phys. Chem. C 2013, 117 (48), 25381–25389, https://doi.org/10.1021/jp409765w.

[4] Das, S.; Attia, P. M.; Chueh, W. C.; Bazant, M. Z. Electrochemical Kinetics of SEI Growth on Carbon Black: Part II. Modeling. J. Electrochem. Soc. 2019, 166 (4), E107–E118. https://doi.org/10.1149/2.0241904jes.

[5] Attia, P. M.; Das, S.; Harris, S. J.; Bazant, M. Z.; Chueh, W. C. Electrochemical Kinetics of SEI Growth on Carbon Black: Part I. Experiments. J. Electrochem. Soc. 2019, 166 (4), E97–E106. https://doi.org/10.1149/2.0231904jes.

[6] Smith, R. B.; Bazant, M. Z. Multiphase Porous Electrode Theory, J. Electrochem. Soc. 2017, 164 (11). https://doi.org/10.1149/2.0171711jes.

[7] Bazant, M. Z., Theory of chemical kinetics and charge transfer based on nonequilibrium thermodynamics, *Accounts of Chemical Research*, *46*(5), 1144–1160. https://doi.org/10.1021/ar300145c

[8] T. Gao, Y. Han, S. Das, T. Zhou, D. Fraggedakis, N. Nadkarni, C. N. Yeh, W. Chueh, J. Li, M.Z. Bazant, Interplay of lithium intercalation and plating on graphite using in-situ optical measurements, submitted.

[9] Huang, W.; Attia, P. M.; Wang, H.; Renfrew, S. E.; Jin, N.; Das, S.; Zhang, Z.; Boyle, D. T.; Li, Y.; Bazant, M. Z.; McCloskey, B. D.; Chueh, W. C. and Cui, Y.; Nano Letters 2019 19 (8), 5140-5148. DOI: 10.1021/acs.nanolett.9b01515

589787 Model Development of Solid Oxide Fuel Cell Thermal Performance Using Artificial Neural Network

Abstract:

NASA's Johnson Space Center has recently begun efforts to eventually integrate air-independent Solid Oxide Fuel Cell systems, with landers that can be propelled by LOX-CH4, for long duration missions. Using landers that utilize such propellants, provides the opportunity to use such systems as a power option, especially since they will be able to process methane into a reactant through fuel reformation. To ensure fuel reformation in the systems, Steam Methane Reformation (SMR) will be employed. Various lead-up activities, such as hardware testing and computational modelling, have been initiated to assist with this developmental effort including the development of dynamic empirical models using artificial neural network. Such a modeling approach has shown to be accurate ($R^2>0.9$) to predict the thermal and electrical performance for SOFC systems. Hence, such models could be used for system optimization and control design.

Impact of Spacer Nature on Rheological Properties of Cationic Gemini Surfactant

Abstract ID: 589650

Shams Kalam[1], Muhammad Shahzad Kamal[2], Shirish Patil[1] and S. M. Shakil Hussain[2]

(1) KFUPM, Dhahran, Saudi Arabia

(2) Center for Integrative Petroleum Research, King Fahd University of Petroleum & Minerals, Dhahran, Saudi Arabia

Surfactants play a vital role in Chemical Enhanced Oil Recovery (cEOR) techniques. Gemini surfactants are a unique style of surfactants having more water-soluble hydrophilic head and oil-soluble hydrophobic tail units. Both groups are connected by a spacer at or near to the head groups. Polymers are added into the surfactants to increase the sweep efficiency of the flooding processes. This paper presents the rheological properties of Surfactant Polymer (SP) solutions in which two recently synthesized cationic gemini surfactants with different spacer nature are used.

The rheological properties of novel SP solutions were determined using Discovery hybrid rheometer (DHR-3) from TA Instrument. The geometric configuration set for this work was concentric cylinder. All experiments were carried out at 80 °C, while the shear rate for steady shear viscosity measurements ranges from 0.001 to 1000 s^{-1}. 0.25 weight % of a polymer FO 4290 SSH was used with different concentrations of newly synthesized gemini surfactants for Rheological measurements. Two hours magnetic stirring and twenty hours retention time were provided to make uniform surfactant polymer solutions.

Both surfactants exhibited good thermal stability in seawater with no precipitation. Shear viscosity and storage modulus were measured as a function of shear rate and angular frequency respectively at different surfactant concentrations ranging from 0 to 0.5 mM, where 0 mM shows pure polymer. We discovered that counter ions also influence the rheological properties significantly. Research outcomes revealed that the novel SP solution with a benzene ring and chlorine counterion produces better performance in comparison with benzene ring and bromine counterion. Moreover, the effect enhances if we increase one more benzene ring in the spacer.

This study reports two recently synthesized SP solutions for cEOR processes in carbonate reservoirs. It presents the effect of counterion on rheological properties and improves the understanding of tuning the surfactant structure in making SP solutions to increase the flooding sweep efficiency.

590257 Real Time Energy Management Systems Implementation for Continually Optimized Performance

Abstract:

It is well known that energy management is key for the optimal operation of refining and petrochemical sites. Industrial energy systems are constituted by complex utilities networks that are interconnected and have complex interactions that in order to achieve it best performance need to comply with increasing environmental rules, push against operating constraints and adapt to altering power prices and rapidly changing equipment availability.

After the initial installation, management and continuous improvement of these systems present challenges not only from a technological but also from an organizational point of view.

The implementation of an advanced digital Energy Management System (EMS) which includes capabilities for multi-period optimized scheduling, real time optimization and monitoring, deployed either as cloud-based or on-premise solutions that could be acting on open or closed loop is known to provides large benefits.

Critical is the sustainability of the EMS in the ever-evolving energy supply and demand of refining and petrochemical sites. Commercial, asset and organizational factors all change and the EMS needs to adapt to the developing environment of each enterprise meaning the type of implementation, including appropriate management structure, culture, best practices and Key Performance Indicators (KPIs) needs to be continually improved.

The EMS implementation at the large and complex Eastman site in Longview, TX, will be presented as an example. It will provide insight on the initial successes achieved in real time energy management for both, cost optimization and energy related KPIs performance tracking. The case study evolves to discuss how technical asset improvements, once replicated in the EMS, provided alternative operating modes, large additional benefits and sustained optimal

Andrew Lammers, NN2

Eastman Longview

Pablo Montagna*, Cristian Rojas

KBC Advanced Technologies, Inc

589818 Analysis of Fuel Cell System through Project Based Learning for Application of Classical Thermal Fluid Concepts in Mechanical Engineering Laboratory Course

Abstract:

A hands-on project involving analysis of air noncircular channels in Fuel Cell to measure the pressure was introduced in a thermal fluid laboratory course. Although the students studied fluid mechanics as a theoretical course, this project gave them the opportunity to build an experimental setup from scratch to help showcase their hands-on skills and to apply their understanding of relevant concepts of fluid sciences. The results of the project included a successful test of a conceptual procedure, designed to measure and calculate pressure drop and flow rates across the anode and cathode sides of a fuel cell. The students enjoyed completing and learning from the project. They also provided very useful feedback on the project by incorporating a functional experimental system for the future laboratory project. This project based learning opportunity allowed students to build their confidence in what they are learning as well as provide a glimpse of what the real world may hold.

589438 Measured Versus Calculated Sour Gas Properties

Abstract:

Measured versus calculated sour gas properties
Abstract:
Recent Environmental agencies have restricted standards for sulfur emission from gas chemical processing. Sour gas can extremely impact the delivery and processing by corroding the equipment. Gas properties such as compressibility, density and viscosity are a major factor controlling prediction of future production pipelines, and production tubing and pipelines in gas industry.
Sour gas effect on all gas industry aspects usually gas properties are measured in laboratories but occasionally experimental values are not available due to complexity and time consuming hence a reliable method would be proposed.
We are assessing the accuracy of the industry standard methods in calculating sour gas properties, compressibility factor using the following correlations:
Drunchuck - Abukkasem, Hall – Yarbarough, Drunchuck- Purvis- Robinson, Londono et-al, Alanazi et-al and Azizi et-al.
Viscosity using empirical correlations such as Carr-Kobayashi Burrows, Lea-Gonzalez –Eakin and Dean- Stiel.
Density using equations of state Redlich Kwang and Soave -Redlick - Kwang.
This study shows that the best correlation for calculating compressibility factor and density of high concentration sour gas is Londono-et al, all other considered methods are not accurate. All other methods investigated include high error especially for high H2S concentration.
However the error in estimation of Z factor for most of the considered methods has great impact on estimated sour gas density. Therefore we recommend that sour gas properties be measured experimentally at high pressure and high temperature conditions.

587967 The Study of a S-CO2 Oxy-Fuel Combustion Power Plant for Low Grade Coals

Abstract:

Recently, it is very important to have reliable solutions that can maximize power generation efficiency of current power generation processes and decrease emissions of CO2 and other pollutants because there are increasing concerns about the global warming and the depletion of fossil fuels. Among the available carbon capture technologies and power generation technologies, a S-CO2 (supercritical carbon dioxide) cycle based oxy-fuel combustion process appears to be a promising mid-term solution for the production of electricity from fossil fuels. Since supercritical carbon dioxide (S-CO2) involves a rapid density change at small temperature and pressure ranges, the S-CO2 cycle can extract large amounts of energy from relatively small equipment and it is possible to minimize or simplify a power generation block and to obtain higher thermal efficiency than existing steam cycles used in power generation plants. Therefore, an indirect S-CO2 oxy-fuel combustion power plants for low grade coals was designed and analyzed by using ASPEN-PLUS. Also, the operating characteristics of the indirect S-CO2 Oxy-fuel combustion power plant were compared with those of conventional air combustion power plants.

590099 Characterization and Performance Assessment of Oil Entrapment By Commercially Available Surfactants

Abstract:

Oil spills occur frequently and are a threat to the health and safety of human society and the well-being of the environment and ecosystem. Remediating oil spills is an ongoing challenge. Surfactant based dispersing agents, such as the Corexit® product line, are widely used for remediation of oil spills mainly through direct application on top of the oil slick. This work encompasses the investigation of variables that influence the emulsification process and therefore the encapsulation of oil within the surfactant micelles. Of central importance are the conditions under which oil entrapment in emulsions even occurs. A central finding is that there appears to be a threshold shear rate for emulsifying oil with surfactants, below which no emulsification occurs. Different schemes for applying shear stress in this work included simply stirring, using a high energy colloid mill, and flowing the mixture through a high shear capillary. Also, the sequence in which the oil/surfactant/water (O/S/W) components are mixed impinges strongly on whether and how much oil entrapment occurs. To try to correlate shear stress and velocity profiles throughout the sample volume with entrapment efficiency and kinetic stability of these O/S/W systems, Computational fluid dynamics (CFD) simulations were conducted. Another important factor is the ionic strength of the aqueous medium, which greatly impacts kinetics of these emulsions, but with a smaller impact on the threshold at which any emulsion is obtained at all. Surfactants tested include Corexit® EC9500A, sodium dodecyl sulfate (SDS), Triton X-100, and cetyltrimethyl ammonium bromide (CTAB). The quantification of oil entrapment was performed through either thermal gravimetric analysis (TGA) or differential scanning calorimetry (DSC).

Designing a Native Model for Assessment of the Effectiveness of Advertising

Sorush Niknamian
PhD, military Medicine Department, Liberty University, USA
Email: so.niknamian@gmail.com

Abstract

Reviewing and evaluating the effectiveness of advertising can be used as a criterion to convert weaknesses into strengths and budgeting. Also, Identification of the factors influencing the effectiveness of advertising can lead to more professional design and implementation. In the banking industry, with the increasing number of banks and increased right of customers to choose, the role of advertising has become increasingly important to succeed in this field of competition; so it should be said that advertising can make the bank close to its goals when it is sufficiently effective. The current research aimed to identify and rank the effective factors on the design of a native model for evaluation of the effectiveness of advertising of the Eghtesad Novin bank in Iran. The statistical population of this research is all managers and experts of the Eghtesad Novin bank. According to the research literature, the criteria and sub-criteria affecting the effectiveness of advertising have been extracted. By using the fuzzy Delphi technique and based on the experts' opinion, the final indices were extracted. Then, using the fuzzy AHP method, the rank of criteria has been addressed. The derived criteria are demographic characteristics, individual emotions, and attitudes of the customer, apparent features of the message, message content, market characteristics, cultural, economic features, design and nature of the service, advertising methods. The results of the ranking showed that the methods of advertisement, the design, and nature of the service, the apparent characteristics of the message, the emotions and attitudes of the customer, the demographic characteristics, the content of the message, economic, market characteristics, were placed in the first to ninth ranks, respectively.

Keywords: banking services, banking services marketing, advertising, advertising effectiveness,

Introduction:

Advertising is one of the promotional mix elements of marketing mixes, which play an important role in attracting customers. Advertising is a bridge between the organization and customers and can introduce the organization and its products to customers, inform consumers about the goods and services of the organization, convince them about the benefits of the product, and ensure them to have the best choice even after the purchase when the customer is uncertain (Buck, 2016: 172).

Given the increasing competition among the brands in the marketing sector, advertising is a necessary tool to introduce the characteristics of goods and services to customers. Leong et al. (1999) pointed out the fact that consumers overcome advertising information, and many of them do not remember the advertisements that they are exposed to, and it is necessary to consider the advertising and its attractive methods to attract customers' attention. The ability to attract customers' attention is one of the main indicators of advertising effectiveness that should be taken into consideration. The attention of customers is a key component of the overall effectiveness of advertising (Pileliene et al., 2017: 2). Any professional marketing without advertising will not be effective. Despite relatively high advertising costs, advertising is mentioned as an investment rather than the cost. Since many advertisements cause that the advertising companies do not benefit their advertising because of the wrong way of advertising, the competitive companies benefit from the advertising of that company (Hartnet, 2016: 23)

Advertising should be in line with the goals and plans, as well as the strategies of the organization. Therefore, we should see to what extent the advertising has met such goals. Allocating funds without evaluating the effectiveness of past advertising is the waste of resources of the organization. By evaluating the effectiveness of advertising, changes can be made to the form, the content of messages, type of media, budget, communication channels, and even the time and conditions for conducting advertising to create more useful and effective advertising than the past (Yazgirudo-Yasteh, 2015: 359). In the current competitive world, the importance of advertising is increasing. As a result, businesses have devoted a great deal of money to customers' attraction, but it is important to see whether the company advertising has been designed and implemented in such a way to meet the goals of the company. This issue has been often neglected in our country. (Kim, 2014: 2107).

The amazing effect of advertising on the lives of individuals requires that this scientific, social and communicative branch be carefully studied and advertising is made with a scientific, artistic, specialized and expert view. Professional advertisement can be a sharp sword for sales promotion, both inside and outside the country, and otherwise, it is a waste of time, money and energy. A professional advertising lever can, with the help of new technology, penetrate humans, and cause more desire for consumption (Arribeh, 2016: 463). Today, banks have turned their attention to the widespread use of mass media advertising, given the highly competitive structure governing on them. Also, given the relationship among a wide range of businesses of organizations and households and banks through various deposit accounts, loans, and other financial services, banks can benefit from advertising in a proper way to meet their goals, namely the introduction of their services, the communication with customers, their effectiveness on them in choosing a bank to do their banking business and ultimately attracting them. Unfortunately, in most banks, despite the massive advertising and the allocation of large amounts of energy and budget, the effectiveness of such advertisements is less evaluated. Since advertising is one-way, it is necessary to receive feedback from the customers (Rabiei, 2011: 19).

Regarding marketing, the effectiveness of advertising depends on the wide range of coverage of various consumers and the cost of advertising campaigns. Economists have defined the effectiveness of advertising as a change in the profit derived from promotional activities. Therefore, the effectiveness of bank advertising is the potential quality of information reported to consumers, and in promotional campaigns, it aims at creating a bank image and informing banking products and services, as well as obtaining a specific economic result before and after the transmission of the message (Denis et al., 2012: 85).

Eghtesad Novin Bank (EN bank), which is one of the oldest banks in Iran, due to the addition of new banks, should use new tools to protect and even increase its market share for the assessment of the effectiveness of its advertising. For this purpose and given that the bank uses various advertisements, it is necessary to design a model to evaluate the effectiveness of the bank advertising; therefore, the present study seeks to answer the following questions:

- What indicators are used to assess the bank native advertising model?

- How are the indicators for bank advertising evaluation weighed and ranked?

In this research, we have tried to identify the criteria using the Fuzzy Delphi process; then we will use the Fuzzy AHP method to rank the identified criteria. In the next section of the research, the literature has been studied. Then the research design has been expressed, and in the last section the conclusion has been presented.

1-Research literature:

1-1-Marketing of banking services

The growing trend in service activities has led to the emergence of new markets and large service companies. The complexity of service industries, such as information systems, management consultancy, and logistics, has increased, as service providers have entered the market with a variety of business backgrounds and capabilities to provide a broader range of services. Facing this complexity requires knowledge and skills to understand the services. The definition of the service includes all activities that have been done (for example, writing document specifications) (Selviaridis et al., 2013: 1398).

In the industrial literature, there are various ways to define the service and product requirements. Van Weele (2010) has distinguished technical and applied definitions from each other. The first requirement expresses the exact technical specifications of the goods with the required services and inputs while it focuses on performance demands and demand-oriented performance orientations. In this regard, Axelsson and Wynstra (2002) have proposed four methods for definition of services: input definitions are based on the sources and capabilities of the service provider while the process-oriented approach focuses on how services are presented regarding activities and processes. Output-oriented definitions focus on service capabilities and performance levels, as well as the interaction between the suppliers and recipients. Ultimately, they focus on stress-oriented results in the economic value of services (Selviaridis et al., 2013: 1399).

The financial sector forms an essential part of the service sector. In this regard, banks have used marketing to improve the strategic performance of banks. In particular, since the 1980s, this trend has been confirmed and criticized. Marketing researchers and experts are interested in researching the various aspects of marketing in banking. The quality of services is essential for customers who use banking services, and the provision of accurate and high-quality services leads to the survival of banks in the competitive market. Improving the quality of services will increase the access and response to the demands of the bank's customers, which will make the bank more distinct than other competitors. Therefore, measuring the quality of services in the banking industry is essential. Providing quality services is a hard and difficult task, in this regard, the features of services in general and banking services, in particular, should be explained. Banking services require a high degree of appropriate communication between service providers and customers. If services are not provided properly, they will be at high risk.

About the five features that distinguish them from material services (intangibility, non-allocation of materials for production, diversification, non-corruption, and lack of ownership), financial services have two other specific features, that have been firstly developed by McKechnie (1997). These characteristics include trusteeship and two-way information. These special features mean that the evaluation of services is more difficult than material products, and they not only create a challenge in the development of marketing activities, but also they are the source of the problem in providing an optimal level of quality services. Banking services require a high level of communication between service providers and customers, as well as a high degree of customer participation in the process of providing services. Angurr et al. (1999) emphasized that due to the high standardization and the development of technologies in the banking sector, human factor has been considered as the most important one in making distinctions among competitors. The expertise and motivation of the employees, as well as the level of customer involvement in the provision of services, significantly affect the quality of services. Due to the wide range of banking activities, attention should be paid to cultural and social changes that affect the attitudes of customers (Durdana et al., 2015: 100).

Marketing of banking services has three components as follows: if any of them were properly executed, banking services would change the country.

1- Market study: Identification of the banking services market system, including the status of current banking services and their weaknesses, strengths, expectations, demands, ideas and opinions of customers, the conditions of environmental instruments affecting the market system, in particular government policies and governing laws, and even global changes in the banking system and successful models in this system. Indeed, market study is the foundation of the evolution of the banking system, because market study (which is the result of constant communication with the market and conducting marketing studies and research in various forms) leads to comprehensive, up-to-date, relevant and necessary information for decision-makers and bank managers (Munsung Rheea & Satish Mehrab, 2006: 508).

2. Market creation: It involves the use of a variety of simple and complex marketing techniques and competitive strategies, as well as the growth, development and market creation in the banking system; it needs to understand the internal customers (employees) and external

customers, to communicate with them and meet their needs and demands. Market creation in the competitive environment requires appropriate strategies (Moch, 2013: 2908).

3. Market protection: One of the most important tasks of marketing in the banking system of the country is to maintain market share, a stable relationship with customers, and to convert ordinary customers to loyal and supportive ones. The more intense the competition, the more valuable the loyal customers will be, as marketing research in the world has shown that the cost of finding a new customer in firms is 3 to 33 times higher than the cost of maintaining current customers regarding the type of products and services. Therefore, market protection prevents the costly advertising, and attraction of new customers and provides a kind of confidence and serenity for market survival, especially in competitive conditions (Fredriksson & Moro, 2014: 69).

1-2-Effectiveness of bank advertising

One of the indicators of success in any organization is the extent to which the organization achieves its goals. Here, the success of any organization is determined by its own goals. This approach has been called effectiveness in the management literature. The effectiveness is the degree to which the organization achieves its goals. Goals must first be defined to measure effectiveness in any organization. Then, plans are set up to fulfill these goals, and thus the plans will be implemented. In the end, the results are compared with predetermined goals (Manjegani et al., 2015: 124).

About 100 years ago, William Heketh statement, "Half of the advertisements were wasted, we just do not know which half it is" attracted everyone to the accountability of advertising. What matters to advertisers is that how much the amount of money spent on advertising has returned? Has the cost spent on making and broadcasting the advertising been effective? Has the effect of advertising been strong or weak? Knowing the answer to these questions is important to the advertiser because today companies pay much money to make and broadcast advertising. According to Jack Trout, the money we use to advertise is very much. Every second of the World War II was $ 9,000, and Vietnam cost $ 22,000 per minute. But in the American Football Final, every minute of advertising cost $ 2 million, so the advertiser has the right to know about the effectiveness of his advertising (Aksoy, Atilla, 2005, Yeni Reklamcilik, Istanbul Bilgi Universitesi, 95). It is very difficult to measure the impact of long-term advertising on many occasions. Also, academically in the world of advertising, it is impossible to create a comprehensive theory that can be used to determine the effectiveness of advertisements on people at different levels under different circumstances. The effectiveness of media is its capacity to create desired results, or special conditions in a particular position and content, that is, attaining the desired communication goals (Wery, 2012)

Evaluating the effectiveness of advertising indicates how much advertising has succeeded in achieving the goals. The evaluation of the effectiveness of advertising acts as a feedback, informs us about the weaknesses and strengths of our advertising by which we find our shortcomings in the advertising. The marketer may choose an appropriate stage regarding product life or brand position, and use it to define his advertising goals and ultimately evaluate its effectiveness (Eckler & Bolis, 2011: 2).

The effectiveness of advertising is a function of its content (message), the execution (how it is transmitted), and repetition (the number of times each consumer sees the message) (Cutler, 1997, quoted by Moorthy & Hawkins, 2005). Some of the effectiveness criteria are the call and recognition of the brand name (Park, 2006), awareness, willingness and intention to purchase (Decrop, 2007), attitude towards the ad, attitude toward the brand name and intent to purchase (Muda et al., 2014)

The control of the communication element has a direct effect on customer perception of different dimensions of the brand name, such as service quality and loyalty. It means that the elements of communication, advertising and promotion must be carefully implemented and controlled (Suvittawat, 2015). Informative advertising is one of the factors that shape the behavior of the bank customers. Studies conducted both on attracting, retaining and loyalty of customers, show the need to examine the impact of advertising on attracting and loyalty of customers (Ioannou, Boukas and Skoufari, 2014: 25).

An ideal model of world-class advertising can: 1. transfer a consistent meaning to the consumer all over the world. 2. Transfer positive global capacity for the brand. That is, the results of advertising among different peoples should produce a positive effect and transfer high-quality signals (Strebinger et al., 2018: 410).

Advertisers want to prepare consumers to like goods or services, but sometimes it harms them through unwanted and intrusive advertising. It leads to the creation of a series of negative emotions in advertising. The purpose of ads is to create positive feelings in consumers about advertised goods and services (Bell & Buchner, 2018: 1).

On the other hand, research shows that commercial advertising can be positively evaluated and considered, and this positive evaluation leads to a positive evaluation of the brand with the product. However, advertising can have negative effects that lead to the formation of negative attitudes towards that product and service. Therefore, by increasing the volume of messages, it can have positive, negative or neutral effects. Therefore, marketers should evaluate advertising messages and their behavioral outcomes. With this regard, the effectiveness of advertising, based on a hierarchical approach, should be fully investigated (Tafani et al., 2018: 125).

There are various factors which influence effectiveness of advertisement like media, time, market life stage, advertisement life stage (new or old advertisement) and also advertisement creative signs (like content and administrative elements) (Talis et.al, 2005). Also using popular cartoon personages (Ranjbarian, 2009), size (Rousat, Venous & Ebrahimi, 2004, Baltas, 2003), color (Park, 2008, Moriarty, Mitchell, Wells, 2009, Cheng & Kao, 2011), complexity or simplicity of the advertisement (Morrison & Dainoff, 1972, Di Res, 2008), using irony factor (Mukherjee, Hober, 2004), using famous artists (Rousta, Venous, Ebrahimi , 2004) and consumer features (Matrin, Phaymi, Agee, 2002, Park, 2008) are among these variables. Advertisement repetition is also a factor which helps in reminding its content to the audiences (Rousta, Venous, Ebrahimi, 2004). One of considered factors in the debate of evaluating effectiveness of advertisement is the attitude towards advertisement. Attitude towards the advertisement includes public interest or lack of interest in an advertising stimulus in encountering a special topic (Moaen & Minor, 2009). Mental imagery (B one & Elen, 1992), content transparence and emotions called through the advertisement are among variables influencing the attitude towards the advertisement. Some of these factors may lead to strong impacts while some of them may create weak impacts or don't include any impact. Awareness of the impact of each of these factors makes a considerable help in making management decisions (Talis et.al, 2005).

Fotini et.al (2009) provided a model of effective advertisements for advertising websites of big companies. The results of the study showed that measurements, activities, priorities and intentions of people have a considerable impact on evaluation and qualitative factors of relation-oriented marketing have a great impact on advertisements effectiveness indicators. The extracted criteria include:

- Advertisement policies: product information, company information, product general display, consistency with traditional advertisement

- Medium features: message features, information structure, responding time, facilitating research, beauty, interactions, communications in real-time, amusement, information control

- Relation- oriented marketing policy: communications, feedback, supporting customers

Today, organizations for competition in the business world and selling their goods and services require purposeful and efficient marketing activities. One of topics which has a very important role in the process of marketing of an organization and assign a high volume of budget to it is advertisement. Advertisement causes some changes in the knowledge, attitude and behavior level of audiences, so attention to evaluating advertisement effectiveness is a necessity for the organizations that consider advertisement as requirement of continuous life, growth and development. By assessing advertisement effectiveness and investigating consistency with the organization objectives and its expected impacts, some changes may be created in its dimensions, content and methods so that a more effective and useful advertising is performed. Regarding the research literature and internal and external background which has been stated, here a general categorization of the model proposed criteria has been provided in the following table:

Table 1- summation and extracted variables

scholar	sub standard	criterion
) Dargi & Mohammadian (2010), Ashouri (2016), Mansouri Moayed (2005), Aghapour (2016), Mirzadh & Mehrmanesh (2016), Keramforn (2011), Mac Ki- Nesbiat et.al (2011), Rosemary et.al (2012), Wang (2015), Maryion et.al(2017)	awareness of sexuality differences	individual factors
Dargi & Mohammadian (2010), Ashouri (2016), Samsam Shariat Keramforn (2011)(2007), Mirzadeh & Mehrmanesh (2016), Mac Ki- Nesbiat et.al (2011), Rosemary et.al (2012), Wang (2015), Maryion et.al(2017)	attention to age requirements	
Dargi & Mohammadian (2010), Dehdashti Shahrokh (2009)	various consumers (permanent, ordinary and case consumers)	

Khashizadeh (2010), Danaei & Dasar (2016), Pillay et.al (2012), Marion et.al (2017)	advertisement richness	
Danaei & Dasar (2016), Dargi & Mohammadian (2010), Pillay et.al (2012), Olson et.al (2012)	advertisement accuracy	
Ashouri (2016), Rabiei (2011), Taeini Harari (2014)	positive mentality of the organization	
Dargi & Mohammadian (2010), Shahmohammadi (2011), Olson et.al (2012)	honesty feeling in advertisement	
Fakhari et.al (2012)	threatening feeling	
Ashourizadeh (2016), Hamidizadeh (2013)	attitude towards brand	
madhoushi (2013), Olson et.al (2012)	being convinced	
Ahmadi (2013), Mac Ki – Nesbiat et.al (2011), Olson et.al (2012)	convicting audiences	
Ashouri (2016), Hosivar (2014), Olson et.al (2012)	mental imagery	
Siadat & Sakhi Jourashk (2014)	positive feeling	
Navaei et.al (2015), Rozmari et.al (2012)	audience being active	
Mosavi Esmaeilzadeh (2015), Olson et.al (2012)	knowledge level	
Dargi & Mohammadian (2010), Ashouri (2016), Shafiei Nikabadi & Zarei (2017)	attitude towards advertisement	
Dargi & Mohammadian (2010), Hosseinzadeh et.al (2016), Sanei et.al (2016)	familiarity with methods and forms of advertisement	message
Ashouri (2016), Olson et.al (2012), Alan (2012), Patric Ro (2014)	attention to reduction of expectation time of the user for observing advertisement	
Danaei & Dasar (2016), Forentebury & Mac Gldric (2011)	beautiful and appropriate	

	combination of colors	
Aghapuor (2016), Sanei et.al (2016)	site and location of advertisement placement	
Hassanzadeh & Danaei (2016), Sanei et.al (2016)	access to sufficient and required information	
Dargi & Mohammadian (2010), Darzban (2016)	starting the advertisement with a challenge	
Navaei et.al (2015), Salimi Fard & Pourdehghan (2016), Bargar et.al (2016)	starting the advertisement with a statement and explanation	
Dargi & Mohammadian (2010), Mirzadeh & Mehrmanesh	using helping words and terms for decision making	
Dargi & Mohammadian (2010), Daneshvar et.al (2016), Alan (2012)	using stimuli with diversions more than usual and expected extent	
Sanei et.al (2016)	using images of women and children	
Sohrabi Ranani (2016), Fortenbury & Mac Goldric (2011)	creativity in the message	
Shafei Nikabadi & Zarei (2017), Denis et.al (2012)	the text of advertisement message being a predicate sentence	

Khashizadeh (2010), Dargi & Mohammadian (2010), Sohrabi Ranani (2016), Patric Ro (2014), Abdolsalam et.al (2016)	continuation and repetition of message	
Daneshfar & Nabizadeh (2015), Fortenbury & Mac Goldric (2011)	using trustable and famous people	
Salimifard & Pourdehghan (2016)	common dialects and terms	
Daneshfar & Nabizadeh(2015), Forentnburi & Mac Goldric (2011)	loveliness of the popular person for the observer	
Daneshfar & Nabizadeh (2015), Forentnburi & Mac Goldric)(2011	familiarity of famous person with the observer	
Mansouri Moayed (2005), Aghapour (2016)	emotional music	
Khonsazadeh (2010), Hosseinzadeh et.al (2016), Aghapour (2016)	using various methods	
Siadat & Sakhi Jourashk (2014), Sohrabi Ranani (2016)	high coverage and being universal	
Navaei et.al (2016), Mosavi Esmaeilzadeh (2015), Hosseinzadeh et.al (2016), Mirzadeh & Mehrmanesh (2016), Rosemary et.al (2012), Mikit et.al (2012), Spenkar (2013)	media advertisement	
Navaei et.al (2015), Mosavi Esmaeilzadeh (2015), Hosseinzadeh et.al (2016), Ashouri (2016), Hassanzadeh & Danaei (2016), Mikit et.al (2012)	multimedia advertisement	
Navaei et.al (2015), Mirzadeh & Mehrmanesh (2016)	radio advertisement	
Navaei et.al (2015), Mosavi Esmaeilzdeh (2015), Ashouri (2016), Hassanzadeh & Danaei (2016)	cell phone advertisement	
Navaei et.al (2015), Hosseinzadeh et.al (2016), Mirzadeh & Mehrmanesh (2016)	creating TV advertisement	
Ahmadi (2013), Salimi Fard & Pourdehghan (2016), Hoang et.al (2013)	attraction of rewards	
siadat & Sakhi Jourashk (2014), Koumra et.al (2016)	being benevolent	

Darzaian (2016), Zabkouich & Louksten (2011), Abdolsalam et.al (2016)	coordination in speaking and behavior	
Ahmadi (2013), Salimi Fard & Pourdehghan(2016), Honari et.al (2015)	reality of messages	
Madhoushi (2013), Salimi Fard & Pourdehghan (2016), Houang et.al (2013)	graphic-designing	
Shahmohammadi (2011), Madhoushi (2013), Salimi Fard & Pourdehghan(2016)	psychological – geographical division	
Mirzadeh & Mehrmanesh (2016), Zabkouich & Lokesten (2011), Menroz et.al (2012), Choubhai et.al (2013), Mokan & Oltz (2014), Ayono et.al (2014)	advertisement on web	
Dehdashti Shahrokh (2009), Shahmohammadi (2011), Mirzadeh & Mehrmanesh (2016), Menroz et.al (2012), Choubhai et.al (2013), Mokan & Oltz (2014), Ayono et.al (2014), Henri et.al (2015), Yazghiro- Basteh (2015), Lin (2016)	internet & cell phone advertisement	
Shahmohammadi (2011), Sanei et.al (2016), Menroz et.al (2012), Chobai et.al (2013), Ayono et.al (2014), Henri et.al (2015)	selecting advertising media	
Shahmohammadi (2011), Sanei et.al (2016), Howang et.al (2013), Henri et.al (2015), Yazghirdo-Yasteh (2015), Martina Santa et.al (2017)	artistic aspect of advertisement	
Kashizadeh (2010), Hoshivar(2014), Pelpin et.al (2017)	gender of speaker in the advertisement	
Hosseinzadeh et.al (2016), oribeh (2016), Madra et.al (2017)	text repetition	
Shahmohammadi (2011), Hassanzadeh & Danaei (2016), Abri & Joumar (2014), Ayouno et.al (2014), Oribeh (2016), Pachman (2016), Bak (2016), Madra et.al (2017)	advertisement	
Samsam Shariat (2007), Sohrabi Ranani (2016), Daneshvar et.al (2016), Bak (2016), Kiji-Kali (2017)	factors relating to usage	environment

Fakhari et.al (2012), Mirzadeh & Mehrmanesh (2016), Howang et.al (2013), Bak (2016), kiji-Kali (2017)	factors relating to facilitation	
Dargi & Mohammadian (2010), Hamidizadeh (2013), Darzban (2016), Belman et.al (2012)	awareness of different status of the environment status	
Dargi & Mohammadian (2010), Mirzadeh & Mehrmanesh (2016)	awareness of access location of objective group to the product	
Dargi & Mohammadian (2010), Salimifard & Pourdehghan (2016)	awareness of norms	
Dargi & Mohammadian (2010), Hamidizadeh (2013), Darzban (2016), Belman et.al (2012), Koumra et.al (2016)	awareness of religions and religious values	
Dargi & Mohammadian (2010), Hassanzadeh & Danaei (2016)	awareness of method of evaluating services in relation to other services	
Hassazadeh & Danaei (2016), Radmiker (2015)	attention to security	
Samsam Shariat (2007), Mirzadeh & Mehrmanesh (2016), Belman et.al (2012)	economic beliefs	
Salimifard & Pourdehghan (2016), Pilay et.al (2012), Howang et.al (2013), Komra et.al (2016)	ethical beliefs	
Salimifard & Pourdehghan (2016)	Islamic teaching	
Samsam Shariat (2007), Daneshvar et.al (2016), Radmiker (2015)	strategic look to advertisement plan	
Dargi & Mohamadian (2010), Ansari & Rasei (2016), Abri & Joumar (2014)	market share	
Dargi & Mohamadian (2010), Darzain (2016)	obviating primary needs of audiences	

Salimi Fard & Pourdehghan (2016), Kim et.al (2012), Howang et.al (2013)	reference groups	
Sohrabi Ranin (2016), Fontini et.al (2009), Belman et.al (2012)	awareness of services	factors relating to services
Salimi Fard & Pourdehghan (2016), Abri & Joumar (2014), Radmiker (2015)	stating potential benefits of purchase ad services consumption	
Dargi & Mohamadian (2010), Hosseinzadeh et.al (2016), Fontini et.al (2009)	rate of services complexity	
Samsam Shariat (2007), Darzaian (2016), Fontini et.al (2009), Radmiker (2015)	display of expected experience	
Khashizadeh (2010), Darzain (2016)	rate of familiarity of consumer with services	
Ansari & Rasaei (2016), Belman et.al (2012)	market researches	
Hosseinzadeh et.al (2016), Mirzadeh & Mehrmanesh (2016)	record of exploiting services	
Dargi & Mohammadian (2010), Ansari & Rasaei (2016), Lousi (2012), Park (2015), Marion et.al (2017)	being unique	
Dehdashti Shahrokh (2009), Daneshvar et.al (2016), Radmiker (2015), Jankoski et.al (2016)	price, benefits	
Daneshvar et.al (2016), Fontini et.al (2009), Jankoski et.al (2016)	relation with customer	
Samsam Shariat (2007), Daneshvar et.al (201), Fontini et.al (2009), Jankoski (2016)	supporting customer	
Dehdashti Shahrokh (2009), Ansari & Rasaei (2016), Fontini et.al (2009), Radmiker (2015), Jankoski (2016)	feedback	

2- Research plan

The present study is of descriptive applied type which has been performed with survey method. The statistical population of this study includes all managers and experts of Eghtesad

Novin Bank. The statistical population of the present study contains all people who are exposed to advertisements of Eghtesad Novin Bank across the country. For selecting samples, Cochran formula is used for unlimited societies that based on this formula, the samples number were determined 400 persons for each province. The sample people were selected based on the following sampling method. At first, using stratified sampling method , the samples were randomly selected from five geographical regions of the country (north, south, east, west and center) of nine provinces (Mazandaran, Khorasan Razavi, Zanjan, Chahar Mahal Bakhtiari , Kerman, East Azarbaijan , Kermanshah, Ilam and Tehran).

3- Variables and research model

In this study, fuzzy Delphi technique has been used for identifying main indicators. For ranking the final criteria and sub standards, multi-criteria decision making models and AHP model have been used with fuzzy approach. For analysis of obtained data, Excel software and coding in the environment of this software have been used. The results obtained from data analysis have been provided in the present chapter.

3-1 Identifying criteria and sub standards of factors effective on advertisement effectiveness

At the first step, the primary indicators were identified and selected. In the first stage, by reviewing the literature and performed studies, a large number of indicators were extracted from the literature. For screening the indicators and identifying the final indicators, fuzzy Delphi approach has been used. The expert views have been collected in the field of rate of significance of main indicators. Fuzzy average and defuzzification output of rates relating to indicators are presented in table 2. The rate of defuzzification greater than 7 is acceptable and every indicator which has the score higher than 7 is confirmed (Wu& Fang, 2011).

Table 2- fuzzy average and fuzzy screening of indicators of factors effective on advertisement effectiveness

Result	Crisp	mean	U	M	L	
acceptance	7.15	(2, 7.97, 9)	9	7.97	2	Q01
acceptance	7.84	(6, 8.01, 9)	9	8.01	6	Q02

Result	Crisp	mean	U	M	L	
acceptance	7.92	(6, 8.13, 9)	9	8.13	6	Q03
acceptance	7.39	(4, 7.83, 9)	9	7.83	4	Q04
acceptance	7.70	(6, 7.8, 9)	9	7.80	6	Q05
rejection	6.68	(2, 7.28, 9)	9	7.28	2	Q06
acceptance	7.71	(6, 7.82, 9)	9	7.82	6	Q07
acceptance	7.66	(6, 7.75, 9)	9	7.75	6	Q08
acceptance	7.02	(2, 7.78, 9)	9	7.78	2	Q09
acceptance	7.27	(2, 8.15, 9)	9	8.15	2	Q10
acceptance	7.09	(2, 7.88, 9)	9	7.88	2	Q11
rejection	6.85	(2, 7.52, 9)	9	7.52	2	Q12
acceptance	7.43	(4, 7.9, 9)	9	7.90	4	Q13
rejection	6.71	(3, 7.07, 9)	9	7.07	3	Q14
acceptance	7.17	(2, 8.01, 9)	9	8.01	2	Q15
acceptance	7.13	(2, 7.95, 9)	9	7.95	2	Q16
rejection	6.85	(4, 7.02, 9)	9	7.02	4	Q17
acceptance	7.43	(4, 7.89, 9)	9	7.89	4	Q18
acceptance	7.70	(6, 7.8, 9)	9	7.80	6	Q19
rejection	6.80	(3, 7.21, 9)	9	7.21	3	Q20
acceptance	7.75	(6, 7.88, 9)	9	7.88	6	Q21
rejection	6.62	(3, 6.93, 9)	9	6.93	3	Q22
acceptance	7.61	(6, 7.66, 9)	9	7.66	6	Q23
acceptance	7.22	(2, 8.08, 9)	9	8.08	2	Q24
acceptance	7.04	(2, 7.81, 9)	9	7.81	2	Q25
acceptance	7.71	(6, 7.82, 9)	9	7.82	6	Q26
rejection	6.65	(3, 6.97, 9)	9	6.97	3	Q27
acceptance	7.98	(6, 8.22, 9)	9	8.22	6	Q28
acceptance	7.70	(6, 7.79, 9)	9	7.79	6	Q29
rejection	6.54	(3, 6.81, 9)	9	6.81	3	Q30
acceptance	7.19	(4, 7.53, 9)	9	7.53	4	Q31
acceptance	7.79	(5, 8.18, 9)	9	8.18	5	Q32

Result	Crisp	mean	U	M	L	
acceptance	7.53	(4, 8.05, 9)	9	8.05	4	Q33
acceptance	7.05	(2, 7.82, 9)	9	7.82	2	Q34
acceptance	7.43	(4, 7.89, 9)	9	7.89	4	Q35
rejection	6.92	(3, 7.38, 9)	9	7.38	3	Q36
acceptance	7.02	(2, 7.78, 9)	9	7.78	2	Q37
rejection	6.91	(3, 7.37, 9)	9	7.37	3	Q38
rejection	6.96	(2, 7.68, 9)	9	7.68	2	Q39
rejection	6.99	(2, 7.74, 9)	9	7.74	2	Q40
acceptance	7.42	(4, 7.88, 9)	9	7.88	4	Q41
acceptance	7.10	(2, 7.9, 9)	9	7.90	2	Q42
rejection	6.59	(2, 7.13, 9)	9	7.13	2	Q43
rejection	6.55	(2, 7.08, 9)	9	7.08	2	Q44
acceptance	7.77	(6, 7.9, 9)	9	7.90	6	Q45
rejection	6.91	(2, 7.62, 9)	9	7.62	2	Q46
acceptance	7.09	(2, 7.88, 9)	9	7.88	2	Q47
acceptance	7.68	(6, 7.78, 9)	9	7.78	6	Q48
acceptance	8.19	(6, 8.53, 9)	9	8.53	6	Q49
rejection	6.88	(2, 7.57, 9)	9	7.57	2	Q50
acceptance	7.75	(5, 8.12, 9)	9	8.12	5	Q51
acceptance	7.03	(2, 7.8, 9)	9	7.80	2	Q52
rejection	6.96	(4, 7.19, 9)	9	7.19	4	Q53
acceptance	7.04	(2, 7.8, 9)	9	7.80	2	Q54
rejection	6.96	(3, 7.45, 9)	9	7.45	3	Q55
acceptance	7.02	(2, 7.78, 9)	9	7.78	2	Q56
acceptance	7.92	(5, 8.38, 9)	9	8.38	5	Q57
acceptance	7.00	(2, 7.76, 9)	9	7.76	2	Q58
rejection	6.90	(1, 7.86, 9)	9	7.86	1	Q59
acceptance	7.10	(4, 7.4, 9)	9	7.40	4	Q60
acceptance	7.86	(6, 8.04, 9)	9	8.04	6	Q61
rejection	6.97	(2, 7.71, 9)	9	7.71	2	Q62

Result	Crisp	mean	U	M	L	
acceptance	7.02	(2, 7.79, 9)	9	7.79	2	Q63
acceptance	7.73	(6, 7.84, 9)	9	7.84	6	Q64
acceptance	7.01	(2, 7.76, 9)	9	7.76	2	Q65
rejection	6.65	(3, 6.98, 9)	9	6.98	3	Q66
acceptance	7.01	(2, 7.77, 9)	9	7.77	2	Q67
acceptance	7.90	(6, 8.11, 9)	9	8.11	6	Q68
acceptance	7.55	(5, 7.83, 9)	9	7.83	5	Q69
acceptance	8.02	(6, 8.28, 9)	9	8.28	6	Q70
acceptance	7.95	(6, 8.17, 9)	9	8.17	6	Q71
acceptance	7.03	(2, 7.79, 9)	9	7.79	2	Q72
acceptance	7.65	(5, 7.98, 9)	9	7.98	5	Q73
acceptance	7.00	(1, 8, 9)	9	8.00	1	Q74
acceptance	7.04	(2, 7.8, 9)	9	7.80	2	Q75
acceptance	7.03	(2, 7.8, 9)	9	7.80	2	Q76
acceptance	7.30	(4, 7.7, 9)	9	7.70	4	Q77
rejection	6.88	(2, 7.57, 9)	9	7.57	2	Q78
acceptance	7.63	(4, 8.19, 9)	9	8.19	4	Q79
acceptance	7.65	(5, 7.98, 9)	9	7.98	5	Q80
rejection	6.59	(3, 6.88, 9)	9	6.88	3	Q81
acceptance	7.09	(2, 7.88, 9)	9	7.88	2	Q82
rejection	6.83	(2, 7.5, 9)	9	7.50	2	Q83
acceptance	7.12	(2, 7.93, 9)	9	7.93	2	Q84

From the 84 indicators which were assessed, 24 indicators which have acquired the average score less than 7 are omitted. So, 60 indicators are finally confirmed (Adelazaret.al, 1995).

The remained indicators will be finally examined in the format of a 60-question questionnaire with analytical hierarchy process.

3-2 determining the priority of factors effective on effectiveness of advertisement using fuzzy AHP technique

For determining the priority of major criteria, fuzzy analytical hierarchy (FAHP) has been used. The analysis process is as follows.

1- Dual comparison of major criteria based on objective and determining the weight of major criteria

2- Dual comparison of sub standards of each criterion and determining sub standards of each cluster

3- Multiplication of the weight of sub standards in the weight of respective criterion and determining the final weight of sub standards

For dual comparison of elements, 9-degree Saaty scale has been used. 9-degree Saaty scale has been provided by Thomas Saaty , codifier of analytical hierarchy theory. Also, in this study fuzzy approach has been used for making the values quantitative. So, Saaty fuzzy range has been used.

Table 3- fuzzy range equal to 9-degree Saaty scale in AHP technique. Habibi et.al , 2014, 77

reversed fuzzy equivalent	fuzzy equivalent	the word term of comparison status of i to j
(1,1,1)	(1, 1, 1)	Preferred Equally
$\left(\frac{1}{3},\frac{1}{2},1\right)$	(1, 2, 3)	intermediate
$\left(\frac{1}{4},\frac{1}{3},\frac{1}{2}\right)$	(2, 3, 4)	Preferred moderately
$\left(\frac{1}{5},\frac{1}{4},\frac{1}{3}\right)$	(3, 4, 5)	intermediate
$\left(\frac{1}{6},\frac{1}{5},\frac{1}{4}\right)$	(4, 5, 6)	Preferred Strongly
$\left(\frac{1}{7},\frac{1}{6},\frac{1}{5}\right)$	(5, 6, 7)	intermediate
$\left(\frac{1}{8},\frac{1}{7},\frac{1}{6}\right)$	(6, 7, 8)	very strongly Preferred
$\left(\frac{1}{9},\frac{1}{8},\frac{1}{7}\right)$	(7, 8, 9)	intermediate
$\left(\frac{1}{9},\frac{1}{9},\frac{1}{9}\right)$	(9, 9, 9)	Extremely Preferred

(source in papersL Khorshid & Zabihi, Lee my H.I; Wen-Chin Chen, Ching-Jan, Chang, 2008, 101)

The pattern of criteria and sub standards has been drawn in the figure using AHP technique. Also, the criteria and sub standards of the study have been named with numeric index so that the research flow is easily detectable.

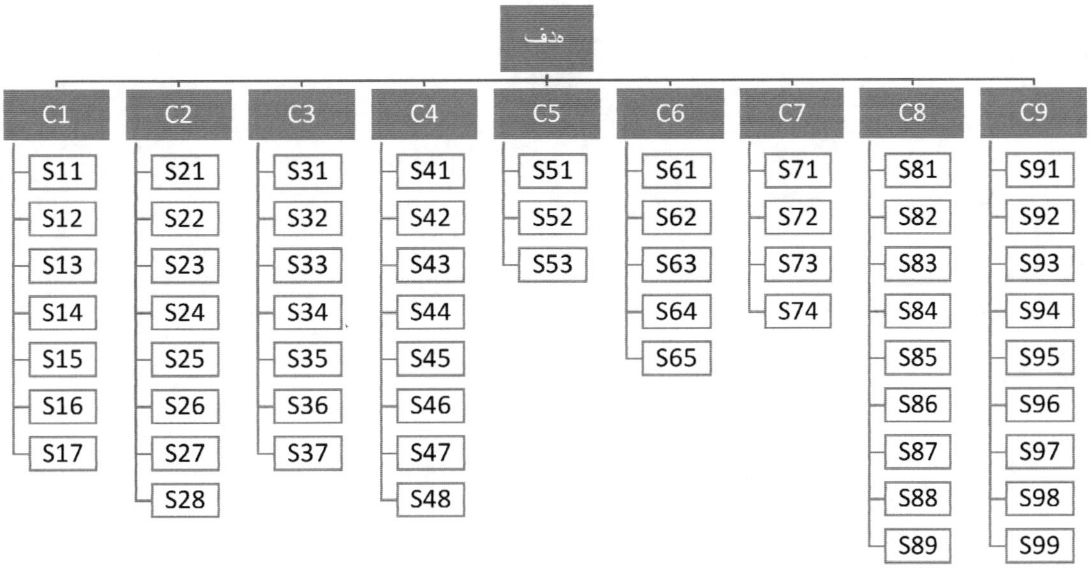

Figure 1- hierarchical model of criteria and sub standards of factor effective on advertisement effectiveness

1- Determining the priority of major criteria based on objective

In the first step, the major criteria based on the objective have been compared dually. Dual comparison is very simple and all elements of each cluster should be compared two by two. Therefore, if there is n element in a cluster, $\frac{n(n-1)}{2}$ comparisons will be performed. Since there are four criteria, so the number of performed comparisons is equal to:

$$\frac{n(n-1)}{2} = \frac{9(9-1)}{2} = 36$$

So, 36 dual comparisons have been performed from group perspective of experts. The experts view has been made quantitative using fuzzy scale. Collecting experts view has been collected with 9-degree Saaty range. So, 6 dual scales of four major four criteria have been performed based on the experts view. Using geometric average the experts views have been aggregated. Fuzzy aggregation will be calculated as follows:

Equation 1

$$F_{AGR} = (\prod(l), \prod(m), \prod(u))$$

After matrix formation of the obtained dual comparisons, the special vector has been calculated. At first, the fuzzy extension of each row is calculated. Every entry of dual comparison matrix $\widetilde{X}$

Is displayed as $\tilde{x}_{ij}$.

The fuzzy extension of every row is also displayed with symbol $\tilde{S}_i$. Therefore, the fuzzy extension of each row is calculated as follows:

Equation 2

$$\tilde{S}_i = \sum_{j=1}^{n} x_{ij}$$

Then, the fuzzy extension of every row elements will be as follows:

Fuzzy extension of row 1 (1,1,1) ⊕ (1,27,1,47,1,75) ⊕ (0.47, 0.55, 0.66) ⊕ (1.52, 1.81, 2.14) ⊕ (0.93, 1.14, 1.39) ⊕ (1.26, 1.61, 2.04) ⊕(1.25, 1.51, 1.84) ⊕(0.78, 0.97, 1.24) ⊕ (0.59, 0.75, 0.96) ⊕ (9.08, 10.8, 13.02)

Fuzzy extension of row 2 (0.57, 0.68, 0.79)⊕(1, 1, 1)⊕(0.58, 0.7, 0.89)⊕(1.84, 2.28, 2.75)⊕(0.68, 0.82, 1.02)⊕(1.42, 1.81, 2.34)⊕(1.64, 2.04, 2.53)⊕(0.71, 0.88, 1.12)⊕(0.73, fuzzy extension of row 3 (1.52, 1.82, 2.11)⊕(1.13,)0.9, 1.1)⊕ = (9.17, 11.11, 13.55 1.42, 1.72)⊕(1, 1, 1)⊕(1.89, 2.34, 2.88)⊕(0.7, 0.86, 1.09)⊕(1.45, 1.78, 2.22)⊕(1.55, 1.88, 2.29)⊕(0.57, 0.71, 0.93)⊕(0.67, 0.8, 0.96)⊕ = (10.47, 12.61, 15.2)

Fuzzy extension of row 4 (0.47, 0.55, 0.66)⊕(0.36, 0.44, 0.54)⊕(0.35, 0.43, 0.53)⊕(1, 1, 1)⊕(0.97, 1.21, 1.51)⊕(0.96, 1.33, 1.8)⊕(2.14, 2.7, 3.37)⊕(0.94, 1.21, 1.56)⊕(1.3, 1.69, 2.16)⊕ = (8.49, 10.57, 13.14)

Fuzzy extension of row 5 (0.72, 0.88, 1.07) ⊕ (0.98, 1.22, 1.48)⊕(0.92, 1.17, 1.43)⊕(0.66, 0.82, 1.03)⊕(1, 1, 1)⊕(0.67, 0.83, 1.04)⊕(0.23, 0.29, 0.36)⊕(0.28, 0.32, 0.39)⊕(0.3, 0.38, 0.48)⊕ = (5.76, 6.91, 8.29)

Fuzzy extension of row 6 (0.49, 0.62, 0.79)⊕(0.43, 0.55, 0.7)⊕(0.45, 0.56, 0.69)⊕(0.55, 0.75, 1.04)⊕(0.96, 1.2, 1.48)⊕(1, 1, 1)⊕(0.34, 0.41, 0.51)⊕(0.27, 0.34, 0.44)⊕(0.24, 0.31, 0.41)⊕ = (4.74, 5.75, 7.08)

Fuzzy extension of row 7 (0.54, 0.66, 0.8)⊕(0.4, 0.49, 0.61)⊕(0.44, 0.53, 0.64)⊕(0.3, 0.37, 0.47)⊕(2.81, 3.5, 4.3)⊕(1.95, 2.43, 2.96)⊕(1, 1, 1)⊕(0.34, 0.42, 0.54)⊕(0.29, 0.36, 0.47)⊕ = (8.06, 9.76, 11.78)

Fuzzy extension of row 8 (0.8, 1.03, 1.28)⊕(0.89, 1.14, 1.41)⊕(1.08, 1.41, 1.76)⊕(0.64, 0.82, 1.06)⊕(2.55, 3.1, 3.63)⊕(2.25, 2.97, 3.72)⊕(1.86, 2.39, 2.94)⊕(1, 1, 1)⊕(0.58, 0.69, 0.83)⊕ = (11.66, 14.56, 17.64)

Fuzzy extension of row 9 (1.04, 1.33, 1.69)⊕(0.91, 1.11, 1.37)⊕(1.04, 1.25, 1.5)⊕(0.46, 0.59, 0.77)⊕(2.08, 2.63, 3.3)⊕(2.42, 3.23, 4.1)⊕(2.15, 2.74, 3.43)⊕(1.2, 1.46, 1.72)⊕(1, 1,)1)⊕ = (12.29, 15.35, 18.88

Then, the fuzzy sum of total priorities column elements is calculated

Equation 3

$$\sum \tilde{S}_i = \sum_{i=1}^{n} \sum_{j=1}^{n} x_{ij}$$

The sum of elements of the priorities column of major criteria will be as follows:

$$\sum_{i=1}^{4} \sum_{j=1}^{4} M_g^j = (79.72, 97.41, 118.59)$$

For normalization of each criterion priorities, that criterion total rates sum should be divided into sum of all priorities (column elements), since the rates are fuzzy, so fuzzy sum of every row is multiplied in summation reverse. The summation reverse should be calculated.

Equation 4

$$if\ \tilde{F} = (l, m, u)\ then\ \tilde{F}^{-1} = \left(\frac{1}{u}, \frac{1}{m}, \frac{1}{l}\right)$$

Then, based on the equation 4, we will have

$$\left(\sum_{i=1}^{n} \sum_{j=1}^{n} M_g^j\right)^{-1} = (0.01, 0.01, 0.01)$$

$$P_k = \sum_{i=1}^{n} M * \left(\sum_{i=1}^{n} \sum_{j=1}^{n} M_g^j\right)^{-1}$$

So, the results obtained from normalization of obtained rates will be as follows:

$\widetilde{W}_{C1} =$ (0.08, 0.11, 0.16)

$\widetilde{W}_{C2} =$ (0.08, 0.11, 0.17)

$\widetilde{W}_{C3} =$ (0.09, 0.13, 0.19)

$\widetilde{W}_{C4} =$ (0.07, 0.11, 0.16)

$\widetilde{W}_{C5} =$ (0.05, 0.07, 0.1)

$\widetilde{W}_{C6} =$ (0.04, 0.06, 0.09)

$\widetilde{W}_{C7} =$ (0.07, 0.1, 0.15)

$\widetilde{W}_{C8} =$ (0.1, 0.15, 0.22)

$\widetilde{W}_{C9} =$ (0.1, 0.16, 0.24)

Each of the obtained rates is fuzzy and normalized weight relating to major criteria. There are various methods for defuzzifcation of the obtained rates. In this study, Bojadziev proposed method has been used for defuzzification.

Table 4- defuzzification and computation of final and normal weight of criteria of factors effective on advertisement effectiveness

Normal	Deffuzy	X3max	X2max	X1max	Crisp
0/111	0/117	0/113	0/115	0/117	C1
0/114	0/120	0/115	0/118	0/120	C2
0/103	0/137	0/133	0/135	0/137	C3
0/108	0/113	0/112	0/113	0/113	C4
0/070	0/073	0/072	0/073	0/073	C5
0/060	0/063	0/062	0/063	0/063	C6
0.101	0/107	0/103	0/105	0/107	C7
0.149	0/157	0/153	0/155	0/157	C8
0.158	0/167	0/163	0/165	0/167	C9

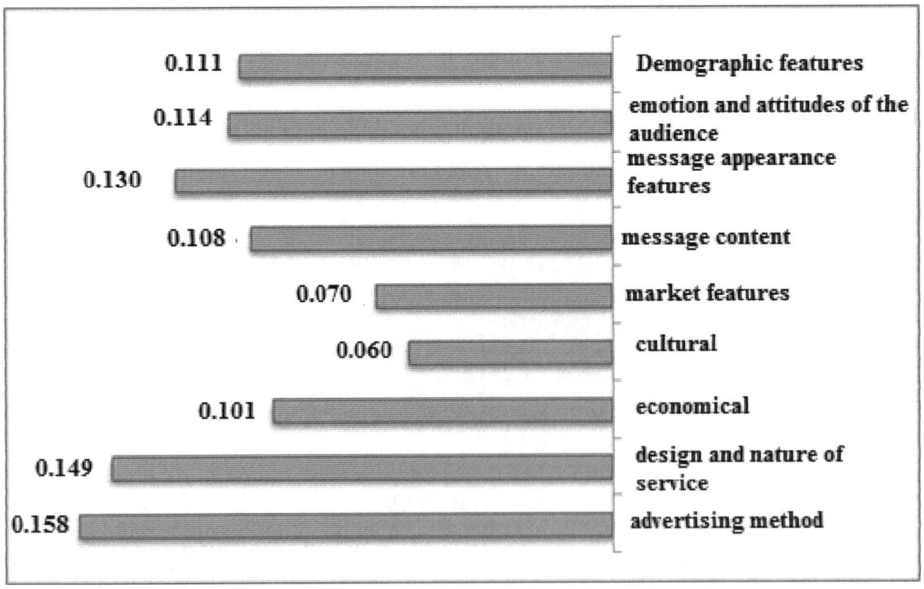

Figure 2- graphical display of significance of criteria of factors effective on advertisement effectiveness

Based on the obtained special vector:

- Demographic features with weight 0.111 have been located in rank five

- Individual feelings and attitude of the addressee with weight 0.114 has been located in rank four.

- The message surface features with weight 0.13 has been located in rank three.

- The message content with weight 0.108 has been located in rank six

- The market feature with weight 0.07 has been located in rank eight.

- Cultural with weight 0.06 has been located in rank nine.

- Economic with weight 0.10 has been located in rank seven.

- Plan and nature of services with weight 0.149 has been located in rank second.

- Advertisement methods with weight 0.158 has been located in rank first.

The inconsistency rate of performed comparisons has been obtained 0.030 which is smaller than 0.1 and so the performed comparisons is sufficed.

2- Comparison and determining priority of sub standards

At first, the experts' views have been collected in the dual comparison of sub standards. Then, the performed calculations for fuzzification of the experts views average have been provided for determining the priority of sub standards.

Priority of sub standards of demographic features

The audience job with weight 0.151 has been located in the third rank.

The audience income with weight 0.093 has been located in the seventh rank.

The audience religion with weight 0.16 has been located in the second rank.

Social class with weight 0.141 has been located in the fifth rank.

Awareness of gender differences with weight 0.173 has been located in the first rank.

The individual knowledge level with weight 0.135 has been located in the sixth rank.

Priority of sub standards of the audience individual feeling and attitude

Positive mentality with weight 0.124 has been located in the fifth rank.

Feeling of honesty in advertisements with weight 0.089 has been located in the eighth rank.

Threatening feeling with weight 0.14 has been located in the third rank.

Mental image with weight 0.102 has been located in the seventh rank.

Attitude towards brand with weight 0.121 has been located in the sixth rank.

Convincing audiences with weight 0.149 has been located in the first rank.

Attitude towards advertisement with weight 0.143 has been located in the second rank.

The individual perception of accuracy of advertisement with weight 0.133 has been located in the fourth rank.

Priority of sub standards of the message surface features

Then, based on the results of performed dual comparisons:

Beautiful and appropriate combination of colors with weight 0.143 has been located in the fourth rank.

Attention to reduction of the user waiting time with weight 0.159 has been located in the second rank.

Graphics- design with weight 0.131 has been located in the sixth rank.

Creativity in the message with weight 0.137 has been located in the fifth rank.

Emotional music with weight 0.099 has been located in the seventh rank.

Text repetition with weight 0.153 has been located in the third rank.

Access to required and sufficient information with weight 0.178 has been located in the first rank.

Determining the priority of sub standards of the message content

Benevolent with weight 0.093 has been located in the eighth rank.

Coordination in speech and behavior with weight 0.125 has been located in the fourth rank.

Reality of the messages with weight 0.119 has been located in the sixth rank.

Using women and children images with weight 0.131 has been located in the third rank.

Gender of speaker in the advertisement with weight 0.106 has been located in the seventh rank.

Using trustful and famous people with weight 0.16 has been located in the first rank.

Loveliness of famous people for the observer with weight 0.122 has been located in the fifth rank.

Familiarity of the famous person with the observer with weight 0.144 has been located in the second rank.

Priority of sub standards of the market features

Then, based on the results of the performed dual comparisons:

Factors relating to its application in the market with weight 0.326 have been located in the second rank.

Awareness of different aspects of status and environment with weight 0.283 has been located in the third rank.

Awareness of evaluation method of service receiver with weight 0.391 has been located in the first rank.

Priority of cultural sub standards

Then, based on the results of performed dual comparisons:

Awareness of norms with weight 0.199 has been located in the third rank.

Awareness of religion and religious values with weight 0.238 has been located in the first rank.

Ethical beliefs with weight 0.18 have been located in the fourth rank.

Obviating primary needs of the audiences with weight 0.164 has been located in the fifth rank.

Reference groups with weight 0.219 have been located in the second rank.

Determining the priority of economic sub standards

Then, based on the results of the performed dual comparison:

Economic beliefs with weight 0.261 have been located in the second rank.

Strategic look at an advertising program with weight 0.215 has been located in the fourth rank.

Market share with weight 0.274 has been located in the first rank.

Attention to economic security with weight 0.251 has been located in the third rank.

Priority of the plan and services nature sub standard

Awareness of services with weight 0.104 has been located in seventh rank.

Expressing potential benefits of purchase and services consumption with weight 0.117 has been located in the third rank.

Services complexity rate with weight 0.089 has been located in the ninth rank.

Uniqueness with weight 0.108 has been located in the sixth rank.

Rate of familiarity of the services consumer with weight 0.12 has been located in the second rank.

Price and allowances with weight 0.114 has been located in the fourth rank.

Relation with the customer with weight 0.111 has been located in the fifth rank.

Displaying the expected experience with weight 0.098 has been located in the eighth rank.

Supporting customers with weight 0.139 has been located in the first rank.

Priority of sub standards of the advertising methods

Beginning the advertisement with a challenge with weight 0.079 has been located in the ninth rank.

Beginning an advertisement with a statement and explanation with weight 0.107 has been located in the fifth rank.

Using stimuli with diversions more than usual and expected with weight 0.097 has been located in the seventh rank.

The advertisement message text being predicating with weight 0.126 has been located in the third rank.

Using diverse methods with weight 0.088 has been located in the eighth rank.

Media advertisement with weight 0.113 has been located in the fourth rank.

Dimensions of TV advertisement with weight 0.151 have been located in the first rank.

Psychological, geographical division with weight 0.107 has been located in the fifth rank.

Advertising media selection with weight 0.132 has been located in the second rank.

Determining the final priority of major criteria with FAHP technique

For determining final priority of major criteria using FAHP technique, the weights relating to major criteria (W1) and weight of indicators based on each criteria (W2) are needed. The results of comparison of the research sub criteria relating to them forms matrix W2. For determining the final priority of indicators with AHP technique, it is sufficient that the weight of indicators based on every criterion (W2) is multiplied in the major criteria (W1). Any of these matrices has been calculated in the previous steps. Based on the obtained output, it is obvious that the market share item, one of economic factors with weight 0. 277 is the most important indicator in the existing indicators. Item of awareness of evaluation method of service receiver in relation to other services with weight of 0.0273 has the second priority. Item of economic believes, one of economic factors, has the third priority.

So, the research model is as follows:

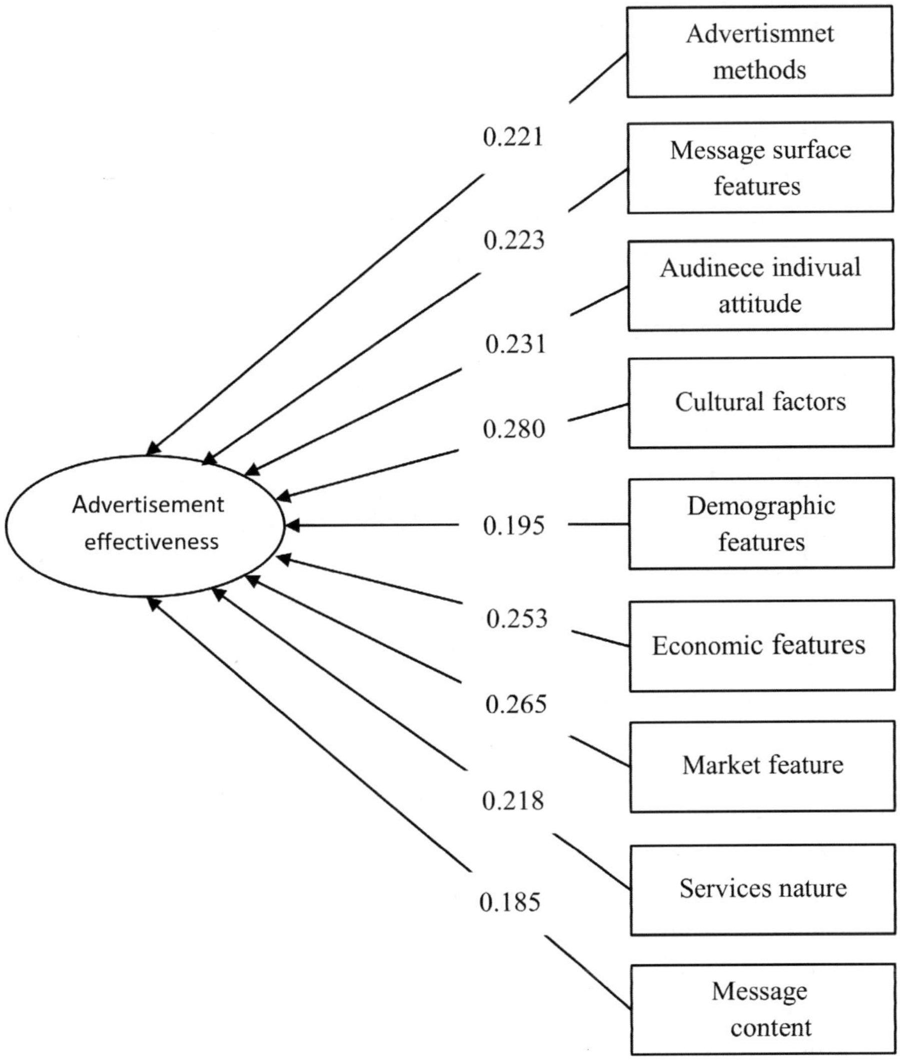

4- Conclusion:

In this study, identification and ranking of factors effective on designing the native model of advertisement in Eghtesad Novin bank has been addressed. In this first step, the primary indicators were identified and selected. In the first stage, by reviewing the literature and performed studies, a great number of indicators were extracted from the literature. For screening the indicators and identifying final indicators, fuzzy Delphi approach has been used. The experts' views have been collected in the field of significant rate of major indicators. From 84 indicators which were assessed, 24 indicators which have acquired score less than 7 are omitted. So, 60 indicators were finally confirmed. Then, regarding the performed calculations, the final weight of any of the model indicators have been calculated with fuzzy AHP. Based on the obtained output, it is obvious that the market share of economic factors with weight of 0.027 is the most important indicator in the existing indicators. The item of awareness of evaluating services in relation to other services is one of market features factors with weight 0.0273 which has second priority. The item of economic beliefs with weight 0.0264 is in the third priority.

Today, considering extension of competition in the international level and formation of world competition, all countries of the world may be successful in this arena to coordinate them with this trend and create more value for their customers by services and products they provide. In the today modern world, almost all people are exposed to advertisements and other communicational activities. Effectiveness of advertisement is one of important issues in the advertisement domain. In this regard, Mansouri Moyed (2005) showed that attention attraction, commercial advertisements of Maskan Bank are effective but in respect of other variables, that is, creating interest, stimulation of tendency and measurement to purchase, commercial advertisements of Maskan Bank are not effective. Samsam Shariat (2007) showed that TV channels and cartoon animations , attractive package, colors relating to the product and innovation, noticing safety feeling and associated symbols, informing of quality, background and price, dialects and common phrases, diversity in the message content, in TV, broadcasting from 18-23 and before beginning programs (in media) have been effective factors in effectiveness of commercial effectiveness from people viewpoint. Mosavi Esmaeilzadeh (2015) showed that oral advertisements (in 4 dimensions) and after that TV advertisement (just in 2 dimensions) are effective. Also, radio, internet and billboard advertisement are not effective. And also Friedman test was used for ranking advertisement tools. Salimi Fard & Pourdehghan (2016) showed that among factors effective on attitude to cell phone advertisement, components of general attitude towards advertisement, knowledge level and value have respectively the greatest significance and components of time of presentation and reference groups with the least significance in effecting the individuals attitude towards cell phone advertisement. Ansari & Rasaei (2016) showed that the creativity of advertisement message , market research, competition, market share, uniqueness and relation with customer are effective on the effectiveness of brand advertisement. Louis (2012) showed that advertising in the urban public spaces will have a great impact in the consumers' mentality. Since some advertisements are transitory and are often displayed in a short period of time and just in several spaces, but advertisement in public spaces exist everywhere and is similar to advertising through print and media, but since they have lower cost for production and distribution, have lower cost in this respect. Zabckowich & Loukston (2011) showed that in necktie advertisements use more of sensory requests and information signs. Findings showed that in western and European countries, people establish more contact due to globalization increase. They declared that content of printed advertisements, includes visual and text advertisements which are effective in establishing relation with many audiences. Also, ethnic identity as an important cultural variable increases attention to consumer purchase behavior. Besides that, they found out that the image of people, models, ceremonies and managers constitute an important part of images and background image plays an important role in effectiveness of visual factors in advertisement. Mac Ki- Nesbiat et.al (2011) showed that younger adults remind emotional messages specially negative ones better than others and older adults more prefer positive and logical messages to negative and emotional ones, but attitude towards advertisement among younger adults is not different. Also, they found that older adults have the most positive attitude towards advertisements.

Perzouti (2015) showed that advertising models are different in various ages and each age is consistent with a special model and using various and consistent models with people conditions may make the advertisement more effective. Abri & Joumar (2014) showed that advertising messages are effective on the consumer behavior and the advertiser may use various media for communicating consumer and media schedule is effective on the advertisement success. Attention to environmental factors where the customer lives are important factors in the advertisement success. Madra et.al (2017) showed that the advertisement text, text repetition impact the advertisement effectiveness. Maryion et.al (2017) showed that simultaneous use of several media like TV and internet may increase effectiveness of the advertisement and considering two variables of gender and multimedia create a better individual attitude towards the message effectiveness.

In ranking the identified factors, advertisement methods have assigned the first rank among factors. It is suggested that in designing advertisement and selecting advertising instruments, the dimensions of TV advertising, selecting median and predicative message text should be noticed. Using advertising instrument like internet advertisement, adverting on public vehicles like buses and metro due to great number of people who use them can increase stability of the advertisements in the customers mind.

In codifying and designing bank advertisements, the design and services nature should be noticed. It is proposed that supporting customer, familiarity of consumer with services are considered and during the advertisement process, it is required that the potential benefits of purchase and services consumption are expressed.

In designing the advertisement and content of bank advertising message, it is required to use famous and trustful people for introducing the services. Also, children as future customers and audiences should be noticed since they are future potentials. Total advertisements provided by the bank through various media should move in a unit line so that the audience can establish a logical relation between the observed advertisements in various media that is total advertising efforts lead to synergy of advertisement effectiveness. This issue won't be acquired unless through a media planning and efficient advertisement.

In advertisements, the message surface features should be noticed. It is required that in advertisement, access to required and enough information for the customers, attention to reduction of expectation time of user for observing the advertisement and suitable repetition of the text are considered. Providing large information in an advertisement which has a short time and providing information that require special mathematical analysis should be avoided.

References:

Hosseinabadi, Maryam, Mahmoud Nematian, Mohammad Besharati, investigating the relation between mixed elements of marketing and customers satisfaction with Tehran

Sarmayeh bank branches, management and entrepreneurship studies, fall 2016, 2nd period, No. 3 (page 162-169)

Dargi Parviz, 2008, subjects and topics of marketing management with Iran market attitude, Tehran, Rasa publication

Sarmad, Zohreh, Bazargan Abbas, Hejazi, Elaheh, 2014, research methods in behavioral sciences, Tehran, Agah. 27-13

Salimi Fard, Khodakaram, Pourdehghan, Adel, 2016, identifying and ranking factors effective on attitude towards cell phone advertisement using AHP, Brand management, period 3, no. 6, summer 2016, page 133-160

Seyed Akbar Manjgani, Vahideh Tabasi Lotfabadi, 2015, factors effective on effectiveness of advertisement on financial stoop of legal clients of Mashhad City with AHP approach, finance journal, fall 2015, No. 75 (page 121-146)

Sanayei, Ali, Majid Mohammad Shafiei, Maedeh Amini Valashani, 2016, effectiveness of bank advertisement in customer attraction, commercial management searches, spring and summer 2016, No. 15 (page 185-210)

Mahmoudian, Mahmoud, Rabiei, Ali, Jamili, Bita. 2011, evaluation of effectiveness of Parsian bank advertisement and identifying the most important factors in increasing its effectiveness in Tehran city, modern marketing studies, summer 2011, No. 2 (page 17-40).

Manjagani, Seyed Akbar, Tabasi Lotfabadi, Vahideh, 2015, factors effective on advertisement effectiveness on financial stoop of legal clinets of Mashhad city with AHP approach, finance journal, fall 2015, No. 75, 1-121

Venous, Davar, Safaeian, Mitra, (2005), marketing bank services in Iran and bank services features, Tehran, Negah Danesh, 4th edition

Angulo, N. (2016). 10 digital marketing trends to watch in 2016 and beyond. Retrieved from: http://www.marketingdive.com/news/10-digital-marketing-trends-to-watch-in-2016-./and-beyond/411457

Bell Raoul, Buchner, Axel.(2018). Positive Effects of Disruptive Advertising on Consumer Preferences, Journal of Interactive Marketing 41 (2018) 1–13

Corvi. Elisabetta.(2010).The effectiveness of advertising: a literature review, 10th Global Conference on Business & Economics, ISBN : 978-0-9830452-1-2

Cramphorn Michael F.,(2011), Forum Gender effects in advertising, International Journal of .Market Research Society Vol. 53 Issue 2

Cronin, J. J. & Fox , G. L. (2010). "The implications of third party customer complaining for .advertising efforts". Journal of advertising. 39(2): 21-23

Decrop, A. (2007). The influence of message format on the effectiveness of print advertisements for tourism destinations. International Journal of Advertising, 26(4), 505–.525

Denis .A. Shakhov and ,Anna A. Panasenko.(2012).Evaluating Effectiveness of Bank Advertising in the Internet: Theory and Practice, World Applied Sciences Journal 18 (Special Issue of Economics): 83-90, 2012 ISSN 1818-4952

Denis 1,2 A. Shakhov and 3Anna A. Panasenko.(2012).Evaluating Effectiveness of Bank Advertising in the Internet: Theory and Practice, World Applied Sciences Journal 18 (Special Issue of Economics): 83-90, 2012

Durdana Ozretic-Dosen Ines Zizak , (2015),"Measuring the quality of banking services targeting student population", EuroMed Journal of Business, Vol. 10 Iss 1 pp. 98 - 117

Eckler, P., and Bolls, P. (2011). Spreading the Virus: Emotional Tone of Viral Advertising and Its Effect on Forwarding Intention and Attitudes. Journal of Interactive Advertising, .11(2): 1-11

Eric W.T. Ngai Ka-leung Karen Moon S.S. Lam Eric S. K. Chin Spencer S.C. Tao , (2015),"Social media models, technologies, and applications", Industrial Management & Data Systems, Vol. 115 Iss 5 pp. 769 – 802

Fotini Patsioura Maro Vlachopoulou Vicky Manthou, (2009),"A new advertising effectiveness model for corporate advertising web sites", Benchmarking: An International Journal, Vol. 16 Iss 3 pp. 372 - 386

Frederic Crane, , G. (2010). «Marketing FO Entrepreneurs: ConceptandApplication for new .Venture. New York: SAGE Pablication, Inc

Frolova. Svetlana.(2014).THE ROLE OF ADVERTISING IN PROMOTING A PRODUCT, Thesis CENTRIA UNIVERSITY OF APPLIED SCIENCES ,Degree Programme in Industrial Management May 2014, p"1-61

Henry Boateng Abednego Feehi Okoe , (2015),"Consumers' attitude towards social media advertising and their behavioural response", Journal of Research in Interactive Marketing, Vol. 9 Iss 4 pp. 299 -312

Henry Boateng Abednego Feehi Okoe , (2015),"Consumers' attitude towards social media advertising and their behavioural response", Journal of Research in Interactive Marketing, Vol. 9 Iss 4 pp. 299 -312

Huang. Jinsong, Song Sub, , Liuning Zhouc, , , Xi Liu .(2013).Attitude Toward the Viral Ad: Expanding Traditional Advertising Models to Interactive Advertising, Journal of Interactive Marketing, Volume 27, Issue 1, February 2013, Pages 36–46

Idris, I., M.S.A. Yajid and A. Khatibi, 2009. Personal and Psychological Factors-Does it .Impact the Choice of Advertising Medium? Journal of Social Sciences, 5(2): 104-111

Ioannou Myria, Nikolaos Boukas, Evi Skoufari .(2014).Examining the role of advertising on the behaviour of co-operative bank consumers, Journal of Co-operative Organization and Management, Volume 2, Issue 1, June 2014, Pages 24-33

Jankowski. Jarosław, Przemysław Kazienko, Jarosław Wątróbski, Anna Lewandowska, Magdalena Ziolo.(2016).Fuzzy multi-objective modeling of effectiveness and user experience in online advertising, Expert Systems with Applications, Volume 65, 15 December 2016, Pages 315-331

Kim, C., Park, S., Kwon, K., & Chang, W. (2012). An empirical test to measure the effectiveness of online advertising in online marketplaces using a hierarchical Bayes model. .Expert Systems with Applications (39), 117–128

Kim, ji-Hern., Hyun, yong j., (2010). " A model to investigate the influence of marketing – mix efforts and corporate image on brand equity in the IT software sector " . industrial . Marketing Management. No 40,2011, pp 424-438, Emerald Group Publishing Limited

Kim, S., & Jun, J. (2016). The impact of event advertising on attitudes and visit intentions. .Journal of Hospitality and Tourism Management, 29, 1-8

Kirby, J. and Marsden .(2006). Connected marketing : the viral, buzz and word of mouth .revolution, Oxford : Butterworth-Heinemann, 2006

Madera. Quetzali, Oscar Castillo ⁎, Mario García-Valdez, Alejandra Mancilla..(2017). A method based on Interactive Evolutionary Computation and fuzzy logic for increasing the effectiveness of advertising campaigns, Information Sciences 414 (2017) 175–186

Marion Garaus and Udo Wagner , Anna-Maria B¨ack.(2017). The Effect of Media Multitasking on Advertising Message Effectiveness, Psychology & Marketing, Vol. 34(2):)138–156 (February 2017

Martín-Santana J.D., E. Reinares-Lara , P. Reinares-Lara .(2017).Influence of radio spokesperson gender and vocal pitch on advertising effectiveness: The role of listener gender, Spanish Journal of Marketing – ESIC, Volume 21, Issue 1, February 2017, Pages 63-71

McKay-Nesbitt, J., V. Manchanda, R., C. Smith, M., & A. Huhmann, B. (2011). Effects of age, need for cognition, and affective intensity on advertising effectiveness. Journal of Business Research (64), 12–17

Moch, Nils (2013). Competition in fragmented markets: New evidence from the German banking industry in the light of the subprime crisis. Journal of Banking &; Finance, Volume 37, Issue 8, Pp 2908–2919

Moorthy, S. and Hawkins, S. (2005). Advertising repetition and quality perception. Journal of .Business Research, 58, 354– 360

Mucan. Burcu, Özeltürkay. Eda Yaşa.(2014).Social Media Creates Competitive Advantages: How Turkish Banks Use This Power? A Content Analysis of Turkish Banks through Their Webpages, Procedia - Social and Behavioral Sciences, Volume 148, 25 August 2014, Pages 137-145

Muda, M.; Musa, R.; Niana-Mohamed, R. and Borhan, H. (2014). Celebrity Entrepreneur .Endorsement and Advertising Effectiveness. Social and Behavioral Sciences, 130, 11-20

Oladele. Kehinde Joseph, Ogunnaike, Akinbode Mosunmola, Aka Deborah .(2016). Effective Advertising: Tool For Achieving Client-Customer Relationships, Researchjournali's Journal of Media Studies Vol. 2 | No. 1 February | 2016

Olsen, G., Pracejus, J. W., & O'Guinn, T. (2012). Print advertising: White space. Journal of .Business Research (65), 855–860

Park, M. (2006). The compensatory effects of pictorial and verbal information for haptic information on consumer responses in non-store shopping environments. Unpublished Ph.D, .Ohio, Ohio state University

Park, T. (2008). Effective design of advertisement in handheld devices (Doctoral dissertation .)PURDUE UNIVERSITY

Park, T. (2008). Effectiveness Design of Advertisement in Handheld Devices. Dissertation for .doctoral degree, Purdue University, West Lafyette India

Pergelova, A., D. Prior and J. Rialp, 2010 Assessing Advertising Efficiency Does the Internet Play a Role? Journal of Advertising, 39(3): 39-54

Pilelienė Lina Viktorija Grigaliūnaitė , (2017)," The effect of female celebrity spokesperson in FMCG advertising: neuromarketing approach ", Journal of Consumer Marketing, Vol. 34 - .Iss 3 pp

Rosemary J. Averya, Matthew D. Eisenbergb, Kosali I. Simon .(2012).The impact of direct-to-consumer television and magazine advertising on antidepressant use, Journal of Health Economics, Volume 31, Issue 5, September 2012, Pages 705–718

Selviaridis. Kostas, Martin Spring b, Luis Araujo b .(2013). Provider involvement in business service definition: A typology, Industrial Marketing Management 42 (2013) 1398–1410

Strebinger .Andreas, Xiaoling Guo, Ferdinand Klauser, Peter Grant-Hayd .(2018).Is Multi-Ethnic Advertising a globally viable strategy for a Western luxury car brand? A mixed-method cross-cultural study, Journal of Business Research 82 (2018) 409–416

Suvittawat, A., (2015). "Commercial Bank Service Quality: Chonburi Province, Thailand". .International Journal of Management Sciences and Business Research. 4(2): 2226-8235

Tafani. Eric, Elyette Rouxa, Rainer Greifeneder .(2018).In the mood for action: When negative program-induced mood improves the behavioral effectiveness of TV commercials, Journal of Business Research 84 (2018) 125–140

Te'eni-Harari, T. (2014). "Clarifying the relationship between involvement variables and .advertising effectiveness among young people". Journal of Consum Policy. 37: 183–203

Thomas, S. and Kohli, S.C. (2011). Can Brand Image Move Upwards after Sideways? A .Strategic Approach to Brand Placements. Business Horizons, 54 (1): 41-49

Yahyaoui . Hamdi, Own .Hala S..(2018). Unsupervised clustering of service performance behaviors, Information Sciences 422 (2018) 558–571

Yuksel, A., Yuksel, F. & Bilim, Y. (2010) "Destination Attachment: Effects on Customer Satisfaction and Cognitive, Affective and Cognitive Loyalty", Journal Of Tourism Management, Vol.31, pp: 274-284.

STUDY OF NATURAL ANTIOXIDANT INHIBITORS FOR CORROSION PROTECTION OF MILD STEEL IN ACIDIC MEDIUM

Ms. Ibtihal Hassan AL-Bahrani, National University of Science and Technology, Sultanate of Oman.

ABSTRACT

Corrosion is a natural phenomena and it is a very big loss for structure as well as business; so corrosion minimization is one of the biggest areas of research. In this project; corrosion inhibition efficiency of berries juice extract was tested in mild steel sample against the sulfuric acid medium which was prepared from four types of berries; blackberries, blueberries, cranberries, and raspberries.

The experiment was designed by using Design Expert V 11 where immersed time, acid and berries juice extract concentrations were set as factors and a limited range was selected. The corrosion rate was set as a response. 20 runs were under taken and the experiments were done in the lab with maximum care.

The corrosion rate of 1020.374 mpy minimized to 597.659 mpy in presence of 5ml of extract and the efficiency of berries juice extract reached to 41.427%. The RSM gave 232.022 mpy corrosion rate as a minimum value and the optimum equation given by RSM was used for Genetic Algorithm analysis which gave 188.866 mpy corrosion rate at the optimizing conditions.

The surface and chemical characterization of the mild steel sample in presence and absence of berries juice extract were analyzed along with the relations between temperature, immersed time, acid and berries juice extract concentrations with response. The mixing tank was designed, calculations and drawing were presented.

KEYWORDS

Corrosion, Mild steel, Berries, Inhibitor, Weight loss, Sulfuric acid, RSM, SEM.

INTRODUCTION

Mild steel is used in most of our daily life applications as well as in industrial applications. In other hands, mild steel has low corrosion resistance which corrodes easily once it reacts with the surrounding environment; corrosion is a phenomenon where the mild steel reacts with the deteriorates and surrounding environment. Corrosion of mild steel is a continuous and constant problem where it severe financial and life loses as it leads to loss in the product, efficiency of equipment and failures in some cases (Roberge,2012).

The corrosion phenomenon controlled by adding a small amount of component to the corrosive environment which is a corrosion inhibitor. There are organic and inorganic inhibitors. The inorganic inhibitors are used in limit amount as they are toxic such as nitrites, molybate, nitrate, silicate, chromate, and phosphate (Cafferty,2010). Now a day, the corrosion engineers looking for cheap, renewable, friendly, readily available and ecologically acceptable inhibitors which are organic inhibitors which comes from the plants such as antioxidant components consisting of tannins, saponins, flavonoids, alkaloids and amino acids. Organic inhibitors consist of some heteroatoms such as oxygen, nitrogen, multi-bound, and sulphur atoms with the conjugated system which allows the absorption in the metal surface. Those atoms attract the absorption of some components on the metal surface and that done by the interaction of metals and by the electron of the conjugated system or by forming a bond with the free electron of heteroatom which forms a layer in the metal surface, as a result, it decreases the corrosion rate (Roberge,2012).

Berries are small pulpy fruits which have up to 9.2 mmol of antioxidant per 100g of berries. Berries can be used as a corrosion inhibitor as they are rich in antioxidant. The studies show that the berries juice extract of blueberries, cranberries, raspberries, and blackberries rich with antioxidant where the antioxidant capacity of them reach to 10 mg/ 100 g of blueberries (Michalska&Lysiak,2015), 58 mg/100g of cranberries (Blumberg,J.,et al,2013), 26.2mg/ 100 g of raspberries (Bobinaite,R.,Viskelis&Venskutonis,2015), and 21 mg/100 g of blackberries (Kaume,L.,et al,2012).

OBJECTIVES

To prepare anticorrosive extract (berries), to study corrosion in sample in presences and absences of the inhibitors, to determine the optimum conditions to inhibit corrosion and to design mixing tank.

METHADOLOGY, DESIGN AND EXPERIMENTATION

PREPARATION OF MILD STEEL SAMPLES

The mild steel sheets were cut into 2.5*3 cm^2 by using steel cutter and cleaned by surface cleaning machine to remove any corroded products in the metal surface then 4mm hole done by using drilling machine. Bunches for all metal samples (1-20) were made to differentiate between them. The drilled samples were washed by distilled water then rinsed with benzene and acetone respectively and dried with a hot air dryer. They were weighted using sensitive balance and kept in the desiccator until the time of usage.

PREPARATION OF CORROSIVE MEDIUM

The corrosive medium of different concentrations of H_2SO_4 prepared by using 36.8 N of the sulfuric acid.

PREPARATION OF BERRIES JUICE EXTRACT

137g, 209g, 282g, and 200g of raspberries, cranberries, blueberries, and blackberries respectively have been used to produce 309.5 ml of berries juice extract. The different types of berries washed by cold water, crashed in the electrical crasher and compressed mechanically by hand then filtered by cloth filter. The filtrate filtered again by using filter paper. The berries juice extract kept in the freezer at 0°C till the time of usage.

The efficiency of berries juice extract has been tested by running three experiments. The cleaned and weighted mild steel samples immersed in 0ml(blank), 2ml and 5ml of berries juice extract respectively with 50 ml of 1M H_2SO_4. After 2 hours; the samples have been taken out and cleaned by distilled water, benzene and acetone then dried by air dryer and weighted again. The corrosion rate and inhibitor concentration have been calculated by using equations (1) and (2) respectively.

$$Cr(mpy) = 22300\times\frac{\Delta W}{\rho\times A\times t} \quad \textbf{equation(1)}$$

$$\mathrm{IE}(\%) = \frac{\text{initalcorrosionrate-finalcorrosionrate}}{\text{initalcorrosionrate}}\times 100 \quad \textbf{equation(2)}$$

DESIGN THE EXPERIMENT CONDITIONS

The experiments conditions have been designed by using Design Expert software-version 11 and following Response Surface Module(RSM); the range of the parameters have been given as inputs to the software which are inhibitor concentration (5-15 ml), immersed time (180-240 min) and acid concentration (1-2 M) and the response set as corrosion rate in mpy. Table(1) shows the experiment conditions given by the software.

Table(1): Designed runs given by RSM.

Run.No	Factor(A): inhibitor, ml	Factor(B): Time, min	Factor(C): Acid concentration, M
1	10	180	1
2	15	240	2
3	5	120	1
4	5	240	1
5	10	180	1.5
6	15	180	2
7	10	120	1.5
8	5	120	1.5
9	10	180	2
10	10	180	1.5
11	10	180	1.5
12	15	240	1
13	15	120	1
14	5	240	2
15	10	180	1.5
16	15	180	1.5
17	5	120	2
18	10	180	1.5
19	10	180	1.5
20	10	240	1.5

APPLYING THE DESIGNED RUNS IN THE LABORATORY

The weighted mild steel sample has been hanged by using a plastic thread in a 50 ml of sulfuric acid in presence of berries juice extract of known concentration and contact time as per the software; the batch has been kept in the shaker. After the contact time the mild steel sample cleaned by using 120 grade emery paper to remove any corroded products, washed by distilled water, rinsed by benzene and acetone respectively and dried by hot air dryer, weighted by using sensitive balance. The corrosion rate in mpy was calculated by using equation(1) where the weight losses results have been used.

STUDYING THE IMPACT OF TEMPERATURE IN CORROSION RATE

To study the impact of temperature in the corrosion rate; mild steel samples weighted and hanged for 2 hours in 50 ml of 1M sulfuric acid with 10ml of berries extract at different temperatures(30, 40, 50 and 60 °C) as shown in Table(2); each batch has been kept in the water bath shaker with heater to provide perfect contact between the solution and mild steel sample and to maintain the required temperature. Corrosion rate has been calculated by equation(1).

Table(2): Experiment conditions of temperature into corrosion rate.

SI.NO	Temperature (°C)	Inhibitor concentration (ml)	Acid concentration (M)	immersed time, (hours)
1	30	10	1	2
2	40	10	1	2
3	50	10	1	2
4	60	10	1	2

SURFACE AND CHEMICAL CHARACTERIZATION ANALYSIS

The mild steel surface of blank, absence and presence of berries juice extract samples have been studied by using Scanning Electron Microscopy (SEM) and Energy Dispersive Spectroscopy (EDS).

OPTIMIZATION STUDY BY GENETIC ALGORITHM

Matlab has been used for optimization where the genetic algorithm method has been followed. The actual equation form Design expert(RSM) has been used where "Optimtool" followed. The same limit of the conditions from RSM has been selected. The same previous procedure has been followed in the lab with maximum care by using those optimum conditions which given by the software, 1.001M of H_2SO_4 with 15 ml of berries juice extract immersed for 237.208 minutes.

DESIGN OF MIXING TANK

Design of mixing tank is a part of the project where the industrial scale mixing tank has been designed, the tank will be formed from high resistance material such as S.S to be able to resist the corrosive medium of sulfuric acid which will be mixed with berries juice extract. It is a container which used to mix various material (heterogeneous materials) together to form homogenous materials even the different viscosity materials can be mixed easily which provides continuous process, good control, easy cleaning, and simplest construction. It can be made from S.S material which provides a smooth surface and it can be easily cleaned (Petropedia,2018).

Parameters to be considered to design mixing tank are length, torque, weight, stiffness and hydraulic loading.(Dynamixianc,2015). The important factors in design mixing tank are tank type, tank volume, specific gravity, process and environment.(Ibryamoval,Y.et al,2009)

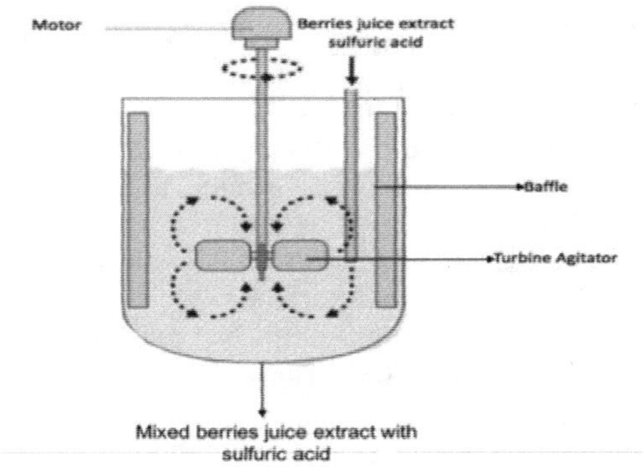

Figure(1): Mixing Tank(*Mixing Tank,*2018)

From the graph in Figure(2) Turbine agitator has been selected as it is suitable with the viscosity of the fluid and the tank volume. (Sinnot&Towler,2009)

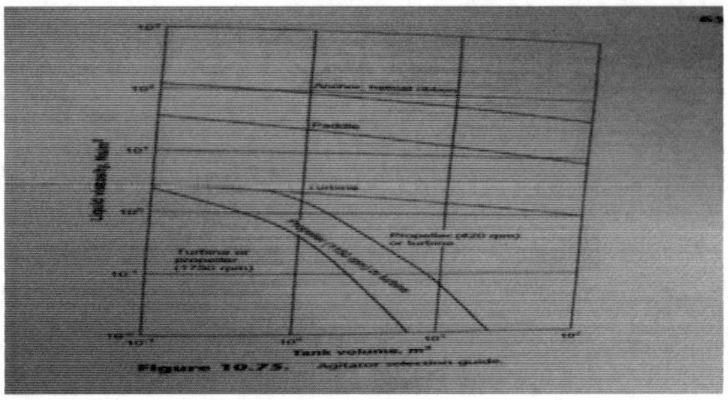

Figure(2): Graph of liquid viscosity Vs tank volume. (Sinnot&Towler,2009)

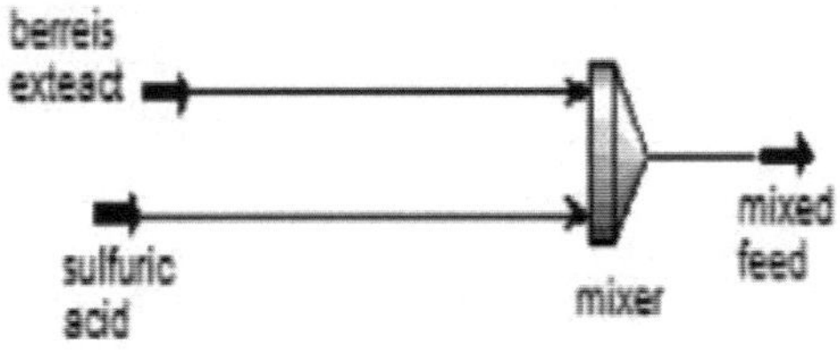

Figure(3): Result of Mixing Tank by Aspen Hysys.

Table(3): Calculated parameters for designing mixing tank

Tank diameter	m	3.5
Tank Height	m	1.81
Across sectional area	m^2	9.62
Volume of the tank	m^3	17.5
Impeller diameter	m	1.16
Length of paddle	m	0.29
Width of paddle	m	0.23
power	watt	204085
Reynolds Number	-	136486.6391

RESULTS AND ANALYSIS

EFFECT OF BERRIES JUICE EXTRACT CONCENTRATION INTO CORROSION RATE AND INHIBITOR EFFICIENCY

As shown in Table(4) the corrosion rate in the absence of berries juice extract inhibitor reach to the maximum (1020.374mpy); the corrosion rate has been minimized to 898.117mpy and 597.659mpy in presence of 2 and 5 ml of berries juice extract inhibitor respectively which means that berries juice extract inhibits the corrosion as the concentration of the berries juice extract increases. As well as the inhibitor efficiency reaches to 41.427% which obtained at 5ml of berries juice extract inhibitor in 1M of sulfuric acid immersed for 2 hours. The adsorption of the organic compounds such as ascorbic acid that present in the berries juice extract is blocking the active sites on the mild steel samples as it is forming a protective film on the metal surface as what kumar.A,2012 mention in his studies so the corrosion rate keeps decreasing as the amount of the berries juice extract is increasing. (Kumar.A,2012) Those compounds have heteroatoms such as oxygen which is a center of adsorption. In addition; It forms a coordinate bond as it has the ability to transfer the electrons to the vacant d-orbital of mild steel from the heteroatoms. (Ostovari, L. et al,2009)

Table(4): Experimental results of testing the efficiency of berries juice extract.

SI.NO	Inhibitor concentration (ml)	Corrosion rate (mpy)	Inhibitor efficiency (%)
1	0	1020.374	-
2	2	898.117	11.976
3	5	597.659	41.427

EFFECT OF IMMERSED TIME INTO CORROSION RATE AND INHIBITOR EFFICIENCY

The maximum corrosion rate reaches to 765.005mpy which obtained at 5ml of berries juice extract in the presence of 50 ml of 2 M of H_2SO_4 at 2 hours. Since the minimum corrosion rate of 232.022mpy obtained in presence of 5ml of berries juice extract with 50 ml of 1 M of H_2SO_4 immersed for 4 hours. The effect of immersed time into corrosion rate can be observed at low acid concentration and in the presence of berries juice extract, the corrosion rate is minimized at high immersed time. The decrease in the corrosion rate with the increase in immersed time is related to the adsorption of the corrosion product ($FeSO_4$) as shown in equation(3); According to Xianghong,L. et al ; the corrosion product of $FeSO_4$ becomes thicker with the immersed time, so it form a protective film (barrier) in the surface of mild steel (Xianghong, L. et al,2009).

$$Fe_{(s)} + H_2SO_{4(aq)} \rightarrow FeSO_{4(aq)} + H_{2(g)}$$, **equation(3)**

EFFECT OF ACID CONCENTRATION INTO CORROSION RATE

The effect of acid concentration into corrosion rate has been studied by using 1, 1.5 and 2 M of H_2SO_4 at different inhibitor concentrations and immersed time where the corrosion rate increase with the acid concentration at low inhibitor concentration and immersed time which reach to 765.005mpy at 2M H_2SO_4, 5ml of berries juice extract which immersed for 120 minutes as a maximum corrosion rate as shown in Table (3). In presence of the corrosive medium of H_2SO_4 with less amount of berries juice extract; the mild steel surface was exposed to H_2SO_4 medium where the corrosion reaction started as shown in equation(3) and $FeSO_4$ was formed with H_2 bubbles.

EFFECT OF TEMPERATURE INTO CORROSION RATE

From Table(5) can be noted that the rate of corrosion increases with temperature where it reaches to maximum 2407.994mpy at maximum temperature of 60°C. According to Ambrish, S. et al, at high temperature the berries juice extract molecules were partially absorbed in the mild steel surface so it was exposed to the corrosive medium of H_2SO_4, as a result, increasing the corrosion rate (Ambrish,S. et al,2012).

Table(5): Results of studying the effect of temperature into corrosion rate

SI.NO	Temperature,° C	Corrosion rate, mpy
1	30	701.507
2	40	1666.404
3	50	2008.861
4	60	2407.994

ANALYSIS BY RSM

Once the experimental results and corrosion rate calculated the RSM analysis has been done where the corrosion rate has been given to the software as a response. The response must be modified where the results fit with linear. The fit summary shows that the model F-value and P-value of 12.66 and 0.0001 respectively imply that the model is significant where A, B, C, and B^2 are the significant terms in this model only. 1.48 is a lack of fit P-value which indicates that the fit is not significant as there are pure errors.

0.5562 of the predicted R^2 is a reasonable agreement with 0.7105 of the adjusted R^2. The signal to noise ratio of 13.7252 of Adeq precision that the ratio grater than 4 is desirable. (Mycourses,2016)

The process variables are studied using ANOVA. The estimate coefficients in term of coded factors are A-Inhibitor, B-time, C-Acid and B^2. It is creating an accompanying condition for the sample with 1.0000 correlation coefficient. The coded and actual equations given by software are shown in equation(4) and (5) respectively.

- Corrosion rate(mpy)= 370.52-40.52(A)-126.97(B)+ 80.85(C) +66.58$(B)^2$ **equation(4)**
- Corrosion rate(mpy)= 1189.15487-8.10479(Inhibitor) -8.77421(Time) +161.69302 (Acid Conc)+0.018495 $(Time)^2$ **equation(5)**

The various relations can be studied from one factor graph; it can be noted that the corrosion rate increases with acid concentration and decreases with the increase in the amount of berries juice extract and immersed time as can be seen from the plots in Figure(3):Plot pf corrsoion rate Vs inhibitor concntration, Figure(4): plot of corrosion rate Vs acid concentration and Figure(5):Plot of corrosion rate Vs immersed time respectively.

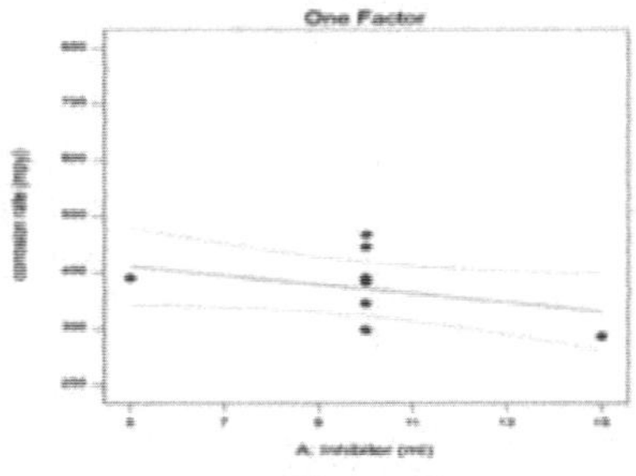

Figure(3)

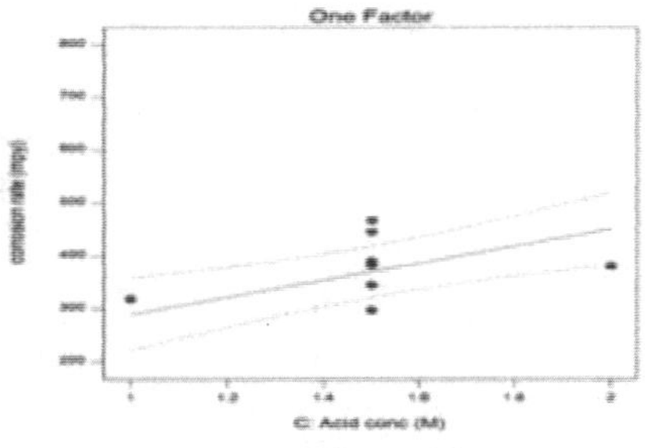

Figure(4)

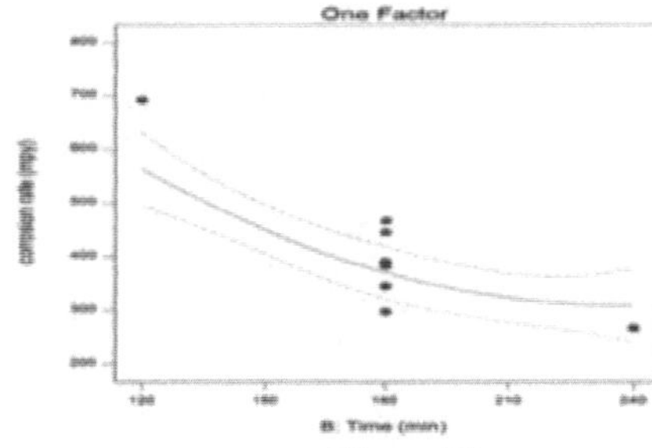

Figure(5)

The effects of two factors into the corrosion rate can be studied from the 3D plots as shown in Figure(6): 3D plot of corrosion rate Vs inhibitor concntration and acid concntration and Figure(7): 3D plot of corrosion rate Vs inhbitor concntration and immersed time.

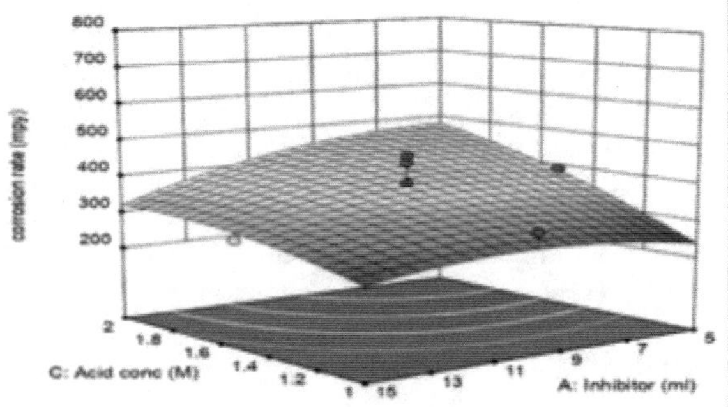

Figure(6)

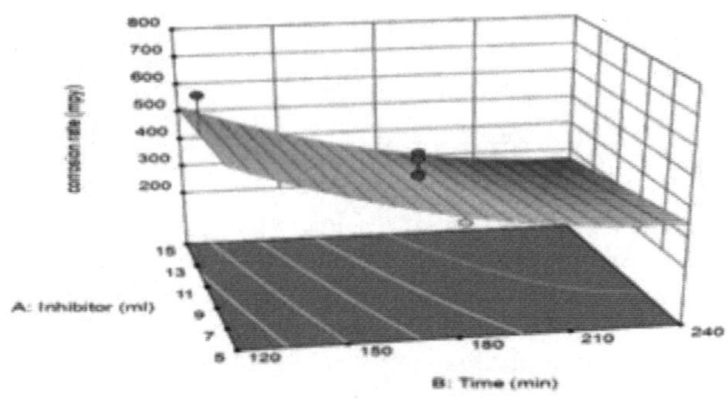

Figure(7)

The optimum conditions given by the RSM of 1M of H_2SO_4 with 5ml of berries juice extract inhibitor runs for 240 minutes gives 213.402mpy as a minimum corrosion rate but the experimental results at the same conditions gives 232.022mpy which means that they are too close as the suggested solution considered as an excellent optimum conditions.

SEM ANALYSIS

The SEM has been done for the blank mild steel sample for facilitating comparison as shown in Figure(8) which shows smooth mild steel surface compare to others. Rough surface in absence of berries juice extract is shown in Figure(9); where corrosion starting with the formation of H_2 bubbles as per the equation(3) which means that the surface of mild steel is exposed to the H_2SO_4 medium. In presence of 15 ml of berries juice extract, the surface of mild steel is protected by the adsorption of the organic compounds that are present in the berries juice extract where fewer cracks and pits can be noted and smooth surface is shown in Figure (10), which means that the mild steel surface is protected. (Raja, P. et al, 2008)

Figure(8)

Figure(9)

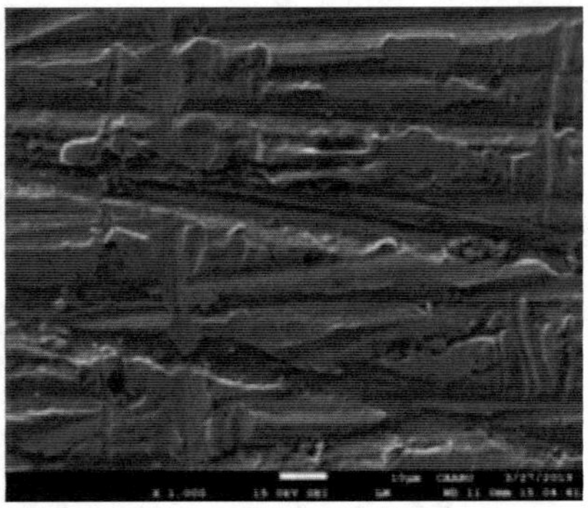

Figure(10)

EDS ANALYSIS

EDS has been used to study the chemical characterization or elemental analysis; Figure (11) shows the blank mild steel sample before being immersed in 1M H_2SO_4 in absence of berries juice extract where high Fe peak is shown, which indicates the presence of high amount of iron which reaches to 88 Wt% in the mild steel surface. Figure (12) shows the mild steel surface that has been immersed for 2 hours in 50 ml of 1M H_2SO_4 in absence of berries juice extract; the mild steel surface contains low amount of Fe (87.3 Wt%) compared to the blank mild steel sample. In Figure (13) the mild steel sample immersed for 2 hours in 50 ml of 1M H_2SO_4 in the presence of 5ml of berries juice extract, it shows high spark of Fe where the mild steel surface contains lower amount of Fe (84.8 Wt%) compared to Figure(11) and (12) which indicate lower iron dissolution rate in 1M H_2SO_4 medium containing berries juice extract (Alaneme, K. et al,2016). A high amount of oxygen and carbon are observed in presence of berries juice extract compared to the absence of berries juice extract. Si is shown in the absence and presence of berries juice extract which means that's it comes out to the surface during the reaction with sulfuric acid.

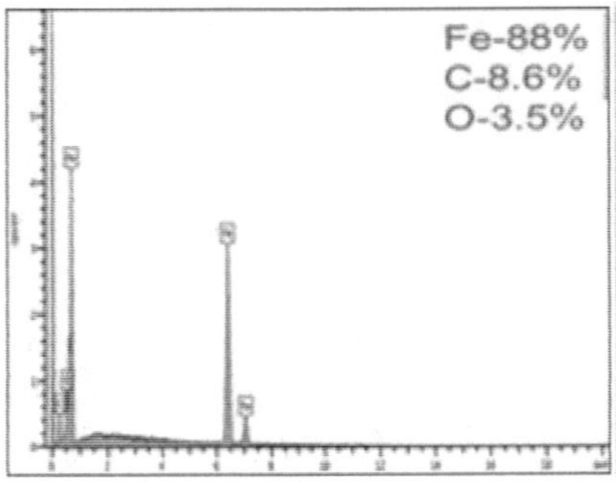

Figure(11)

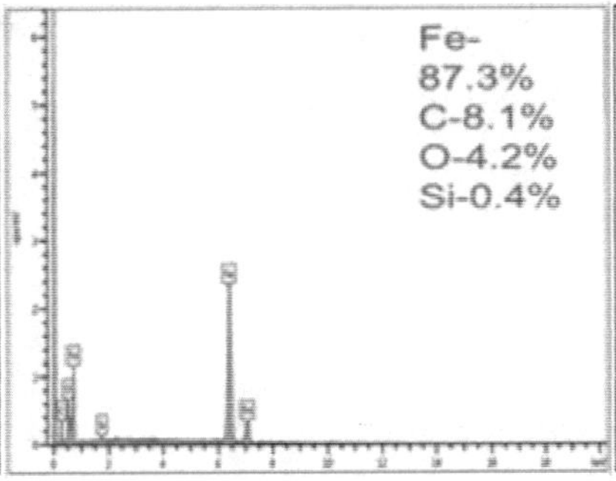

Figure(12)

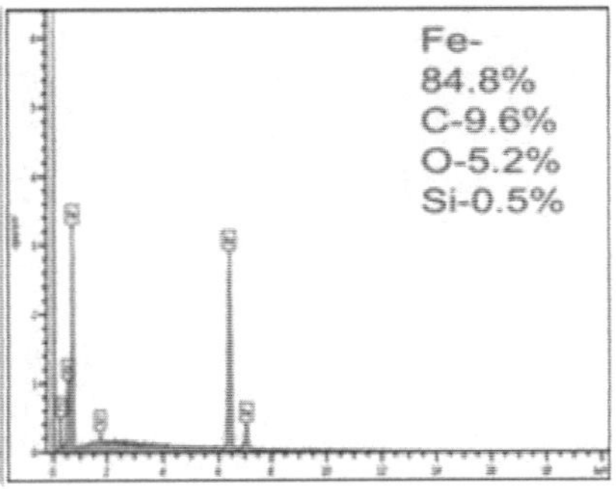

Figure(13)

GENETIC ALGORITHM OPTIMIZATION

The experimental result gives 288.367mpy as a corrosion rate. The software gives the minimum corrosion rate value of 188.866mpy from 50 runs as shown in Figure(14). A small difference can be noted as the genetic algorithm is not the exact fit with linear equations so the experimental and software results are too close to each other.

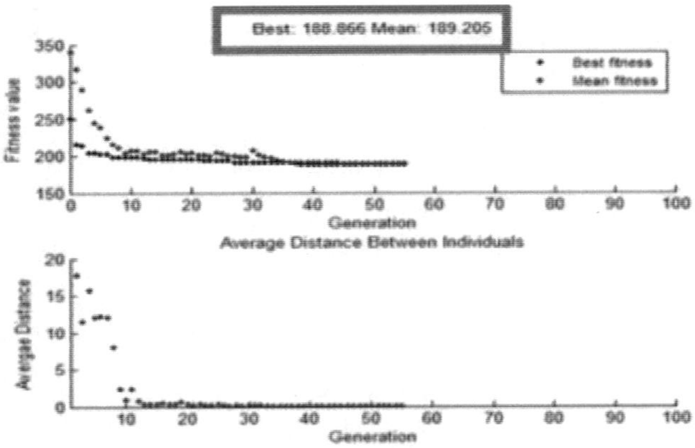

Figure(14): Genetic Algorithm optimization.

CONCLUSION

In this project, all the objectives have been achieved by designing and experimenting.

Firstly; the anticorrosive extract (berries) has been prepared, the prepared extract which contains 4 types of berries namely blueberries, blackberries, cranberries, and raspberries inhibit the mild steel corrosion in sulfuric acid medium with 41.427% efficiency in presence of 5 ml of berries extract.

Secondly, the mild steel corrosion in the presence and absence of berries juice extract were studied where the corrosion rate reaches to 1020.374mpy in absence of berries juice. By using 5 ml of berries juice extract the corrosion rate minimized to 232.022mpy.

Thirdly, the optimum conditions to inhibit the mild steel corrosion were studied by Design Expert V11. At the optimum conditions of 1M of H_2SO_4. 5ml of berries juice extract, 240 minutes and at room temperature the corrosion rate reaches to 232.022mpy as a minimum value.

The actual equation obtained from Design Expert V11 was later optimized using Genetic Algorithm. Genetic Algorithm shows that at 1.001M of H_2SO_4, 15 ml of berries juice extract and 237.208 minutes at room temperature, the corrosion rate is minimized to 188.866mpy.

Finally, mixing tank has been designed with 1.18m height and 2.5m diameter and a turbine agitator has been selected.

ACKNOWLEDGEMENT

I would like to thank god for giving me patience and power to work in my most important part of my life, my father, mother, my brother and my friends for their help and support. I would also like to express my sincere gratitude to Dr. Priy Brat Dwivedi for his support and help in developing my project idea, Mr. M.J.Varghese for his help in RSM analysis, Mr. Hatem Yazidi and Dr. Moammar Ali AlTubi for providing me the technical knowledge in Matlab optimization, Mr.Mohammed Salim Adeeb in providing me the technical help in the laboratory, Ganesh Nagraj Patil, Dr.Lakkim Setty

Rao, and Dr.Santosh Madhav Walke for their help in the calculations of design mixing tank. Last but not the least, I would like to thank National University of Science and Technology-College of Engineering for giving me this opportunity to complete the project in their laboratories and supporting me by providing me with the required chemicals.

REFERENCES

1.Alaneme,K. et al .,2016.Corrosion Inhibitory Properties of Elephant Grass(Pennisetum Purpureum) Extract Effect on Mild Steel Corrosion in 1M HCl Solution :*Alexandria Engineering Journal*.[e-journal]. 55 (8). p. 1069-1076. Available from: https://www.sciencedirect.com/science/article/pii/S1110016816300242.[Accessed: 8th May 2019].

2.Ambrish,S. et al .,2012. The Effect of Environmentally Bening Fruit Extract of Shahjan(Moringa Olifera) on the Corrosion of Mild Steel in Hydrochloric Acid Solution:*Journal of Chemical Engineering Communications*.[e-journal]. 199 (6). p. 63-77. Available from:https://www.tandfonline.com/doi/abs/10.1080/00986445.2011.570390.[Accessed: 6th May 2019].

3.Blumberg,J.,et al.,2013.Carnberries and Their Bioactive Constituents in Human Health:*Journal of Advances in Nutrition*.[e-journal]. 12 (7). p. 618-632. Available from: https://academic.oup.com/advances/article/4/6/618/4595566.[Accessed: 1th May 2019].

4.Bobinaite,R.,Viskelis,P.&Venskutonis,P.,2015.Chemical Composition of Raspberry Cultivars:*International Journal of Molecular Sciences*.[e-journal]. 16 (8). p. 18642-18663. Available from: https://www.researchgate.net/publication/306158828_Chemical_Composition_of_Raspberry_Rubus_spp_Cultivars.[Accessed: 1th May 2019].

5.Cafferty,E.,2010.*Introduction to Corrosion Science*.United States:Springer.
Roberge,P.,2012.*Handbook of Corrosion Engineering*.2nd edition.United States:McGraw-Hill company.
6.Dynamixianc,2015.*Optimal Mixer Design* .[Online]. Available from: https://www.dynamixinc.com/optimal-mixer-design. [Accessed: 14th December 2018]

7.Ibryamoval,Y. et al .,2009.Design of Mixers for Chemical and Biochemical Industry: *Trakia Journal of Sciences*.[e-journal]. 8 (6). p. 6-11. Available from: http://tru.uni-sz.bg/tsj/Vol8.Suppl.3.2010/Y.Ibryamova.pdf.[Accessed: 13th December 2018].

8.Kaume, L. et al.,2012. The Blackberry Fruit A Review on Its Composition and Chemistry, Metabolism and Bioavailability and Health Benefits:*Journal of Agricultural and Food Chemistry*.[e-journal]. 37 (7). p. 5716-5727. Available from: https://www.ncbi.nlm.nih.gov/pubmed/22082199.[Accessed: 6th May 2019].

9.Kumar.A,2012. Inhibition of Mild Steel Corrosion in HCl Solution Using Pipali(Piper Longum)Fruit Extract:*Arabian Journal for Science and Engineering*.[e-journal]. 38 (9). p. 85-97. Available from: https://link.springer.com/article/10.1007/s13369-012-0409-9.[Accessed: 9th May 2019].

10. Michalska,A.& Lysiak.G .,2015.Bioactive Compounds of Bluberries:*International Journal of Molecular Sciences*.[e-journal]. 16 (8). p. 18642-18663. Available from: https://www.ncbi.nlm.nih.gov/pmc/articles/PMC4581264/.[Accessed: 1th May 2019].

11.*Mixing Tank*,2018.[Online]. Available from: https://www.conceptdraw.com/examples/agitator-tank-symbol.[Accessed: 19th November 2018].

12. Mycourses,2016. *Response Surface Modelling-RSM*.[Online]. Available from:https://mycourses.aalto.fi/pluginfile.php/364340/mod_resource/content/1/BOS_Lecture3_2016.[Accessed: 4th May 2019].

13. Ostovari, L. et al.,2009.Corrosion Inhibition of Mild Steel in 1M HCl Soluition by henna extract:*Journal of Corrosion Science*.[e-journal]. 32 (9). p. 1935-1949. Available from: https://www.sciencedirect.com/science/article/pii/S0010938X09002157. [Accessed: 6th May 2019].

14. Petropedia,2018. *Mixing Tank*.[Online]. Available from: https://www.petropedia.com/definition/7742/mixing-tank.[Accessed: 11th December 2018].

15.Raja,P. et al.,2008.Inhibitive Effect of Black Pepper Extract on the Sulphuric Acid Corrosion of Mild Steel:*Journal of Materials Letters*.[e-journal]. 62 (9). p. 1977-2979. Available from: https://www.sciencedirect.com/science/article/abs/pii/S0167577X08001316.[Accessed: 8th May 2019].

16. Sinnot,R.&Towler,G.,2009.*Chemical Engineering Design*.5th edition.India:Elsevier.

17. Xianghong,L. et al.,2009.Inhibition Effect of 6-benzylaminopurine on the Corrosion of Cold Rolled Ssteel in H_2SO_4 Soluition:*Journal of Corrosion Science*.[e-journal]. 51 (9). p. 620-634. Available from: https://www.sciencedirect.com/science/article/pii/S0010938X09000080. [Accessed: 6th May 2019].

590434 Verification of Firefighters' Heuristics and Predictive Model Development through Big Data Analysis

Abstract:

The heuristics accumulated in the field activities of firefighters are reviewed through big data analysis of fire occurrences and emergency responses in Gyeonggi-do Province, Korea and expanded as predictive models to be utilized for proper fire prevention activities according to time, day, and target through quantitative modeling. Empirical rules with high sympathy are collected through direct interviews with firefighters. Among these, three types of high-priority rules are compared and analyzed for big data such as fire occurrence and damage history in Gyeonggi-do in 2018. Big data comparison analyses are conducted, including the number of fires and damages that occurred in Gyeonggi-do in 2018. Furthermore, fire occurrence patterns by region, day of the week, time of day, and building type are derived. In addition to using empirical rules that have been validated through research, relatively inexperienced firefighters also can make better decisions by relying on refined quantitative predictive models replacing empirical rules, including local government and time-based factors that reflect big fire occurrence data.

CONVERTING SHELTER-IN-PLACE LOCATIONS TO ZERO VULNERABILITY SAFE HAVENS

J. A. Rau
MineARC Systems America, LLC
4850 W. Ledbetter Drive
Dallas, Texas 75236
USA

A. Young
MineARC Systems, Pty Ltd
274 Welshpool Rd
Welshpool, WA 6106
Australia

M. McDermott
High Performance Building Solutions
7415 Folger Dr.
Charlotte, NC 28226
USA

CONVERTING SHELTER-IN-PLACE LOCATIONS TO ZERO VULNERABILITY SAFE HAVENS

ABSTRACT

While it is not uncommon to find shelter-in-place or safe haven locations designated across chemical facilities, what varies is the level of protection offered at these sites. Functionality can range from merely a muster point (shelter-in-place) to a positively pressurized room (safe haven). Either way, when determining the room's acceptable performance criteria, the industry generally applies ERPG-3[1] or a vulnerability factor. An acceptable vulnerability factor is often deemed the probability of a fatality being less than one percent.

In 2017, MineARC Systems and High Performance Building Solutions (HPBS) began working with a leading energy company to convert five shelter-in-place spaces to safe havens, which included additional life support systems that would ensure zero percent vulnerability. The objective of this paper is to examine the process to harden the space against contaminated air intrusion and what critical life support systems are essential for converting a shelter-in-place location to a safe haven.

INTRODUCTION

Emergency shelter-in-place (SIP) forms an integral part of any plant's broader emergency response plan. Explosions, fires, and the release of smoke or other toxic gases are some of the types of incidents that can occur despite high levels of planning and safety precautions[2].

Numerous government agencies and other organizations recommend shelter-in-place as a measure to reduce harm in the event of a chemical release. Any building may be used as a temporary measure to reduce health risk from exposure to the toxic materials by simply closing windows and doors and turning off ventilation fans and air conditioning/heating systems.

Three factors govern the effectiveness of a SIP location: 1) the tightness of the building or leakage rate, 2) the concentration of the toxic gas, and 3) the release duration. The tightness of the building is primarily based on building construction; however, temperature differential and wind speed also play a factor. Identification of the maximum toxic concentration occurs during a Facility Siting Study (FSS). FSSs are a hazard analysis that defines Maximum Credible Event (MCE) scenarios using a consequence-based approach to produce a quantitative risk assessment. Determining potential consequences is a necessary step in the process of developing a comprehensive safety plan. The FSS determines the expected release duration based on stored inventories of toxic materials. The higher the toxic concentration and longer the release duration, the less effective sheltering in place becomes.

While occupants may feel safe sheltering-in-place from the external toxic hazard, typical buildings have high leakage rates. A typical wood framed constructed building can have 3-5 air changes per hour (ACH50[3]), rendering it ineffective against most chemical releases. Additionally, sealing to an acceptable airtightness can be time consuming and expensive and even with professional air sealing it is very difficult to air seal existing structures to an acceptable tightness.

A case study for the conversion of five shelter-in-place spaces to zero percent vulnerability safe havens is presented. There are considerations for effectively sealing existing infrastructure and ensuring positive pressure to ensure contaminants do not infiltrate safe haven locations along with the critical element of life support. The critical elements of life support that are of primary importance to human survival in a safe haven are:

- Positive pressure
- Breathable air supply
- Supplemental O_2
- Removal of accumulated CO_2

[1] Emergency Response Planning Guidelines (ERPGs) are values developed by the AIHA Guideline Foundation's Emergency Response Planning Committee. ERPG-3 is the maximum airborne concentration below which it is believed nearly all individuals could be exposed for up to 1 hour without experiencing or developing life-threatening health effects.

2 The CSB is an independent federal agency charged with investigating industrial chemical accidents. Over the last 10 years they have completed investigations into 15 separate chemical release incidents.

[3] ACH50 is the number of times the air volume in a building changes per hour at 50 Pascal of pressure. Commercial air leakage rates are commonly reported at 75 Pascals, but this leakage study was conducted at 50 Pascals based on the targets included in the FSS report.

- Cooling and dehumidifying

CASE STUDY

In 2017 MineARC Systems and its project partner HPBS began working alongside a major energy company to convert five shelter-in-place locations into zero percent vulnerability safe havens.

The process started with the energy company completing its FSS, which included plume modeling to define the chemical release concentrations at each building location along with ideal shelter-in-place locations[4] and internal minimum positive pressure[5]. This information was used by MineARC to calculate indoor concentrations and occupant vulnerability to highlight that shelter-in-place alone would result in the vulnerability of greater than zero percent.

From there, HPBS conducted air infiltration testing following ASTM E779-10 to determine an air leakage baseline (see Table 6) for rooms that were designated to be sealed (pre-weatherization). A site assessment was then completed to review each location and determine its suitability for sealing and conversion to a safe haven. Firstly, the measured baseline infiltration rates were compared to the FEMA 453 estimates based on construction type[6] (see Figure 2). All surfaces of the room were examined for ease of conventional air sealing and mechanical equipment sealing. Rooms that included false floors, large amounts of electrical conduit, cabinetry, computer servers or windows etc. were deemed difficult to seal (i.e. control rooms). Secondly, the building or rooms open floor space was checked to ensure it could accommodate the specified number of people in an emergency. FEMA 453 section 1.6 recommends 20ft^2 per person for durations of less than 24 hours. With the FSS recommending 4 hours, it was deemed acceptable to have an unrestricted floor space of 6ft^2 per person.

Manual weatherization (air sealing) of the rooms included installing a cleanroom suspended ceiling system with gaskets and tiles that met the air permeance rate for an air barrier material, cleanroom airtight light fixtures with gaskets, sealing around electrical and plumbing, air sealing all visible air pathways to make sure the interior gypsum board was functioning as the air barrier, installing low-leakage HVAC dampers, installing an emergency shut off for the mechanical system and adding door weather-stripping and sweeps. Post-weatherization air leakage tests were conducted to determine the air infiltration rate of the SIP spaces after completion of the manual air sealing, cleanroom suspended ceiling system and HVAC work. The weatherization air sealing methods improved the air barrier integrity for all the proposed SIP spaces by between 57% and 65% (refer to Table 6).

Aerosol-based envelope sealing was used to reduce leakage rates even further. AeroBarrier® is a building envelope sealing technology that simultaneously measures and seals buildings. AeroBarrier® works by forcing an atomized sealant into orifices under pressure gradually sealing leaks. Rooms were sealed during night shift with personnel able to re-access the rooms by morning. The AeroBarrier® air sealing methods improved the air leakage for all the SIP spaces between 7% and 88% over the post manual weatherization leakage rates (refer to Table 7). AeroBarrier® is not effective for sealing large holes so manual air sealing is required for certain locations prior to using AeroBarrier®.

MineARC engineers once again calculated infiltration concentration rates along with the vulnerability rate. As the probability of a fatality was still higher than zero, MineARC then specified its AirBANK Pressure System and necessary carbon dioxide and oxygen control equipment.

SAFE HAVEN LIFE SUPPORT EQUIPMENT

Primary Life Supporting Atmosphere

A safe haven needs to provide a reliable breathable atmosphere independent of external toxic gas concentrations. Human breathing consumes oxygen (O_2) and expels carbon dioxide (CO_2), carbon monoxide (CO), water vapor, and heat. At complete rest, an average person exhales air at 17% O_2 and 3.2% CO_2. This breathing rate results in a 'resting' O_2 uptake of 0.3L/min of O_2 and a resting CO_2 discharge of 0.24L/min (Brake, Fellows, & Bates, 1999). Taking into consideration the above values, a person in a closed system such as a shelter-in-place succumbs to CO_2 poisoning well before O_2

4 Shelter-in-place location is beyond the scope of this paper, but should be determined based on the chemical hazard, facility layout, and location of personnel.

5 Positive pressure requirements are based on average wind speed at the facility. As wind speed increases so does the SIP leakage rate.

6 The FEMA 453 estimates are intended to estimate filter sizing when ASTM E779 air leakage testing is not conducted.

asphyxiation. Venter et al. (1998) found an average oxygen consumption for 12 entrapped individuals of 0.44 liters per minute, at STP, over 24 hours (including sleeping). A figure of 0.5 liters per minute over a non-sleeping entrapment period of 12 hours is a prudent, conservative design value. This corresponds to a metabolic rate of about 160W or 80W/m^2 for a typical person with a body skin area of about 2m^2, which is equivalent to a breathing rate of about 12.5 liters per minute of fresh air.

Consequently, life support systems are needed within a safe haven that can ensure positive pressure keeps contaminants outside, along with replacing the O_2 consumed, and removing exhaled CO_2, moisture, and heat to maintain the atmosphere within acceptable threshold limits.

Positive Pressure

Positive pressure is integral to prevent the ingress of airborne contaminants that may compromise the breathable atmosphere for occupants. Establishing a Class 1 Safe Room (FEMA 453) requires the sufficient release of filtered air from an external source, to form a positively pressurized zone.

Buildings are typically constructed to provide natural or forced ventilation to allow permanent occupancy and potentially combined with large exposed windows and surfaces which can present challenges for ensuring protection during a chemical release. When determining adequate positive pressure within a space, several structural factors need to be considered, including:

- Shelter volume
- HVAC (Heating, Ventilating, and Air Conditioning)
- Door and window tightness
- Existing opening such as vents and elevator shafts
- Indoor and outdoor temperature
- Wind speed and building orientation
- The external screening of the structure

Building design and multiple ingress points suggest the conversion of a smaller interior room within a building can provide a higher level of protection, than the whole building, as it is less subject to wind and buoyancy forces which induce the infiltration of external hazards.

Breathable air from compressed air cylinders venting directly into the SIP forms a positive internal pressure. Determining the amount of air required to reach the necessary level of positive pressure involves measuring the areas internal volume and the relationship between the internal and external atmospheric conditions.

After reaching required ACH (air changes per hour) or tightness, a positive-pressure management system must be sized for the expected shelter duration.

Breathable Air Supply

If the shelter-in-place is airtight, a supply of breathable air is necessary to support occupants over the required shelter duration. Unventilated SIPs that are tightly sealed cannot be occupied for long periods without the risk of high carbon dioxide levels. A secondary source of breathable air must be circulated within the shelter to meet occupancy over the required amount of time.

Clean, breathable air is supplied to a chemical, biological and radiological (CBR) safe room in a few ways: by trapping air inside the room and minimizing the air exchange, passing contaminated air through a filter (not practical for high concentration chemical releases or for chemicals that cannot be filtered) to purify it as it is supplied to the room, and the supplemental supply of oxygen through high pressure compressed air cylinders.

The air supply must contain enough oxygen in its mix, and not contain contaminants like carbon monoxide, carbon dioxide, water, or any other toxins that poison safe haven occupants during entrapment.

Table 1 lists international standards for flowrates for breathable compressed air for self-contained breathing apparatus (SCBA), supplied air respirators (SAR) and refuge chambers.

Table 1 – Standards for acceptable air supply requirements

Air supply requirements	Relevant Standard
Open-circuit positive-pressure SCBA, including all components and accessories except the air cylinder (shell) must resist the permeation and penetration of HD and GB chemical agents when tested for a period of 6 hours on an upper-torso manikin connected to a breathing machine operating at an airflow rate of 40 liters per minute (lpm), 36 respirations per minute, 1.1 liters tidal volume.	NIOSH 42 CFR Part 84 Respiratory Protective Devices
The flow rate should be at least 85L/min (3cfm) per person to remove expired CO2 and CO but at least 255L/min (9cfm) to provide adequate cooling and maintain comfortable conditions. The air inflow is assumed to be at a temperature of 26.5°C and 65% relative humidity	ITA Report No. 14 – Guideline for the provision of refuge chambers in tunnels under construction, May 2014
For manually operated blower type air-hose respirators, the blower shall be capable of delivering air to the space enclosed by the facepiece at a flow rate of at least 85 L/min.	AS/NZS 17:16:2003 Respiratory Protective Devices

The NIOSH standard of 40L/min is for SCBA apparatus (i.e. mouthpiece or face mask) where dilution of CO_2 is not a concern. Conversely, the ITA and Australian standards specify 85L/min and are concerned with both oxygen consumption and carbon dioxide dilution. It is important to note that three times the amount of air is required to maintain comfortable conditions inside a refuge chamber, when compared to simply breathing (Health & Safety Executive UK, 2006).

Unless the positive pressure system is supplying breathable air at a rate that maintains oxygen and compressed gas levels within acceptable ranges (dependent on safe haven occupancy and duration), a CO_2 scrubber and supplementary oxygen supply are required.

Supplemental O_2

O_2 is an essential requirement for cellular activity, and therefore, the absence of O_2 for relatively short periods results in death. Humans within an enclosed environment consume O_2 and if not replaced, would result in an irrespirable atmosphere. An O_2 concentration of less than 11% poses a high risk of death.

Retaining the oxygen levels between acceptable limits is critical. Various international standards outline acceptable O_2 limits for a breathable environment; examples are shown in Table 2.

Table 2 - Standards for allowable O_2 levels

Minimum O_2 Content (% by volume)	Maximum O_2 Content (% by volume)	Relevant Standard
18.5%	23.0%	Mines Safety and Health Administration 30 CFR Parts 7 and 75 Refuge Alternatives for Underground Coal Mines; Final Rule
19.5%	23.5%	Occupational Safety and Health Administration 29 CFR 1910.146 Permit-Required Confined Spaces for General Industry
19.5%	23.5%	Australian Standard AS 2865–2009 Safe working in a confined space

OSHA and Australian Standards for working in a confined space specify an acceptable lower limit of 19.5%, which is consistent with many other standards around the world. The slightly lower MSHA limit of 18.5% is based on a person at rest, not working, therefore, not likely to have trouble in breathing; conditions that are likely to be present in a refuge chamber.

O_2 itself is not flammable but supports and accelerates combustion. OSHA defines O_2 enriched mixtures or atmospheres as those containing more than 23.5% O_2 by volume (OSHA, 2013). In O_2-enriched atmospheres, the reactivity of O_2 significantly increases the risk of ignition and fire.

Within a safe haven, pure O_2 must be utilized in conjunction with a CO_2 removal system.

Removal of Accumulated CO_2

Protection within a safe haven is substantially increased by filtering CO_2 from the air. CO_2 is being produced continuously by the cells of our body and is a byproduct of respiration. The inability to expel this gas or an environment with high concentrations of CO_2 leads to CO_2 poisoning. Breathing air with more than 3% CO_2 can induce dizziness, headaches, and vision disturbance. Prolonged exposure at greater than 6% could result in unconsciousness and death. Listed in Table 3 are examples of international standards of satisfactory CO_2 removal requirements for a breathable environment.

Table 3 – Acceptable CO_2 removal requirements

CO_2 Removal Requirements	**Relevant Standard**
CO_2 removal components shall remove CO_2 at a rate of 1.08 ft^3 (30.5 L)/h/person.	Mines Safety and Health Administration 30 CFR Parts 7 and 75 Refuge Alternatives for Underground Coal Mines; Final Rule
OSHA has established a Permissible Exposure Limit (PEL) for CO2 of 5,000 parts per million (ppm) (0.5% CO2 in air) averaged over an 8-hour workday (time-weighted average or TWA.)	Occupational Safety and Health Administration
5.2.8 Atmospheric monitoring. Atmospheric monitoring There should be the capability to monitor temperature continuously, and for carbon monoxide, carbon dioxide, and oxygen in the chamber during use in both externally supported or standalone modes. The concentration of carbon dioxide should not exceed 10000 ppm.	ITA Report No. 14 – Guideline for the provision of refuge chambers in tunnels under construction, May 2014

The CO_2 that is exhaled by the occupants within the safe haven must be removed by, constant flushing by an external breathable air source (e.g. compressed air cylinders used for positive pressure) or by the regeneration of the internal atmosphere with a CO_2 absorbent (scrubber).

Different regulations around the world, state varied human production rates between 30–50L per hour per person. When considering an acceptable production rate; it must be noted that the rate of breathing is primarily triggered by the CO_2 content of inspired air with higher CO_2 levels triggering faster respiration rates (Brake, Fellows, & Bates, 1999) (i.e. if the maximum allowable CO_2 concentration inside the safe haven is 1%, a lower production rate is acceptable compared to a 3% CO_2 limit).

If CO_2 removal mechanisms are inadequate, the occupants can develop CO_2 toxicity by re-breathing their own exhaled CO_2.

Cooling and Dehumidifying

Thermal tolerability with a safe haven should be addressed during the design phase. Assumptions must be made in preparation for safe haven design in regard to the number of personnel, activity level, available hydration and heat production. The minimum airflow to maintain a comfortable environment is set by taking these assumptions into account as well as the input temperature and humidity of supplied air.

The healthy human body maintains its internal temperature around 37°C. An essential requirement for continued normal body function is that the deep body temperature is maintained within the narrow 1°C limit (Epstein, Moran, 2006). A change of body temperature exceeding 1°C occurs only during illness or when environmental conditions surpass the body's ability to cope with extreme temperatures.

When a person is subject to extreme heat and humidity over some time, the person may experience a very rapid progression of symptoms or exhibit a few of the less significant symptoms before falling victim to more severe forms of heat-related illness.

The acceptable cooling requirements for underground refuge chambers are defined in various international standards, examples of which are listed in the below table.

Table 4 – Standards for acceptable cooling requirements

Cooling Requirements	Relevant Standard
Simulated emergencies, in which a full complement of people has occupied a refuge chamber for a significant period, indicate that humidity and temperature can increase very rapidly to potentially heatstroke-inducing levels. Refrigerated air conditioning is strongly recommended for both externally supported and stand-alone refuge chambers to counter this potentially dangerous problem. Inevitably, this places a heavy demand on the stand-alone power supply, but there are systems available that can cope, and at an acceptable cost.	Department of Mines, Industry Regulations and Safety, 2005 Refuge Chambers in Underground metalliferous mines – guideline: Safety and Health Division, Department of Industry and Resources Western Australia, 32 pp
Apparent temperature in the fully occupied refuge chamber must not exceed 95 °F.	Mines Safety and Health Administration 30 CFR Parts 7 and 75 Refuge Alternatives for Underground Coal Mines; Final Rule
It should be possible to maintain an internal temperature not exceeding 30 °C wet bulb globe temperature for at least 24 hours	ITA Report No. 14 – Guideline for the provision of refuge chambers in tunnels under construction, May 2014

MSHA regulations specify a maximum internal apparent temperature[7] of 95°F (35°C). NIOSH specifies an equivalent maximum wet bulb temperature of 28.9°C (NIOSH, 2007).

The temperature maintained inside the safe haven is determined by the metabolic heat liberated by the occupants, dissipated to the external environment by the heat transfer characteristics of the room or building. The external temperature primarily influences the heat transfer while the humidity is the combination of moisture produced by the occupants (via exhaled air and sweat) and water liberated by CO_2 scrubbing chemicals.

It is important to remember that in most cases, before sheltering-in-place, the air conditioning system will need to be shut down (unless it is a split system). For short durations, artificial cooling is not a critical requirement unless external temperatures are extraordinarily high, and/or the safe haven's volume is small compared to its occupancy.

INFILTRATION TESTING

Background

High Performance Building Solutions Inc. (HPBS) was contracted to perform an air infiltration study for spaces identified as candidates to be used as shelter-in-place (SIP) in the event of an accidental chemical release. This SIP strategy uses the existing structure and its indoor atmosphere to separate individuals from a hazardous outdoor atmosphere temporarily.

A series of pre-weatherization air leakage tests were conducted to determine the baseline air infiltration rate of potential SIP spaces. Manual air sealing, a cleanroom suspended ceiling system and low-leakage HVAC dampers were installed to reduce the overall air infiltration of the SIP spaces and to close off air leakage pathways from non-protected spaces in the building through the HVAC system. Post-weatherization air leakage tests were conducted to determine the air infiltration rate of the SIP space after the completion of weatherization and HVAC work. The SIP spaces were additionally air sealed using the proprietary AeroBarrier® system to further reduce infiltration rates.

Scope

Air sealing work was completed, and low-leakage HVAC dampers were installed to reduce the overall air infiltration of the SIP spaces and to isolate the HVAC from non-protected spaces in the building. Post-weatherization air leakage tests were conducted to determine the air infiltration rate of the SIP space after the completion of air sealing, suspended ceiling and HVAC work. AeroBarrier® air sealing was conducted on SIP spaces as a pilot project on September 18 through September 21, 2017. Post AeroBarrier® air leakage tests were conducted on September 21 and September 22, 2017, to determine the air infiltration rate of the SIP space after the completion of the AeroBarrier® sealing.

7 Apparent temperature is the temperature equivalent perceived by humans, caused by the combined effects of air temperature and relative humidity.

Following ASTM E 779-10, pressurization air leakage tests were conducted on the following spaces recommended in the Facility Siting Study (FSS).

Table 5: SIP specifications

SIP Space	Occupancy	Floor Area (Square Feet)	Volume (Cubic Feet)
SIP Space 1	40	990	10,481
SIP Space 2	6	955	8,210
SIP Space 3	5	391	3,917
SIP Space 4	10	1,697	16,968
SIP Space 5	40	293	2,637

Air Leakage Test Results

Positive pressure air leakage test results are reported in four different ways:

CFM4 is the quantitative volume of air passing through an enclosure under a pressure difference of 4 Pascals (equivalent to a 5mph wind). This is a measurable physical quantity and the value recommended in the FSS to eliminate toxic gas ingress into the SIP spaces.

CFM25 is the quantitative volume of air passing through an enclosure under a pressure difference of 25 Pascals (0.1-in w.g.). FEMA 453 requires Class 1 Safe Havens be pressurized to 25 Pascals. This is a measurable physical property.

CFM50 is the quantitative volume of air passing through an enclosure under a pressure difference of 50 Pascals (0.2-in w.g.).

ACH50 is the quantitative volume of air passing through an enclosure under a pressure difference of 50 Pascals (0.2-in w.g.) normalized by the enclosure volume. ACH is the abbreviation for air changes per hour. It is the number of times the air volume in a building changes per hour at 50 Pascals of pressure.

Table 6 below shows the CFM50 and ACH50 for each of the SIP spaces pre-weatherization and post-weatherization along with the percentage reduction in leakage. SIP Space 1 and SIP Space 5 did not have pre-weatherization results as initially the whole building was tested. Smaller spaces were selected as the whole building was too large to feasibly convert to a SIP space (re-testing of the new SIP spaces was not completed). The control room, SIP space 4, was abandoned in favor of a smaller adjacent room due to high leakage post manual weatherization and the computers that could not be shut down, which prevented the use of AeroBarrier® in the space.

Table 6: Post-weatherization leakage results

Safe Haven Space	CFM50		ACH50		% Reduction
	Pre-Weatherization	Post-Weatherization	Pre-Weatherization	Post-Weatherization	
SIP Space 1	N/A	1,871	N/A	10.7	N/A
SIP Space 2	3,623	1,560	26	11.4	57%
SIP Space 3	385	135	6	2.1	65%
SIP Space 4	N/A	N/A	N/A	N/A	N/A
SIP Space 5	N/A	655	N/A	19.4	N/A

The AeroBarrier® air sealing was prescribed a target. According to the AeroBarrier® simultaneous air leakage calculations, all the spaces met and surpassed this airtightness target; AeroBarrier® sealing was halted at this point. Table 7 shows the CFM50 and ACH50 for each of the SIP spaces as tested by HPBS post AeroBarrier® along with the percentage reduction. The HPBS results were utilized to calculate the amount of positive pressure required to convert the rooms to positive pressure safe havens.

Table 7: Post AeroBarrier® leakage results

Safe Haven Space	CFM50		ACH50		% Reduction
	Post-Weatherization	Post AeroBarrier®	Post-Weatherization	Post AeroBarrier®	
SIP Space 1	1,871	455	10.7	2.6	76%
SIP Space 2	1,560	182	11.4	1.3	88%
SIP Space 3	135	125	2.1	1.9	7%
SIP Space 4	N/A	173	N/A	3.3	N/A
SIP Space 5	655	138	19.4	3.1	79%

The AeroBarrier® air sealing methods improved the air barrier integrity for all the proposed SIP spaces between 7% and 88% over the post manual weatherization leakage rates. Air leakage was still observed behind many electrical penetrations including HVAC thermostats, and fire and smoke alarms. These were covered with plastic to protect them from the AeroBarrier® Sealant but had hidden holes behind them.

Figure 1 displays the post AeroBarrier® leakage rate in comparison to the FEMA 453 estimates for air leakage rates based on construction types. FEMA 453 estimates are reported in square feet of floor area at 25 Pascals and do not take into consideration the ceiling height and volume of the space.

Figure 1: Post AeroBarrier® results compared to FEMA 453 estimates

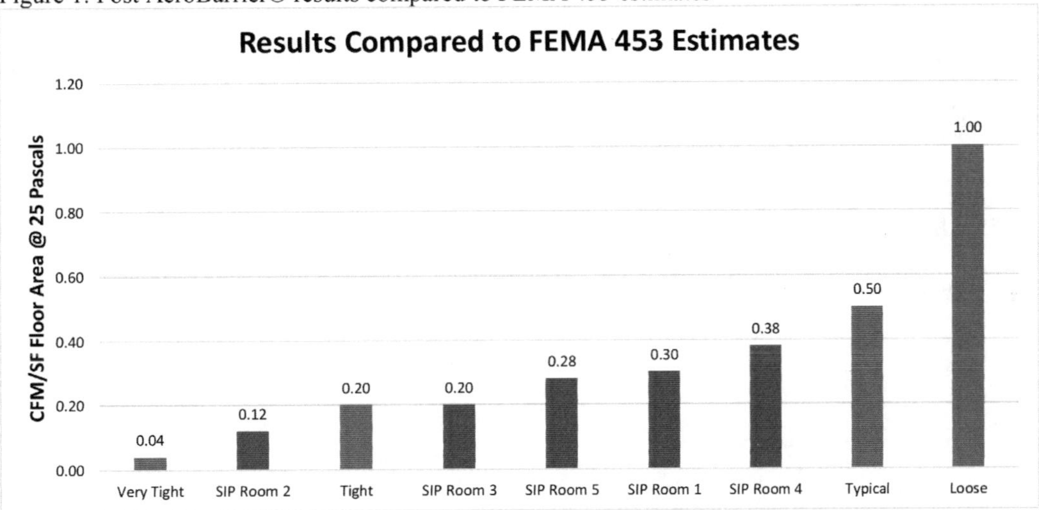

Figure 2 is an estimated leakage rate based on construction type to estimate the filtration and positive pressure system size under the FEMA 453 standard. All the SIP locations were between tight and typical except for SIP Room 2, which was a small prefabricated metal constructed portable building.

Figure 2: Estimated leakage rate per square foot of floor space to achieve 0.1 iwg (25Pa) overpressure

Construction Type	cfm per square foot of floor area
Very tight: 26-inch thick concrete walls and roof with no windows	0.04
Tight: 12-inch thick concrete or block walls and roof with tight windows and multiple, sealed penetrations	0.20
Typical: 12-inch thick concrete or block walls with gypsum wall board ceilings or composition roof and multiple, sealed penetrations	0.50
Loose: Wood-frame construction without special sealing measures	1.00

Table 8 shows the combined AeroBarrier®, and manual air sealing methods improved the airtightness for all the proposed SIP spaces between 68% and 95% over the original leakage rates.

Table 8: Air infiltration improvement for SIP spaces

Safe Haven Space	ACH50			% Reduction
	Pre-Weatherization	Post-Weatherization	Post AeroBarrier®	
SIP Space 1	N/A	10.7	2.6	76%
SIP Space 2	26.5	11.4	1.3	95%
SIP Space 3	5.9	2.1	1.9	68%
SIP Space 4	N/A	N/A	3.3	N/A
SIP Space 5	N/A	19.4	3.1	84%

Figure 3 compares the ACH50 pre-weatherization, post-weatherization and AeroBarrier® air infiltration results.

Figure 3: Pre-weatherization, post-weatherization and AeroBarrier® comparison

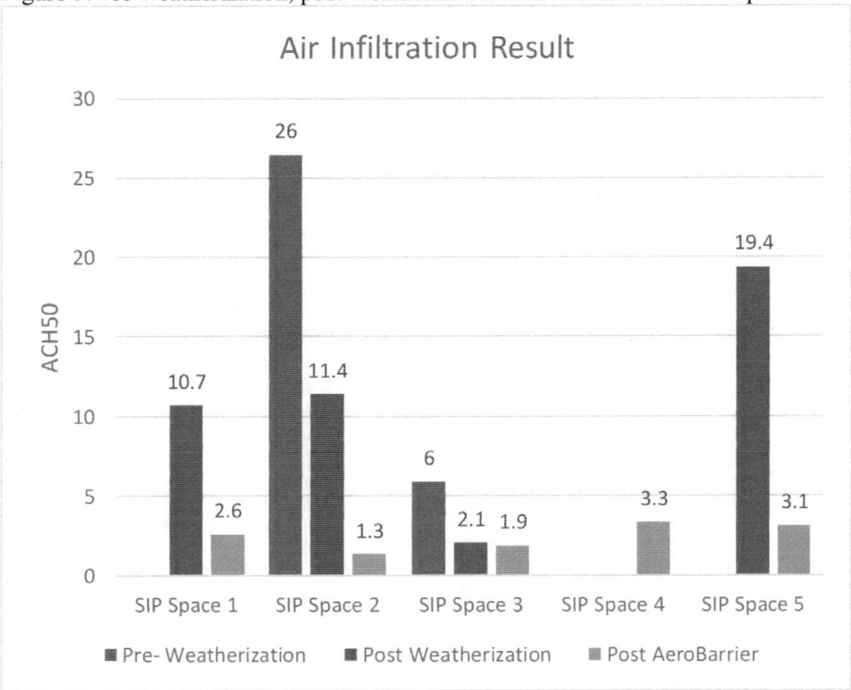

<u>Life Support Systems</u>

MineARC re-calculated the vulnerability for each SIP space based on the post AeroBarrier® leakage rates. All SIP spaces still had a vulnerability of 100% and therefore required the installation of independent life support systems to be deemed zero percent vulnerability safe havens.

Figure 4 shows the toxic gas concentration, duration of the release, and vulnerability for each SIP space. As the largest MCE was a full tank rupture the toxic concentrations were extremely high, but the duration was short at one minute.

Figure 4: SIP space vulnerability post AeroBarrier®

	TOXIC GAS				
Safe Haven Location	Worst Case Threat Scenario	Toxic External [ppm]	Duration of Reslease [min]	ACM	Vulnerability
SIP Space 1	Tank Rupture	585,000	1	0.032	100%
SIP Space 2	Tank Rupture	652,000	1	0.042	100%
SIP Space 3	Tank Rupture	660,000	1	0.007	100%
SIP Space 4	Tank Rupture	412,000	1	0.042	100%
SIP Space 5	Tank Rupture	600,000	1	0.028	100%

The leakage rate at 4 Pascals of pressure difference is slightly more than would occur under a 5 mile per hour wind used in the dispersion model in the FSS report based on climatic data for the facility (to maintain positive pressure and neutralize air infiltration). Post-weatherization infiltration rates in cubic feet per minute at 4 Pascals of pressure for each of the SIP spaces is shown in Table 9.

Table 9: Post AeroBarrier® air leakage results at CFM4

Safe Haven Space	Volume (Cubic Feet)	CFM4
SIP Space 1	10,481	99
SIP Space 2	8,210	32
SIP Space 3	3,917	24
SIP Space 4	3,110	45
SIP Space 5	2,637	21

The air leakage test results provide a mathematical formula that can be used to calculate the amount of breathable compressed air cylinders (437ft^3) that must be added to each candidate SIP space to maintain a 4 Pascal positive pressure to overcome a 5mph design wind coincident with a toxic release.

MineARC Systems was contracted to install life support systems to convert the SIP spaces to positive pressure safe havens. The MineARC AirBANK Positive Pressure System provides rapid pressurization, which is activated and maintained using the AirBANK Control via a simple human-machine interface (HMI) touch screen. MineARC's integrated Aura-FX Gas Monitor ensures breathable air automatically remains within acceptable limits. Additionally, the installation of supplementary oxygen and carbon dioxide scrubbing systems in safe havens, are as needed.

Figure 5 shows the number of 437 cubic feet cylinders required for each safe haven location. The minimum breathable air per person is 2.5cfm (70L/min) per person based on MineARC's experience and testing within the refuge chamber industry.

Figure 5: Quantity of cylinders required for 4Pa positive pressure

	SAFE HAVEN SPECIFICATIONS		POSITIVE PRESSURE (@ 4Pa)						
Safe Haven Location	Occupancy	Duration	CFM	CFM per Floor Space	ACH	No. Cylinders	No. Cylinder Racks	Breathable Air per Person [CFM]	Breathable Air per Person [LPM]
Safe Haven Space 1	40	4	99	0.10	0.6	55	5	2.5	70.8
Safe Haven Space 2	6	4	32	0.03	0.2	18	2	5.4	152.9
Safe Haven Space 3	5	4	24	0.06	0.4	14	2	4.8	135.9
Safe Haven Space 4	10	4	45	0.14	0.9	25	3	4.5	127.4
Safe Haven Space 5	40	4	21	0.07	0.5	12	2	0.6	17.0

The maximum CO_2 concentration is at 1% and O_2 between 19.5% and 23.5% based on OSHA limits (refer to Table 2). Due to the volume and number of people in Safe Haven Space 5, it required supplemental oxygen and carbon dioxide scrubbing to maintain the internal gas concentrations within specified limits. Three sodium chlorate oxygen candles and two AirGEN CO_2 scrubbers were included in the design.

Figure 6: Internal gas concentrations in safe havens

	SAFE HAVEN SPECIFICATIONS		POSITIVE PRESSURE (@ 4Pa)			02 & CO2 CONCENTRATIONS (@ 4Pa)			
Safe Haven Location	Occupancy	Duration	CFM	CFM per Floor Space	ACH	Indoor O_2 %	No. O_2 Candles	Indoor CO_2 %	No. AirGEN Scrubbers
Safe Haven Space 1	40	4	99	0.10	0.6	20.45	0	0.50	0
Safe Haven Space 2	6	4	32	0.03	0.2	20.84	0	0.17	0
Safe Haven Space 3	5	4	24	0.06	0.4	20.77	0	0.23	0
Safe Haven Space 4	10	4	45	0.14	0.9	20.66	0	0.31	0
Safe Haven Space 5	40	4	21	0.07	0.5	18.67	3	2.04	2

In 2018 MineARC completed installation and training on the five safe haven spaces. Each safe haven was tested during the commissioning process to ensure it could maintain acceptable positive pressure.

CONCLUSIONS

The conversion of shelter-in-place (SIP) locations to zero percent vulnerability safe havens is a four-step process consisting of the following:

1) Identifying the correct SIP location based on ease of sealing and small volume based on the required occupancy.
2) Manual air sealing of SIP spaces.
3) AeroBarrier® sealing of SIP spaces.
4) Installation of life support equipment.

Identifying SIP spaces that can seal from the outset is crucial to a successful result. In the case study, three of the five original locations were abandoned in favor of rooms in adjacent locations due to high leakage rates or reducing the size to make the project more cost effective. Control rooms pose the most difficulty in trying to seal due to the false floor and a myriad of cable tray and pass-throughs installed for the computer systems. The smaller the room and the more robust the construction (i.e. prefabricated metal or concrete buildings), the better it seals.

Manual air sealing needs to be overseen by someone with familiarity in building tightness. While most of the tasks are general construction, standard contractors are not trained to work at the level of precision needed to ensure proper sealing. Even a tiny orifice or ceiling tile not sealed correctly can ruin a leakage test result. HPBS and MineARC Systems have developed a system for ensuring airtightness in construction methodology and by avoiding un-sealable spaces or spaces that are not cost effective to seal.

For future construction projects, incorporation of SIP spaces should occur during the design phase. Air barrier commissioning is standard in new construction and should include the following tasks at a minimum: design review of the plans with specifications for the air barrier and mechanical system - specifically as these relate to spaces designated for use as safe havens; quality assurance inspections during construction to ensure a continuous air barrier is in place; and final air leakage testing of the continuous air barrier to ensure the leakage rate of the safe haven spaces are within acceptable limits.

Complementary life support equipment is essential to sustain life within an airtight environment. For personnel to remain within the sealed safe haven, the environment must be habitable. Air is to be free of toxins such as carbon dioxide and acceptable levels of oxygen maintained. Positive pressure systems reduce the ingress of toxins and provide some or all of the required source of breathable air. Additional equipment such as compressed air, oxygen cylinders and air scrubbers are used to meet the shortfall. Furthermore, internal conditions must consider the human condition, supporting workable body temperature and humidity.

Reducing the risk of harm to personnel is a priority for emergency response planning. While the likelihood of an MCE occurring may be low, the risk to on-site personnel in the event of an incident is high. Converting shelter-in-place (SIP) locations to zero vulnerability safe havens improves emergency response.

REFERENCES

ASTM E779-10, *Standard Test Method for Determining Air Leakage Rate by Fan Pressurization*, ASTM International, West Conshohocken, PA, 2019, www.astm.org

Brake, Fellow, & Bates (1999). *Criteria for the design of emergency refuge stations for an underground metal mine*. Aus. IMM Proceedings, No.2, 1999, pp 1–7.

Department of Mines and Petroleum, 2013, *Refuge chambers in underground mines — guideline: Resources Safety*, Department of Mines and Petroleum, Western Australia, 41 pp.

Epstein, Y., & Moran D. (2006). *Thermal Comfort and the Heat Stress Indices,* Industrial Health, 44, 388–398.

FEMA 453. (2006). *Design Guidance for Shelters and Safe Rooms.* Federal Emergency Management Agency

Health & Safety Executive UK (2008). *Guidance and Information on the role and design of safe havens in arrangements for escape from mines.* Retrieved from http://www.hse.gov.uk/pubns/mines08.pdf

ITA Working Group 5 (2016). *Provisions for refuge chambers in tunnel projects under construction.* ITA Report N°014 - V2 / March 2018

Mines Safety and Health Administration (2008). *30 CFR Parts 7 and 75 Refuge Alternatives for Underground Coal Mines; Final Rule.*

National Institute for Occupational Safety and Health (2007). *Research Report on Refuge Alternatives for Underground Coal Mines.* Retrieved from http://www.cdc.gov/niosh/docket/archive/pdfs/NIOSH-125/125 ResearchReportonRefugeAlternatives.pdf

Occupational Safety & Health Administration (2013). *Permit-required confined spaces.* Retrieved from http://www.osha.gov/pls/oshaweb/owadisp.show_document?p_table=STANDARDS&p_id=9797

Suming Du. (2014). *On the effectiveness of shelter-in-place as a measure to reduce harm from atmospheric releases. Journal of Emergency Management, Vol. 12, No, 3, May/June 2014.*

Venter. J. (1998). *Portable Refuge Chambers: Air or Tomb in Underground Escape Strategies.* School for Mechanical and Materials Engineering, Potchefstoom University, Republic of South Africa

Worksafe Australia National Standard (2005). *Australian Standard AS: NZS 1716:2003 Respiratory protective devices.*

Yantosik. G. (2006). *Shelter-in-Place Protective Action Guide*. The Chemical Stockpile Emergency Preparedness Program (CSEPP). U.S. Department of Energy

589008 How the Field of Human Factors Helps Solve the Riddle of Serious Injuries and Fatalities

Abstract:

Process and occupational safety are often treated as distinct endeavors, with the former tending to focus on facility system function and the latter on individual worker wellbeing. Though other differences certainly exist between the two, both endeavors are generally purposed to prevent incidents that can cost lives, health, time, and money, in addition to employee morale and public trust. The overarching thesis of this paper is that the field of Human Factors is uniquely positioned to effect positive change in both of these major safety categories.

The backdrop for our argument is the recent finding that while injury and illness incidence rates published by the Bureau of Labor statistics have decreased in recent years, the occurrences of serious injuries and fatalities have declined at a significantly slower rate. We discuss possible explanations for this finding but note that regardless of the preferred explanation, the trend should motivate scientists, engineers, and other safety professionals to revisit long-held accident causation theories holding that near misses – these are incidents that could have resulted in serious injuries and fatalities but did not – are precursors of future serious incidents. While this idea still has merit, the just-mentioned statistics warrant reconsideration of the relations between serious and less-serious incidents to explain why the rates of the two are not decreasing in lockstep. We suggest these data indicate additional (human) factors are involved.

To address this riddle and to support our general thesis, we (a) review core tenets of process safety, occupational safety, and the field of Human Factors, (b) point out that the human is always an element common to both process and occupational safety contexts, and (c) discuss how preventing or mitigating human error can benefit both. We offer specific examples from both safety contexts to show how Human Factors perspectives and techniques can help. Examples touch on standards, policies, and other administrative controls (e.g., procedures), human-machine interfaces of both workplace hardware (e.g., valves) and software (e.g., SCADA systems), environments (e.g., physical, regulatory, and cultural), and communications between personnel (e.g., elements of team resource management). We conclude that fundamental Human Factors considerations can help prevent and mitigate human performance gaps and therefore increase the rate of reduction in serious injuries and fatalities in both process and occupational safety.

589233 Initial Conditions Used in Atmospheric Dispersion Models:Issues with Commonly Used Assumptions for Gas Jets and Plumes

Abstract:

When simulating jets and plumes, all atmospheric dispersion models require initial conditions that appropriately parameterize the behavior of a gas release to the atmosphere. Often, process conditions are such that loss of primary containment of a gas will result in choked flow conditions. A great deal of experimental and theoretical consideration has been given to the flow from stagnation conditions to choked flow conditions, but less experimental work has centered on describing the flow through the depressurization process to atmospheric pressure where atmospheric dispersion models would be applicable. Traditional modeling of the depressurization process has taken the simplified approach of assuming a (depressurized) "top hat" profile (laterally constant) with application of the principles of conservation of mass, momentum, and energy. Such assumptions are known to result in predicted depressurized velocities that exceed the (sonic) velocity at the point of loss of containment, and because of the energy balance, the associated temperatures are significantly less than temperatures at the choked conditions. Physically, these effects could possibly self-cancel because as jet velocity slows, the decrease in kinetic energy of the jet will increase the jet temperature. However, no atmospheric dispersion model used for routine hazard assessment purposes accounts for kinetic energy in the energy balance. The traditional modeling approaches of the depressurization process are reconsidered based on analysis of available data, and an alternative approach to modeling the depressurization process is proposed. Example cases showing the important differences between the two approaches are discussed.

Fault Tree Uncertainty Analysis

Raymond "Randy" Freeman
S&PP Consulting
12303 Lake Shore Ridge
Houston, TX 77041
Cell: 713 408 0357
Rafree@yahoo.com

For Presentation at the 2020 Spring National AIChE Meeting
Houston, Texas

March 30, 2020

ABSTRACT

Fault tree analysis (FTA) is a widely used methodology in the process industries. FTA is used for the development of failure mechanisms, computation of failure frequencies and the determination of the probability of failure on demand of safety systems. Much of the data used in a FTA study are uncertain. For example, the failure rate of a pump is often not known with great precision. Likewise the failure rates of instrumentation are often known only within some defined limits. The common practice, used by analysts in the quantification of a fault tree, is to use the most likely or best guess as to the needed failure rate data. The use of best guess values as data inputs to the quantification of a fault tree creates uncertainty in the computed results.

This paper presents a general methodology for the determination of the impact of uncertainty on the results of a fault tree study. The general methodology is based on the mathematics of propagation of error and variance contribution analysis. For a fault tree that contains 10 event minimal cut sets, the methodology will compute the variance in the probability (or frequency) of the top event with a maximum error of 5%. For 5 event minimal cut sets the error in variance of the top event probability (or frequency) will be no more than 2%. An example is presented to illustrate the application of the fault tree uncertainty analysis methodology to a real world problem.

Keywords

Quantitative Risk Assessment, Fault Tree Analysis, Uncertainty Analysis

1. Background

Fault tree analysis is a widely used method for representation of failure mechanisms in chemical plants. The methodology is described in the CCPS CPQRA guideline book [Ref. 1] and is extensively discussed in Chapter 9 of Lees [Ref. 2]. Lacking in these discussions is a systematic method to quantify the uncertainty of the results of the fault tree calculations.

Fault tree analysis has also been used in the aircraft and aerospace industries, nuclear power plant review and is recognized by the OSHA Process Safety Management (PSM) standard [Ref. 3] as a method for process hazards analysis.

2. Review of Error Propagation and Variance Contribution Analysis (VCA) Methodology

The mean and variance of a function of random variables can be approximated using the method described by Haugen [Ref. 4] and applied by Freeman [Ref. 5, 6, 7]. Define an arbitrary function of a set of random variables, x_i, as:

Let

$$Y = F(x_i) \qquad \text{(Eq 1)}$$

The mean of Y can be estimated using the following approximation:

$$E(Y) = F[\, E(x_i)] \qquad \text{(Eq 2)}$$

Where:

E(Y) = expected value of random variable Y = mean of Y

$E(x_i)$ = expected value of random variable x_i = mean of x_i

The variance of Y can likewise be estimated as:

$$V(Y) = \sum_{i=1}^{n} \left[\frac{\partial Y}{\partial x_i} \right]^2 V(x_i) \qquad \text{(Eq 3)}$$

Where:

$V(Y)$ = variance of random variable Y as defined above in Equation 1

$V(x_i)$ = variance of random variable x_i as defined above in Equation 1

Note that the variance is simply the square of the standard deviation. Using the variance will simplify the mathematics that is described below. The contribution of each independent random variable to the overall variance in the function is:

$$V(Y \text{ from } x_i) = \left[\frac{\partial Y}{\partial x_i}\right]^2 V(x_i) \qquad \text{(Eq 4)}$$

The relative contribution of each term to the overall variance V(Y) is a measure of the importance in the uncertainty in the particular random variable, x_i. In effect, this is a sensitivity analysis combined with a uncertainty evaluation. The variance contribution combines the sensitivity in the answer to changes in the uncertain random variable, x_i, with a measure of the uncertainty in the random variable, x_i. The overall variance in Y is found by summing the sensitivity weighted variances from each random variable.

Numerical Estimate of the Sensitivity

Let us return to the fundamental definition of the derivative.

As before, set y = F(xi)

$$\frac{\partial y}{\partial x} = \lim_{\Delta xi \to 0} \left[\frac{F(xi + \Delta xi) - F(xi)}{\Delta xi}\right] \qquad \text{(Eq 5)}$$

We are interested in a numerical estimate of the derivative. Therefore, we will make the following approximation.

$$\frac{\partial y}{\partial x} \sim \left[\frac{F(xi + \Delta xi) - F(xi)}{\Delta xi}\right] \qquad \text{(Eq 6)}$$

Linear Functions

Let us examine the approximation for a simple linear function.

Let Y = AX + B (Eq 7)

What is the sensitivity of Y to a change in variable X? By simple calculus:

$$\frac{\partial y}{\partial x} = A \qquad \text{(Eq 8)}$$

We can also find the sensitivity using the numerical approximation as:

$$\frac{\partial y}{\partial x} \sim \left[\frac{F(xi+ \Delta xi)- F(xi)}{\Delta xi}\right] \qquad \text{(Eq 9)}$$

$$\frac{\partial y}{\partial x} \sim \left[\frac{(A(xi+ \Delta xi)+ B)-(A(xi)+ B)}{\Delta xi}\right] \qquad \text{(Eq 10)}$$

$$\frac{\partial y}{\partial x} \sim \left[\frac{(A(xi+ \Delta xi))-(A(xi))}{\Delta xi}\right] \qquad \text{(Eq 11)}$$

$$\frac{\partial y}{\partial x} \sim \left[\frac{((A\, xi+A\, \Delta xi))-(A(xi))}{\Delta xi}\right] \qquad \text{(Eq 12)}$$

$$\frac{\partial y}{\partial x} \sim \left[\frac{A\, \Delta xi}{\Delta xi}\right] \qquad \text{(Eq 13)}$$

$$\frac{\partial y}{\partial x} = A \qquad \text{(Eq 14)}$$

Note that for a simple linear equation the numerical result for the sensitivity is exactly the same as the analytical expression. There is no restriction on the size of the perturbation (Δxi) used in the calculations. Previously, Freeman [Ref 7] has suggested using a 10% perturbation in xi to do the sensitivity calculations. However any size of perturbation will work and generate the correct sensitivity.

This is a general result. As long as the function can be expressed as a linear function of a number of variables, xi, the sensitivity can be directly calculated by a simple perturbation of the function. The resulting sensitivity will be exactly the same as would be obtained using analytical methods.

Non-Linear Expressions

In the quantification of a fault tree, the top event frequency is computed using failure rates of the component devices and time that measures the interval between system validations. In a minimal cut set of two or more basic events, the time variable may be the same. For example the test interval for a valve may be the same for a pressure transmitter. A minimal cut set involving both the valve and the pressure transmitter would create a 2rd order term in the test interval. If the test interval is uncertain, how do we compute the sensitivity of the top event frequency with respect to the test interval? We will now explore how to deal with this type of problem when completing the uncertainty analysis.

As before, let

$$y = F(xi) = Axi^n + B \qquad \text{(Eq 15)}$$

What is the sensitivity of Y to a change in variable X? By simple calculus:

$$\frac{\partial y}{\partial x} = nAxi^{n-1} \qquad \text{(Eq 16)}$$

Let us evaluate the numerical approximation for this case. If n=2, the function of interest is:

$y = F(xi) = Axi^2 + B$ (Eq 17)

From basic calculus, the exact analytical expression for the sensitivity is:

$$\frac{\partial y}{\partial x} = 2\ A(xi) \qquad \text{(Eq 18)}$$

If xi = 1

$$\frac{\partial y}{\partial x} = 2\ A \qquad \text{(Eq 19)}$$

Again the approximation of the derivative is:

$$\frac{\partial y}{\partial x} \sim \left[\frac{(A(xi+\Delta xi)^2+B)-(Axi+B)}{\Delta xi}\right] \qquad \text{(Eq 20)}$$

$$\frac{\partial y}{\partial x} \sim \left[\frac{F(xi+\Delta xi)-F(xi)}{\Delta xi}\right] \qquad \text{(Eq 21)}$$

Inserting the equation for F(xi) we obtain:

$$\frac{\partial y}{\partial x} \sim \left[\frac{(A(xi+\Delta xi)^2+B)-(Axi+B)}{\Delta xi}\right] \qquad \text{(Eq 22)}$$

$$\frac{\partial y}{\partial x} \sim \left[\frac{(A\ (xi^2+2\Delta xi+\Delta xi^2)\ +B)-(Axi+B)}{\Delta xi}\right] \qquad \text{(Eq 23)}$$

$$\frac{\partial y}{\partial x} \sim \left[\frac{A\ (xi^2+2\Delta xi+\Delta xi^2)\ -Axi}{\Delta xi}\right] \qquad \text{(Eq 24)}$$

$$\frac{\partial y}{\partial x} \sim A\left[\frac{(xi^2+2\Delta xi+\Delta xi^2)\ -xi}{\Delta xi}\right] \qquad \text{(Eq 25)}$$

Setting xi =1, and using a 10% perturbation as $\Delta xi = 0.1\ xi = 0.1$

$$\frac{\partial y}{\partial x} \sim A\left[\frac{(1^2+(2)(0.1)1+0.1^2\ )\ -1}{0.1\ (1)}\right] \qquad \text{(Eq 26)}$$

$$\frac{\partial y}{\partial x} \sim A\left[\frac{(2)(0.1)+0.01}{0.1}\right] \qquad \text{(Eq 27)}$$

$$\frac{\partial y}{\partial x} \sim A\left[\frac{0.21}{0.1}\right] \qquad \text{(Eq 28)}$$

$$\frac{\partial y}{\partial x} = 2.1\ A \qquad \text{(Eq 29)}$$

Note that the numerical estimate (Eq 29) is approximately 5% greater than the analytical value (Eq 19). As shown in Table 1, for higher order equations the error will increase as the size of the perturbation becomes more important. Table 2 shows the impact of reducing the perturbation size to 1%. For 10th order variables (say a minimal cut set with 10 basic events) the error is 5%. For almost all engineering evaluations, a maximum error of 5% should be adequate.

For cases where the test frequency, repair time, or mission time appears in a minimal cut set 3 or more times, I suggest that the following perturbation in time be used to compute the sensitivity of the top event:

$$\Delta time = 0.01\ time \qquad \text{(Eq 30)}$$

The sensitivity of the top event frequency to all uncertain variables can be computed using a perturbation of 1%.

$$\Delta xi = 0.01\ xi \qquad \text{(Eq 31)}$$

3. Fault Tree Analysis

Fault trees are often used to analyze potential failures of an engineered system. A fault tree is a failure logic diagram that shows the relationship of device or system failures leading to an accident. Fault tree analysis is covered in detail in the CCPS CPQRA book [Ref. 1]. A fault tree is constructed by starting with a top event or condition and asking the question:

What conditions lead to this outcome?

Each condition in the next layer down is then analyzed by asking the same question. What conditions lead to this outcome? The logic of how these intermediate conditions (termed intermediate events) are related is expressed using AND and OR logic gates. An AND gate output is true if and only if (iff) all of the inputs to the AND gate are all true simultaneously. An OR gate output is true iff one of the inputs to the OR gate is true.

Once the fault tree logic structure is created, numerical data can be used to calculate the frequency or probability of the top event of concern occurring. For fault trees where a basic event is found in multiple branches of the tree, such as power failure, the logic structure of the tree must be simplified to remove the impact of the repeated events on the top event of concern. This simplification is done using Boolean Algebra. Each resulting grouping of basic events can cause the top event to occur. These groupings of basic events are called minimal cut sets. As a simple example consider the Boolean Algebra representation of a fault tree :

$$T = BE11 + BE1 \bullet BE2 + BE1 \bullet BE3 + BE2 \bullet BE9 + BE3 \bullet BE7 \qquad \text{(Eq 32)}$$

Where:

T = Fault Tree Top Event

BE11 = Basic Event 11

BE1 = Basic Event 1

BE2 = Basic Event 2

BE3 = Basic Event 3

BE7 = Basic Event 7

BE9 = Basic Event 9

+ = Boolean Algebra Addition Symbol

• = Boolean Algebra Multiplication Symbol

For this example the top event T, the minimal cut sets are:

BE11

BE1 BE2

BE1 BE3

BE2 BE9

BE3 BE7

For example the existence of basic events BE1 and BE3 guarantees that the top event T will occur. Publically available standard computer software such as SAPHIRE [Ref. 8] may be used to find the minimal cut sets of a large fault tree.

The representation of the top event in equation 32 can now be used to compute the probability or frequency of occurrence of the top event. If the devices describe by the basic events are non-repairable and are tested after a period of time TI, the probability of top event, T, may be calculated as:

$$\text{Prob (T)} = \lambda_{11}\ TI/2 + (\lambda_1\ TI/2)\ (\lambda_2\ TI/2) + (\lambda_1\ TI/2)\ (\lambda_3\ TI/2) + (\lambda_2\ TI/2)\ (\lambda_9\ TI/2) + (\lambda_3\ TI/2)\ (\lambda_7\ TI/2) \qquad \text{(Eq 33)}$$

Where:

λ_{11} = failure rate of device 11 in basic event BE11, failures per unit time

λ_1 = failure rate of device 1 in basic event BE1, failures per unit time

λ_2 = failure rate of device 2 in basic event BE5, failures per unit time

λ_3 = failure rate of device 3 in basic event BE3, failures per unit time

λ_7 = failure rate of device 7 in basic event BE7, failures per unit time

λ_9 = failure rate of device 9 in basic event BE9, failures per unit time

TI = test interval, time

Equation 33 assumes that the device failure rates are small and the product λ TI < 0.1. The other major assumption is that the values of all of the failure rates and test intervals are known. In reality the failure rate data are not known exactly with only a range or probability distribution representing the state of knowledge.

4. Example Problem

The following example is taken from the ISA technical report on fault tree analysis [Ref. 9]. This example should not be considered as a recommendation by either ISA or AIChE. The example is presented to allow for the demonstration of the methods that can be used in a fault tree study to quantify the uncertainty in the resulting calculated failure frequency or failure probability.

Consider the interlock on the intermediate storage tank T101 shown on Piping and Instrument Diagram in Figure 1. The interlock block diagram is shown in Figure 2. The interlock is intended to prevent an abnormal condition in the tank which could lead to a uncontrolled release of the tank contents. The process hazards analysis team has recommended that the interlock be designed to safety integrity level 2 (SIL 2) with a target probability of failure on demand (PFD) of no greater than 1E-2 or a risk reduction factor (RRF) of 100. Does the proposed interlock shown on Figure 1 meet the SIL 2 target? What is the uncertainty in the predicted PFD of the interlock?

A fault tree for the failure of this interlock is shown in Figure 3. A minimal cut set analysis has been completed and the resulting minimal cut sets are presented in Table 3. We can now write the Boolean Algebra equation that represents this fault tree as:

$$T = PE + TS1 \cdot TS2 + LS1 \cdot LS2 + FT1 \cdot FT2 + FT2 \cdot FT3 + FT1 \cdot FT3 + BV1 \cdot BV2 + BV1 \cdot SOL1 + BV2 \cdot SOL2 + SOL1 \cdot SOL2 + PT1 \cdot PT2 \qquad \text{(Eq 34)}$$

Assuming that the interlock is non-repairable until tested at time TI and using lamda-time (λT) approximation of the failure rate, the Boolean representation of the fault tree may now be converted to a failure probability model as:

$$\begin{aligned}\text{Prob}(T) = {} & \text{Prob}(PE) + (\lambda_{TS1}\, TI/2)(\lambda_{TS2}\, TI/2) + (\lambda_{LS1}\, TI/2)(\lambda_{LS2}\, TI/2) + (\lambda_{FT1}\, TI/2)(\lambda_{FT2}\, TI/2) + \\ & (\lambda_{FT2}\, TI/2)(\lambda_{FT3} TI/2) + (\lambda_{FT1}\, TI/2)(\lambda_{FT3}\, TI/2) + (\lambda_{BV1}\, TI/2)(\lambda_{BV2}\, TI/2) + \\ & (\lambda_{BV1}\, TI/2)(\lambda_{SOL1} TI/2) + (\lambda_{BV2}\, TI/2)(\lambda_{SOL2}\, TI/2) + \\ & (\lambda_{SOL1}\, TI/2)(\lambda_{SOL2}\, TI/2) + (\lambda_{PT1}\, TI/2)(\lambda_{PT2}\, TI/2) \qquad \text{(Eq 35)}\end{aligned}$$

Where:

Prob(PE) = failure probability of electronic logic solver, assumed to be constant at 5E-3.

λ_{TS1} = failure rate of temperature switch 1, failures per unit time

λ_{TS2} = failure rate of temperature switch 2, failures per unit time

λ_{LS1} = failure rate of level switch 1, failures per unit time

λ_{LS2} = failure rate of level switch 2, failures per unit time

λ_{FT1} = failure rate of flow transmitter 1, failures per unit time

λ_{FT2} = failure rate of flow transmitter 2, failures per unit time

λ_{FT3} = failure rate of flow transmitter 3, failures per unit time

λ_{BV1} = failure rate of block valve 1, failures per unit time

λ_{BV2} = failure rate of block valve 2, failures per unit time

λ_{SOL1} = failure rate of solenoid valve 1, failures per unit time

λ_{SOL2} = failure rate of solenoid valve 2, failures per unit time

λ_{PT1} = failure rate of pressure transmitter 1, failures per unit time

λ_{PT2} = failure rate of pressure transmitter 2, failures per unit time

TI = test interval, time

The electronic logic solver PE is assumed to have a constant failure probability of 5E-3.

Failure rate data for the other devices are present in Table 4. The mode (most likely value) of the failure rate is set equal to the point value taken from the ISA example [Ref. 9]. The triangular probability distribution (Appendix A) was assumed to represent the failure rate data for the equipment items. The upper and lower limits are based on the failure rate data of Appendix 4 of Smith [Ref. 10].

Using the device failure rates represented by the Table 4 column labeled as the mode and the minimal cut sets presented in Table 3, the ISA technical report [Ref.9, page 35] computes an interlock PFD as 7.5E-3 (RRF of 133). This would satisfy the requirement for a SIL 2 interlock {PFD of 1E-2 or a RRF of 100}. Now what is the uncertainty in the interlock PFD?

The first thing to remember is that the PFD should be calculated based on the mean value of the failure rate lambda (λ), not the mode of the failure rate. For a data set, the mean is the best single point representation of the data set. Table 5 presents the PFD of each minimal cut set using the mean of the device failure rate lambda (λ). The total of the PFD becomes 1.47E-2 (RRF of 68). This would not satisfy a SIL 2 interlock requirement.

Now we will compute the variance of the predicted probability of top event, T. To compute the variance of the probability of the top event, T, we will need to compute the sensitivity of the top event to each of the variables.

Sensitivity Using Numerical Perturbation Calculation

We will use the ISA interlock example previously analyzed by Freeman (Ref. 7). Using public data sources Freeman found that the probability of failure on demand (PFD) was 1.49E-2. To use the perturbation method for the evaluation of the sensitivities requires evaluating the impact of small changes in the uncertain input parameters. To enable these calculations to be completed numbers using 8 significant digits were used. The calculations were completed using Microsoft Excel which uses double-precision floating-point arithmetic compliant with IEEE 754 specification. Excel nominally carries 15 significant figures in calculations. We will complete the calculations with Excel displaying 5 significant digits and round the result at the end of the calculations to 4 significant digits. We will illustrate the calculations by looking at the Flow Transmitter, FT1.

a. First we recalculate the PFD of the example and find it to be 1.4594E-02. This is the base number that we will use in the perturbation calculations.
b. We now make a small change in mean failure rate of FT1. The mean failure rate is 8E-6 failures/hour. A 1% perturbation is used or an increase of 0.08E-6 failures/hour.

c. The perturbed value of the failure rate of FT1 becomes (8 + 0.08) * 1E-6 failures/hour or 8.08E-6 failures/hour.
d. The perturbed value of the failure rate of FT1 is now used to re-calculate the PFD of the interlock. The resulting re-calculation finds that the PFD is 1.4618E-02
e. The change in the PFD from the base number due to the perturbation of the failure rate of FT1 is now computed as:

   delta PFD = 1.4618E-02 - 1.4594E-02 = 2.4556E-05

f. The sensitivity of the PFD to a change in the failure rate of FT1 is now computed by dividing the change in the PFD by the change FT1 failure rate as:

   sensitivity = (2.4556E-05)/(0.08E-6 failures/hour) = 307.0 hours
   Result is rounded to 4 significant digits.

We now note that in the paper by Freeman (Ref. 7) that equation 32 presents the sensitivity for FT1 as:

$$\frac{\partial Prob(T)}{\partial \lambda \mathrm{FT1}} = 307.0 \text{ hr} \qquad \text{(Eq 32 of Ref 7)}$$

The answers are exactly the same. The sensitivity calculations for all of the devices are presented in Table 6. Note that in every case the sensitivity found using the analytical expression and the numerical estimate are the same. This is a general result which will be true for all basic event equipment failure rates found in a fault tree minimal cut set.

Once found, the sensitivities can now be used to compute the variance contribution of each particular device to the overall variance of the top event PFD in the same manner as Freeman previously presented.

The sensitivities of the top event probability are summarized in Table 6. We may now compute the variance contribution of each uncertain variable using Equations 1 – 4 in the methodology section of this paper. The relative variance contribution is expressed as a percent of the total variance of the top event probability. The top event variance is found to be 8.24E-6 (failures per hour)^2 or equivalently the top event standard deviation is 2.78E-3 failures per hour.

In this example the variance (measure of uncertainty) is dominated by the flow transmitters. The flow transmitters account for almost 2/3 of the uncertainty in the top event probability. Failures of the flow transmitters appear in 3 of the minimal cut sets. Since the uncertainty in the transmitter failure rates is large, the flow transmitters should contribute a significant amount to the variance of the fault tree top event.

We can use the variance in the top event to define the likelihood of achieving SIL 2 performance of the interlock. The calculations are done using the normal probability distribution. The normal distribution is tabulated as the standard normal distribution using a normalization factor Z.

The standard normal factor, Z, [Ref. 11] is defined as:

$$\mathrm{Z} = \left[\frac{xi - E(x)}{\sigma}\right] \qquad \text{(Eq 36)}$$

Where:

σ = standard deviation. Note that the variance of a random variable is the square of the standard deviation of the random variable.

$E(x)$ = Expected value of the random variable x_i

For the example, the expected value of the probability of the top event of concern was calculated as

$E[Prob(T)] = 1.47E\text{-}2$ (Eq 37)

From Table 4, the standard deviation in the probability of the top event is

$\sigma_T = 2.78E\text{-}3$ (Eq 38)

The target PFD for a SIL 2 interlock must not be worse than 1E-2 (RRF of 100).

We compute the normal distribution Z factor at the SIL 2 target value as

$Z = (0.01 - 0.0147)/2.78E\text{-}3$ (Eq 39)

$Z = -1.69065$ (Eq 40)

Tabulations of standard normal distribution are presented in most statistics books such as Meyer [Ref. 11]. For this value of Z, the corresponding probability that the interlock will achieve SIL 2 performance is 4.55 percent. We can also compute the risk reduction that the proposed interlock design will achieve.

What is the interlock probability of failure on demand (PFD) that we can be 95% certain that the interlock will provide? At the 95% level, the corresponding Z factor is 1.65 [Ref 11]. We now compute the corresponding PFD of the interlock that corresponds to this Z factor.

$$Z = \left[\frac{x95\% - E[Prob(T)]}{\sigma}\right]$$ (Eq 41)

In this case:

$x_{95\%}$ = the PFD that we can be confident that the interlock will achieve

$\sigma_T = 2.78E\text{-}3$ (Eq 42)

$E[Prob(T)] = 1.47E\text{-}2$ (Eq 43)

$Z = 1.65$ (Eq 44)

Rearranging equation 40 to find $x_{95\%}$ we obtain:

$x_{95\%} = E[Prob(T)] + Z\,\sigma_T$ (Eq 45)

$x_{95\%} = 1.47E\text{-}2 + (1.65)(2.78E\text{-}3) = 0.0193$ (Eq 46)

or a 95% certain risk reduction factor of

$RRF = 1/PFD = 1/0.0193 = 52$ (Eq 47)

By the same logic there is 5% chance that the risk reduction is a PFD of 0.01 or an RRF of 99. In summary, this interlock is likely to perform as a mid-range SIL 1 interlock, not the SIL 2 interlock desired by process hazards analysis team.

5. Methodology for Fault Tree Uncertainty Analysis

We can now generalize the methods used in the above example to a proposed method for the evaluation of the uncertainty in a fault tree. The following step-by-step procedure is modified from that that previously published by Freeman (Ref 7) to indicate the use of numerical methods to compute the sensitivity. Step 8 shown in *italics* indicates the modification..

1. Create the fault tree using standard methods outline in the CPQRA book [Ref 1].
2. Determine minimal cut sets by hand or using standard computer software such as SAPHIRE [Ref 12].
3. Define the needed failure rate data for each basic event.
4. Define those basic events that are considered to be uncertain.
5. For the basic events that are uncertain, define the probability distribution and associated parameters needed to numerically define the probability distribution. Appendix B of this paper presents the needed description of 4 commonly used probability distributions. In many cases the triangular distribution will be selected. The minimum, maximum and mode (most likely) parameters will be needed for the triangular distribution.
6. Compute the mean of each of the uncertain variable
7. Compute the probability or frequency of the fault tree top event of interest using the mean value for each uncertain variable.
8. *Compute the sensitivity of the top event probability or frequency using the numerical methods outlined in this paper.*
9. Compute the variance of each of the uncertain variables
10. Compute the variance contribution of for each of the uncertain variables to the top event probability or frequency using equation 6.
11. Compute the total variance of the top event of interest probability or frequency by summing all of the contributions determined in Step 10.
12. Compute the variance contribution percent of each uncertain variable by dividing the variable contribution (Step 10) by the total variance of the top event of interest (Step 11).
13. Define the level of risk the project is willing to take. What chance will the project management accept for potential failure of the interlock to achieve the desired risk reduction? In the above example, I have used a 5% risk of failure or a 95% certainty that the interlock will achieve a risk reduction factor of 52. In the example, there is a very low probability that the interlock will achieve a SIL 2 PFD of 0.01 or an RRF of 100.

Alternately you can report the 90% range for the top event of interest. For the example the 90% range starts at the 5% RRF of 99 and there is a 95% certainty that that an RRF of at least 52 will be achieved. The expected risk reduction for this interlock is an RRF of 68.

6. Conclusions

A general procedure has been presented to quantify the uncertainty in calculations of failure frequencies using the fault tree methodology. The uncertainty analysis procedure is based on the application of propagation of error and variance contribution analysis techniques to the minimal cut sets created during a fault tree study. The procedure is simple and can be incorporated into standard fault tree analysis programs such as SAPHIRE [Ref 8]. This procedure is based on perturbation calculation of the frequency of the top event from changes in the uncertain variables. Incorporation of a numerical perturbation method into standard fault tree software such as SAPHIRE would be easy to implement and would provide the means for evaluation of the uncertainty in the quantified fault tree results.

For the example interlock presented in this paper, the uncertainty in the device failure rates degrades the predicted interlock SIL from meeting SIL 2 requirements to becoming a mid range SIL 1 interlock. This is a general result. Uncertainty in design parameters will reduce the likelihood that the equipment will achieve desired results. The design engineer has two choices:

- Accept the conclusion that the interlock will not perform as SIL 2
- Re-design the interlock using more reliable components to achieve the SIL 2 target

The variance contribution analysis shown in Table 7 indicates that the major contributor to the uncertainty in the interlock performance is the Flow Transmitters. Flow transmitters certified to SIL-2 or SIL-3 performance with lower failure rates could improve the calculated PFD of the interlock.

7. References

1. Center for Chemical Process Safety , *Guidelines for Chemical Process Quantitative Risk Analysis – 2rd edition*, American Institute of Chemical Engineers, New York, 2000.
2. Frank P Lees, Loss Prevention in the Process Industries – 2rd Edition, Butterworth, ISBN 0-7506-1547-8, Oxford, UK, 1996
3. US Department of Labor, Occupational Safety and Health Administration, CFR 1910.119 - Process Safety Management of Highly Hazardous Chemicals, 1910.119(e)(2)(vi).
4. Edward B. Haugen, *Probabilistic Approaches to Design*, John Wiley, New York, 1968
5. Raymond. Freeman and Angela Summers, "Evaluation of Uncertainty in Safety Integrity Level Calculations," *Process Safety Progress,* Published online in Wiley Online Library (wileyonlinelibrary.com) DOI 10.1002/prs. 11805, 2016
6. Raymond. Freeman, "General Method for Uncertainty Evaluation of Safety Integrity Level Calculations – Part 2 Analytical Methods," *Process Safety Process*, Published online in Wiley Online Library (wileyonlinelibrary.com) DOI 10.1002/prs 111915, October 2017
7. Raymond. Freeman ,"Error Propagation and Uncertainty Analysis: Application to Fault Tree Analysis", Process Safety Progress, Published Wiley Online Library (wileyonlinelibrary.com/journal/prs) DOI: https://doi.org/10.1002/prs.12080, June 29, 2019

8. US Nuclear Regulatory Commission, "Systems Analysis Programs for Hands-on Integrated Reliability Evaluations (SAPHIRE) – Version 8," NUREG/CR-7039, Vol 3, June 2011.
9. ISA, "Safety Instrumented Functions (SIF) – Safety Integrity Level (SIL) Evaluation Techniques Part 3: Determining the SIL of a SIF via Fault Tree Analysis, ISA Technical Report ISA-TR-84.00.02-2002 – Part 3, Approved 17 June 2002.
10. David J. Smith, *Reliability Maintainability and Risk – 8th Edition*, ISBN 978-0-08-096902-2, Elsevier, London, 2011
11. Paul L. Meyer, Introductory Probability and Statistical Applications, John Wiley, New York, 1972

Table 1. Comparison of Analytical Sensitivity with Numerical Approximation with Perturbation of 10% in xi for the equation: $y = F(xi) = Axi^n + B$

Order of Equation n	Equation for Analytical Sensitivity $\frac{\partial y}{\partial x} = nAxi^{n-1}$	Sensitivity with xi = 1 and A=1	Numerical Sensitivity with $\Delta xi = 0.1\ xi$	Error %
1	$\frac{\partial y}{\partial x} = 1A(xi)^0$	1	1	0
2	$\frac{\partial y}{\partial x} = 2A(xi)^1$	2	2.1	5
3	$\frac{\partial y}{\partial x} = 3A(xi)^2$	3	3.31	-10%
4	$\frac{\partial y}{\partial x} = 4A(xi)^3$	4	4.64	-16%
5	$\frac{\partial y}{\partial x} = 5A(xi)^4$	5	6.11	-22%
6	$\frac{\partial y}{\partial x} = 6A(xi)^5$	6	7.72	-29%
7	$\frac{\partial y}{\partial x} = 7A(xi)^6$	7	9.49	-36%
8	$\frac{\partial y}{\partial x} = 8A(xi)^7$	8	11.44	-43%
9	$\frac{\partial y}{\partial x} = 9A(xi)^8$	9	13.58	-51%
10	$\frac{\partial y}{\partial x} = 10A(xi)^9$	10	15.94	-59%

Error = (Analytical- Numerical)/Analytical

Table 2. Comparison of Analytical Sensitivity with Numerical Approximation with Perturbation of 1% in xi for the equation: y = F(xi) = $Axi^n + B$

Order of Equation n	Equation for Analytical Sensitivity $\frac{\partial y}{\partial x} = nAxi^{n-1}$	Sensitivity with xi = 1 and A=1	Numerical Sensitivity with $\Delta xi = 0.1\, xi$	Error %
1	$\frac{\partial y}{\partial x} = 1A(xi)^0$	1	1	0
2	$\frac{\partial y}{\partial x} = 2A(xi)^1$	2	2.01	-1%
3	$\frac{\partial y}{\partial x} = 3A(xi)^2$	3	3.03	-1%
4	$\frac{\partial y}{\partial x} = 4A(xi)^3$	4	4.06	-2%
5	$\frac{\partial y}{\partial x} = 5A(xi)^4$	5	5.10	-2%
6	$\frac{\partial y}{\partial x} = 6A(xi)^5$	6	6.15	-3%
7	$\frac{\partial y}{\partial x} = 7A(xi)^6$	7	7.21	-3%
8	$\frac{\partial y}{\partial x} = 8A(xi)^7$	8	8.29	-4%
9	$\frac{\partial y}{\partial x} = 9A(xi)^8$	9	9.37	-4%
10	$\frac{\partial y}{\partial x} = 10A(xi)^9$	10	10.46	-5%

Error = (Analytical- Numerical)/Analytical

Table 3. Basic Events in Minimal Cut Sets For Example

Cut Set No	BE 1	BE 2
1	PE	-
2	TS1	TS2
3	LS1	LS2
4*	FT1	FT2
5*	FT2	FT3
6*	FT1	FT3
7*	BV1	BV2
8*	BV1	SOL1
9*	BV2	SOL2
10*	SOL1	SOL2
11	PT1	PT2

* Minimal cut Set with same basic event as another cut set

Table 4. Lamda (λ) Failure Rate Data With Uncertainty Limits

Device	Minimum Failures per 1E6 Hours	Mode Failures per 1E6 Hours	Maximum Failures per 1E6 Hours	Mean Failures per 1E6 Hours	Lamda Variance (Failures per 1E6 Hours)^2
Flow Transmitters FT1, FT2 and FT3	1.0	2.9	20.0	8.0	18.3
Pressure Transmitters PT1 and PT2	1.0	2.3	20.0	7.8	18.8
Temperature Switch TS1 and TS2	3.0	7.6	20.0	10.2	12.9
Level Switch LS1 and LS2	2.0	4.6	20.0	8.9	15.8
Block Valves BV1 and BV2	0.2	2.3	10.0	4.2	4.4
Solenoid Valves SOL1 and SOL2	1.0	2.3	8.0	3.8	2.3

Table 5. Probability of Minimal Cut Sets Based on Mean Failure Rates

Cut Set No	BE 1	BE 2	PFD based on mean λ
1	PE	-	5.00E-03
2	TS1	TS2	2.00E-03
3	LS1	LS2	1.52E-03
4	FT1	FT2	1.23E-03
5	FT2	FT3	1.23E-03
6	FT1	FT3	1.23E-03
7	BV1	BV2	3.38E-04
8	BV1	SOL1	3.38E-04
9	BV2	SOL2	3.38E-04
10	SOL1	SOL2	2.77E-04
11	PT1	PT2	1.17E-03

Table 6. Sensitivity of Example System PFD to Basic Event Failure Rates Based on a 1% Perturbation

Basic Event Label	Device Type	Mean Failures per 1E6 Hours	Failure Perturbation of 1% of Mean Failures per 1E6 Hours	Perturbed Failure Rate Failures per 1E6 Hours	Total System PFD Using Perturbed Failure Rate of Basic Event	Change in Total System PFD	Sensitivity of PFD to Change in Failure Rate Hours
FT1	Flow Transmitter	8	0.08	8.08	1.461848E-02	2.455603E-05	307.0
FT2	Flow Transmitter	8	0.08	8.08	1.461848E-02	2.455603E-05	307.0
FT3	Flow Transmitter	8	0.08	8.08	1.461848E-02	2.455603E-05	307.0
PT1	Pressure Transmitter	7.8	0.078	7.878	1.460560E-02	1.167179E-05	149.6
PT2	Pressure Transmitter	7.8	0.078	7.878	1.460560E-02	1.167179E-05	149.6
TS1	Temperature Switch	10.2	0.102	10.302	1.461389E-02	1.995945E-05	195.7
TS2	Temperature Switch	10.2	0.102	10.302	1.461389E-02	1.995945E-05	195.7
LS1	Level Switch	8.9	0.089	8.989	1.460912E-02	1.519596E-05	170.7
LS2	Level Switch	8.9	0.089	8.989	1.460912E-02	1.519596E-05	170.7
BV1	Block Valves	4.2	0.042	4.242	1.460037E-02	6.445958E-06	153.5
BV2	Block Valves	4.2	0.042	4.242	1.460037E-02	6.445958E-06	153.5
SOL1	Solenoid Valves	3.8	0.038	3.838	1.459976E-02	5.832058E-06	153.5
SOL2	Solenoid Valves	3.8	0.038	3.838	1.459976E-02	5.832058E-06	153.5

Table 7. Variance of Top Event Probability Due to Uncertain Device Failure Rates

Device No	Device Type	Label	Sensitivity of Top Event Probability to Device Failure Rate Hour^{-1}	Variance of Device Failure Rate (Failures per 1E6 Hour)^{2}	Variance Contribution to Top Event Variance	Contribution to Top Event Variance %
1	Logic Solver	PE	-	0	0.00E+00	0%
2	Temperature Switch	TS1	195.7	12.9	4.94E-07	6%
3	Temperature Switch	TS2	195.7	12.9	4.94E-07	6%
4	Level Switch	LS1	170.7	15.8	4.60E-07	6%
5	Level Switch	LS2	170.7	15.8	4.60E-07	6%
6	Flow Transmitter	FT1	307	18.3	1.72E-06	21%
7	Flow Transmitter	FT2	307	18.3	1.72E-06	21%
8	Flow Transmitter	FT3	307	18.3	1.72E-06	21%
9	Block Valve	BV1	153.5	4.4	1.04E-07	1%
10	Block Valve	BV2	153.5	4.4	1.04E-07	1%
11	Pressure Transmitter	PT1	149.6	18.8	4.21E-07	5%
12	Pressure Transmitter	PT2	149.6	18.8	4.21E-07	5%
13	Solenoid valve	SOL1	153.5	2.3	5.42E-08	1%
14	Solenoid valve	SOL2	153.5	2.3	5.42E-08	1%

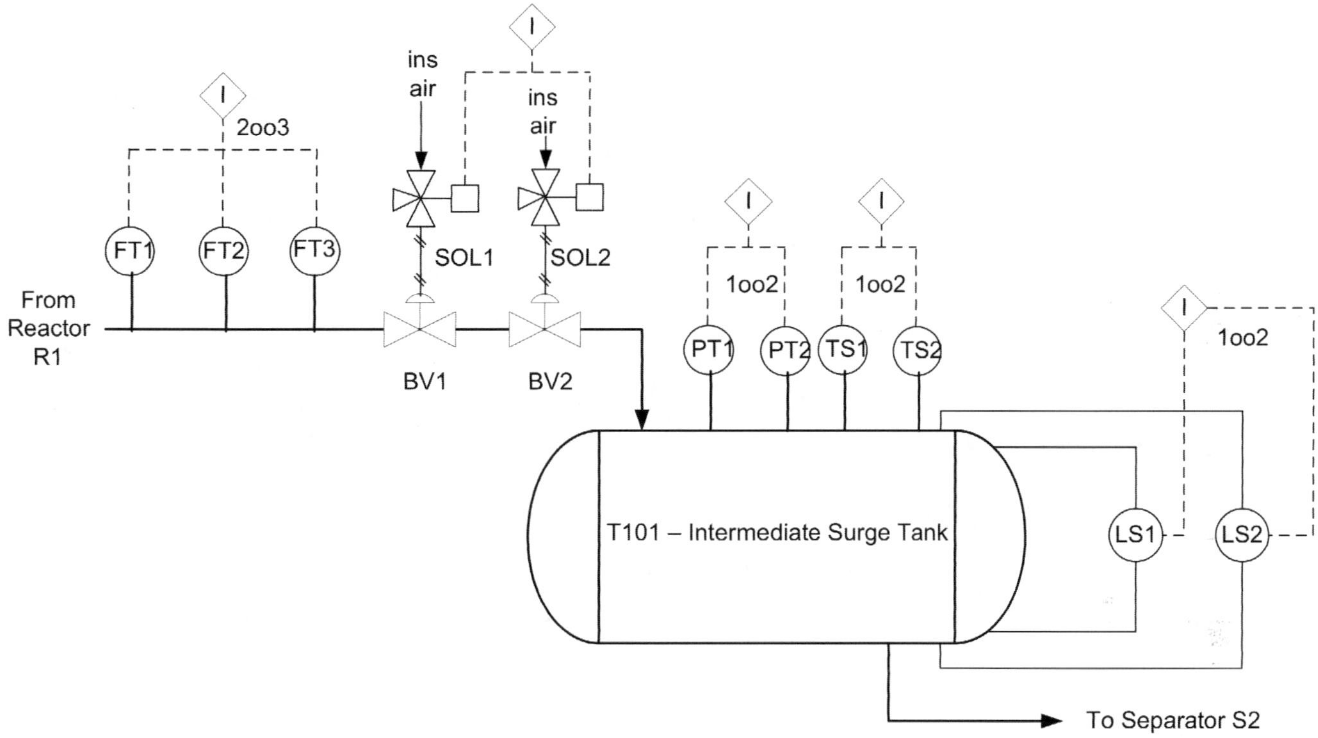

Figure 1. P&ID Diagram for Example

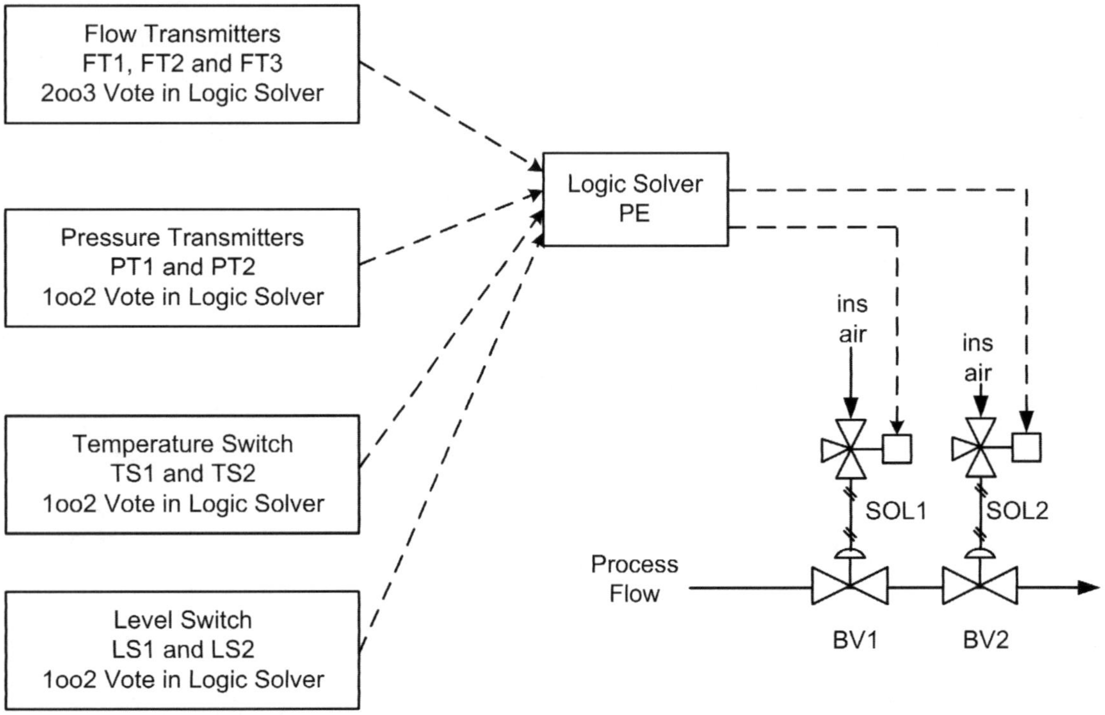

Figure 2. Interlock Block Diagram

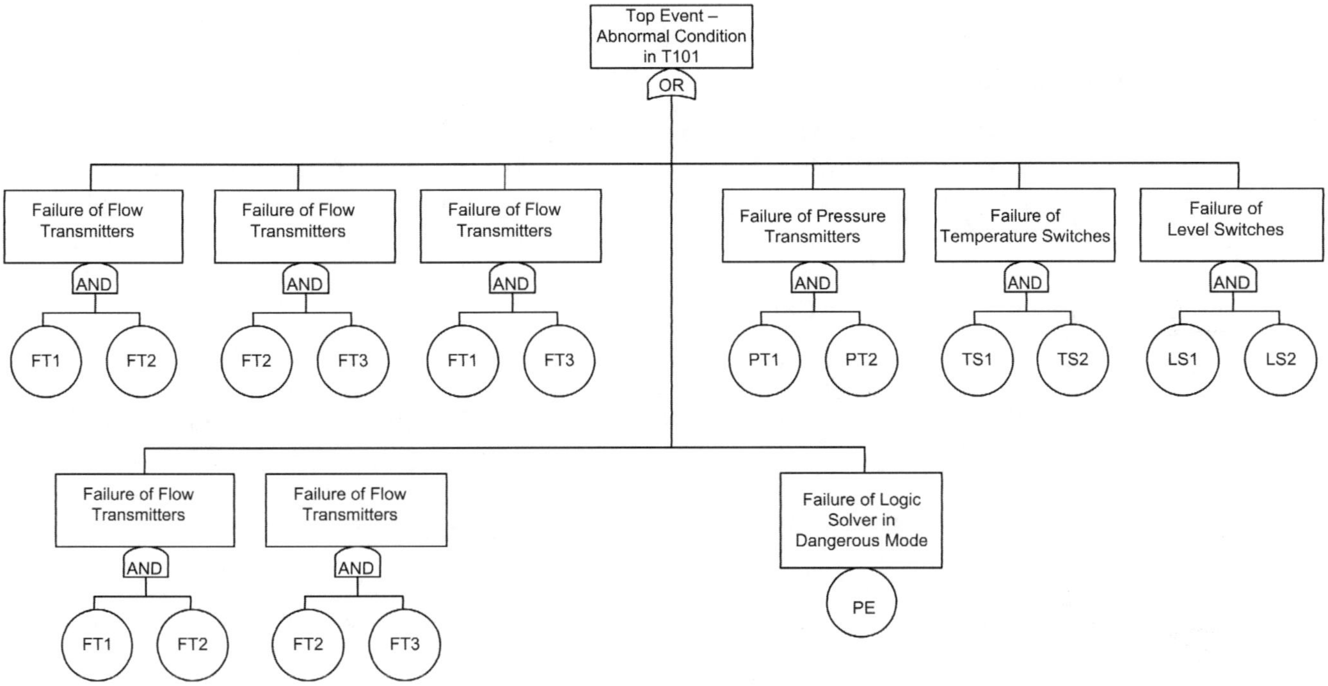

Figure 3. Fault Tree for Example Interlock

Appendix A – Commonly Encounter Probability Distributions

There are four probability distributions used to represent the failure rate of equipment items. These are:

1. Uniform Distribution
2. Triangular Distribution
3. Normal Distribution
4. Log-Normal Distribution

This appendix reviews the properties of these four probability distributions.

1. Uniform Distribution

The uniform distribution [Ref. A-1, pg 74; A-2, pg 687-688] is used when a variable is only known within a defined range. Figure A-1 presents a plot of the probability density function (pdf) for the uniform distribution. The distribution is defined using two parameters, A and B.

Where:

A = minimum value of the random variable

B = maximum value of the random variable

The mean or expected value of a random variable, x, that follows the uniform distribution is:

Mean = Expected Value – $E(x) = (A+B)/2$ (Eq A-1)

The variance of a random variable, x, that follows the uniform distribution is:

Variance = $V(x) = (B-A)^2/12$ (Eq A-2)

The standard deviation of a random variable, x, that follows the uniform distribution is:

Standard Deviation = $(\text{Variance})^{1/2} = (B-A)/(12)^{1/2}$ (Eq A-3)

2. Triangular Distribution

The triangular distribution [Ref. A-2, pg 686-687] is used when a variable is known to lie within a defined range and a "best guess" or estimate can be made as to the most likely value to the variable. Figure A-2 presents a plot of the probability density function (pdf) for the triangular distribution. The distribution is defined using three parameters, A. B and C.

Where:

A = minimum value of the random variable

B = maximum value of the random variable

C = most likely (mode) of the random variable

The mean or expected value of a random variable, x, that follows the triangular distribution is:

Mean = Expected Value – $E(x) = (A+B+C)/3$ (Eq A-4)

The variance of a random variable, x, that follows the triangular distribution is:

Variance = $V(x) = [A^2 + B^2 + C^2 - AB - AC - BC]/18$ (Eq A-5)

The standard deviation of a random variable, x, that follows the triangular distribution is:

Standard Deviation = $(\text{Variance})^{1/2} = \{[A^2 + B^2 + C^2 - AB - AC - BC]/18\}^{1/2}$ (Eq A-6)

3. Normal distribution

The normal distribution [Ref A-2, pg 665-666] is used when a variable when data analysis finds the normal distribution is the best model to describe the spread in the measured variable. Figure A-3 presents a plot of the probability density function (pdf) for the normal distribution. The distribution is defined using two parameters, μ and σ.

Where:

μ = mean of the measured data for variable

σ = standard deviation of the measured data for variable

The expected value of a random variable, x, that follows the normal distribution is:

Expected value = mean = $E(x) = \mu$ (Eq A-7)

The variance of a random variable, x, that follows the normal distribution is:

Variance = square of standard deviation = $V(x) = \sigma^2$ (Eq A-8)

The standard deviation of a random variable, x, that follows the normal distribution is:

Standard deviation = σ (Eq A-9)

4. Lognormal Distribution

The lognormal distribution [Ref. A- 2, pg 658-659] is used when a variable when data analysis finds the lognormal distribution is the best model to describe the spread in the measured variable. Figure A-4 presents a plot of the probability density function (pdf) for the lognormal distribution. Many variables in nature are lognormally distributed (for example the height of people). The log normal distribution is used when the variable of interest has a known physical lower limit of zero. A variable, x, is lognormally distributed if ln(x) is normally distributed. The lognormal distribution is defined using two parameters, μ_y and σ_y.

Let the original data be defined by the variable x with mean and standard deviation as:

μ_x = mean of the measured data for variable x

σ_x = standard deviation of the measured data for variable x

Define a new variable y as:

$y = Ln(x)$ (Eq A-10)

Then variable x is said to be lognormally distributed if y is normally distributed. We can now write the mean and standard deviation of y (μ_y and σ_y.) [Ref. A-3, A-4 and A-5] as

$$\mu_y = Ln[\mu_x^2 /((\sigma_x^2 + \mu_x^2)^{1/2})] \quad \text{(Eq A-11)}$$

$$\sigma_y = [Ln(\sigma_x^2 / \mu_x^2 + 1)]^{1/2} \quad \text{(Eq A-12)}$$

Where:

Ln is the natural logarithm (base e) of the argument

Appendix A References:

A-1. P. L. Meyer, *Introductory Probability and Statistical Applications*, 2nd Edition, Addison-Wesley, Reading Mass, Library of Congress Catalog no. 75-104971, 1972, pp. 74

A-2. D. Vose, *Risk Analysis–A Quantitative Guide*, 3rd edition, Wiley, Chichester, West Sussex, England, ISBN 978-0-470-51284-5, 2008, pp. 686-687

A-3. "Lognormal mean and variance - MATLAB lognstat". www.mathworks.com. Retrieved 27 December 2018

A-4. M. Mood, F. A. Graybill, and D. C. Boes. *Introduction to the Theory of Statistics.* 3rd ed., New York: McGraw-Hill, ISBN 0-07-042864-6, 1974. pp. 117.

A-5. E.L. Crow and K. Shimizu, Lognormal Distributions – Theory and Applications, Marcel Dekker, New York, ISBN 0-8247-7803-0, 1988, pp. 9

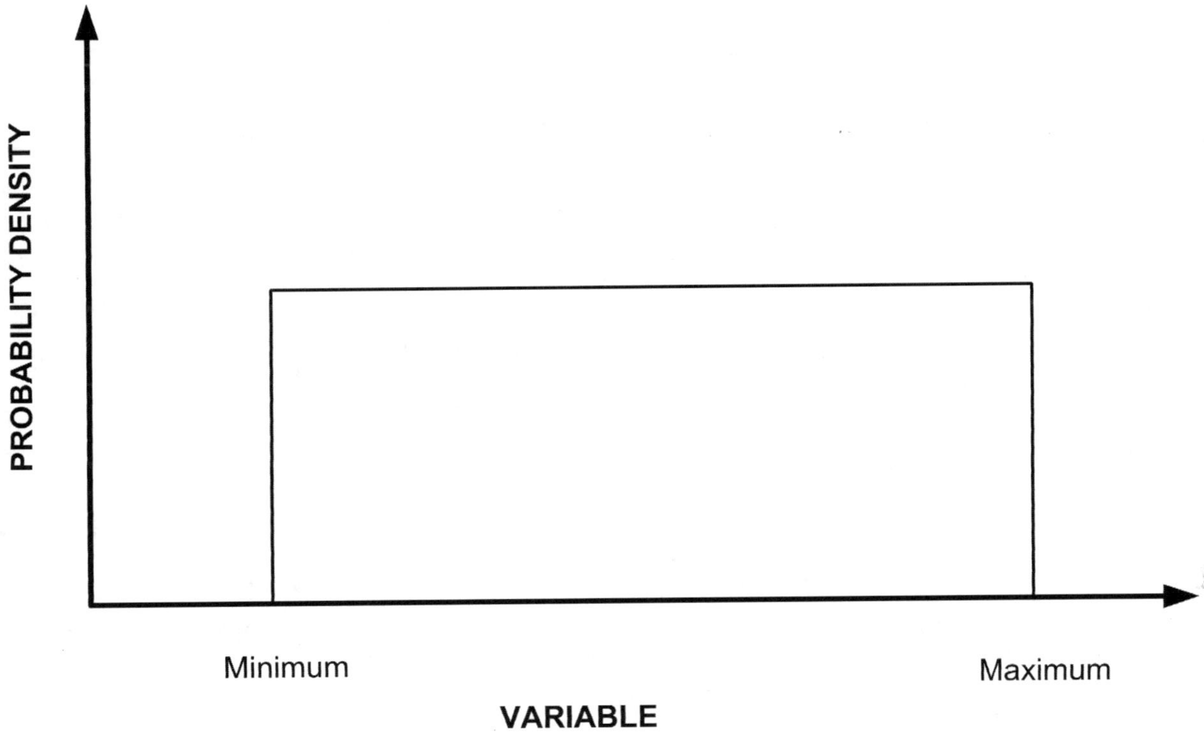

Figure A-1. Uniform Probability Distribution

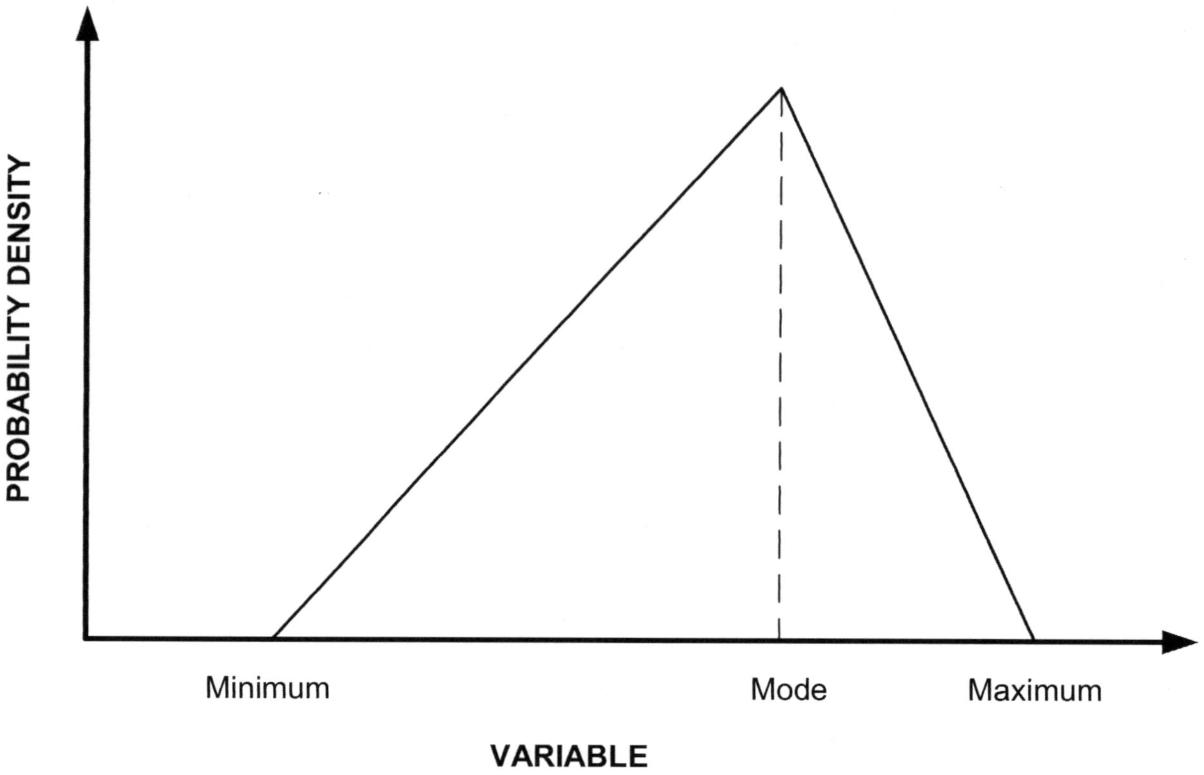

Figure A-2. Triangular Probability Distribution

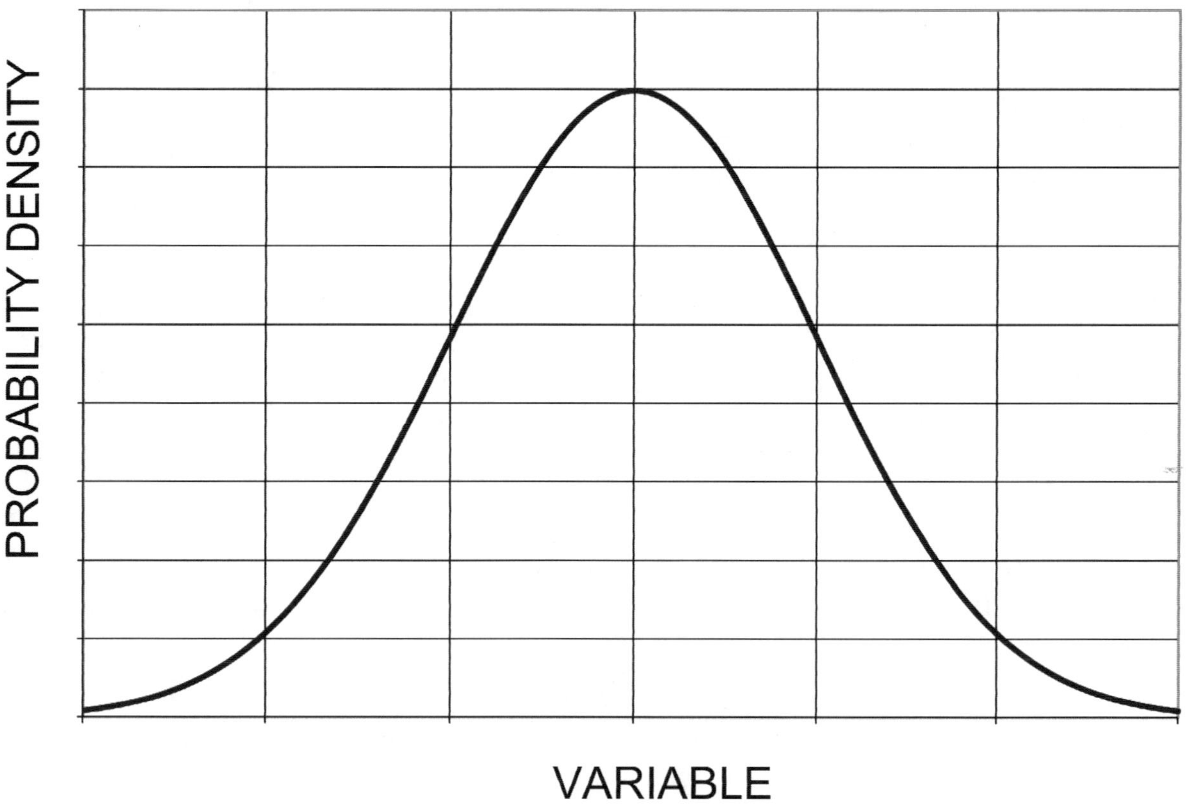

Figure A-3. Normal Probability Distribution

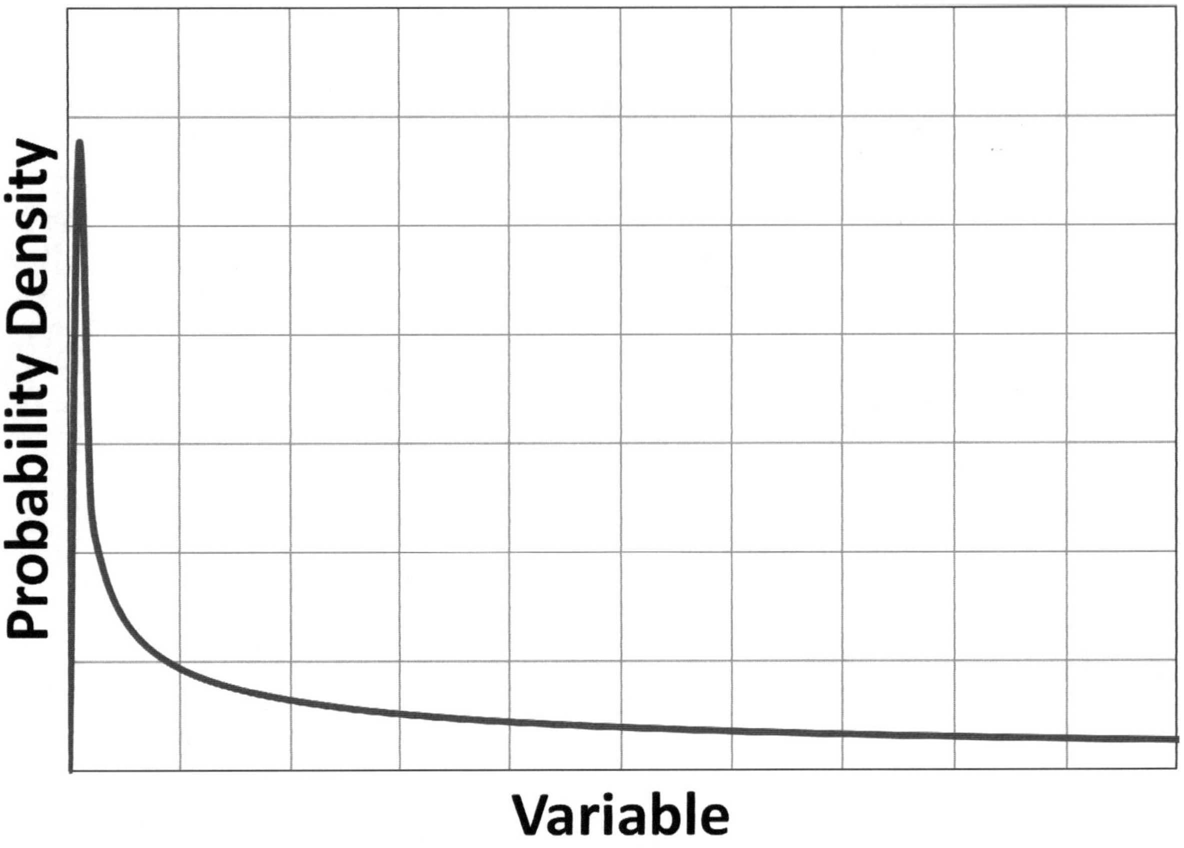

Figure A-4. Lognormal Probability Distribution

590424 Making the Most of Your PHA

Abstract:

A Process Hazard Analysis (PHA) is required to address OSHA Process Safety Management (PSM) and EPA Risk Management Program (RMP) regulations. Understanding the potential process hazard is widely recognized as a foundational element of PSM and RMP, and in addition to simply meeting regulatory requirements, can be a critical mechanism for flushing out potential vulnerabilities early in the design process for capital projects. Although an important activity, PHA is also recognized as a sizeable investment in time and energy for an Owner/Operator. So why not make the most of it? Other important activities at plants such as Mechanical Integrity, Alarm Rationalization, and Management of Change are able to stay focused and possibly be streamlined (i.e., reduced cost) based on the ability to use the results of a quality PHA having the necessary information and usability of documentation. So why not structure the PHA and documentation to be able to capitalize on potentially significant cost-savings and possible program integration?

The thrust of this paper and conference presentation will be to identify tips for conducting and documenting the PHA to maximize its ability to be used for supporting other important programs such as MOC, MI, and Alarm Rationalization. The objective will be to allow PHA Coordinators to be able to readily apply these tips to bring about improvements and cost-savings on these interrelated programs.

589814 Theoretical Analysis of the Viability of Isopropanol-Cyclohexane Solvent Systems for Biomatter Lipid Extraction

Abstract:

A theoretical alternative solvent combination for the extraction of lipids from biomatter is the Water, Isopropanol, and Cyclohexane instead of the more common combination used in the Bligh and Dyer Method (Water, Methanol, and Chloroform). While the Bligh and Dyer Method is considered effective, toxicological concerns have been raised over its use of both Chloroform and Methanol. The alternative method utilizes solvents that do not pose as high as a toxicological concern while theoretically maintaining a similar level of practical viability. The alternative method's extraction efficacy was judged against the performance of the Bligh and Dyer method using the simulation software Aspen Plus V10®. The equilibrium data generated from Aspen Plus V10® was then graphed into tertiary diagrams for data analysis using the graphing software SigmaPlot.

592166 Techno-Economic Analysis of the Conversion of CH4 and CO2 Intoacetic Acid Using Microwave Plasma

Abstract:

Methane and carbon dioxide together represent 90% of greenhouse gases [1], and one way of diminishing their emissions is through the development of pathways that convert them into added value chemicals. Acetic acid is one of such chemicals, and efforts at the laboratory scale have been carried out to obtain alternative ways of producing it. However, there are very few studies that evaluate the possibility of implementing these processes at a larger scale. In this paper, techno-economic and sustainability analyses were used to assess the feasibility of upscaling two pathways to produce acetic acid: (1) by direct conversion of methane and carbon dioxide and (2) by indirect conversion, using syngas (obtained from dry methane reforming) as an intermediate. For the direct method a microwave non-catalytic plasma reactor was considered as the main driver in the process simulation [2]. For the indirect method, a zeolite catalytical single step reaction unit was considered [3]. APEN Plus was used to simulate the reaction and purification systems of both methods, and alongside literature values and plant cost estimation techniques, the capital and operational expenses were calculated. Finally, the feasibility of both projects is assessed on carbon sequestration mass balances and economic metrics. The main sensitivity variables that need to be improved in order to obtain a minimum ROI threshold value were also evaluated.

Key words: carbon management, process design, microwave plasma reactor

[1] EPA. *Overview of Greenhouse Gases*. United States Environmental Protection Agency. 2017.

[2] Wang, L., Yi, Y., Wu, C., Guo, H., Tu, X. *One-Step Reforming of CO2 and CH4 into High-Value Liquid Chemicals and Fuels at Room Temperature by Plasma-Driven Catalysis.* Angewandte Chemie. 2017.

[3] Lopez, L., Velasco, J., Montes, V., Marinas, A., Cabrera, S., Boutonnet, M., Jaras, S. *Synthesis of Ethanol from Syngas over Rh/MCM-41 Catalyst: Effect of Water on Product Selectivity*. Catalysts. 2015.

589668 Model Predictive Control Design for the Allam Power Plant Integrated with an Air Separation Unit

Abstract:

Since electricity and heat generation account for around 25% of this greenhouse gas emission, high efficiency power generation technology such as the Allam cycle which emits essentially no CO_2 and NOx is much needed. Being an oxy-fuel power cycle, the Allam power plant is inherently integrated with an air separation plant (ASU). In this study Aspen Plus Dynamics was used to model the integrated Allam power plant/Air Separation Unit (ASU) with a high degree of heat and work integration based on the steady state work of Net Power. In our previous work, regulatory and advanced PID controls were implemented and tested for various set point tracking and disturbance rejection scenarios. However, PID controllers have limitations for various reasons. Consequently, a model predictive control (MPC) was designed in this study to optimize and improve the performance of the integrated unit and to test its performance in terms of dynamic behavior and robustness to model uncertainties.

589672 Explore the Function of Ion-Exchange Membrane (IEM) in Membrane Capacitive Deionization (mCDI) Cell, an Alternative Technique in Brackish Water Desalination Field

Abstract:

In the arid west, the freshwater supply of many communities is limited, leading to increased interest in tapping brackish water resources. Although reverse osmosis is the most common technology to upgrade saline waters, there is also interest in developing and improving alternative technologies. Here we focus on Capacitive Deionization (CDI), which has attracted broad attention as an energy efficient desalination technology, especially for ion extraction from low salinity brackish water. The objective of this work is to explore the function of ion-exchange membrane (IEM) in enhancing the performance of CDI cell. In this study, a fully coupled two-dimensional model describing ion transport and adsorption behavior in Membrane Capacitive Deionization (mCDI) cell was developed and solved numerically, taking advantage of Darcy's Law, Nernst-Planck equation, and Donnan equilibrium considering hydraulic dispersion in channel, non-ideal membrane, and modification of ion diffusivity in porous media. The model behavior was supported by the experimentally observed variations in transient effluent concentration under different operating conditions. The results showed that ion flux was significantly improved because of the IEM, leading to a faster desalination rate. The equilibrium adsorption amount is increased due to the permselectivity of the IEM, resulting in extra ion accumulation in electrode's macropores during desalination process. The cut-off time (optimal time to switch polarity to generate highest energy efficiency) of mCDI was shorter with a higher salt removal percentage compared to CDI under the same operating conditions. A sensitivity study of the thickness and the fixed charge density (starting from 500mM) of the IEM was done separately, and neither of them were found to be the dominant influencing factors under a low concentration range of brackish water (below 100mM) in this study.

590103 An Integrated Process for the Production of Propylene Glycol from Shale Gas

Abstract:

Propylene glycol (PG) is produced on a large scale primarily for the production of environmentally safest and least toxic industrial and domestic products such as polyester resins, thermoset plastics, polyurethanes, food additives and antifreeze. In this research, a novel integrated plant is developed for the production of PG from shale gas. This integrated approach has the benefit of safer operating conditions since the intermediate propylene oxide, which is explosive, does not need to be stored and transported. Furthermore, there are potential economic benefits from integration. The overall plant is simulated in the Aspenplus environment, and a variety of process conditions are tested at steady-state to optimize the production of PG. Heat-integration tools are utilized for energy-saving and capital cost reduction opportunities. A comparative economic assessment based on existing plant information indicates that the use of process integration techniques has the potential to reduce costs significantly.

Keywords: Process Integration, Heat Integration, Propylene Glycol

592456 Monitoring Methods for Fault Detection and Diagnosis: Algorithms and Applications

Abstract:

Many industrial processes collect an abundance of data from different sensors. These sensors often measure a wide variety of physical properties in order to ensure that these parameters are adhering to expected values. This is essential to ensure plant and operator safety, increase economic benefits, and maintain product quality.

A number of algorithms have been developed in order to improve existing fault detection and diagnosis performance. These algorithms integrate a number of different data-driven driven tools and methods. Multiscale wavelet-based representation of data can be used in order to handle data that is autocorrelated, non-Gaussian, and noisy. Hypothesis testing methods such as the Generalized Likelihood Ratio (GLR) technique can be used in order to provide the best possible detection for a fixed false alarm rate. Moreover, certain model-based methods have been developed in order to monitor process drifts and degradations in the process model, even when a process is operating under control.

This work will discuss the different tools utilized in development of the algorithms, their features, and practical applications, including the benchmark Tennessee Eastman Process (TEP).

589902 Selective Separation of Industrial Relevant Metal Ions from High Tds Water Using Wafer-Enhanced Electrodeionization (WE-EDI)

Abstract:

There is a need for an efficient and economical process of ion-selective separation. In hydraulic fracturing, for instance, several million gallons of water are used in a site, most of which are disposed of by deep well injection taking the water from the ecosystem and causing earthquakes. However, due to the high amount of ions in wells at fracking sites, pressure driven processes for water recovery are impossible to use. This project will demonstrate an emerging environmentally friendly Wafer Enhanced Electrodeionization (WE-EDI) technology which attempts to eliminate the shortcomings that come with normal EDI and separates effluent streams into concentrated ions and clean water. In WE-EDI, ion-exchange resins are replaced by wafers that consist of cation and anion exchange beads and a polymer binding agent. In this study, selective separation of these ion exchange wafers and the effect of electroosmotic drag on the membrane structure have been performed for various anions and cations (Na^+, K^+, Ca^{2+}, Cl^-) and improved selectivity and higher water recoveries have been observed while significant modeling has taken place to predict the selective separation as a strong function of the equilibrium constants between the ions and the beads. This project is supported by the Membrane Science and Technology (MAST) Center and is conducted with the collaboration and guidance of industry mentors.

593303 Influence of Particles on Amine Losses during CO_2 Capture: A Process Simulation Coupled Aerosol Dynamics Model _

Abstract:

Influence of Particles on Amine Losses during CO_2 Capture: A Process Simulation Coupled Aerosol Dynamics Model

David Dhanraj and Pratim Biswas

Aerosol and Air Quality Research Laboratory

Department of Energy, Environmental and Chemical Engineering

Washington University in St. Louis

St. Louis, MO 63130, USA

Efforts to reduce the concentration of atmospheric CO_2 have been initiated due to escalating concerns of global warming and climate change. Amine-based post-combustion CO_2 capture offers an economically viable retrofit for power-plants in combating these emissions. However, solvent losses have been a major challenge to commercializing this technology. Recent experiments[1] show that vapor-based and aerosol-based solvent losses are the key mechanisms of amine solvent loss during CO_2 capture. While vapor-based amine losses can be abated by employing techniques like water-washing, there has been little work to curtail aerosol-based solvent losses. This is primarily because the underlying mechanisms governing these losses are still not well understood. Although there are a few numerical models proposed for this system, there are no models that consider aerosol phenomena such as coagulation or particle loss. Moreover, these models have not been fully validated through experiments.

Furthermore, there are a few techniques that are being developed to tackle solvent losses. It is known that reducing particle number concentration in the flue gas will reduce amine emissions. Conventional Electrostatic Precipitators (ESP) are however inefficient in capturing particles in the size range 0.1 – 1.0 μm. Amongst different approaches demonstrated, introducing a pre-charging stage and a photo-ionization stage has shown potential in increasing the charging efficiency.

Therefore, the broad objectives of this work are primarily focused on two facets of the amine loss problem. The first is the development of a model that integrates aerosol phenomena with a process engineering model that accounts for the reactive mass transfer process in a packed-bed absorber. The second is the design, fabrication and testing of a bench-scale Photoionization Enhanced Two-staged Electrostatic Precipitator (PI-ESP) at Abbott Power Plant. In essence, with the PI-ESP aimed at reducing the particulate concentration in the flue gas, and the model relating the influence of particulate concentration in the flue gas to amine emissions, this work aims to illuminate the underlying mechanisms governing amine emissions, and demonstrate a solution to tackle the problem.

References

1. International Journal of Greenhouse Gas Control, vol. 85, pp. 86-99, 2019.
2. International Journal of Greenhouse Gas Control, vol. 19, pp. 138-144, 2013.

588702 A Real-World Case Study: Distillation Tower Constraint Control

Abstract:

Distillation operations constitute a significant portion of energy usage and unit operations of every Hydrocarbon Processing Industry (HPI) facility. While the principles of thermodynamic and mechanical design of these unit operations are well understood, often times the design of the instrumentation and controls becomes the rate-limiting factor affecting their reliable performance. Further, as ultimately happens, economic drivers compel operators to maximize production from these units, thus requiring the controls to keep the units' operations within mechanical limitations, which further complicates the controls' design.

Historically, Model Predictive control (MPC) or "APC" was successfully implemented to ensure reliable and optimal operations at the mechanical limits of existing equipment. Sometimes, for a variety of reasons, however, MPC fails to generate the expected outcomes. A common observation made across multiple HPI sites and companies is that constraint-based regulatory controls are, poorly understood and, consequently, either not implemented at all or implemented with limited success.

This presentation uses a real-world example of a distillation tower that was subject to numerous operating constraints and illustrates the thought process behind designing the regulatory cont

589872 Simultaneous Design and Control of a Distillation Column Under Disturbance for Cost, Inherent Safety, and Controllability

Abstract:

Traditional inherent safety strategies in the early stages of design focus on reducing the overall hazard of a process without considering its operability. The process is first designed to be inherently safer with respect to a nominal, steady state case. Then, after the process is designed, layers of protection are added and operability issues are addressed.

However, the way a process is designed not only affects its safety but also heavily impacts its operability. An intensified process may contain less of a hazardous substance and thus be inherently less hazardous, but if the design restricts the controllability of the process, then the design may have a higher risk and be less safe overall. Therefore, in the early stages of design, it is necessary to consider of both the inherent hazard contained within the process and the ease by which these hazards can be controlled. With the dynamic behaviour of process systems becoming increasingly complex, the consideration of operability issues in the design stage becomes even more necessary to prevent incidents.

Simultaneous design and control is an effective method to predict and reduce operability issues that result from an uncontrollable process design. Incorporating traditional inherent safety metrics into simultaneous design and control subsequently allows for industry to investigate inherent safety in the early design stage along with an analysis of error tolerance and ease of control.

The objective of this research is to implement a strategy to simultaneously design and control an inherently safer distillation column. The PARametric Optimization and Control (PAROC) framework is used as a basis for the simultaneous design and control of a distillation column. gPROMS is used to perform dynamic optimization of a distillation column design. Multiparametric Model Predictive (mp-MPC) control is used to constrain the control limits of the control response, and the Safety Weighted Hazard Index (SWeHI) is incorporated into the PAROC framework to measure and constrain the safety of the system. Therefore, the distillation column is optimized for minimal cost under control and safety constraints.

A Corresponding States Approach to Liquid Heat Capacity QSPRs

Joey Bloxham
joeybloxham@byu.edu

BYU

Map

1. Define QSPRs
2. Discuss regression training and testing
3. Physical interpretations of correlated descriptors
4. Suggest a new equation to fit liquid heat capacity from corresponding states methods
5. Discuss success of methods

BYU

Motivations

- DIPPR 801
- Research group focus is physical properties (prediction, measurement, and understanding)
- This work was undertaken to get a qualitative understanding of liquid heat capacity (LCP), but quickly discovered predictive capabilities

BYU

Uses in Process Design

- Liquid heat capacity is an important design consideration
- Needed for design of heat exchangers, reactors, etc.
- Accurate prediction of LCP allows better process design

BYU

Defining a QSPR

- Quantitative-Structure-Property-Relationships (QSPRs, sometimes QSARs) use regression techniques to relate geometric, topological, functional group counts, and other descriptors to chemical properties
- Always expressed as single value
- Thousands of descriptors
- QSPRs accurate predictions of many properties (boiling point, viscosity, solubility, etc.)

BYU

Van der waals volume v. LCP

- Naef (2019) show linear correlation between molecular volume and LCP at 298.15 K
  - Also note this trend is possible at other temperature ranges
- Molecular volume hard to define, other descriptors needed

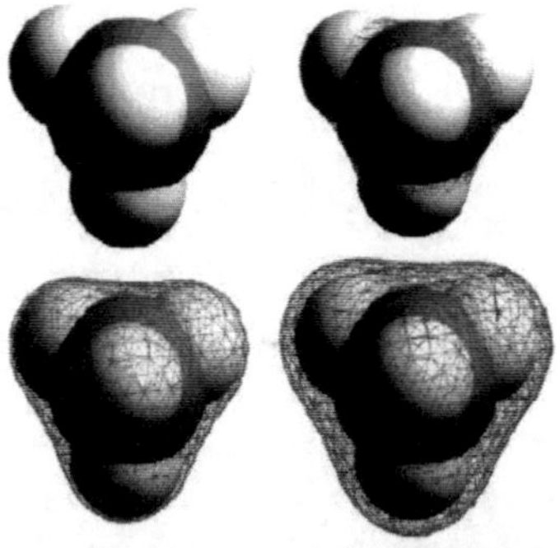
Methane molecular surface with 4 different electric cloud densities. Relick (1997)

BYU

QSPR Design

- DIPPR database has property data for 2730 chemicals
- Database was searched for chemicals with experimental data collected at 200, 300, and 400 K
  - If experimental data within 25 K of target T, correlation value at target T was used
  - Total studied species
    - 165 at 200 K
    - 639 at 300 K
    - 876 at 400 K
  - Species come from many chemical families, organic and inorganic

BYU

QSPR Design

- 5270 descriptors were calculated using Dragon 7
  - Optimized chemical structures used, from *Gaussian 09* using B3LYP/6-11+G(3df,2p)
- Species divided into training (80%) and testing (20%) groups
- Species at or above .8 T_r were eliminated

BYU

Model Choice

- Each descriptor then regressed to several different models
  - Various hyperbolic, logarithmic, polynomial, and exponential models were used, at least 15 unique models attempted
  - Both sum squared error and sum of absolute error objective functions tested
- Linear models were found to give best average absolute error and standard deviation in error
- Single-parameter linear models chosen
  - At all temperatures, meaningful descriptors had high collinearity (VIF well over 10)

BYU

Parameter Optimization

- Parameters were found 15 times from random testing/training arrangements, then averaged
- Accuracy values are using optimized parameters, averaged over 15 random arrangements
  - This method used to avoid using a "preferred fit" for training and testing data

BYU

Significant Descriptors – General Trend

- Most highly correlated descriptors tended to describe:
  - Size
  - Atomic electronegativities
  - Molecular volume
  - Ionization potentials
- These trends make sense, as there are known trends between LCP and molecular size and intermolecular forces

BYU

Data on 200 K

Training Set	%
Average Error	0.03
Average Absolute Error	10.38
Testing Set	**%**
Average Error	-0.50
Average Absolute Error	9.97

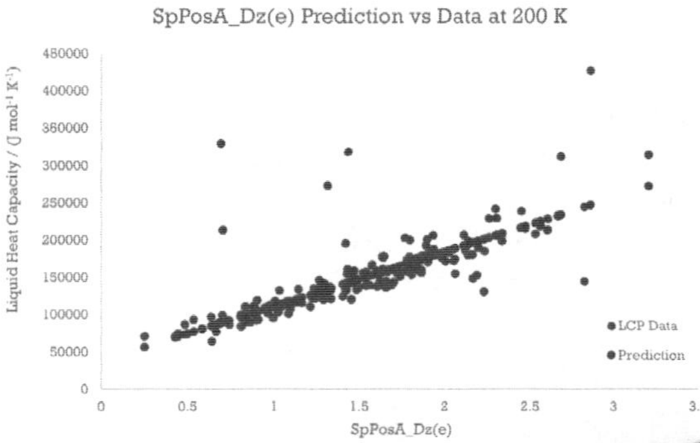

BYU

Data on 300 K

Training Set	%
Average Error	0.78
Average Absolute Error	8.18
Testing Set	%
Average Error	0.46
Average Absolute Error	8.43

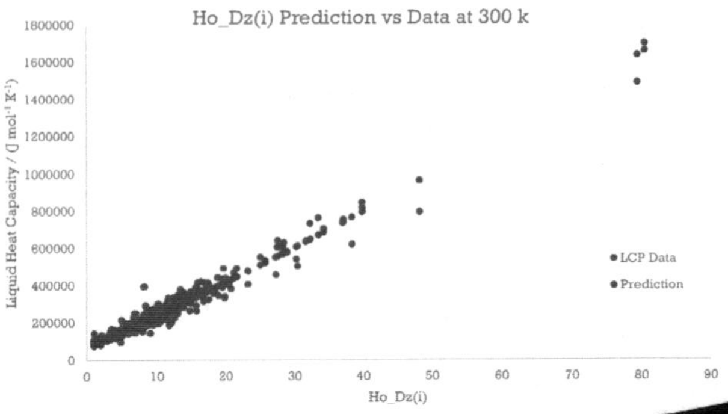

Data on 400 K

Training Set	%
Average Error	0.10
Average Absolute Error	9.43
Testing Set	%
Average Error	0.23
Average Absolute Error	9.77

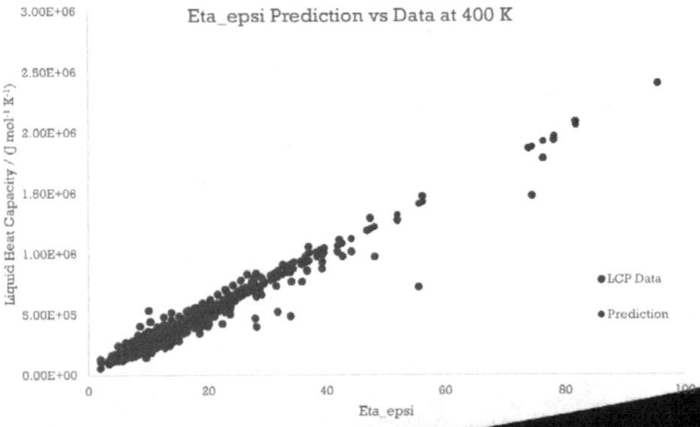

Problems with Other LCP Predictions

- Many are group contribution methods
  - This creates problems if groups are unavailable

Benefit of QSPR - atypical molecules

- Group contribution methods often skip inorganics, salts, etc.
- Descriptors available for these
- Siloxanes cannot be used on many group contribution methods
  - With Eta_Epsi QSPR at 400 K, silicon-containing compounds have average absolute error of 12.63%, and average error of -0.76%

Problems with Other LCP Predictions

- Often, methods only give LCP at single T
- If there is T-dependent portion, often ignores critical temperature effects
  - Popular methods like Ruzicka-Domalski use 2nd order polynomial fit
  - This leads to poor extrapolation at high reduced-T

BYU

Can QSPR Overcome T-Dependence?

- This work gives QSPRs with good performance at several temperatures, and capability to quickly create QSPRs at any temperature with sufficient data
- How to extrapolate between different QSPRs?

BYU

Potential Solution

- Use QSPRs to inform a LCP model inspired by corresponding states
- Proposed model – modified Lee-Kesler residual heat capacity expression
- $C_{pL} = A + \frac{B}{(1-T_r)} + C\frac{(1-T_r)^{\frac{1}{3}}}{T_r}$
  - (Since constant pressure CP used in regression, equation gives constant P LCP)
- Fitting this equation to QSPRs theoretically gives appropriate behavior near critical temperature
  - Also provides flexibility to use other prediction methods for well-suited species

BYU

Theoretical Basis of Eqn

- Rowlinson residual heat capacity function is:
- $\frac{C_{pL}-C_p^{\circ}}{R} = 1.45 + \frac{.45}{(1-T_r)} + .25\omega\left[17.11 + \frac{25.2(1-T_r)^{\frac{1}{3}}}{T_r} + \frac{1.742}{(1-T_r)}\right]$
- Treating the constants as variables:
- $\frac{C_{pL}-C_p^{\circ}}{R} = a + \frac{b}{(1-T_r)} + c\omega\left[d + \frac{e(1-T_r)^{\frac{1}{3}}}{T_r} + \frac{f}{(1-T_r)}\right]$
- And combine like terms
- $\frac{C_{pL}-C_p^{\circ}}{R} = (a + c\omega d) + \frac{(b+c\omega f)}{(1-T_r)} + \frac{c\omega e(1-T_r)^{\frac{1}{3}}}{T_r}$

BYU

Theoretical Basis of Eqn

- $\frac{C_{pL}-C_p^\circ}{R} = (a + c\omega d) + \frac{(b+c\omega f)}{(1-T_r)} + \frac{c\omega e(1-T_r)^{\frac{1}{3}}}{T_r}$
- Rearranging
- $C_{pL} = C_p^\circ + R(a + c\omega d) + \frac{R(b+c\omega f)}{(1-T_r)} + \frac{Rc\omega e(1-T_r)^{\frac{1}{3}}}{T_r}$
- Let $\mathrm{A} = C_p^\circ + R(a + c\omega d)$, $\mathrm{B} = \mathrm{R}(b + c\omega f)$, and $\mathrm{C} = Rc\omega e$
- $C_{pL} = A + \frac{B}{1-T_r} + C\frac{(1-T_r)^{\frac{1}{3}}}{T_r}$

BYU

Ideal Gas Assumption

- Ideal gas heat capacity (C_p°) was treated as constant
- In temperature range studied (200-400 K) most species C_p° well represented by first-order linear model
- Early tests used $C_{pL} = A + \frac{B}{1-T_r} + C\frac{(1-T_r)^{\frac{1}{3}}}{T_r} + DT_r$ to account for C_p°, but no improvement in fit was found
- Equation has enough flexibility to fit changes in C_p° in B and C terms

BYU

Example

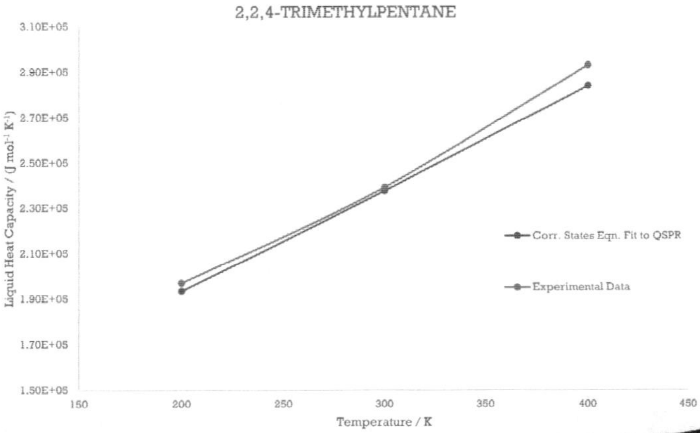

BYU

Performance at .9 T_r

- Need to test performance at high reduced temperatures
- Unfortunately, not enough experimental LCP data at .9 Tc to get estimates on accuracy

BYU

Future Work

- Test high reduced-temperature fit
- Additional fitting constraints for high reduced temperature
- Splitting data sets into number of oxygens increases accuracy – easy way to improve method

BYU

Acknowledgements

- Daniel Hill
  - PhD Student, PRISM Group, BYU

- Dr. W. Vincent Wilding
  - Associate Dean of Engineering and Technology, BYU

- Dr. Knotts
  - Professor of Chemical Engineering, BYU

- Dr. Giles
  - DIPPR 801 Project Coordinator

BYU

Bibliography

- Naef, R. Calculation of the Isobaric Heat Capacities of the Liquid and Solid Phase of Organic Compounds at and around 298.15 K Based on Their "True" Molecular Volume. *Molecules* **2019**, *24*, 1626.
- Rellick, L. M.; Becktel, W. J. Comparison of van der Waals and semiempirical calculations of the molecular volumes of small molecules and proteins. *Biopolymers* **1997**, *42*, 191-202.
- Rowlinson, J. S. Liquids and Liquid Mixtures, 2nd ed., Butterworth, London, **1969**.

BYU

ABSTRACT

As part of our ongoing work to improve thermodynamic consistency in property prediction for associating fluids, an investigation into molecular descriptor-based prediction methods was undertaken. Quantitative Structure-Property Relationships (QSPRs) have found applications in many areas of chemistry and engineering as effective prediction methods. QSPRs use molecular descriptors to simplify complex molecular properties to a single value. These values can then be related through linear or non-linear models to physical properties. Melting point, viscosity, solubility, and other properties have been well represented using molecular descriptors in QSPRs.

Liquid heat capacity (LCP) is another property where QSPRs can be helpful prediction tools. Researchers have shown strong correlation between the LCP and various molecular descriptors, but these predictions are limited to a single temperature, usually 298.15 K. Engineers often need LCP prediction methods that can be extrapolated through a large temperature range for safe process design, and currently available QSPR methods for LCP lack this ability. Another important consideration is the steep increase in heat capacity as the liquid nears its critical temperature, an issue that is not accounted for in traditional polynomial-based predictions, such as the Ruzicka-Domalski group contribution method.

In this work, multiple QSPRs at various temperatures are combined using a modified corresponding states approach. LCP data was compiled or collected at several temperatures and QSPR predictions were made at each temperature point. Then, using statistical analysis to combine uncertainties, a modified corresponding states method was used to link the individual QSPRs while maintaining the known behavior as the critical point is approached.

This method improves on existing QSPRs for LCP by using unique descriptors and adding a temperature dependence that has not been previously implemented. This functionality gives process engineers a thermodynamically rigorous prediction method that can be used to extrapolate up to the critical temperature.

BYU

589934 Computational Tools for Physical Property Estimation and Solvent Selection

Abstract:

We will present the details of a computational, physical property-based solvent selection methodology that we have used in numerous process development projects. We will use the example of selecting a solvent for a liquid extraction process to explain the methodology's steps. Specifically, we will discuss:

- Identification of physical property constraints, e.g., the high solubility of the key component, immiscibility with the feed solvent, proper volatility for regeneration by distillation, etc.
- Identification of required or disallowed substructures, e.g., restricting chemically or thermally unstable substructures, requiring reactive substructures, restricting environmentally undesired substructures, etc.
- Estimation of required physical properties, e.g., liquid-liquid equilibrium for extraction, vapor pressures for distillation, surface tensions and densities for settling, etc.
- Screening and ranking of candidate solvents, e.g., selecting a database of candidate solvents, identifying general trends in recommendations, iterating on the entire selection process, etc.

We will demonstrate how this selection methodology has been implemented in a software tool and show how such software tools are essential for conducting a rapid and thorough selection process.

SwIFT (SwRI Interdependent Flexible Time-Driven) Process Development
Paper #181c

Monica Brinkman; Eloy Flores, III; Kyle March; Southwest Research Institute, San Antonio, TX

Abstract

For many new technologies and processes, the timeframe to go from bench-scale to production can take multiple years. Companies are often racing to build and develop a new technology with limited information about how their innovation will translate from their lab experiments to pilot units or full-scale production. Southwest Research Institute uses its SwIFT (SwRI's Interdependent Flexible Time-Driven) technique in pilot plant development for rapid design, fabrication, and testing of integrated pilot plant units to collapse the overall schedule for the project. SwIFT process development, similar to Agile Programming, uses several small design and redesign cycles throughout the project to increase feedback and uncover unknown risks and information as early as possible in the project cycle allowing them to be corrected in the first version. This greatly reduces the project schedule and allows for flexibility in design, fabrication, and operations as the design team discovers information regarding the new technology's implementation.

The SwIFT Process

SwRI's Interdependent, Flexible, Time-Driven (SwIFT) Process Development centers on quickly and efficiently brining new processes to commercialization. The SwIFT process focuses on the following factors:

- Variables and designs in the process are ***interdependent***. Design for unit operations and systems within the process are continuously being reviewed, reevaluated, and improved as new information is discovered.
- ***Flexibility*** is built into the design, so that changes can be made to the system when new information is identified or refined without rebuilding the entire unit.
- The project goals remain ***time-driven***. Schedule is critical for its effect on the budget, scope, delivery, and schedule constraints.

These concepts are combined to result in a honeycomb-like design-build structure. Design-build is a technique in which several small design, build, and test cycles operate in parallel instead of in series in a project in order to accelerate schedule. As each design-build-test cycle completes, more information is learned about the overall process and can be incorporated in other aspects and design cycles.

Step 1: Categorize Data

In the development of new technologies, there is a large amount of unknown information at the beginning of a project. At the start of the project, SwRI evaluates parameters needed for the design and categorizes the parameters as known information or unknown/assumed information. If there are any design-build-test cycles that can be completed with the known data, these are started at this time.

The way that SwRI does this is by generating simple engineering design documents such as input-output diagrams, block flow diagrams, and process flow diagrams. This creates a starting place for engineers to identify what data is needed and then begin to further categorize that data into known and unknown data.

Like other aspects of the SwIFT process, categorizing data is an iterative process that must often be revisited. As more data becomes known and the engineers can dive deeper into the design process, more unknown data may become apparent. For example, in a block flow diagram, it may be identified that a previously unrecognized separation process is needed. Initially, the type of separation process needed is unknown. Once the type of separation process required is identified, for example a crystallization unit, there are more unknowns around the design of the crystallization unit that have to be resolved such as nucleation rates, particle size and distribution, residence time, effect on stream composition, separation efficiency, and other process parameters.

Step 2: Minimize Unknown Data

Once the data is categorized, it is prudent to move as much of the unknown data as possible into the known data category. This is often done through as-needed laboratory testing.

SwRI evaluates any small-scale experiments or easily-tested properties to move the unknown/assumed information into the known category. Any tests with quick turnaround times, such as ASTM property tests or small-nonintegrated bench-scale tests are completed. This minimizes the unknowns and reduces the number of assumptions being made in the design, or worse, discovered in operation.

Step 3: Bound Unknown Data

Next, SwRI evaluates impact of the information that is unknown or has assumed values and bounds the information into a range with upper and lower limits instead of a single point. Engineers and scientists then design the pilot unit to handle the broad range of values so that the precise value to be used for plant design can be determined during pilot plant testing and operations. Where possible, flexibility is built into the test system to make it as robust as possible, such as using parallel components.

For example, based on screening tests, one may determine that a reactor will likely operate best at 475°C, 10 psig, and 1.5 seconds residence time. However, a full characterization of the product may not be able to be completed until the material is tested in a large-scale pilot unit, so optimization runs may need to be run once the pilot unit is complete. Additionally, the reactants may act differently in a larger unit or in an integrated system perhaps because, for example, of a temperature gradient in larger equipment. Perhaps new byproducts or transport issues not seen at the bench-scale screening test may be observed. By bounding design parameters and designing a flexible system that can operate between

400 – 600°C, at pressures 5-20 psig, and with a residence time of 1 – 3 seconds, the system now has flexibility to complete optimization runs at larger scale and to make changes in the process parameters without rebuilding equipment.

What if it is not feasible to run a screening test or lab-scale experiment? Sometimes, the complexity or cost of a screening test makes it impractical to run a bench scale test. If tests cannot be run, literature data might be used to supplement unknown design data and to guide where the bounding should be. Occasionally, depending on how far you are on the edge of technology, there is no information available at all and the lab-scale experiments give minimal information. In this case, multiple design paths may need to be considered, as discussed in the next section. Alternatively, correlation or theoretical estimates may suffice until operating measurements are made. Having equipment with enough leeway to accommodate this unknown can save refits and thus time on the schedule.

Step 4: Evaluate the Impact of Unknown Data

At this stage, SwRI determines the dependencies of the design information. Unknowns or assumptions that affect multiple parts of the pilot unit are given more weight/attention than those that have less impact on the overall process. Occasionally, for high risk or long lead time items that can only be tested in a fully integrated system, parallel design options can be fabricated so that they can be quickly and easily switched during testing.

Additionally, if a small part of the pilot unit can be built and tested, this is done concurrently with construction prior to full commissioning. Priority is generally given to the high-impact unknowns. Often, fabrication and testing of different pieces in the unit begin prior to completion of the full design with the focus on eliminating or validating as many of the unknowns/assumptions as possible. If an assumption proves incorrect, the design team will use the information gathered from the singular equipment testing to modify the design, a step more preferable than revising the finished pilot plant.

However, some assumptions can only be validated or disproved in a fully integrated system. In this case, the design and fabrication team must be prepared to respond quickly with a new design based on the information gathered during the early operations and testing period.

Step 5: Create a "Critical" Design Path

Prioritize which units in a pilot system need to be designed first and identify which are dependent on the designs/outputs of others. Traditionally, one might design/fabricate a system in the process flow order. For the SwIFT process design, after laying out the known and unknown data in a process flow diagram and evaluating the impacts of the unknown data, you would then generate a "critical" design path to determine the equipment design order. Once this initial "critical" design path has been identified, a P&ID is started concurrently with the unit operation/equipment design and a second more detailed "critical" design path is generated in parallel. This allows engineers to start design work on the large, long lead-time items in parallel with the detailed P&ID and other design work. Due to the complexity of the larger unit operations, final approvals for the single unit operations are ready in the time it takes to put together the design documents for the rest of the pilot unit.

Just like in scheduling, design has dependencies. By determining what design parameters are affecting the greater number of unit operations, one can determine which properties most urgently needed to be determined and calculate them first. In addition to which unit operations have been affected, lead-time for unit operations should also be taken into account when prioritizing design parameters. If parameter A affects more unit operations in total, but the unit operations only take a few days or a couple weeks to arrive, then it will have less impact than parameter B which only affects a single unit operation that may have a 12-week lead time.

Step 6: Continually Update the Design

The design can be continually updated during the design-build-test cycles; especially for items or design parameters with multiple dependencies. The design-build-test cycle is completed as many times as needed throughout the project for small and large items. Flexibility is key, particularly if the number of unknowns at the start of the project is large. Implementing a first-of-kind technology advances the “cutting edge” and will benefit from the SwIFT approach.

The SwIFT goal is to quickly produce a robust solution that can adapt to developing technical challenges as they are identified, shakedown and troubleshoot problems early as they arise, and bring the number of technical unknowns down to zero, derisking the client’s process and making it ready for commercialization.

Advantages and Pitfalls of the SwIFT Process over Traditional Development

Although the SwIFT technique likely has more cycles of design-build-test, it allows the design team to discover and shakeout failures more quickly than a traditional prototype creation cycle. This rapid development approach can take years off a development cycle and get the company into production faster.

Advantages of the SwIFT Process

Time: While the traditional linear design process may take years to get to commercialization, the SwIFT Process Development generally starts to produce pilot-plant data that can be used for commercial design in 1-2 years with some pilot plants up and running in less than 6 months.

Data Density: Information is revealed quickly and efficiently. This allows the client to quickly make informed decisions about the process and the overall system, say including utilities or product handling. If a system has a fatal flaw, the client is in a better position, realizing this at month 6 rather than year 3. Stage gates for smaller tests often begin within a few months of the project start as data is generated even before the main fabrication of the pilot unit begins.

Flexibility: Since new information is quickly available, the team has the ability to adapt the design based on the new information as it becomes available. This sometimes completely changes the direction of the project, but allows for a greater chance of success.

Innovation: With a high data density and flexibility available, innovative solutions can be achieved for unique technical challenges. The iterative process may reveal aspects that were not evident in the process conception.

Pitfalls of the SwIFT Process

Cost: When rapid schedule and increased scope are higher priorities of the triple constraints of time, cost, and quality, cost will be unavoidably higher. Sometimes this is a matter of "pay me now or pay me later". The broader SwIFT approach avoids costly surprises that kill projects by resolving them early.

Aesthetics: Although safety is still the number one priority, rapid installation of quickly fabricated items is often "functional and neat, but not very pretty". For example, operators may grind down welds so that they will not cut any passersby, but not until they are polished enough to see your reflection. Another example may be when wire is being pulled or added to the control box because a new sensor was added to the system; it is safely in the cable tray but may not be bundled with cable ties since it may have been rerouted several times depending on how many times the design has changed.

Rapid Decision Making: The client and SwRI staff have to make quick decisions after running tests and analyses. In order to keep schedule requirements, there is precious little time to mull over decisions, analyze data, or change your mind. This may require more client engagement than originally considered.

SwRI Case Study – SwIFT Process Development for Gas to Liquids

SwRI worked with a client using the SwIFT Process Development for a pilot unit to produce olefins from natural gas. The main project goals were 1) rapid time-frame development of the pilot unit to make up for time lost on the schedule prior to starting the project at SwRI, and 2) a proven inline analytical system. To help expedite the process, the pilot skid was built at SwRI at a "brown site", a site which already had utilities, which previously had been used for a different technology platform.

For the first step, identifying unknowns, SwRI engineers and operators first keyed-in on the site work and utilities. Many of the skids, such as a molecular sieve, natural gas, and compressor skids, had not been used for several years and needed to be recommissioned or upgraded. Additionally, for this pilot unit, new systems requiring oxygen-clean equipment would need to be added. Another unknown, which was a top priority for this client (and most of our clients) was the analytical system setup. For pilot plants, the process data and the analytical data are the outputs used to evaluate the process and design for the subsequent commercial step. The client needed inline analytical data that would tie to the process data. Although there were many other unknowns, a fourth large unknown was heat transfer and heat loss in the system. Both the client and SwRI modeled the heat transfer, but these needed to be validated and tested in hardware to ensure that the temperature range for testing could be reached.

In the second step, minimizing unknowns through testing, SwRI began to complete initial testing on components where possible. For starters, an initial heater function test was completed on the methane and nitrogen heaters in order to validate the heat-up rate, heat-up duration, and ability to reach the set point temperature at the desired flow rates. Although the initial model suggested the heat of reaction would provide sufficient heat to the system, a larger internal heater had to be installed during the testing process to make up for heat losses throughout the unit. Additionally, onsite testing was done to find a special high-temperature seal appropriate for the application. In the reactor itself, insulation types were tested in addition to the validation of the heat loss models. With each run, improvements were made to the design until a three-layer insulation design was finalized, which allowed the reactor to maintain the desired run temperature. Another challenge, which included several unknowns, included the inline analytical system. There was a great distance from the reactor to the analytical instrument, which could greatly impact the temperature profile in the analytical lines. Additionally, water entrainment was a concern. These problems were addressed as encountered yielding a faster result.

For the third and fourth steps, bound unknowns and evaluate impact, engineers sized equipment, such as flowmeters, in the system to give the most flexibility. However, even though the unit operations allowed for a high degree of flexibility, the final operating regime ended up drastically different, and one of the heat exchangers required a redesign. Since engineers were constantly monitoring the impact of changes in the system, the project team was prepared for the redesign and was able to quickly turnaround the design, fabrication, and installation to get the system back up and running during operations.

For the fifth step, create a critical design path, the project team ordered long lead time items first, and made good use of long lead times by working many projects in parallel. When considering which items to finalize in the design and to order first, priority was given to unit operations which were found to have the longest lead times. Additionally, for specialty equipment such as the oxygen-clean equipment, the training and site preparation for installation were completed faster in-house (compared to awaiting a contractor) while awaiting equipment delivery.

Also, as part of the critical design path, there were several operations within the project that were scheduled in parallel. The natural gas skid recommissioning occurred at the same time as the Hazards and Operability (HazOp) review since the natural gas skid had been previously operated, had its own existing HazOp, and there were no modifications made to it. The molecular sieve media replacement, compressor upgrade, and reactor design were all completed simultaneously. Controls programming and system buildup occurred at the same time. Analytical commissioning occurred while the wiring and functional testing of instrumentation was being completed. All of these allowed the project to safely compress schedule.

As a result of using the SwIFT Process Design method, SwRI engineers and scientists were able to condense this 15-month build-up of the system to 6 months. During operations, SwRI engineers and scientists successfully tested multiple catalysts, multiple reactor inserts, reactor supports, heat retention, and validated the heat balance. There was room for modifications throughout operations as new information was learned about the process and SwRI engineers were able to quickly adapt the pilot system to generate the data needed for the client.

Conclusion

Overall, SwIFT Process Development reduces schedule, reveals unknowns in new systems earlier in the development process, discovers potential technical missteps or misunderstandings at an early stage, and saves money in the long run, although more upfront cost is required at the start for testing and to build flexibility into the system.

Using the SwIFT Process Development for clients, SwRI has greatly reduced the time for clients to move their process from bench-scale to commercialization, with project timelines ranging from 6 months to 3 years. SwRI's 1500 acre site in San Antonio, Texas is uniquely suited to pilot plant design, buildup, and operations having adequate buffer space and operating safety systems. Southwest Research Institute possesses pilot plant emission permits, multiple fabrication and machine shops for rapid construction and modification, 24/7 operations and analytical services available, and multidisciplinary expertise for any unexpected technical challenges in engineering, chemistry, analysis, automation, fabrication, or operations.

591038 Three Green Projects from a Gulf Coast Company

Abstract:

This presentation describes three fuel- and emissions-free waste heat-to-power industry solutions developed by Gulf Coast Green Energy, Bay City, Texas. These include fuel efficiencies for large engines, process fluid cooling, and using flare gas for a beneficial use (making power). All three applications use wasted heat to produce on-site power. This can dramatically reduce emissions and power costs.

The examples include:

• a successful Dept. of Defense funded project to reduce emissions through reduced fuel consumption in large engines;

• a successful flare reduction trial on a Hess Corp. North Dakota well pad.§

• an overview of power from produced water.

The presentation will include an overview of the projects from design to installation/commissioning, and through the successful sustained operations.

§ This project was funded by the Houston Advanced Research Center (HARC) and its Environmentally Friendly Drilling Systems (EFD), the U.S. Dept. of Energy, and Hess Corp.

591048 Large Scale Industrial Waste Heat-to-Power/Cogeneration: A Case Study

Abstract:

The benefits and challenges of a large scale industrial waste heat-to-power/cogeneration facility are presented in this case study of the Port Arthur Steam Energy Project. The project recovers 2000ºF flue gas energy from a petroleum coke calcining plant and converts it to 350,000 lb/hr of steam and 5MW power which are provided to a neighboring refinery, the host plant and the utility grid. The project is a great example of industrial cooperation and energy efficiency, with the capability of offsetting 159,000 tons CO2 annually.

590790 Options for Carbon Capture from Refineries / Chemical Sites Smr

Abstract:

IPCC's special report on the impact of 1.5C warming which concludes that *"early scale-up of industry CCS is essential to achieve the stringent temperature target"*. CCUS technologies (Carbon Capture and Utilization or Sequestration) are especially of interest for hydrogen production units, which emit significant amounts of CO_2 and offer several options to capture this CO_2 from otherwise clean gases, such as Pressure Swing Adsorption (PSA) feed gas, PSA purge gas, or Steam Methane Reformer (SMR) furnace exhaust gas. Hydrogen manufacturing units are thus likely to be among the first sources of CO_2 that refiners and chemical sites in USGC will look at to reduce their carbon footprint.

Amine-based absorption systems offer a proven, reliable and cost-effective method to capture CO_2 from the different SMR gas streams. While the SMR syngas (outlet of shift reactor after heat recovery), characterized by a high CO_2 partial pressure, presents the opportunity to capture CO_2 at the lowest unit cost ($/tCO2), post-combustion (furnace exhaust) capture maximizes the amount of CO_2 captured. Both have been proven at large scale; in particular capture from the high-pressure syngas has been successfully applied at the Scotford refinery (Alberta), where the Quest project, using the ADIP-X technology, has already captured and sequestered more than four million tons of CO_2 and generated valuable learnings for replication in future SMR CCUS projects.

The presentation will discuss the relative merits and costs of both options, and in particular how integration with an industrial gas SMR located at USGC site can help provide the energy required for absorbent regeneration, which is around 2 GJ/ton of CO_2, and for 1,000 KTA CO2 capture unit results in a LPS steam requirement of approximately 110 tonne/hr.

AICHE PAPER NUMBER: 159b

NATURALLY OCCURRING RADIOACTIVE MATERIALS IN ETHYLENE PLANTS REVISITED

Dr. Debby M. Rossana
Senior Technical Lead
Nalco Water

Dr Phillip James
Technology Director
Evonik Performance Materials GmbH

Justin Harris
Radiation Safety Services Manager
Quantum Technical Services

Dr David Slim
Technology Advisor
Evonik Corporation

Prepared for Presentation at the 2020 Spring National Meeting
Houston, TX (March 29 - April 2, 2020)

NATURALLY OCCURRING RADIOACTIVE MATERIALS IN ETHYLENE PLANTS REVISITED

Dr Debby M. Rossana
Senior Technical Lead
Nalco Water

Dr Phillip James
Technology Director
Evonik Performance Materials GmbH

Justin Harris
Radiation Safety Services Manager
Quantum Technical Services

Dr David Slim
Technology Advisor
Evonik Corporation

Abstract

Naturally occurring radioactive materials (NORM) are an occupational health concern in the ethylene plant. NORM (as radon) have been known to be a contaminant of natural gas for nearly 100 years but it was only in 1971 that radon was found to concentrate in the lighter natural gas liquids during processing (ref 1). The presence of radioactivity has not been widely reported in the ethylene industry. A perusal through the EPC archives shows only 6 papers directly related to NORM. A survey carried out in 1994 by Hart, Wolf and Graham (ref 2) looked at 17 US ethylene plants and 53% of the respondents had identified the presence of NORM in their plants. A 2020 survey that looked at 42 ethylene plants shows the number of ethylene plants with identified NORM has increased to 87%. Is this increase due to better monitoring and awareness or could the increase in shale fracking be the root cause?

1 Background

Uranium and thorium and the products of their decay, radium 226 and radium 228, can be found in rock formations where oil and gas are produced (ref 3). Radon gas, which comes from radium, can end up in the natural gas. Therefore, the primary radioactive material that concerns the ethylene plant comes from radon, which can be found in gas feeds such as propane and to a lesser extent ethane (ref 4). The amount of radon in the propane and ethane the ethylene plant receives depends on where the gas originated and how long since it was extracted from the ground. It should be noted that Taylor and Cleghorn (ref 5) measured higher than expected radon in their feeds due to a concentrating effect.

Radon boils between ethane and propane (ref 6) see table 1 and therefore NORM tend to collect in the C2 and C3 areas of the ethylene plant (ref 7), although NORM have been reported in feed pumps, compressors and caustic towers.

Table 1 – key boiling points

	Boiling point °C	Boiling point °F
Ethylene	-103.7	-154.7
Ethane	-88.5	-127.4
Radon	-61.7	-79.1
Propylene	-47.6	-53.7
Propane	-42.2	-43.8

Radon is a colorless, odorless gas, with a half-life of 3.8 days. It decays via alpha particles to polonium to bismuth to lead (ref 8 Hyperphysics site). While radon is likely to be in the gas feeds, because of the short half life of radon, it is usually solid polonium, bismuth or lead that is measured as NORM in ethylene plant equipment. **Figure 1** (ref 9) shows the degradation products of radioactive materials found in rock formations. Radon and its progenies have short half lives, except for lead 210, which has a half life of 22.3 years.

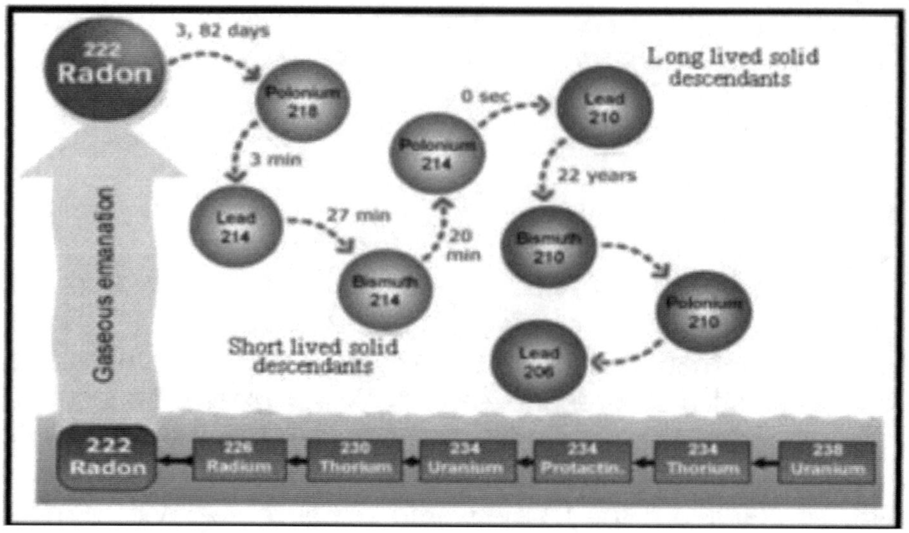

Figure 1: Half-lives of Radon and its progenies

Note: half-life is the average time taken for half the weight of an isotope to disappear: e.g. 1 gram of U-238 becomes 0.5 grams in 4.5 billion years whereas Po-214 disappears almost instantly. **Figure 2** (ref 10) shows the overall time it takes for radon 222 to decay to polonium 218.

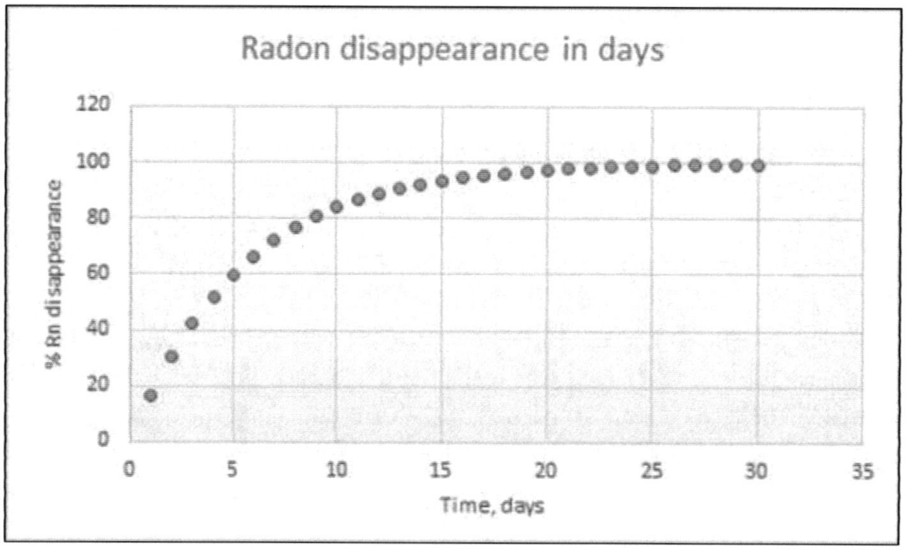

Figure 2: Time for Radon 222 to decay to Polonium 218

As radon is the only gas in the decay pathway, it can escape rock formations with the hydrocarbon gas and depending on how the gas is processed may follow either the ethane or propane. If it follows the ethane, it will enter the steam cracker. If it follows the propane, there are more potential outlets. Propane can be a steam cracker feedstock, go to a propane dehydrogenation unit, or end up in fuel use. For the purposes of this paper we will focus on steam cracking.

Because radon decays so quickly, radioactive contamination of steam crackers is most likely only an issue for plants with a feedstock supply chain from well to plant of less than 1 month. Consequently, it seems unlikely that US shale ethane which is shipped overseas to Europe and Asia will still contain radon by the time it reaches its destination. Instead it will most likely have reverted to solid form, lead 210, as the immediate progenies polonium 218, lead 214, bismuth 214 & polonium 214 are so short-lived. It is expected that the lead 210 will form a solid deposit at the base of the shipping vessel. The risk posed by this material needs further evaluation. Radioactive scales and sludges are discussed in reference 6.

2 NORM in the ethylene plant

As described above, NORM poses the highest risk to steam crackers if it is present in the form of radon gas. Radon is chemically inert, being a noble gas, so it is not expected to react with anything else during its passage through the cracker. The only expected reaction is radioactive decay. The general consensus is that the high temperature to which it is exposed in the cracking furnaces does not affect the rate of decay. After leaving the furnace, the radon will follow ethane and propane through the ensuing units, until reaching the light ends separation unit of the plant. Based on its boiling point (see table 1) it could either end up in the C2 or C3 splitter. In the case of the C2 splitter it will follow ethane and will be recycled to the furnace to complete the loop again. In the case of the C3 splitter, it is likely to follow the propylene and will end up in the propylene product.

Depending on where the radon entering the plant is in its radioactive decay timeframe, will dictate whether it remains as radon during its passage through the cracker. Since it has a half life of just 3.83 days, it is quite likely that at some point it will decay into its immediate progenies. As these are metallic solids, they are much less likely than radon to continue their passage in the gas phase, and therefore will start to form deposits inside the plant. Of these progenies, polonium 218, lead 214, bismuth 214 & polonium 214 are so short-lived that within a few hours, the only remaining species will be lead 210.

If the radon is caught in the ethane recycle loop it will inevitably end up as lead 210 before it can leave the cracker. If it escapes with the propylene, its fate will be decided by how the propylene is handled. If it directly enters a downstream process such as polypropylene, acrylonitrile, acrylic acid or phenol-acetone production, it could also contaminate that process with radiation. If it firstly enters a storage vessel, then the decay may happen at that location, leading to contamination of that vessel.

If the radon in the cracker feedstock has already decayed by the time it reaches the steam cracker, the decay products – primarily lead 210 – may be present as solid deposits at the entry point to the plant.

3 Shale Fracking and NORM

Rock formations contain radioactive materials and radon gas. As fracking increases, there appears to be a likelihood of an increase in NORM in the ethane and propane that is received by the US ethylene plants. Several papers have been published suggesting homes near shale fracking areas have higher than normal levels of radon (ref 11 and 12). In particular, the Marcellus shale play is considered to have higher levels of radioactive materials vs other plays and has been measured at 20 times higher. According to a paper by Frackcident Injury Law (ref 13), the radioactivity is so high that radiation detectors can be used to find deposits.

Figure 3 (ref 14) shows the radon levels measured across the US. The geometric figures superimposed on the map show the approximate locations of shale gas plays. The Bakkan (circle), Niobrara (triangle) and Marcellus (rectangle in the Ohio/Pennsylvania area) are areas of high radon. The odds are greater that the ethylene plants who receive feeds from these areas will receive NORM containing ethane and propane.

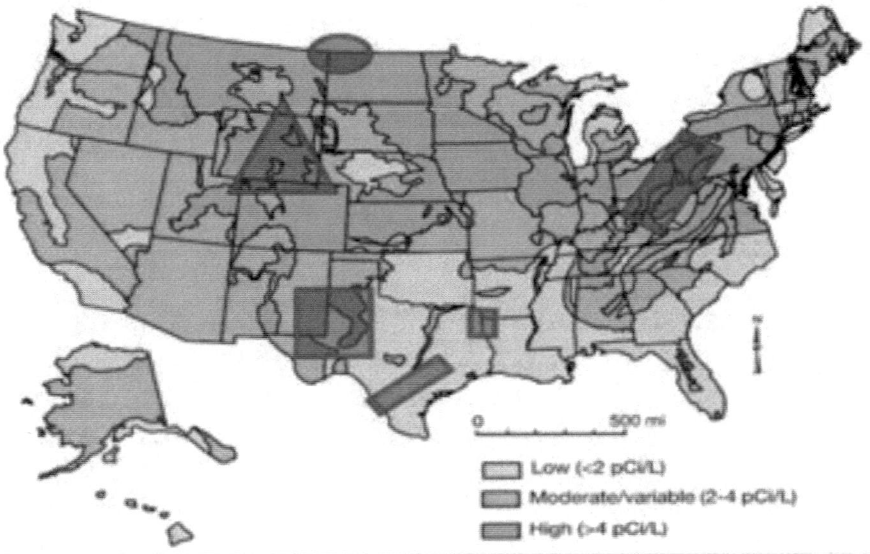

Figure 3: Major shale fracking areas in geometric shapes superimposed on high Radon areas in the US

4 Survey Results

As mentioned earlier, a survey was performed by the authors to determine if the presence of NORM has increased in recent years. All North American ethylene plants were contacted to participate in the 2020 survey. A response rate of 69% was achieved. Of the responses, 6 were new ethylene plants. 2 new plants have already detected NORM in deposits and/or equipment and 4 have not tested yet. In addition, 3 other established ethylene plants responded that they do not test for NORM.

Of the 69% who responded to the survey, 17% do not test for NORM and 83% do test, as shown in **Figure 4**.

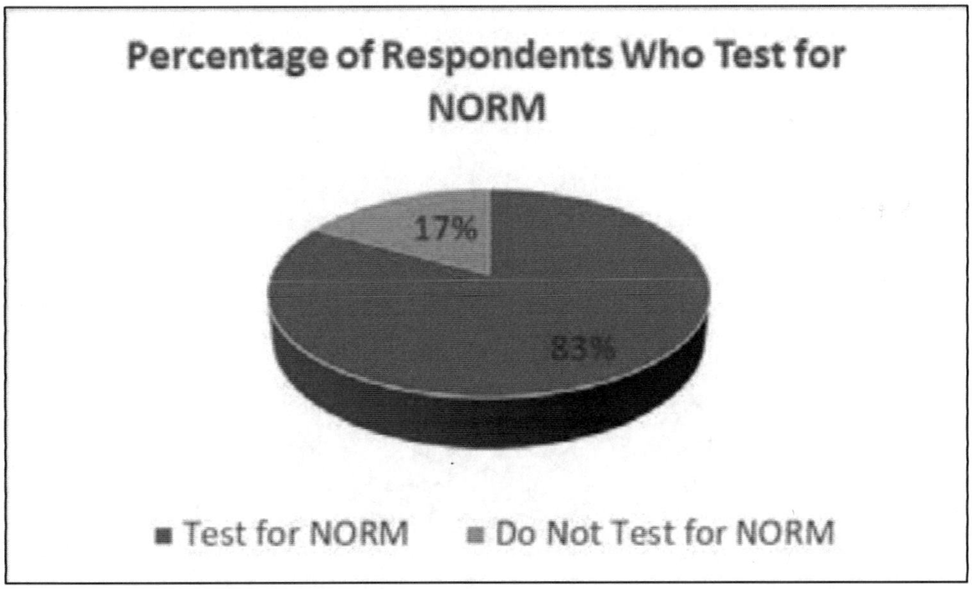

Figure 4: Percentage of respondents who test for NORM in their plants

Of those respondents who do test for NORM, 86% have found NORM in their plant while 14% have not. This is up from 53% detection in 1994, which represents an increase of 160% (**Figure 5**).

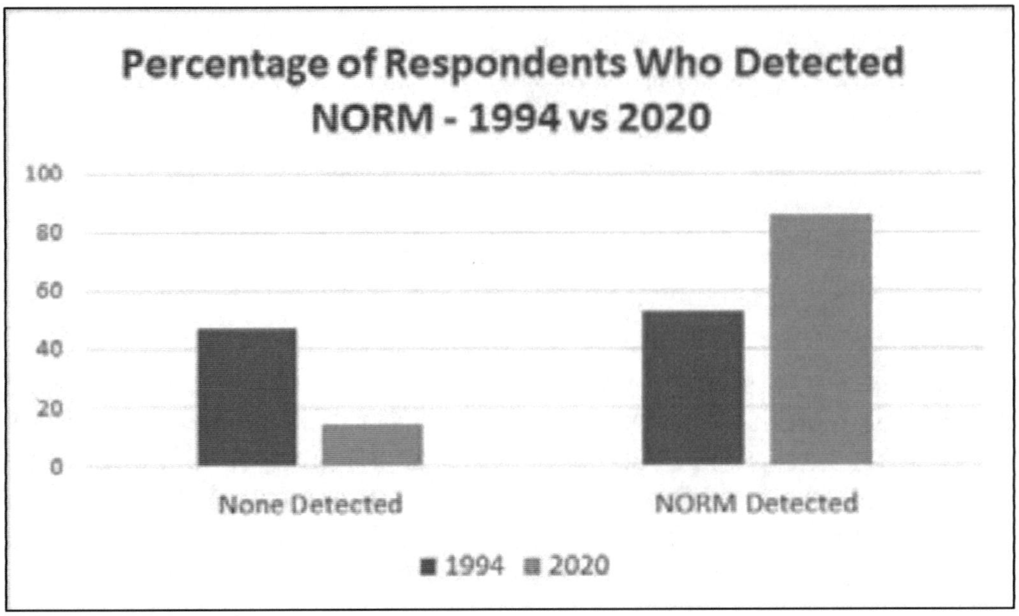

Figure 5: Percentage of respondents who detected NORM in a 1994 survey vs a 2020 survey

Survey data shows that both gas and mixed crackers have detected NORM in their plants (**Figure 6**).

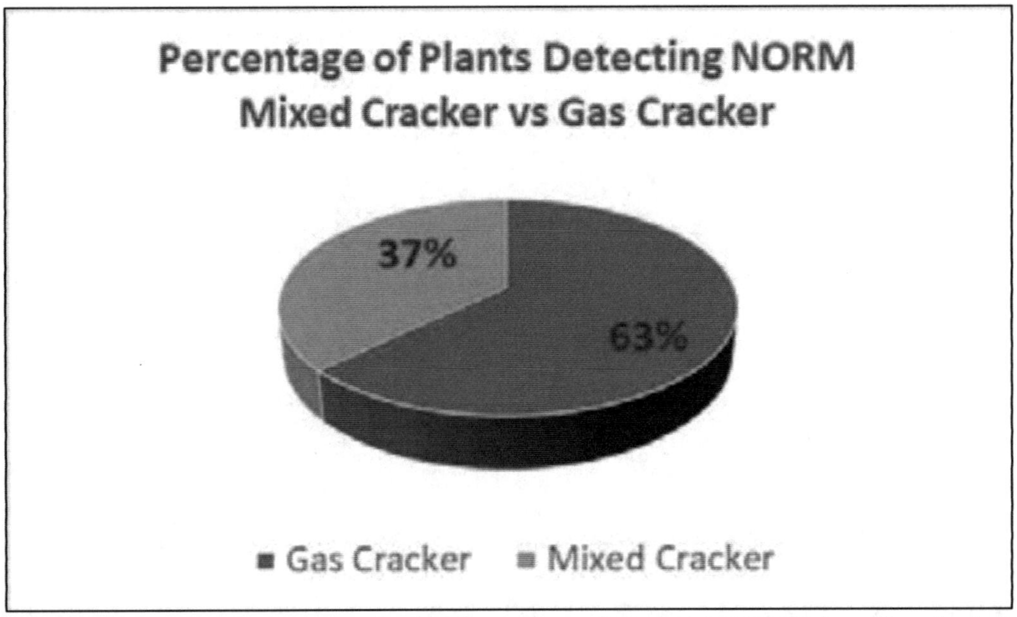

Figure 6: Percentage of plants who detect NORM by feed slate

The survey asked respondents if they felt this increase was due to gas feeds from shale fracking or due to more awareness and better detection. Overwhelmingly, the respondents felt the higher NORM presence was due to better awareness and measurement and not necessarily from shale fracking. A paper by Hlibrich (ref 15), however, suggests this increase is, at least in part, due to shale gas feeds.

Of those who detected NORM in their plant, all found it in the light ends (deethanizer, depropanizer, C3 splitter). Most of the respondents only measure for NORM during maintenance but a number of plants also check their feeds and/or solid waste and some have found NORM in these areas. For some plants NORM have been found in nearly every area of their plant including compressors and caustic towers while other plants see NORM primarily in the light ends towers.

Although most respondents did not know which radioisotopes were detected in the NORM, 2 have identified various amounts of the radon progenies in their plants. Most are only detecting alpha and beta radiation as expected but a few have found gamma radiation as well (ref 15).

The most common measurement tool used by the respondents is the Ludlum produced handheld Geiger counter similar to **Figure 8b**.

The largest variation seen among the surveyed plants was their handling of NORM. Several plants have very detailed procedures with mapping of contaminated areas such as Taylor and Cleghorn discussed in their 1996 EPC paper (ref 5) and others have decontamination stations for workers as they leave NORM areas. Most, if not all, plants use air movers as part of their NORM procedure to blow the NORM out of towers before entering with or without a minimum waiting time before workers can enter. Most require a minimum of a disposable TYVEC™ suit and respirator as shown in **Figure 7** (ref 16).

Figure 7: Proper PPE for NORM exposure

Radioactive waste removal is typically done by an outside company with the proper training to handle these wastes.

5 Handling NORM

During normal operation, there is little concern around exposure to radiation. However, NORM poses a problem to workers particularly during maintenance, waste transport and processing, and decommissioning. However, there may be instances around certain types of equipment (i.e. pumps, filters) where external gamma readings reach a level that need to be posted to maintain compliance with general public dose limits.

Although NORM contamination will be general throughout an ethylene plant, the contamination will usually be greatest in areas of high turbulence, such as in pumps and valves. Thus, the effects of radon progenies are mainly seen in compressors and light ends fractionators. When workers open equipment and vessels, precautions should be taken to prevent contamination. It is recommended that maintenance procedures include the use of respirators, wet grinding, and good hygiene to prevent inhalation of radioactive dust and other materials.

Occasionally, a plant processing ethane and propane is taken out of service and the facility sold or dismantled. Any equipment with internal surface deposits of NORM must receive special consideration when scrapped, sold, transferred or disposed of. In most cases, analysis for lead-210 will be required to verify the extent of contamination and to determine if special handling is required. Care must be used to prevent employee exposure to NORM contamination.

6 Measuring NORM in the Ethylene Plant

External and internal measurements can be performed to measure NORM levels in the ethylene plant (ref 17). External surveys can be done while equipment is running. The primary radioactive component measured externally is the gamma peak from bismuth-214. Internal surveys are done when the equipment is opened for maintenance and inspection. It should be noted that not finding any NORM on the outside of equipment does not mean there are no radioactive components inside the equipment. However, the converse is not true; readings outside of equipment does indicate internal NORM presence.

A means of detecting radioactivity is an electrical device called a Geiger counter. Some examples are shown in **Figures 8a and 8b**. The Geiger counter contains a gas-filled chamber with a thin membrane on one end that allows radiation emitted from radioactive nuclei to enter the chamber and knock electrons off atoms of gas (usually argon). The presence of electrons and positively charged ions causes a small current, which is detected by the Geiger counter and converted to a signal on a meter or, commonly, an audio circuit to produce an audible "click." Detection of alpha and beta radiation is very sensitive to the sample environment, a layer of hydrocarbon on the sample is likely to adsorb the radiation and not be detected by the Geiger counter

Figure 8a – Gamma Scintillator

Figure 8b – Pancake Probe

Internal NORM measurements are done at turnaround or when equipment is opened. During pre-opening purging, some of the radon present will be removed. The solid decay products of radon such as lead-210 and polonium-210 will remain during purging. A pancake probe (**Figure 8b**) is used for checking workers for contamination and for measuring radiation in equipment.

Types of Radiation (refs 18 and 19)

Alpha particles: these are fast moving helium atoms. They have high energy, typically in the MeV (meg electron volt) range, but due to their large mass, they are easily stopped, even by a few inches of air. Alpha particles cannot penetrate most matter. A piece of paper or the outer layers of skin is sufficient to stop alpha particles. Radioactive material that emits alpha particles (alpha emitters) can be very harmful when inhaled, swallowed, or absorbed into the blood stream through wounds.

Beta particles: these are fast moving electrons. They typically have energies in the range of a few hundred keV (kilo electron volts) to several MeV. Since electrons are lighter than helium atoms, they can penetrate further, through several yards of air. Beta particles can be stopped by a layer of clothing or by a few millimeters of a substance such as aluminum. Beta particles can penetrate the skin causing radiation damage, such as skin burns. As with alpha emitters, beta emitters are most hazardous when they are inhaled or swallowed or absorbed into the blood stream through wounds.

Gamma rays and X-rays: these are photons, but have high energy, typically from several keV to several MeV. Depending on their energy, they are capable of greater penetration than other rays. Several feet of concrete or a few inches of lead are required to stop them. Gamma rays and X-rays are a radiation hazard for the entire body. While gamma rays and X-rays can easily pass completely through the human body, some fraction of the energy will always be absorbed by body tissue.

Units of Measurement for Radioactivity (ref 3)

The original unit for measuring the amount of radioactivity was the curie (Ci) which was first defined to correspond to one gram of radium-226 and more recently defined as $3.7x10^{10}$ radioactive decays per second. In the SI system the curie has been replaced by the becquerel (Bq), which equals 1 radioactive decay per second or $2.703x10^{-11}$ Ci.

The absorbed dose describes the amount of radiation absorbed by an object or person (that is, the amount of energy that radioactive sources deposit in materials through which they pass). The units for absorbed dose are the radiation absorbed dose (rad), which is defined as the energy of 100 ergs per gram of irradiated material; and gray (Gy) where 1 Gy = 100 rads.

Equal doses of all types of ionizing radiation are not equally harmful. Alpha particles produce greater harm than do beta particles, gamma rays and x rays for a given absorbed dose. To account for this difference, radiation dose is expressed as equivalent dose in units of Sievert (Sv). The dose in Sv is equal to "absorbed dose" multiplied by a radiation weighting factor (1 for beta and gamma and 20 for alpha). 1 Sievert is equivalent to 100 Roentgen equivalent man (REM).

The effects of being exposed to large doses of radiation at one time (acute exposure) vary with the dose are shown below in **Table 2**.

Table 2 - effects of radiation

Dose (sievert)	Effect
10	Risk of death within days or weeks
1	Risk of cancer later in life (5 in 100)
0.1	Risk of cancer later in life (5 in 1000)
0.05	TLV for annual dose for radiation workers in any one year
0.02	TLV for annual average dose, averaged over five years

7 Regulatory Limits (ref 17)

Currently there are only 8 states in the U.S. that have NORM specific regulations. Since regulations vary slightly per state, Texas will be used as an example to put this into perspective. Regardless of the source of the ionizing radiation, exposures to the general public must be kept below 100 mR/yr and not to exceed 2 mR in any one hour. To be in compliance, facilities must maintain survey data to show these limits are not being exceeded. There are also limits for equipment release, contamination in soil, contamination of media (filters), and inhalation/ingestion to employees. **Table 3** illustrates these limits:

Table 3 – regulatory limits

Limit	Description	Reference
>50 μR/hr at any accessible point	Materials and equipment in recycling process & pipe tubulars and other downhole or surface equipment used in oil production	25 TAC 289.259(d)(2) & 25 TAC 289.259 (d)(3)
5,000 dpm/100cm^2	Contamination that cannot be readily removed from surfaces by casual contact, wiping, brushing or washing	25 TAC 289.259(w)
1,000 dpm/100cm^2	Radioactive material that can be removed from surfaces by non-destructive means, such as casual contact, wiping or brushing	25 TAC 289.259(w)
30 pCi/gram of Ra-226 or Ra-228 (soil/other media)	Oil and gas NORM waste	25 TAC 289.259 (d)(1)(A)(i)
150 pCi/gram any other radionuclide (soil/other media)	Oil and gas NORM waste	25 TAC 289.259 (d)(1)(A)(i)(II)
Derived Air Concentration (DAC) Pb-210 (4x10^{-1} μCi/ml) Po-210 (3x10^{-10} μCi/ml)	Concentration of radioactive material in air that will result in an annual limit of intake (based on 2,000 hours)	25 TAC 289.202 (ggg)(2)(F)
Annual Limit on Intake (ALI) Pb-210 (1x10^{0} μCi) Po-210 (6x10^{-1} μCi)	Derived limit for the amount of radioactive material taken into the body by inhalation/ingestion in a year.	25 TAC 289.202 (ggg)(2)(F)

Note: dpm = disintegrations per minute. 60 dpm = 1 Becquerel. 50 microroentgens/hr = 0.5 microSievert/hr

8 Health Concerns from Radon Gas and NORM

Radiation is a mutagen, which eventually can lead to cancer. Radiation can either kill cells or damage the DNA within them, which damages their ability to reproduce and can eventually lead to cancer. When radiation is present, high energy particles pass through the body. These can collide with atoms in the body and disrupt atomic structure and damage DNA. This can lead to uncontrolled cell division which is the definition of cancer.

Radon gas accounts for 55% of a typical person's radiation exposure (ref 20) as shown in **Figure 9**. Exposure to radon gas is the leading cause of lung cancer for non-smokers (ref 21).

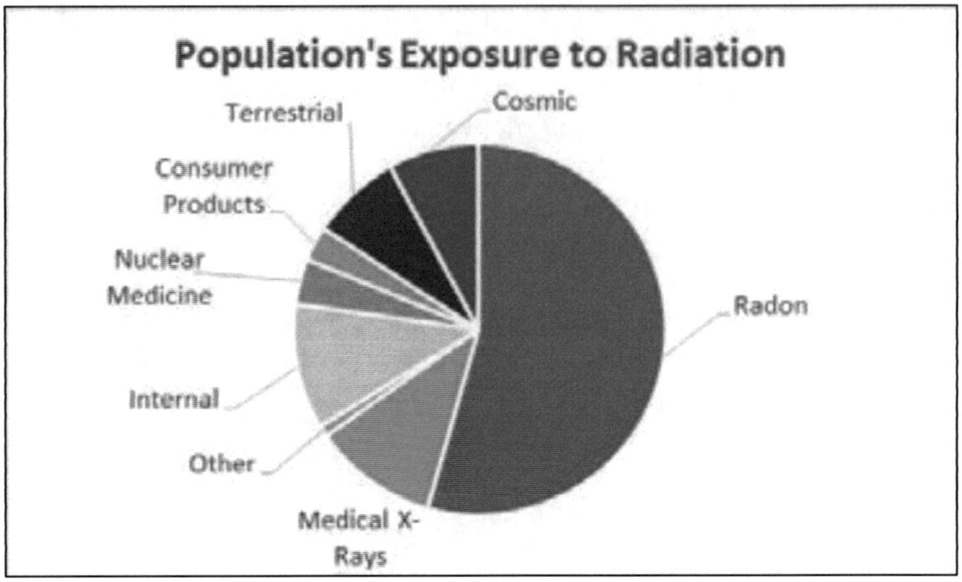

Figure 9 – Break down of typical exposure to radiation

9 Protection of Workers and the Environment

Each location should have a process for managing radioactivity. Considerations include

- A Radiation Safety Officer (RSO)
- Detection equipment (see above)
- Detailed procedures
- PPE (The IAEA has a practical radiation technical manual relating to PPE)
- The IATA held a conference in 2004 describing a methodology for disposal of low active radioactive waste

Any facility containing NORM should have a designated Radiation Safety Officer that is responsible for making decisions relating to NORM maintenance. The main goal of any radiation protection program is keeping all exposures to ionizing radiation **A**s **L**ow **A**s **R**easonably **A**chievable (ALARA). It is impossible to pre-plan for every scenario that may arise. Occasionally, there may be activities that aren't directly outlined in a written procedure. During these times, the RSO will review the work scope and make a determination on the best way to approach the task.

The following safety precautions should be followed when working with NORM contaminated items: (ref 22):

1.) Direct skin contact with radioactive scale, solids, and liquids should be avoided to the maximum extent possible.
2.) Eating, drinking, smoking and chewing shall not be allowed in the NORM work area
3.) Personnel shall thoroughly decontaminate after working with NORM containing equipment, before eating, drinking, or smoking.
4.) Work activities involving NORM should be conducted in areas to which access has been restricted. The number of personnel in the area should be kept to a minimum.
5.) Openings on contaminated equipment shall be sealed or wrapped in plastic or other suitable materials.
6.) NORM containing materials shall be kept wet to limit the possibility of NORM dust becoming airborne.
7.) Plastic ground covers should be utilized to the extent possible to contain NORM and facilitate cleanup.
8.) Workers who may be exposed to NORM must be provided with and wear proper Personal Protective Equipment. (PPE) to reduce the risk of exposure. The most effective PPE for working around NORM includes:
    - Protective eyewear, specifically goggles in high dust areas
    - Disposable or non-porous gloves (two layers of gloves are recommended)
    - Hooded disposable coveralls (i.e., TYVEK™ suits) or non-porous rain suits
    - Rubber or other non-porous protective footwear
    - Respirators with appropriate NIOSH-approved HEPA filters
9.) When respiratory protection is required, the selection, use, and care of any respiratory protective equipment must be in compliance with all company procedures.
10.) Gloves, respirators, coveralls, plastic and rags that have come in contact with NORM should be decontaminated or placed in containers, drums, sealed and held for disposal. Contaminated PPE should not be removed for the restricted work area until it has been decontaminated or placed in approved waste receptacles.
11.) NORM containing material shall be sealed in DOT 17-H steel drums or other suitable containers. Containers of NORM shall be labeled with a durable, clearly visible label bearing the radiation symbol and the words, "Caution Naturally Occurring Radioactive Material (NORM). The label shall also provide applicable information such as the radionuclides present, radiation levels, the quantity of radioactivity (in ρCi/g) if identified, the date for which the activity is estimated, and the type of material. This will assist individuals handling containers or

working near the containers to take precautions to avoid or minimize exposures.

10 Maintenance & Off-Site Repairs

When dealing with NORM, it is imperative that a plan be in place that outlines and controls the maintenance activities of equipment containing NORM. The plan should include the monitoring of work that is performed inside the facility as well as any items being shipped off-site for repair. NORM should be taken into serious consideration while planning for an outage or turnaround. Poor planning could result in a longer downtime as well as a potential liability. NORM contaminated equipment must be decontaminated to levels below the regulatory threshold before being shipped off-site. NORM contaminated equipment exceeding the surface contamination levels can be shipped to maintenance facilities providing they hold a specific license to accept this material. A monitoring program should be in place to identify all NORM contaminated equipment that is being opened or removed during a turnaround. This will ensure that the contaminated equipment does not inadvertently get shipped off-site to an unlicensed company. Having a good understanding of suspect NORM equipment involved in a turnaround results in an improved way of managing it ahead of time. Pre-planning for off-site maintenance will ensure that the facility as well as the employees are not exposed to NORM during repair/maintenance. It is very important to ensure that all decontamination activities are performed by a company holding a specific license to perform that service. The RSO should maintain a current radioactive materials license for any contractor performing decontamination activities at their site.

Summary

A survey carried out in preparation for this paper found that Naturally Occurring Radioactive Materials are present in nearly all ethylene plants. The root cause is known to be radon gas which exists in the rock formations where hydrocarbon feedstocks are often found. Radon is a gas with a short half-life, which poses a challenge to understanding how individual plants will be affected by NORM. In some cases the radon could pass through the plant without leaving a trace, while in others it could decay to non-volatile progenies which are retained in the plant. The known presence of NORM has increased by 160% over the past 25 yrs. This is certainly due to better awareness of the dangers of radioactive materials in our plants. However, the data also suggests shale fracking may be a contributor to higher NORM. How to address the presence of NORM varies plant to plant. It is our recommendation that this topic should be regularly revisited by industry bodies including the US Ethylene Producers Committee, and that guidelines should be developed and shared by the community.

References

1. P.R. Gray. Journal of Petroleum Technology, Vol 45, issue 01, January 1993
2. R.A. Hart, R. Wolf, M. Graham, "Fluoride and Radon Survey Results", EPC article 24B, 1994
3. Texas RRC. NORM (Naturally Occurring Radioactive Material)
4. E. Scarnichia, A. Etchepareborda, M. Arrivere. "Radon in Propylene, Unexpected Influence of NORM in a Chemical Plant.", 2008; SAR; Buenos Aires (Argentina); IRPA 12: 12. International congress of the International Radiation Protection Association (IRPA): Strengthening radiation protection worldwide; Buenos Aires (Argentina);
5. G. Taylor, M. Cleghorn, "Naturally Occurring Radioactive Materials: Contamination and Control in a World Scale. EPC paper 21D, 1996
6. P.R. Gray, "NORM Contamination in the Petroleum Industry." JPT, Jan 1993, pgs 12-16
7. J.G. Barret, "Naturally Occurring Radioactive Material (NORM) Contamination Encountered during Propylene Fractionator Modifications." EPC paper 34B, 1997
8. R. Nave. Georgia State University, HyperPhysics, 2016
9. www.radioactivity.eu.com. Radon Properties.
10. D. Slim, Evonik Industries, private communication
11. "Fracking linked to higher radon levels in Ohio homes." Science Daily, University of Toledo. June 18, 2019.
12. J. Grant. "Fracking is on the rise in Pennsylvania. So are radon levels. Are the two connected?" Living on Earth. August 18,2018
13. Frackcident™ Injury Law. Blue Bell, PA
14. Apex Inspection Service
15. J. Hilbrich, "Experience with NORM", EPC Paper 2020.
16. Picture courtesy of Sheri Snider, Nalco Water
17. J. Harris, "Naturally Occurring Radioactive Material (NORM): Detection and Protection in the Petrochemical Industry", EPC paper 96F, 2012.
18. Howstuffworks.com
19. D. Slim, Radioactivity in ethylene plants, prepared for the Hazardous Chemicals sub-committee of the EEPC, 2019
20. St. Louis Test and Mitigation
21. D. Satterfield. AGU Advancing Earth and Space Science, SEPT 2009 "How Much Radiation Does it Take to Kill You?"
22. BP GoM Region Naturally Occurring Radioactive Material (NORM Manual)

AIChE PAPER NUMBER: 159c

PLASTIC WASTE TO CRACKING FEEDSTOCKS AND THE CIRCULAR ECONOMY

Andre Bernard

NOVA Chemicals

M. David Smith

NALCO Water, an Ecolab Company

Thomas Pickett

Technip Stone & Webster Process Technology

Joe Lally

Evonik Corporation

Prepared for presentation at the 2020 Spring National Meeting

Houston, TX, March 30 – April 4, 2020

PLASTIC WASTE TO CRACKING FEEDSTOCKS AND THE CIRCULAR ECONOMY

Andre Bernard

NOVA Chemicals

M. David Smith

NALCO Water, an Ecolab Company

Joe Lally

Evonik Corporation

Thomas Pickett

Technip Stone & Webster Process Technology

Abstract:

Processing plastic waste into oil that can be use as feedstock to steam cracking is an alternative that is gaining attention. The prospect of this process, although at an embryonic stage, is believed to provide a significant boost in the overall waste plastic recycle yield when compared to existing methodologies. However, this potential new feedstock to steam cracker comes with a number of contaminants, which would likely impact the reliability and performance of ethylene plants. This paper is a high-level discussion on the potential of using waste plastic as feedstock and the contaminants that can be expected.

Introduction

The negative impact of plastic waste is currently shadowing the benefits of plastic products to society. Public perceptions and pressures on governments to act resulted in 60 countries around the globe banning or being in the process of banning single use plastics.[1] Literature reports that only 12 percent of plastic waste is reused or recycled.[2] This low recycling yield is attributed to the current methodology. Processing plastic waste to cracking feedstock is an emerging technology that has the potential to boost the recycling yield and meet the vision of a plastic circular economy model.

Ethylene Value Chain

Figure 1 illustrates the predominant value chain of ethylene crackers leading to different types of plastic materials. The production of polyethylene represents a large fraction of ethylene consumption, where it accounts for about 36 percent of the end products illustrated on Figure 1. The production of ethylene oxide follows and is used to make ethylene glycol. Monoethylene glycol is then used to manufacture polyester fibers and polyethylene terephthalate (PET). PET accounts for 8 percent of the plastic in the value chain. Ethylene dichloride made by chlorination of ethylene is then cracked into vinyl chloride monomer (VCM) for producing polyvinyl chloride (PVC), which represents 18 percent of the plastics fraction of the value chain. Polystyrene contributes to 8 percent of the plastic pool and uses ethylene as feedstock by reacting it with benzene to form ethyl benzene that is further processed with styrene into polystyrene. Polypropylene which constitutes 26 percent of the plastic pool is made from propylene. Although not shown on Figure 1, butadiene recovered from the mixed C_4s stream from ethylene crackers is used for manufacturing rubber, which contributes for about 3 percent of the plastic pool illustrated on Figure 1.

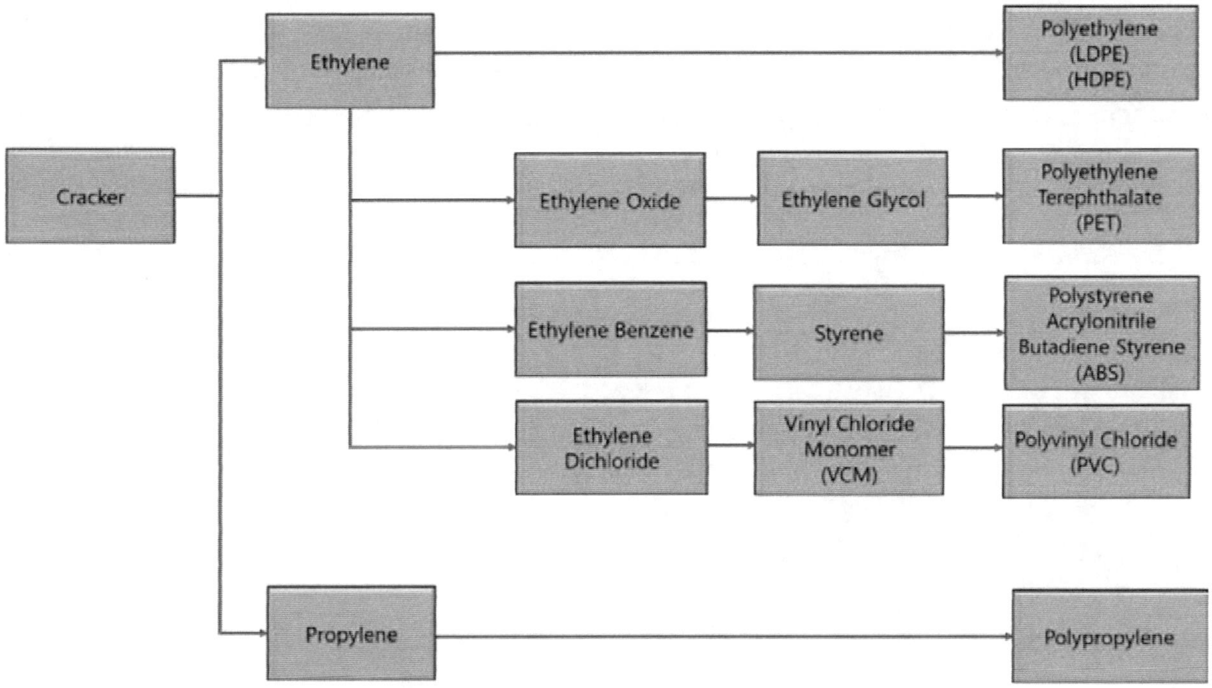

Figure 1: Ethylene Cracker Value Chain

Plastic Production and Usage

Global plastic production is estimated at about 390 million metric tons per year.[3] Although used earlier, the boom of plastic usage started after WWII and started growing exponentially from the 1950s to date. The production of plastic is expected to quadruple by 2050. Figure 2 illustrates the distribution of the different types of plastics produced worldwide. Those categorized as others includes sealants, coatings, polycarbonate, polyamide and bioplastics.

Thirty percent of the plastic produced is use for packaging, which includes a large variety of products. Some are single use such as plastic bags, food wrapping, plastic bottles (water and soft drinks). Others are products intended to be reused like plastic buckets, boxes and multipurpose bottles. About 50 percent of polyethylene production is dedicated to the plastic film market, which is typically single use and rarely accepted by recycling programs as it tends to get stuck in the sorting machinery.

Building and construction consumes about 17 percent of the plastic production through products such as pipes and fittings, siding, flooring and others. Those materials are installed and have a useful life expectancy of many years.

Plastic comprises about 15 percent of the total mass of cars. The use of carbon fiber reinforced plastic is use in the aviation industries to reduce the weight of airplanes allowing gains in fuel efficiencies. Plastic consumption from the transportation sector accounts for about 14 percent.

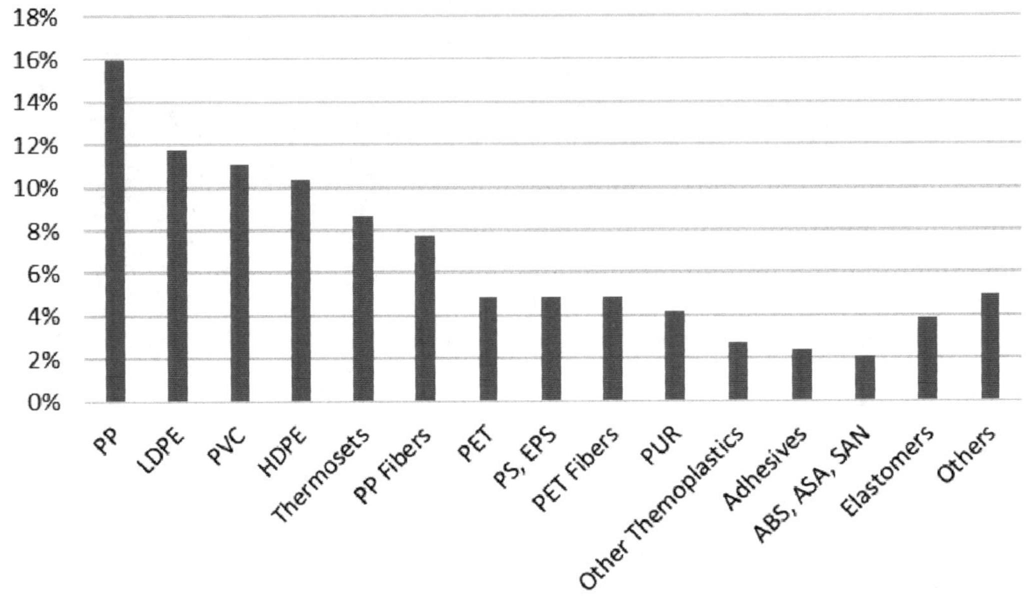

Figure 2: Plastic Production Distribution by Type

Many other products such as toys, healthcare/medical, kitchenware and many other products represent about 11 percent of the total plastic consumption. Textiles and the electric/electronics sector consume 10 and 4 percent respectively of the total plastic production.

Negative Impacts of Plastic Waste

Plastics have largely contributed to the betterment of society. However, their disposal after use creates an environmental problem. The life of plastic material is currently more in alignment with a linear economic model from generation to disposal. The value of the material is lost as waste after single use. A lot of plastic waste finds its way in the ocean. The UN estimated that the equivalent of 1 garbage truck per minute is released to the ocean.[4] It is reported that by 2050 there will be more plastic than fish in the ocean. An increase in synthetic fiber contamination in freshwater systems and terrestrial habitats has also been reported.

Forty years after the launch of the international recycle symbols, only 9 to 15 percent, depending on region,[5,6] of plastic is being recycled. Of that volume, about only 5 percent will be recovered for further use. Ninety percent of plastic ends up in landfills; as such, the plastic content in municipal waste has increased from 1 percent in 1960 to over 10 percent.[7]

The plastic industry is trailing behind paper, metal and glass with regards to recycling. Data from 1990 to 2015 suggest that the recycling rate in the US for paper has increased from 27 to 66 percent, 24 to 34 percent for metal and 20 to 26 percent for glass.[6] The plastic recycle rate has only increased from 3 to 9 percent. The variety of plastics composition and handleability makes it challenging for sorting the different types of plastic. For instance, as mentioned earlier, plastic bags are rarely recycled as they get stuck in the sorting machinery. Recycling facilities focus more on PET, HDPE and PP. Recycling is also limited by the capability of existing methodologies.

Regulatory Response

Many countries around the globe have implemented or are in the process of implementing bans on single use plastic. The European Union voted to ban single use plastic by 2021, targeting products such as straws, plates and cups.[10] In the footstep of the European Union, Canada is proposing a ban on single use plastic to come into effect by 2021. Some states in the US are considering or already have bans on a single use plastic. California and Hawaii already have bans in place, while New York state lawmakers have approved a statewide plastic bag ban to come into effect in March 2020.

Those bans have a large impact on plastic films, which represent a significant fraction of the value chain. The industry must therefore catch up on recycling strategies and methodologies.

Traditional Recycling (Mechanical)

Mechanical recycling is currently the dominating recycling methodology. It applies to polyethylene terephthalate (PET), high density polyethylene (HDPE) and polypropylene (PP). Plastic must first be sorted and separated from other waste (metal, glass, paper). Magnets are used for removal of ferrous metals; an optical sorter is used for removing carton. Eddy current is used for removing non-ferrous metal (mostly aluminum). Plastic is then separated based on its molecular signature using near infrared (NIR)[8] and an air nozzle, which blows the targeted plastic off the conveyor into a recovery bin. The recovered plastic parts are then washed and ground, followed by compounding and pelletizing. Polymer grade quality, however, degrades after multiple cycles due to accumulation of contaminants, primarily color pigments. Higher selectivity relative to pigmentation increases the cycles but at the expense of recovery.

Chemical Recycling and The Circular Economy

In response to the regulatory actions taken against plastics, particularly single-use plastics, and the challenges inherent in mechanical recycling, a new paradigm was developed: the "Circular Economy" for plastics. Rather than mechanically recycling plastic materials to use in increasingly lower-value applications with degrading product quality, the Circular Economy paradigm utilizes technological innovations to chemically break down the polymeric chains in plastic into either fuel, monomer, or feedstock to a steam cracker – a process known as Chemical Recycling.[11]

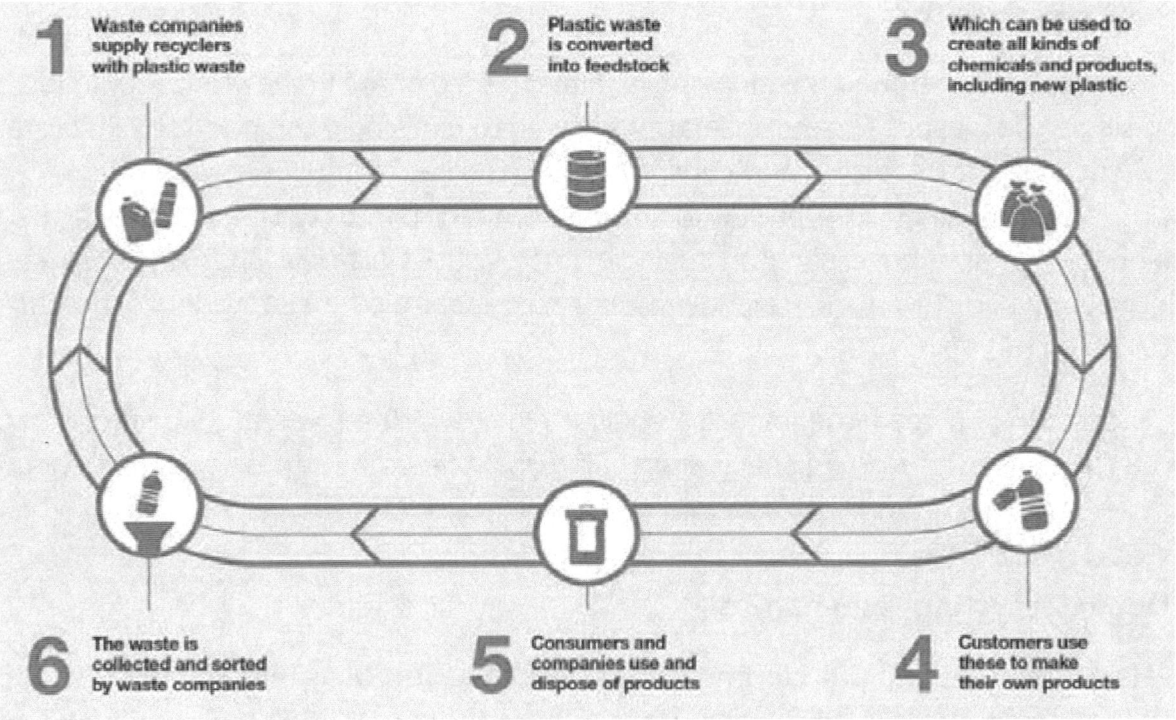

Figure 3: The Circular Economy for Plastics[28]

As shown in Figure 3, there are several steps in the full cycle of chemical recycling. Waste collection facilities provide recyclers with plastic waste. The waste must be sorted carefully, because certain types of plastic have components that are considered feedstock contaminants for petrochemical manufacturers; these will be discussed in more detail below. Next, one of several methods is used to break the polymer chains in the plastic down to smaller molecules that can be used for fuel, or in chemical plants and refineries to produce other consumer products.

Methods of Chemical Recycling

As shown in Figure 4, there are two main subsets of chemical recycling of plastics: decomposition, also known as monomer recycling, and conversion, also known as plastic-to-fuel (PTF) recycling. Each of these subsets utilizes various technologies to accomplish the breakdown of plastic into usable materials.

Monomer Recycling

- Chemolysis
- Hydrolysis
- Methanolysis
- Glycolysis
- Aminolysis
- Other methods

Plastic-to-Fuel Recycling

- Pyrolysis
- Fluid catalytic cracking
- Hydrogen technologies
- Catalytic pressure conversion
- Gasification
- Hydrothermal liquefaction (super-critical water)
- Other methods

Figure 4: Methods of Chemical Recycling

Chemical Recycling Initiatives in the Industry

Interest in chemical recycling within the industry has been high. An organization known as the Alliance to End Plastic Waste is a consortium of companies that manufacture, use, and recycle plastics. Comprised of such companies as BASF, ExxonMobil Dow, NOVA Chemicals, Sabic, Shell, and others, the Alliance has committed to investing $1.5 billion in promoting chemical recycling efforts over a 5-year period. Their goal is to dramatically reduce the estimated $120 billion in plastics they say enter the environment each year. Another industry group called Styrenics Circular Solutions includes such companies as INEOS Styrolution, Repsol, Total, Trinseo, and Versalis and is focused on recycling polystyrene (PS) to styrene monomer (SM).[12]

BASF's Ludwigshafen steam cracker was among the pioneers of recycled plastic-to-feedstock in ethylene production. They announced a successful run of processing pyrolysis oil supplied by the German chemical recycling firm Recenso, which uses a process called catalytic tribochemical conversion that can be applied to both plastics and agricultural biomass waste. The BASF program, which they call "Chemcycling", utilizes both pyrolysis oil and syngas for petrochemicals production.[13]

One of the first reported polymer-to-monomer processes occurred in the United States. American Styrenics, a joint venture of Trinseo and ChevronPhillips, announced in March 2019 that they had received recycled styrene monomer for processing at their St. James, Louisiana site. The supplier of recycled monomer, Agilyx, has been recycling plastic waste to synthetic crude oil for over a decade, processing 8 million pounds of mixed plastics into 800 thousand gallons of crude in 2014. They commissioned their first polystyrene-to-styrene monomer facility in 2018. Agilyx has also signed a joint discovery agreement with INEOS Styrolution and announced a plan to build a processing facility for polystyrene-to-styrene in Europe.[14,15,16,17]

Another successful industrial application of chemical recycling was announced in November 2019 by Shell Chemical. Nexus Fuels supplied Shell's Norco, Louisiana steam cracker facility with liquid feedstock produced from a pyrolysis process that uses high- and low-density polyethylene (HDPE, LDPE), polypropylene (PP), and polystyrene (PS). Shell has a stated goal of using 1 million tons of recycled plastic as feedstock by 2025.[18,19]

Other companies are involved in partnerships as well. In 2018, Sabic and Plastic Energy signed a Memorandum of Understanding (MoU) to build a commercial facility to produce "TACOIL", a feedstock produced via a process known as thermal anaerobic conversion.[20]Plastic Energy currently operates two units in Spain, with the intention of putting more TACOIL units throughout the world, close to centers of plastic supply. Plastic Energy has also signed an MoU with Petronas Chemicals.[21] Another company, UPM Biofuels, has announced agreements with INEOS and Dow Chemical to supply wood-based feedstocks for plastic production.[22,23]

Among the other companies at the forefront of chemical recycling include the following:

- Recycling Technologies, which uses a thermal cracking process that is similar to a refinery fluidized catalytic cracker to produce a variety of "Plaxx" oils and has signed an alliance with chemical manufacturer Total;[24]
- ReOIL, which recycles used tires into pyrolytic oil and carbon black;[25] and
- ReNew ELP, which uses super-critical water in a catalytic hydrothermal reactor to produce original monomer and oils.[26]

Overall, there are a variety of processes for chemical recycling, mostly based around some version of pyrolysis, and many of the largest petrochemicals producers are developing partnerships with chemical recycling companies to develop and implement an effective Circular Economy strategy.

Challenges to Chemical Recycling: Feedstock Contaminants

Contaminants in petrochemical feedstocks can cause a variety of problems in a steam cracker. To provide information on various contaminants and their potential impacts, the Ethylene Producers Committee (EPC) of the American Institute of Chemical Engineers (AIChE) established a Feedstock Contaminants Database that indexes various contaminants, the locations in an ethylene plant or downstream processing units these contaminants might impact, and what the negative consequences could result from the presence of these contaminants.[27]

Because of the deleterious impacts of certain contaminants, plastics being collected for chemical recycling as petrochemical feedstock must be separated carefully. In particular, two types of plastic must be removed prior to processing chemically for use as feedstock: polyethylene terephthalate (PET) and polyvinyl chloride (PVC).[28]

PET is produced by polymerizing ethylene glycol and terephthalic acid and is commonly used for soda bottles and other food and pharmaceutical packaging.[31] As shown in Figure 5, PET contains a considerable amount of oxygen. Oxygenates in petrochemical feed provide a number of negative consequences in an ethylene plant, including increased acetic acid production leading to corrosion, increased carbonyl concentration that can lead to aldol polymer fouling, and concentration of carbonyls in the crude C_4 product.[27]

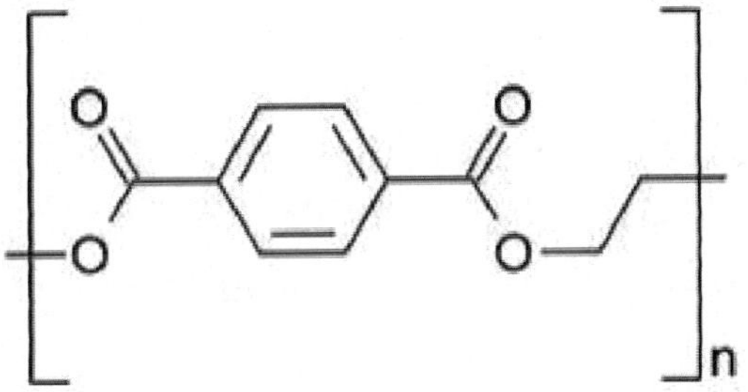

Figure 5: Polyethylene Terephthalate

PVC is produced by the polymerization of vinyl chloride monomer and is used in piping and plumbing, automobiles, raincoats, and a variety of other applications.[32,33] Chlorides in the feed can produce hydrochloric acid (HCl). HCl is highly corrosive, particularly at dew point conditions that can occur in petrochemical plants. Because HCl is a stronger acid than the organic acids typically observed in ethylene plants, it can be especially corrosive.[34]

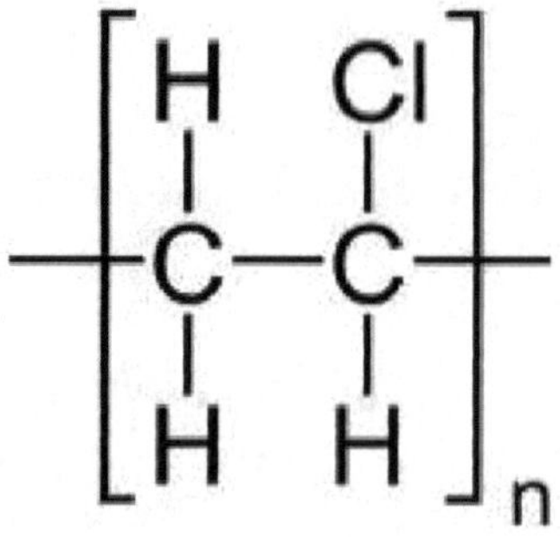

Figure 6: Polyvinyl Chloride

Other potential contaminants could exist as well, such as nitrogen, which can crack to form ammonia, and phosphorous, which can poison the gasoline hydrotreater catalyst. Following the more widespread use of these recycled feedstocks, a more exhaustive survey of observed contaminants and their impacts can be catalogued.[27]

Other Challenges to Chemical Recycling

There are several additional challenges that exist to implementing a widespread program of cracking recycled feedstocks. These include the following:

- Sorting Efficiency. As mentioned above, effective sorting is important to prevent feedstock contaminants. Sorting is currently an inefficient, expensive, and time-consuming operation.
- Geography. In addition to the nature of the sorting itself, sorting centers are still few are far between, requiring transport from collection to the sorting, potentially from the sorting center to the reprocessing facility, and then from the processing facility to the manufacturer.
- Scale. Currently, there are not enough collection, sorting, and processing facilities to provide feedstocks that meet the Circular Economy goals of governments, regulatory agencies, industry groups, and manufacturers.
- Economics. Because of the cost incurred by collecting, sorting, processing, and transporting the plastics and the recycled feedstock, it is still not meritorious from a purely economic standpoint.

Summary

Plastics have become nearly ubiquitous, touching nearly every aspect of our lives. However, the current practice of use-and-dispose has created problems of pollution and waste that are considered unsustainable many consumers, government agencies, and environmentally conscious individuals and groups. As a result, strict limits are being put in place that make recycling of plastics a necessity. In addition to the legacy methods of physical recycling, which recycles products into increasingly less valuable end uses due to product deterioration, chemical recycling promises to provide a method of truly having a circular economy, where new plastics are produced from feedstocks derived from previous goods, without a degradation of product quality. These feedstocks provide challenges that must be overcome for chemical recycling to become a world-scale, economically efficient endeavor.

References:

1. P. Hodges, "Tide will turn against single-use plastic", eChem, January 14th, 2019
2. T. Hundertmark, M. Mayer, C. Mcnally, T.J. Simons, C. Witte, "How plastics waste recycling could transform the chemical industry", Hydrocarbon Processing, April 2019.
3. M.W. Ryberg, A. Laurent, M. Hauscgild, "Mapping of global plastics value chain and plastics losses to the environment", United Nations Environment Programme, 2018
4. L.L. Alexander, "Repackaging the Planet", Chemical Engineering Progress (CEP), May 2018
5. "The new plastics economy. Rethinking the future of plastics", World Economic Forum, Ellen MacArthur Foundation and McKinsey & Company, 2016
6. A. H. Tullo, "Plastic has a problem; is chemical recycling the solution?", Chemical & Engineering News, October 6, 2019
7. R. Geyer, J.R. Jambeck, K.Lavender Law, "Production, use, and fate of all plastics ever made", Science Advances, July 19, 2017
8. "Near Infrared (NIR) Sorting in the Plastics Recycling Process", The Association of Plastic Recyclers, Document number: RES-SORT-2, July 20, 2018
9. K. Ragaert, L. Delva, K. van Geem, "Mechanical and Chemical Recycling of Solid Waste", Waste Management, August 2017.
10. R. Andrews, "The European Union Just Voted Ban on Single-Use Plastics by 2021", Forbes, October 25, 2018.
11. Rubel, Holger, Udo Jung, Clint Follette, Alexander Meyer zum Felde, and Santosh Appathurai (2019), "A Circular Solution to Plastic Waste", Boston Consulting Group.
12. Alliance to End Plastic Waste (2019), "We Have A Clear Vision to Eliminate Plastic Waste: First Anniversary Report, www.endplasticwaste.org.
13. BASF (2018), "BASF for the first time makes products with chemically recycled plastics", Press Release, www.basf.com
14. Recenso (2020), "Plastic Conversion", Press Release, www.recenso.eu
15. Styrenics Circular Solutions (2020), "Teaming Up for Change", www.styrenics-circular-solutions.com/circular-economy
16. American Styrenics (2019), "Agilyx Begins Delivery of Recycled Styrene Monomer to AmSty", Press Release, www.amstyrenics.com
17. Agilyx (2019), "SCS members Trinseo and INEOS Styrolution plan commercial scale polystyrene chemical recycling plant with Agilyx in Europe", Press Release, www.newswire.ca
18. Moore, Stephen (2020), "Shell uses plastic waste to produce resin feedstocks, *Plastics Today,* January, www.plasticstoday.com
19. Shell Chemical (2019), "Shell uses plastic waste to produce chemicals", Press Release, www.shell.com
20. Plastic Energy (2018), "Sabic Signs MoU with PLASTIC ENERGY", Press Release, www.plasticenergy.com

21. Duckett, Alan (2019), "Petronas Chemicals looks to convert waste plastic into feedstock", *The Chemical Engineer*, July, www.thechemicalengineer.com
22. UPM Biofuels (2019), "Dow and UPM partner to produce plastics made with renewable feedstock", Press Release, www.upmbiofuels.com
23. UPM Biofuels (2020), "INEOS and UPM Biofuels announce supply agreement for renewable raw materials to make plastic", Press Release, www.upmbiofuels.com
24. Recycling Technologies (2019), "Citeo, Total, Recycling Technologies, Mars, and Nestle Join Forces to Develop Chemical Recycling of Plastics in France", Press Release, www.recyclingtechnologies.co.uk
25. Reoil (2020), "Tire Recycling by Continuous Pyrolysis", Press Release, www.reoil.us
26. Renew (2020), "We achieve in just 20 minutes what takes nature 200 million years: turning waste plastic into readily usable hydrocarbon feedstock", Press Release, www.renewelp.co.uk
27. Bernard, Andre (2019), "2019 Feedstock Contaminants Database Presentation", presented at the AIChE Spring National Meeting, Ethylene Producers Conference, Session 35.
28. Chitalia, Ashish (2019), "Global Polyolefins: Are the upside risks over and done with?", Wood Mackenzie Americas Olefins Conference, Polymer Focus Session.
29. Editors of Encyclopaedia Britannica, "Polyethylene terephthalate", www.britannica.com
30. Hood, David and Steve Coleman (2007), "Minor Components, Major Issues: An Overview of Contaminants in Ethylene Plants", AIChE Spring National Meeting, Ethylene Producers Conference, Paper 51c.
31. "Polyethylene Terephthalate", www.wikipedia.com
32. Editors of Encyclopaedia Britannica, "Polyvinyl chloride", www.britannica.com
33. "Polyvinyl Chloride", www.wikipedia.com
34. Elder, Sherri, Nicolas Genty, and Wendy Young (2006), "Neutralizer Selection for Control of Acid Corrosion in Ethylene Plants", AIChE Spring National Meeting, Ethylene Producers Conference, Session 218.

589971 Pixel-Based Approach in Combination with GC Ã— GC for Fouling Tendency Prediction in Steam Crackers

Abstract:

Pixel-based approach in combination with GC × GC for fouling tendency prediction in steam crackers

Marko R. Djokic[1,*], Victor Abrahamsson[2], Nenad Ristic[1], Kevin Van Geem[1]

[1] Ghent University, Laboratory for Chemical Technology, Technologiepark 914, 9052 Gent, Belgium.

[2] Jet Propulsion Laboratory, California Institute of Technology, Pasadena, CA 91109, USA

Abstract

Fouling tendencies of a series of gas condensates were evaluated using comprehensive two-dimensional gas chromatography with flame ionization detection and sulfur chemiluminescence detection. A pixel-based approach was applied in order to identify parts of the chromatograms which were associated with the reactor coil fouling. Particular emphasis is given in this work to evaluate several feature selection methodologies along with various data preprocessing procedures. It was found that both aspects were crucial for studying the fouling tendencies and, as part of the subsequent partial least squares model development, predominantly the feature selection. Based on the flame ionization detector chromatograms and using the RReliefF algorithm for feature selection, a partial least squares regression model with one latent variable resulted in a root mean square error of the cross-validation of 0.65 g deposit/6 h (17%) (see Fig 1.). Based on the sulfur chemiluminescence detector chromatograms, the F-statistics feature selection generated a slightly better partial least squares regression model compared to using RReliefF, thus generating a model using one latent variable with a root mean square error of the cross-validation of 0.81 g deposit/6 h (21%) (see Fig. 2). Heavy aromatic compounds and heavy sulfur containing compounds were negatively associated with the fouling rate. Both were crucial in developing a partial least squares model with good prediction power, however, worked independently as predictors.

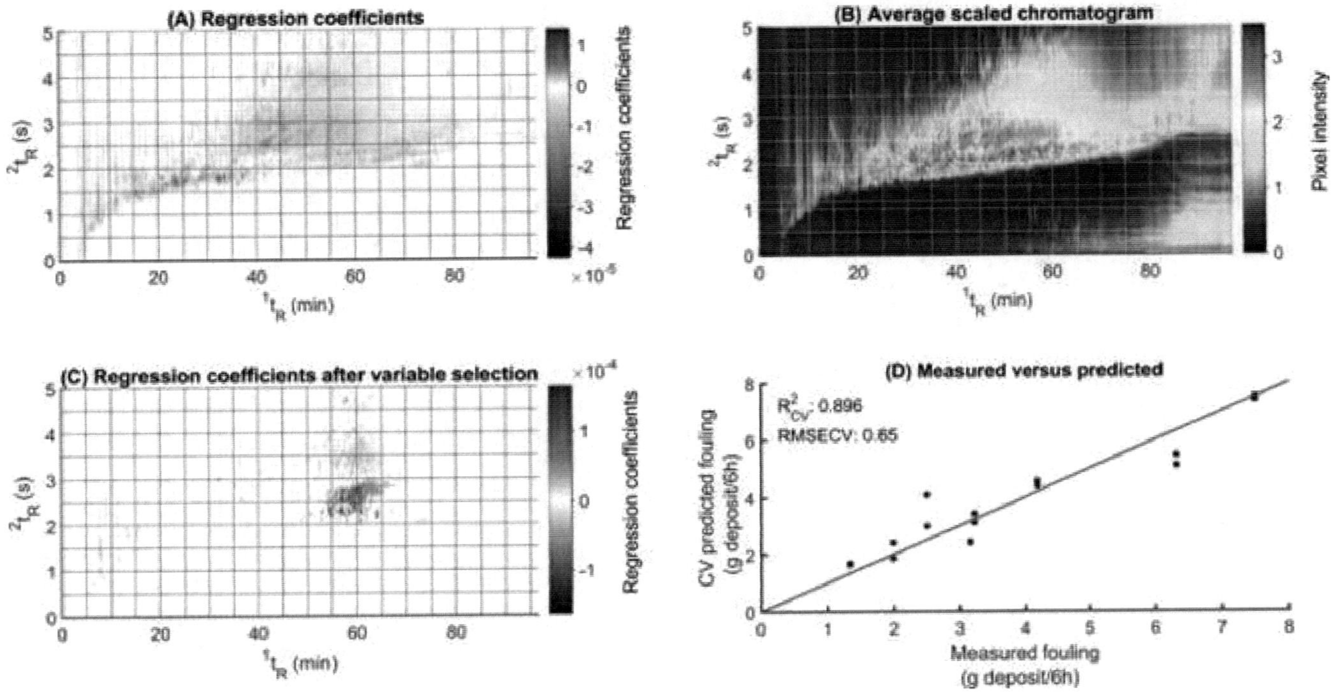

Fig 1: PLSR model based on the scaled FID chromatograms. (A) Presents the regression coefficients prior to feature selection, while Figure C presents the regression coefficients after using RReliefF as feature selection and selecting the optimum cut-off value. (B) Shows an average GC × GC-FID chromatogram based on the arithmetic mean of the scaled and base 10 logarithm transformed chromatograms. The measured versus the cross-validated prediction of the final PLSR model is displayed in (D).

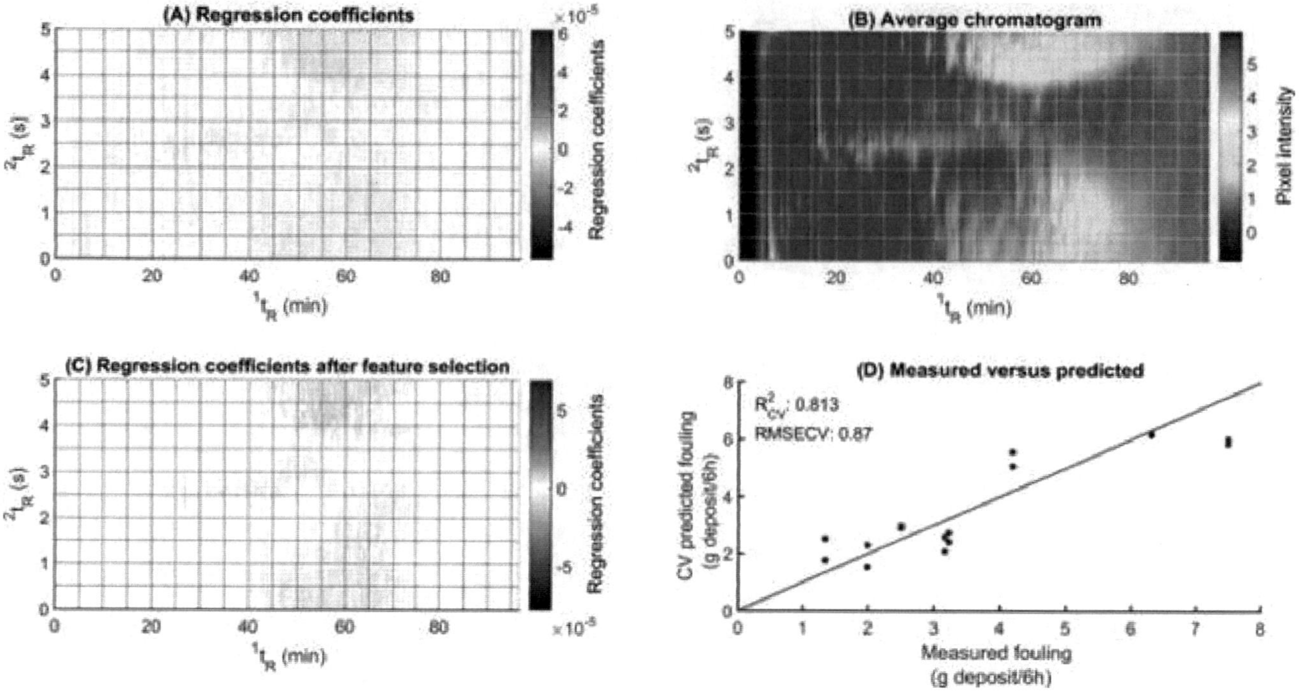

Fig. 2: PLSR model based on the SCD chromatograms. (A) Presents the regression coefficients prior to feature selection, while (C) presents the regression coefficients after using F-statistics as feature selection and selecting the optimum cut-off value. (B) Shows an average GC × GC-FID chromatogram based on the arithmetic mean of the base 10 logarithm transformed chromatograms. The measured versus the cross-validated prediction of the final PLSR model is displayed in (D).

AICHE PAPER NUMBER: 159e

CARBON DISULFIDE AND THIOPHENE: FORMATION AND MITIGATION REVIEW

Dr David Slim

Technology Advisor

Evonik Corporation

Dr Phillip James

Technology Director

Evonik Performance Materials GmbH

Joseph Lally

Technical Service Manager

Evonik Corporation

Prepared for Presentation at the 2020 Spring National Meeting

Houston, TX, March 30 – April 2, 2020

CARBON DISULFIDE AND THIOPHENE: FORMATION AND MITIGATION REVIEW

Dr David Slim

Technology Advisor

Evonik Corporation

Dr Phillip James

Technology Director

Evonik Performance Materials GmbH

Joseph Lally

Technical Service Manager

Evonik Corporation

Abstract

Thiophene and carbon disulfide are two compounds that can cause difficulties in the steam cracking process. This paper describes how these species are formed, the levels that are normally experienced, their impact and how they are normally managed. Thiophene and carbon disulfide are found in cracking furnace effluent. Both species are poisons to hydrotreating catalysts. In addition, thiophene boils very closely to benzene and hence is an issue for benzene purification.

Summary

Thiophene (C_4H_4S) and carbon disulfide (CS_2) have been identified as important contaminants to steam crackers. Both have been observed in cracker feed and carbon disulfide can be made in the cracking furnaces. Both compounds cause problems in product quality and catalyst activity. Both compounds have low chemical reactivity and hence are difficult to remove from product streams. However, technology exists to ensure product quality specifications and catalyst activity targets are met. All ethylene plants can manage these contaminants. Such technologies include polyamines and additives for CS_2 and extractive distillation for thiophene

Introduction

The EPC has reviewed various aspects of sulfur species and their impact on the steam cracking process. It is well known that certain sulfur compounds can produce H_2S in the cracking furnaces. This can retard coke formation and enable optimal run lengths and high olefin yields to be obtained. The mechanism and sulfur levels have been described in previous EPC presentations. In summary H_2S forms at the metal interface and undergoes the following inhibition process

- Catalytic coke: H_2S produces [S] which is adsorbed on the surface and poisons the catalytically active metal sites responsible for hydrocarbon decomposition reactions such as dehydrogenation.
- Radical-induced coke: •HS radicals (from H_2S) inhibits by transfer reactions. Continuous sulfur compound addition is required to maintain an optimum amount of adsorbed S needed to inhibit undesirable reactions.

The types of molecules that liberate H_2S include thiols (RSH e.g. C_2H_5SH) sulfides (R_1SR_2 e.g. $C_2H_5SCH_3$) and polysulfides ($R_1(S\text{-}S)_nR_2$ e.g. CH_3SSCH_3. After the cracking furnaces, the lighter S species must be removed or controlled as they can impact downstream process es and product quality. Examples include acid gas treatment for removal of H_2S, thiols and to some extent COS. Heavy sulfur species that concentrate in gas oil cuts are treated outside of the ethylene plant.

The two contaminants that can "slip through the cracks "are CS_2 and thiophene. This paper discusses these components.

Discussion

Properties of thiophene and its comparison to benzene

The general properties of thiophene are described in the sketch and table below

Table 1: properties of thiophene

Chemical Formula	C_4H_4S
Molecular weight	84.14 g/mol
Appearance	colorless liquid
Specific gravity	1.051
Melting point	−38 °C
Boiling point	84 °C

Thiophene is a heterocyclic compound with the formula C_4H_4S. It consists of a planar five membered ring. Thiophene is pseudo-aromatic like benzene. it is resonance stabilized and thus all the bonds in the ring have partial multiple bond character. The resonance stabilization is only slightly less than benzene (120kJ vs 150kJ). This is expected as the S atom is more electronegative than a C atom which draws electrons away from the pi system. Because of its aromaticity, thiophene does not exhibit the properties seen for conventional sulfides. For example, it does not liberate H_2S in cracking furnaces. It is a colorless liquid with a benzene-like odor.

The C–C bonds to the carbons adjacent to the sulfur are about 134 pm, which is comparable to that of benzene, while the other C-C bond is a little weaker (about 141 pm). A normal C-C bond is about 154 pm. Thus, all the C-C bonds in thiophene have multiple bond character. The C–S bond also has multiple bond character (170 pm vs 183 pm as in dimethyl sulfide). These properties show why thiophene is difficult to decompose in a cracking furnace.

Thiophene and especially its derivatives e.g. benzothiophene occur in petroleum sometimes in concentrations up to 1–3%. Thiophene boils close to that of benzene and is found in furnace effluent stream. Benzothiophene and higher thiophenes boil in the gas oil and tar streams and are not issues in steam cracking.

Benzene is a precursor to heavy chemicals, such as ethylbenzene and cumene. As a common contaminant in benzene, thiophene harms the benzene process industry, for example, eroding metallic equipment, making catalyst poisoning, affecting product quality and causing environmental pollution. It is difficult to separate thiophene from benzene by distillation due to their similar boiling points (4°C difference at ambient pressure). Like benzene, thiophene forms an azeotrope with ethanol.

Table 2: Boiling Points of thiophenes and benzene

Compound	**Boiling point, °C**
Thiophene	84
Benzothiophene	221
Benzene	80

Properties of carbon disulfide

The general properties of carbon disulfide are described in the table below

Table 3: Properties of carbon disulfide

Appearance	Colorless liquid Impure: light-yellow
Density	1.2927 g/cm^3 (0 °C)
Melting point	−111.6 °C
Boiling point	46.2 °C
Water Solubility	2.17 g/L (20 °C)
Solubility in organic solvents	Soluble in alcohol, ether, benzene
Vapor pressure	48.1 kPa (25°C)

Carbon disulfide is a colorless volatile liquid with the formula CS_2. The impure carbon disulfide that is usually used in most industrial processes is a yellowish liquid with an unpleasant odor, like that of rotting radishes. It evaporates at room temperature, and the vapor is more than twice as heavy as air.

CS_2 is a non-polar molecule because it has a linear geometric shape with a symmetric distribution of charge. The CS bonds are double bonds (bond length 155 pm). CS_2 is isoelectronic with carbon dioxide but it is more reactive toward nucleophiles and more easily reduced. These differences in reactivity can be attributed to the weaker π donor-ability of the sulfido centers, which renders the carbon more electrophilic.

The commercial manufacture and uses of CS_2 and thiophene are described in appendix 1

Furnace impact of thiophene and CS_2

The carbon sulfur bonds in both thiophene and carbon disulfide have multiple bond character. This is exemplified in the following table

Table 4: Bond order for CS and thiophene

Molecule	C-S bond length (pm)	C-S bond strength kJ	C-S Bond order
Methane thiol	182	272	1
thiophene	170	~420	1.5
Carbon disulfide	160	573	2

Given the multiple bond character of both thiophene and CS_2, these species and derivatives do not decompose in the cracking furnaces. The C-S bond strengths are significantly stronger than C-C single bonds (~350 Kj). Given this thiophene and CS_2 are termed "non-crackable" and hence will not generate H_2S. Thus CS_2 or thiophene in furnace feed passes through unscathed.

The possibility of CS_2 formation in the radiant section must be considered as well. Up to the 1950s, CS_2 was commercially produced from the elements at about 800°C. However, the partial pressures of sulfur and carbon in the radiant section are mostly regarded as too low and thus formation of CS_2 can only contribute marginally to the overall balance. It has been observed that in many imported feeds the level of CS_2 has recently increased remarkably. This is worth exploring. There have also been reports of low levels of CS_2 being formed as side products from using crackable sulfur compounds.

While mercaptans and organic disulfides are mainly cracked to H_2S in the radiant coils and later washed out in the caustic column, thiophenic, benzothiophenic, and other complex sulfur compounds are reasonably stable and end up in the pyrolysis gasoline fraction. The same happens to carbon disulfide (CS_2), which mostly originates from the cracker feed passing the radiant section without conversion.

There have been reports of a reaction of CS_2 with alkali metal hydroxides (KOH and NaOH). The consensus seems that the rate of dissolution of CS2 is very slow and the kinetics are also unfavorable.

CS_2 reaction with amines

CO_2, COS and CS_2 are iso structural. Since amines e.g. are used to remove high levels of CO_2 from process streams, it is important to compare their effectiveness in treating streams containing COS and CS_2

The reaction of these three components has been established to proceed via a zwitterion formation. Thus, for CO_2

$$CO_2 + RNH_2 = RNH_2^+COO^-$$

$$RNH_2^+COO^- + RNH_2 = RNH_3^+RNHCOO^-$$

For CS_2 the product is termed a dithiocarbamate. The first equation is reversible i.e. the back reaction has to be considered while the second equation is irreversible i.e. the back can be ignored Using conventional kinetics and the measured values of the rate constants, the rate equation can be expressed in the form

$$\text{Rate} = k_3[RNH_2]^2[CO_2]$$

The values of the rate constant k_3 have been measured. The value for CO_2 is approximately 100 times that for COS which in turn is greater than that for CS_2. Note in primary amines, e.g. MEA the thiocarbamate from COS decomposes irreversibly to form a cyclic derivative. It is likely that the dithiocarbamate from CS_2 behaves in a similar manner.

Given the above, conventional amines are relatively ineffective in removing CS_2

However, CS_2 reacts readily with diamines. This forms the basis for CS_2 removal systems. This can be described in the example

$$\underset{\text{Ethylenediamine}}{NH_2CH_2CH_2NH_2} + CS_2 \leftrightarrow \underset{\text{Zwitterion}}{NH_3^+CH_2CH_2NHCS_2^-}$$

Zwitterion $\rightarrow$ 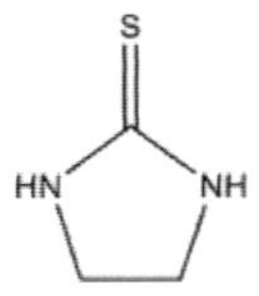

$+ \; H_2S$

Product is 2-imidazolinethione or ethylene thiourea mp 203°C, bp 347°C

In this case, the dithiocarbamate ion decomposes and liberates a molecule of H_2S to forma cyclic derivative. Presumably the H2S is picked up by the amine. This reaction path forms the basis for CS_2 removal, which is described later

Impacts of thiophene and CS2

CS_2 concentrates in the C5+ stream (pyrolysis gasoline) from the quench water settler and debutanizer bottoms in the steam cracker. For further processing the operator has three main options

- Extract the C5's from the pygas in order to process them in a C5 extraction unit (often a so-called isoprene extraction unit)
- Hydrogenate the C5's for recycle cracking
- Send the C5+ stream to the Pygas hydrotreater.

The extraction and hydrogenation units are the units within the steam cracker process and have reported issues with their performance due to the presence of thiophene and CS_2

<u>PyGas Processing</u>

The purpose of hydrogenation of pyrolysis gasoline is to eliminate highly reactive species deemed to polymerise and potentially cause fouling and plugging, such as alkynes, diolefins, styrene and olefin compounds, as well as to remove sulfur, while minimizing losses of aromatics. This is commonly achieved by hydrogenation in two subsequent stages. The first stage is a trickle phase operation employing palladium or sulfided nickel catalyst to remove alkynes as well as diolefinic and styrenic species. The second stage is a vapour phase operation employing sulfided nickel molybdenum (NiMo) and cobalt molybdenum (CoMo) catalysts to remove mainly olefins and convert sulfur compounds to H_2S

The formation of metal sulfides has an impact on the hydrogenation reactions as it retards all hydrogenation rates. It improves the selectivity of converting di-olefins above the selectivity of converting olefins. Unsulfided Ni or Pd catalyst could hydrogenate aromatics and the presence of sulfur strongly suppresses the hydrogenation of aromatics.

CS_2 acts like a "semi-permanent" poison on both nickel and palladium catalyst applied in the 1st stage. One should not neglect the impact on the 2nd stage pygas reactor as underperformance of the 1st stage reactor cascades to the 2nd stage reactor and a substantial shift in exotherm could occur. The customer example shown in this article shows that catalyst developments and service support (modeling and technical support) can help improve performance by shifting the operation to a more pro-active response.

The deactivation of palladium-based catalyst for the hydrogenation of pyrolysis gasoline has been investigated. the content of CS_2 in the feedstocks was high in the range of 1.8-20.2 μg/g. The CS_2 content in feedstock is recommended to be less than 1 μg/g in well-running hydrogenation units. A deactivation test for the hydrogenation catalyst has been carried out by adding CS_2, dimethyl disulfide and thiophene to feedstock. The effects of the catalyst poisoning on the hydrogenation are in order of $CS_2 >>$ dimethyl disulfide > thiophene. Given this while thiophene can be classified as a catalyst poison, it is not regarded as a major poison. Thus, the deactivation of the catalyst can result from CS_2 in the feedstock. The CS_2 content is often linearly related with the C5 content in the feedstock. If the CS_2 content in the feedstock is less than 1 μg/g when the C5 content is less than 0.5%(w), and then the hydrogenation unit can run normally.

Targeting reliable and stable operation, the performance of the catalyst is essential. Knowing the feed composition, operating parameters, and start of run catalyst activity, one can model and predict the catalyst's performance, inclusive of cycle length, for given end of run criteria. This includes adjustments of the model for known feed poisons and inhibitors. If catalyst activity is permanently reduced by a feed contaminant, one classifies such a contaminant as a poison. If catalyst activity is restored after contact with the specific feed contaminant has been stopped, one classifies such a contaminant as a temporary poison or inhibitor.

However, a great advantage is the flexibility to use a broad range of cracker feedstocks. Inherent with the benefit of such flexibility, the composition of the feed and the catalyst poisons and inhibitors in the feed change by nature and quantity.

For stable operation of a pyrolysis gasoline unit, it is of high value to know the effects that various poisons and inhibitors have on the performance of the catalyst as well as to know if and what measures can be taken to counteract and to restore the catalyst's performance.

Other Removal options

Use of polyamines to remove CS_2

Certain polyamines e.g. triethylenetetramine (TETA) are effective in removing CS_2. The structure of TETA is given below

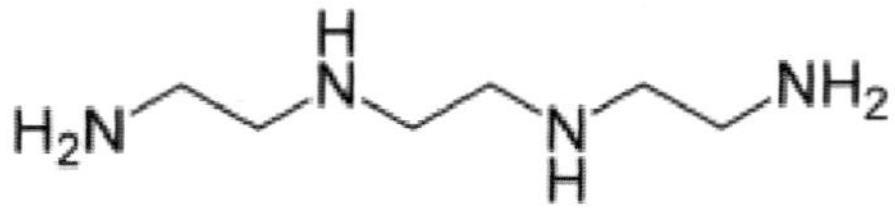

Based on this structure and the chemistry discussed for amines, it is clear that molecules such TETA can be effective in removing CS_2

Separation of thiophene from benzene

There is an urgent need to remove thiophene to ensure the high purity of benzene. Therefore, separation of the benzene-thiophene mixture is of great industrial interest in purification technology. The separation of azeotropic or close boiling binary mixtures is very difficult using ordinary distillation processes. As a result, extractive distillation, a special distillation technique that removes certain compounds from close-boiling mixtures in industrial processes, has received much attention. Such processes are currently being used. Optimization has been achieved recently making the processes very effective. Extractive distillation has effectively replaced the older technology that involved the use of reagents of a group consisting of diuphenylsulfone and dihydroxydiphenylsulfone and distilling the resulting mixture.

Conclusion

Thiophene (C_4H_4S) and carbon disulfide (CS_2) have been identified as important contaminants to steam crackers. They are both found in cracker furnace effluents. Both compounds cause problems in product quality and catalyst activity. Both compounds have low chemical reactivity and hence are difficult to remove. However, technology exists to ensure product quality specifications and catalyst activity targets are met. All ethylene plants can manage these contaminants. Such technologies include polyamines and additives for CS2 and extractive distillation for thiophene

Several references

Sulfur myths – what we know and what we don't know by M Brayden AICHE Proceedings: Volume 23 – 2011 Session: 33

The Influence of Metal Surface and Sulfur Addition on Coke Deposition in the Thermal Cracking of Hydrocarbons by M. Reyniers G.F. Froment AICHE Proceedings: Volume 6 – 1994 Session: 26

Introduction to Sulfur Chemistry as Related to the Olefin Industry by J. Reid AICHE Proceedings: Volume 10 – 1998 Session 23

Carbon Disulfide (CS2) - A New Challenge for Steam Cracker Complexes Pascal Coenen Ernesto Angueira AICHE Proceedings: Volume 28 – 2016 Session: 126

Introduction to Sulfur Compounds and Sulfur Removal as Applied to Olefin Unit Design and Operation by J Reid AICHE Proceedings: Volume 10

Kinetics of removal of carbon disulfide by aqueous monoethanolamine by Sang Cheol Lee, Michael J Snodgrass, Moon Ki Park and Orville C Sandall,; Environmental Science Technology 2001, 35, 2352-2357

Reactions of carbonyl sulfide and methyl mercaptan with ethanolamines by Mahmud A Rahman, Robert N Maddox and C. J. Mains; Industrial Engineering Chemical Research, 1989 28 470-475

Divalent Sulfur Based Functions, by P. J. Ewing Vol 13 Part 2 Treatise on Analytical Chemistry 1966

Method for separating thiophene from mixtures thereof with benzene US patent US40889273 1973-10-23 1973-10-23

APPENDIX 1

COMMERCIAL MANUFACTURE AND USES OF CS_2 AND THIOPHENE

Commercial thiophene synthesis

The first synthesis of thiophene by Meyer, reported the same year that he made his discovery, involves acetylene and elemental sulfur. Thiophenes are classically prepared by the reaction of 1,4-diesters, or dicarboxylates with sulfidizing reagents such as P_4S_{10}. Specialized thiophenes can be synthesized similarly.

Thiophene is produced on a modest scale of around 2,000 metric tons per year worldwide. Production involves the vapor phase reaction of a sulfur source, typically CS_2, and a C4 source, typically butanol These reagents are contacted with an oxide catalyst at 500–550 °C

Thiophene chemistry

General

As a consequence of its aromaticity, thiophene does not exhibit the properties seen for conventional sulfides double bond, giving the thiophene 2,3-epoxide, followed by

Oxidation

Oxidation can occur both at sulfur, giving a thiophene *S*-oxide, as well as at the 2,3-double bond. This gives the thiophene 2-3 epoxide, followed by the subsequent rearrangement. Oxidation of thiophene also demonstrates both reaction pathways. The major pathway forms the *S*-oxide as an intermediate, which undergoes subsequent Diels-Alder type dimerization and further oxidation, forming a mixture of sulphoxide and sulphone products with a combined yield of 83%

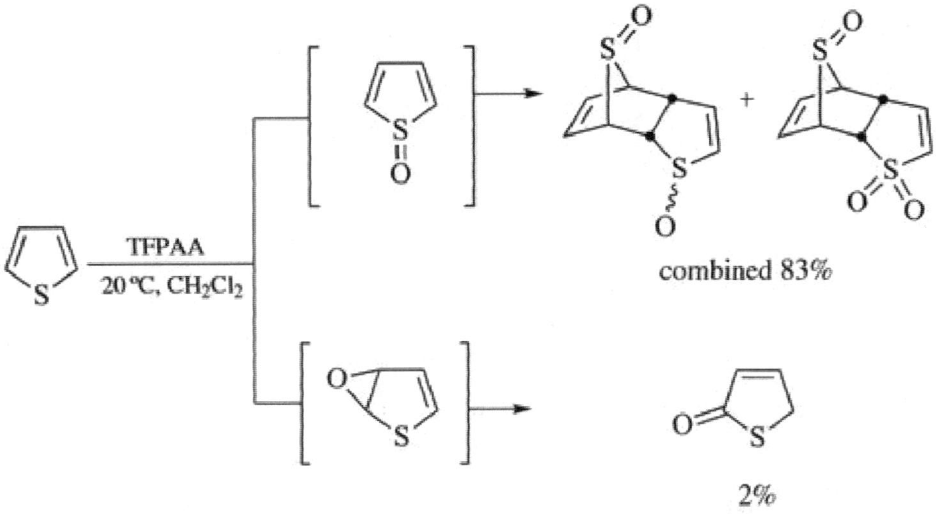

Alkylation

Although the sulfur atom is relatively unreactive, the flanking carbon centers, the 2- and 5-positions, are highly susceptible to attack by electrophiles Halogens give initially 2-halo derivatives followed by 2,5-dihalothiophenes. Thiophene brominates 10^7 times faster than does benzene.

Chloromethylation and chloroethylation occur readily at the 2,5-positions. Reduction of the chloromethyl product gives 2-methylthiophene. Hydrolysis followed by dehydration of the chloroethyl species gives 2-vinylthiophene.

Desulfurization by Raney nickel

Desulfurization of thiophene with Raney nickel affords butane. When coupled with the easy 2,5-difunctionalization of thiophene, desulfurization provides a route to 1,4-disubstituted butanes.

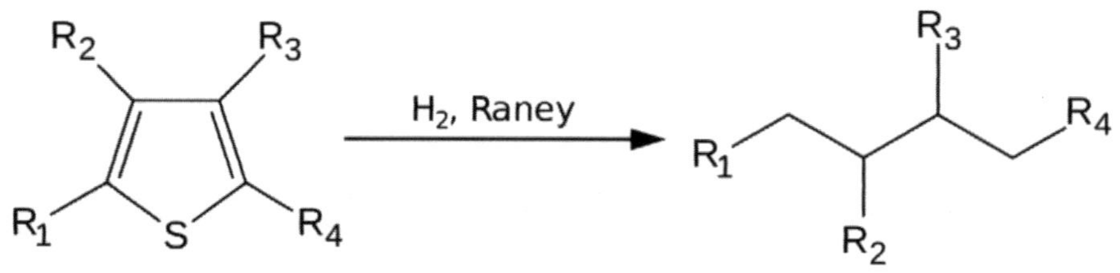

Polymerization

The polymer formed by linking thiophene through its 2,5 positions is called polythiophene Polymerization is conducted by oxidation using electron-transfer reagents. An idealized equation is shown:

$n\ C_4H_4S \rightarrow (C_4H_2S)_n + 2n\ H^+ + 2n\ e^-$

Polythiophene itself has poor processing properties and so is little studied

CS_2 chemistry

At high temperatures CS_2 can be manufactured by combining coke or carbon with sulfur. At lower temperatures, approximately 600°C, in the presence of silica gel or alumina catalysts natural gas can react with sulfur to produce CS_2.

$$2\ CH_4 + S_8 \rightarrow 2\ CS_2 + 4\ H_2S$$

In organic chemistry it is used to manufacture regenerated cellulose, the main ingredient of viscose, rayon and cellophane. It is also used as an industrial and chemical non-polar solvent. Chlorination of CS_2 is the principal route to produce carbon-tetrachloride. From a safety point of view, CS_2 is a toxic and highly flammable component. Its toxicity is close that of H_2S) and as such elevated levels of CS_2 are life threatening as it affects the nervous system. The flash point of CS_2 is -30°C and auto ignition can occur below 100°C

AUTHOR INDEX

AIChE
120 Wall Street, FL 23
New York, NY 10005-4020

ISBN 978-1-7138-2338-4

CURRAN ASSOCIATES INC.
proceedings.com